STARCH INDUSTRIES

VOLUME 3

STARCH INDUSTRIES

Processes and Innovative Products in Food and Non-Food Uses

VOLUME 3

Edited by

MARNEY PASCOLI CEREDA
Retired from São Paulo State University, Botucatu, São Paulo, Brazil
Tecsapp Solutions, Campo Grande, Mato Grosso do Sul, Brazil

OLIVIER FRANÇOIS VILPOUX
Federal University of Rondonopolis, Post-graduate Program in Environmental Management and Technology - PPGTA, Rondonopolis, Mato Grosso, Brazil
Tecsapp Solutions, Campo Grande, Mato Grosso do Sul, Brazil

ELSEVIER

ACADEMIC PRESS
An imprint of Elsevier

Academic Press is an imprint of Elsevier
125 London Wall, London EC2Y 5AS, United Kingdom
525 B Street, Suite 1650, San Diego, CA 92101, United States
50 Hampshire Street, 5th Floor, Cambridge, MA 02139, United States
The Boulevard, Langford Lane, Kidlington, Oxford OX5 1GB, United Kingdom

ISBN: 978-0-323-90842-9

For Information on all Academic Press publications
visit our website at https://www.elsevier.com/books-and-journals

Original cover art and designs created by Carolina Cereda Rafaine.

Publisher: Nikki P. Levy
Acquisitions Editor: Nina Bandeira
Editorial Project Manager: Lindsay C. Lawrence
Production Project Manager: Kumar Anbazhagan
Cover Designer: Christian J. Bilbow

Typeset by MPS Limited, Chennai, India

Contents

List of contributors

Luc Avérous BioTeam/ICPEES-ECPM, University of Strasbourg, Strasbourg, France

Luis A. Bello-Perez National Polytechnic Institute, CEPROBI, Yautepec, Morelos, Mexico

Camila Delinski Bet Food Engineering, State University of Ponta Grossa, Ponta Grossa, Parana, Brazil; UniCesumar, Ponta Grossa, Parana, Brazil

Fabiano P. Branco Dom Bosco Catholic University - UCDB and Federal Institute of Mato Grosso do Sul, Campo Grande, Mato Grosso do Sul, Brazil

Marney Pascoli Cereda Retired from São Paulo State University, Botucatu, São Paulo, Brazil; Tecsapp Solutions, Campo Grande, Mato Grosso do Sul, Brazil

Aina Liz Alves Cesar Una, Bom Despacho, Minas Gerais, Brazil

Ivo Mottin Demiate Food Engineering, State University of Ponta Grossa, Ponta Grossa, Parana, Brazil

André Augusto Gomes Faraco Department of Pharmaceutical Sciences, Faculty of Pharmacy, Federal University of Minas Gerais, Belo Horizonte, Minas Gerais, Brazil

Cleverson Fernando Garcia Department of Chemistry, Federal Center of Technological Education of Minas Gerais, Belo Horizonte, Minas Gerais, Brazil

Maria Aparecida Vieira Teixeira Garcia Food Department, School of Pharmacy, Federal University of Minas Gerais, Belo Horizonte, Minas Gerais, Brazil

Monserrat Hernandez-Bautista National Polytechnic Institute, CEPROBI, Yautepec, Morelos, Mexico

Vivian Cristina Ito UniCesumar, Ponta Grossa, Parana, Brazil; Food Science and Technology Graduate Program, State University of Ponta Grossa (UEPG), Ponta Grossa, Parana, Brazil

Juliana Bonametti Olivato Pharmaceutical Sciences Department, State University of Ponta Grossa, Ponta Grossa, Parana, Brazil

Lucas Felipe dos Ouros DPV-Food Technolgy, São Paulo State University - UNESP, Botucatu, Sao Paulo, Brazil

Hector A. Romero-Hernandez National Polytechnic Institute, CEPROBI, Yautepec, Morelos, Mexico

Apolonio Vargas-Torres Autonomous University of the State of Hidalgo, Institute of Agricultural Sciences, Tulancingo de Bravo, Hidalgo, Mexico

Pricila Veiga-Santos DPV-Food Technolgy, São Paulo State University - UNESP, Botucatu, Sao Paulo, Brazil

Olivier François Vilpoux Federal University of Rondonopolis, Post-graduate Program in Environmental Management and Technology - PPGTA, Rondonopolis, Mato Grosso, Brazil; Tecsapp Solutions, Campo Grande, Mato Grosso do Sul, Brazil

CHAPTER

1

Introduction: starch processing and products—between tradition and future

Marney Pascoli Cereda[1,2] *and Olivier François Vilpoux*[2,3]

[1]Retired from São Paulo State University, Botucatu, São Paulo, Brazil [2]Tecsapp Solutions, Campo Grande, Mato Grosso do Sul, Brazil [3]Federal University of Rondonopolis, Post-graduate Program in Environmental Management and Technology - PPGTA, Rondonopolis, Mato Grosso, Brazil

1.1 Introduction

Starch is an important energy source, with high global availability and relatively low prices. As indicated in Chapter 3 of Volume 1 in this series, the world starch market is essentially concentrated in four starchy raw materials, corn, cassava, wheat, and potato, in order of importance. It is also possible to mention rice and sweet potatoes, with significant productions, mainly in Asia. While the first four compete in the commodity market, the other two essentially occupy niche markets.

The identification of the main raw materials for starch production highlights the importance of underground starchy crops originating from South America, as half of them originate from that region, potato, cassava, and sweet potato. However, South America gave origin to many other underground starchy crops that did not meet the same success. Thus, canna (*Canna indica*) and arrowroot (*Maranta arundinacea*) are only of regional importance, while others have little or no use for starch production (*Xanthosoma sagittifolium*, *Arracacia xanthorrhiza*, *Dioscorea* sp., *Pachyrhizus* sp., *Ullucus tuberosus*, etc.). For more information on these crops, see Volume 2 in this series.

All these crops have high percentages of starch and could be used as commercial raw materials. Thus, the first part of this volume, the third in the series "Underground Starchy Crops of South American Origin: Production, Processing, Utilization, and Economic Perspectives" describes different processes of extraction and modification of starches from these underground crops, as well as starch derivatives, such as tapioca pearls, sour starch, and starch noodles.

However, in addition to their economic value, these crops also have a cultural importance, an aspect that has already been addressed in the first two volumes of this series and will be further explored in the upcoming Volume 4. This importance is translated by the existence of traditional processes of starch extraction in small artisanal units scattered in various parts of South America. These units process mainly cassava (Brazil, Paraguay, and Colombia), arrowroot (Brazil, Colombia, and Venezuela), and canna

Starch Industries
DOI: https://doi.org/10.1016/B978-0-323-90842-9.00001-7

(Colombia and Venezuela). Thus, although the starch industry is present worldwide and occupied by large modern production units, this volume could not fail to address the processes used by small traditional units.

Almost all the underground starchy crops contain high levels of phenols, compounds remain even after washing. This phenomenon is often considered a negative factor in the quality of the products, but new research indicates that it may also be a strength that differentiates these types of raw materials from cereals, with a favorable effect on human health.

Other markets also offer opportunities for starches from underground starchy crops. Chapters 3 and 14 of Volume 1 in this series already covered the main uses of starch and new consumption expectations. In Chapter 3 of Volume 1, the food, paper, and paperboard markets were already well developed. It was therefore decided in the second part of this volume to focus on lesser-known markets that offer good development prospects, such as the pharmacology and cosmetology, the glue, carrier and encapsulation products, and the film and bioplastics industry.

1.2 Processes of starch extraction and production of starch derivatives

Starch extraction basically follows the same steps for all underground starchy crops: washing (with or without peeling), grinding, extraction, concentration and refining, dewatering, and drying. In small units, the concentration (starch decantation), refining (cleaning of the lighter decanted part, made of starch and impurities mixture), and dewatering (removal of the decantation water) operations occur usually in decantation tanks or channels.

Depending on the size of the companies, the degree of mechanization and equipment varies, but these operations are always present. However, even in the most modern enterprises processing underground starchy crops, the efficiency of the extraction process is lower than in cereal starch companies. While in cereal starch enterprises the extraction percentage is over 90% and can reach 98%–99%, in the case of roots, tubers, and rhizomes, except for potatoes, this extraction rate is much lower. In the case of cassava, which is the most produced starch worldwide after corn and is essentially extracted in large modern companies, the extraction percentage is around 70%.

The explanation for the low yield with cassava comes from the greater difficulty in extracting the starch. However, it is interesting to note that the three raw materials with the highest yields, corn, wheat, and potato, had their processes developed initially in more industrial countries, in Europe and the United States. In the case of the other underground starchy crops, processing is done in developing countries, even if, in the case of cassava, part of the production is carried out by international companies.

Chapter 2 presents the main equipment used along the starch processing chain in large processing units of underground starchy crops, comparing large producing countries located in distant places, Thailand and Brazil, and China, a large consumer country, but which also has several production plants. The equipment presented is used mainly in cassava starch companies but could be used in all kinds of large units processing other underground starchy crops, such as sweet potatoes, arrowroot, and canna.

In this chapter, the similarity of the equipment used is evident, despite the distance between the different countries and the differences in size between the Thai and Brazilian companies. This chapter also highlights similar efficiency between units in Brazil and Thailand.

In Chapter 3, Vilpoux and Cereda present the equipment used in smaller units, but still with mechanization of the whole process. This equipment is used for a wide variety of underground starchy crops, such as cassava, canna, yams, arrowroot, and sweet potato. The same processing steps can be identified as in the larger companies, but performed with simpler and less efficient equipment, with higher losses in the bran and in the washing waters. As for the large companies, similar solutions can be found among the different countries and for the different crops identified.

The large processing units work with modern equipment, often copied from European or American companies, which were initially developed for the raw materials processed in these countries, essentially corn, wheat, and potatoes. Although this equipment has been adapted by national companies to the local reality and to the necessary dimensions, several improvements could be implemented. Thus, in Chapter 4, Branco and Cereda show that the changes were made without considering technical efficiency criteria and safety aspects, which impairs the final yields, a situation evidenced by the energy consumption as a function of the processing capacity of the industrial grinders used for cassava, sweet potato, and potato.

Despite some problems, large processing units can count on modern technologies that provide high yields, which is not the case for the small ones, which in many cases use, when mechanized, local confection equipment built without technical criteria. Based on this observation, Branco and Cereda describe the process of developing a machine to disintegrate and extract starch to be used in small units.

If the native starch knows a renewed attraction, due to the consumer demand for healthier products, with fewer chemicals, as well as the search for less harmful processes to the environment, most commercialized starches go through a modification process, whether physical, chemical, or enzymatic. Thus Chapters 5 and 6 describe the main types of modified starches that exist and the modification processes. In Chapter 5, Cereda describes the enzymatic and acid hydrolysis processes, with example of experiments done with cassava and sweet potato. In Chapter 6, Demiate, Bet, and Ito focus on the other types of modifications, physical and chemical.

If the modifications described in Chapters 5 and 6 can be applied to all starchy crops, Chapter 7 deals with a modification specific to cassava and present in several Latin American countries, the sour starch. This product undergoes an acid modification, followed by an ultraviolet treatment, which allows it to acquire expanding characteristics without the need for extrusion. In the traditional process, UV treatment is performed during sun drying, but that could be replaced by UV lamps in a more modern system. In this chapter, Cereda explains that starch, during the process of lactic fermentation and UV treatment, undergoes a photochemical modification reaction, with oxidation and probably etherification.

In its native or modified forms, starch also serves as food for people with celiac syndrome, who have an allergy to wheat gluten. This market has some traditional products, such as Asian noodles made from various raw materials, including sweet potato, arrowroot, and canna. In Latin America, biscuits and cheese bread made from sour starch can also be considered as products for this market. Chapter 8 describes the traditional process of producing noodles in South China from canna and arrowroot starches. It also discusses the development of starch products, including sour starch, for the celiac market.

Finally, Chapter 9 closes the presentation of starches and derivative processes with the production of tapioca pearls and two similar products, the broken tapioca pearl, consumed in Southern Brazil, and the popped tapioca pearl, a product similar to expanded rice and consumed in Northern Brazil. In addition to the process in large, fully mechanized companies, the processing in small companies, with very little mechanization, is described. Thus, in addition to showing the processing of these starch derivatives, the chapter allows discovering the process of starch extraction in small nonmechanized units, with more rudimentary technologies compared to those presented in Chapter 3.

1.3 Starch products and their derivatives

The two most consumed starches in the world are corn and cassava, which have managed to differentiate themselves essentially based on their production costs. The starch market is essentially a commodity market, where price and availability are predominant factors. However, more and more niche markets are emerging where specific characteristics can become a market differential.

In Chapter 10, Belo Perez discusses the complexation of starch with phenolic compounds, a characteristic that differentiate the starches of underground crops from those of cereals, where the complexation occurs with lipids. He highlights the importance of these complexes in obtaining resistant starches (RS), products that are increasingly valued in food production, in search of healthier products. Thus, phenols, which were considered for a long time as negative compounds in starches from underground crops, mainly due to the darker yellow or orange color provided to the final product, can become a strength in their commercialization.

Other markets little addressed in the literature, the cosmetology and pharmacology starch markets, are presented in Chapters 11 and 16, respectively. In these chapters, the products that require starch and the types of starch used are described. However, it is clear from reading these chapters that the selection is essentially made based on the price of the product, and that in this case corn is by far the preferred one, followed by cassava starch. However, some market niches can be occupied by differentiated starches, such as those of underground starchy crops of South American origin. For this to happen, especially in the pharmacological market, these starches should first be registered in the pharmacopoeias of the main consuming countries (United States, Europe, and Japan), which for now has only occurred with potato and cassava starches.

Related to the pharmacology and cosmetology markets, but also in food and beverage uses, the spray-dryer dehydration and encapsulation market is discussed in Chapter 12. This market is little known but presents a growing interest with a consumer looking for new flavors, closer to the flavors of fresh and increasingly natural products. Starch and its derivatives offer several solutions in this regard.

Other markets with great potential, but that remain far away for underground starchy crops, are biodegradable film products and bioplastics, developed as shown in Chapters 13 and 14. These markets have seen great development and are experiencing accelerated growth that is expected to continue in the coming years. However, despite its importance, starch is considered a commodity, where the most sought-after characteristic is price. Some other characteristics are sought for some products, such as gel transparency and amylose content, but the available information on the relationship bet-ween starch characteristics and properties of packaging and biodegradable materials is scarce. For the use of starches from underground crops, this knowledge will be important to allow for the development of niche markets, where price is no longer the main selling point.

Finally, Chapter 15 deals with a small market that has been little addressed in the literature, that of starch-based glues and adhesives. As for most of the markets addressed in this volume, the main characteristic of interest for choosing the raw material is price and the adhesive properties being obtained from the modifications made with the starch. Some competing products to starch are also addressed, which makes price even more important.

1.4 New perspectives in the starch market

In addition to the more widespread uses covered in Volumes 1 and 3 of this series, starch can be used in a wide range of applications such as metal separation by flotation, thermal and electrical insulation, and lubricant for oil drilling, among others. These markets have existed for many years and are already well known by the companies. Thus, in Brazil, some cassava starch units market bran, a residue of the cassava starch production process and containing high levels of starch (see Chapter 2), for lubrication during the drilling of oil wells. The utilization of residues from the extraction process will be the focus of the upcoming Volume 5 in this series.

In addition to these markets, other new ones are being developed that are worth mentioning for their originality, even if they have not been covered in more depth in this volume. The following are some examples of recent publications on high potential uses for starch (more information about new markets for starch can be encountered in Chapter 6).

- Superabsorbent
  - Superabsorbent hydrogels: biodegradable products with high absorption properties Bora and Karak (2022) cite a hydrogel made from starch and itaconic acid, with absorption capacity of 650 g/g. Kolya and Kang (2023) use starch recovered from rice cooking water to produce hydrogel that can be used in seedling planting. Srikaew et al. (2023) produced ecofriendly superabsorbent hydrogel films based on cassava starch and polyacrylic acid. This superabsorbent could be used to remove dye pollutants.
- Encapsulation and porous starch
  - *Controlled release of pesticides and herbicides*: although known for many years (Doane, 1992; Trimnell & Shasha, 1988), the use of starch in the encapsulation of agrochemicals for controlled release gains momentum with growing environmental concerns (Patil et al., 2018). Riseh et al. (2022) studied starch-based capsules used to enclose antimicrobial agents as effective tools against phytopathogens. Chen et al. (2023) developed a slow-release fertilizers using an in-situ radiation-synthesizing monolithic urea-embedded starch-based hydrogels through grafting polyacrylamide.
  - Zhang et al. (2023) used a corn starch-based aerogel with a microporous network structure to encapsulate β-carotene, with good improvement of product stability.
  - Zhang, Wang, Lim et al. (2022) obtained a significant increase of water and oil absorption with porous starch prepared with critical melting and freeze-thawing treatment. Porous starch is widely used in food and pharmaceutical industries. Jiang Junejo et al. (2023) made a starch cryogel with good potential for oil absorption, emphasizing the importance of amylose content. The treatment of corn starch with a combination of microwave and complex enzymatic (α-amylase and amyloglucosidase) hydrolysis increases the specific surface area and pore capacity of porous starch, improving its water and oil adsorp-tion performance (Jiang Wang et al., 2023).
- Purification—separation of material:
  - Sewage and water purification: Kang et al. (2022) developed a starch-derived flocculant for the treatment of wastewaters containing toxic contaminants of organic dyes, heavy metals, and antibiotics.
  - Separation: Dasgupta et al. (2022) used starch nanoparticles for the extraction of bitumen from oil sands
- Battery production: Jothi et al. (2022) utilized corn starch and polyvinylpyrrolidone to produce polymer electrolytes for proton battery.

- 3D printers:

  Zhang, Zheng, Tang et al. (2022) controlled the digestibility and Liu et al. (2022) the starch–lipid interactions of starch in hot-extrusion 3D. In their experiments, Zhang, Wang, Lim et al. (2022), Zhang, Zheng, Tang et al. (2022), and Zhang, Liu, Zhou et al. (2022) obtained a significant increase in slowly digestible starch (SDS) and RS content of rice, wheat, and potato starch-based printing food samples. Liu et al. (2022) obtain an increase of SDS and RS with the addition of stearic acid. These experiments allow for a personalized customization of starch-based foods.

  Zheng et al. (2023) tested annealing and heat-moisture treatment (HMT) on the quality of 3D-printed wheat starch gels and concluded that HMT can improve 3D printing performance of the starch.

  Vicentini et al. (2022) highlighted properties of interest of various types of starch in conjunction with (bio)materials in the chemical modification of sensors and biosensors, focusing on potato and cassava starches. For the authors, the use of 3D printing technology can play an important role, enabling, besides the formation of surface films on 3D-printed electrodes, the mixture of starches in the structure of thermoplastic filaments, attributing flexibility and elasticity to the developed sensors.

- Uses in human medicine:

  Yang et al. (2022) developed a starch-regulated adhesive hydrogel dressing, which provides a facile and effective method for wound management and treatment and has great promise for long-term application of wound dressings. Einipour et al. (2022) prepared a silk dressing containing dialdehyde starch nanoparticles loaded with vancomycin that can cure infection through the controlled release of antibiotics.

  Cespedes et al. (2022) evaluated the antithrombogenic activity of polyurethanes starch composites. They found that polyurethanes composites with zwitterionic starch improved antithrombogenic property and reduced the viability of the endothelial cells. They concluded that these biomaterials are inert and can be used for cardiovascular applications.

  Sivamaruthi et al. (2022) made a revision about the use of starch nanoparticles as micro- and nanobased drug delivery systems in cancer therapy, treatment of neurodegenerative disorder, and infectious disorder. For them, carbohydrate-based polymers have received much interest as drug delivery systems owing to their biocompatible nature, enhanced encapsulation efficiency, ecofriendly nature, low cost, and controlled release of drugs. Abundance, cost-effectiveness, ecofriendly, and biocompatible nature of the starch have turned the focus of researchers toward starch as a suitable candidate, but for the authors, physical, chemical, and enzymatic modification are necessary.

  Zhang, Liu, Zhou et al. (2022) utilized pregelatinized starches as a cohesion promoter to improve mechanical property and surgical performance of calcium phosphate bone cement. Waxy starch gave the best result, and the authors concluded that decreasing amylose content significantly improves the surgical performance and mechanical strength of calcium phosphate cement.

1.5 Final considerations

Starch is a carbohydrate of great versatility, used in numerous markets, from the most traditional to the most modern. Despite this great number of applications, the production of starch is concentrated in four raw materials:

corn with more than 70% of production, followed by cassava, with almost 14%, and wheat and potato with less than 10% production (see Chapter 3 of Volume 1 in this series). Rice and sweet potatoes also have relevant regional productions, although with much lower production volumes.

The market concentration on a few types of starch can be explained by the great importance given by users to availability and price. Thus, these characteristics represent one of the main arguments for the use of starch cited in the literature. However, its poor functional properties mean that for many applications, physical, chemical, or enzymatic modifications are required.

In many uses, the different characteristics of starch are little considered, and researchers end up working with the raw material most available to them. This situation is clear not only in the uses described in this volume (pharmacology, cosmetology, encapsulation, biomaterials, and adhesives), but also in the more recent articles surveyed in Section 1.4. It is not clear what the most important characteristics are for a starch to be used in these markets; the information is often limited to amylose/amylopectin contents.

This lack of information greatly hinders the use of starches from other sources, such as the underground starchy crops from Latin America, which are the subject of this series. The regional production and the limited amount of research on these crops (see Volume 2 in this series) make starch more expensive, making it impossible to compete in the international market, or even in the national or local markets. However, different characteristics of these starches could facilitate the occupation of market niches (see the production of noodles with sweet potato, canna, and arrowroot starches in Chapter 8). However, this would require a very good knowledge of the characteristics sought for each use.

One market seems promising according to the information available and raised in Chapter 10: RS, favored by starch/phenol complexation. The presence of phenols may be important in some roots and tubers (see Chapter 8 in the Volume 1 of this series) and may represent a differential factor in competitiveness.

References

Bora, A., & Karak, N. (2022). Starch and itaconic acid-based superabsorbent hydrogels for agricultural application. *European Polymer Journal, 176*, 111430.

Cespedes, J. F., Arévalo-Alquichire, S., Diaz, L. E., & Valero, M. F. (2022). Assessment of the antithrombogenic activity of polyurethane starch composites. *Journal of Functional Biomaterials, 13*(4), 184.

Chen, F., Miao, C., Duan, Q., Jiang, S., Liu, H., Ma, L., Li, Z., Bao, X., Lan, B., Chen, L., & Yu, L. (2023). Developing slow release fertilizer through in-situ radiation-synthesis of urea-embedded starch-based hydrogels. *Industrial Crops and Products, 191*, 115971.

Dasgupta, N., Wang, J. Z. O., Nguyen, V. T. A., & Gauthier, M. (2022). Thermoresponsive starch nanoparticles for the extraction of bitumen from oil sands. *Colloids and Interfaces, 6*(4), 67.

Doane, W. M. (1992). Encapsulation of pesticides in starch by extrusion. *Industrial Crops and Products, 1*(2–4), 83–87.

Einipour, S. K., Sadrjahani, M., & Rezapour, A. (2022). Preparation and evaluation of antibacterial wound dressing based on vancomycin-loaded silk/dialdehyde starch nanoparticles. *Drug Delivery and Translational Research, 12*(11), 2778–2792.

Jiang, K., Wang, W., Ma, Q., Wang, J., & Sun, J. (2023). Microwave-assisted enzymatic hydrolysis as a novel efficient way to prepare porous starch. *Carbohydrate Polymers, 301*, 120306.

Jiang, Y., Junejo, S. A., Jia, X., Zhang, B., Fu, X., & Huang, Q. (2023). Amylose content and pre-freezing regulate the structure and oil absorption of polyelectrolytes-based starch cryogel. *Carbohydrate Polymers, 302*, 120386.

Jothi, M. A., Vanitha, D., Sundaramahalingam, K., & Nallamuthu, N. (2022). Utilisation of corn starch in production of 'eco friendly' polymer electrolytes for proton battery applications. *International Journal of Hydrogen Energy, 47*(67), 28763–28772.

Kang, S., Liu, W., Wang, Y., Wang, Y., Wu, S., Chen, S., Yan, B., & Lan, X. (2022). Starch-derived flocculant with hyperbranched brush architecture for effectively flocculating organic dyes, heavy metals and antibiotics. *Journal of the Taiwan Institute of Chemical Engineers, 135*104383.

Kolya, H., & Kang, C. W. (2023). Synthesis of starch-based smart hydrogel derived from rice-cooked wastewater for agricultural use. *International Journal of Biological Macromolecules*, *226*, 1477–1489.

Liu, Z., Chen, L., & Zheng, B. (2022). Control of starch–lipid interactions on starch digestibility during hot-extrusion 3D printing for starchy foods. *Food & Function*, *13*(9), 5317–5326.

Patil, M. D., Patil, V. D., Sapre, A. A., Ambone, T. S., Torris AT, A., Shukla, P. G., & Shanmuganathan, K. (2018). Tuning controlled release behavior of starch granules using nanofibrillated cellulose derived from waste sugarcane bagasse. *ACS Sustainable Chemistry & Engineering*, *6*(7), 9208–9217.

Riseh, R. S., Hassanisaadi, M., Vatankhah, M., & Kennedy, J. F. (2022). Encapsulating biocontrol bacteria with starch as a safe and edible biopolymer to alleviate plant diseases: A review. *Carbohydrate Polymers*, *302*, 120384.

Sivamaruthi, B. S., Kumar Nallasamy, P., Suganthy, N., Kesika, P., & Chaiyasut, C. (2022). Pharmaceutical and biomedical applications of starch-based drug delivery system: A review. *Journal of Drug Delivery Science and Technology*, *77*, 103890.

Srikaew, M., Jumpapaeng, P., Suwanakood, P., Kaiyasuan, C., Promarak, V., & Saengsuwan, S. (2023). Rapid synthesis and optimization of UV-photopolymerized cassava starch-based superabsorbent hydrogels as a biodegradable, low-cost, and effective adsorbent for MB removal. *Journal of Industrial and Engineering Chemistry*, *118*, 53–69.

Trimnell, D., & Shasha, B. S. (1988). Entrapment of herbicides in starch for spray applications. *Journal of Controlled Release*, *7*(3), 263–268.

Vicentini, F. C., Silva, L. R., Stefano, J. S., Lima, A. R., Prakash, J., Bonacin, J. A., & Janegitz, B. C. (2022). Starch-based electrochemical sensors and biosensors: A review. *Biomedical Materials & Devices*, 1–20.

Yang, M., Fei, X., Tian, J., Xu, L., Wang, Y., & Li, Y. (2022). A starch-regulated adhesive hydrogel dressing with controllable separation properties for painless dressing change. *Journal of Materials Chemistry B*, *10*(31), 6026–6037.

Zhang, C., Wang, S. Y., Lim, S. T., Wan, K. X., Wang, Z. J., Qian, J. Y., & Liu, Q. Q. (2022). Critical melting assisted freeze-thawing treatment as a novel clean-label way to prepare porous starch: Synergistic effect of melting and ice recrystallization. *Food Hydrocolloids*, *131*, 107730.

Zhang, P., Wang, Y., Liu, Y., Wu, Y., & Ouyang, J. (2023). Improved stability of β-carotene by encapsulation in SHMP-corn starch aerogels. *Food Chemistry*, *406*, 135040.

Zhang, R., Liu, H., Zhou, H., Liang, C., Wang, X., Yang, H., Bai, Y., & Yang, L. (2022). Pregelatinized starch as a cohesion promoter improves mechanical property and surgical performance of calcium phosphate bone cement: The effect of starch type. *Materials Technology*, *37*(14), 3110–3121.

Zhang, Z., Zheng, B., Tang, Y., & Chen, L. (2022). Starch concentration is an important factor for controlling its digestibility during hot-extrusion 3D printing. *Food Chemistry*, *379*, 132180.

Zheng, L., Zhang, Q., Yu, X., Luo, X., & Jiang, H. (2023). Effect of annealing and heat-moisture pretreatment on the quality of 3D-printed wheat starch gels. *Innovative Food Science & Emerging Technologies*, *84*, 103274.

CHAPTER

2

Cassava starch processing units: comparison of industrial technologies from China, Thailand, and Brazil

Olivier François Vilpoux[1,2]

[1]Federal University of Rondonopolis, Post-graduate Program in Environmental Management and Technology - PPGTA, Rondonopolis, Mato Grosso, Brazil [2]Tecsapp Solutions, Campo Grande, Mato Grosso do Sul, Brazil

2.1 Introduction

Starch represents a significant source of carbohydrates. It is used in all the countries, and its consumption increases along with development. With an estimated production of around 12 million tons in 2020, cassava starch is the second most produced starch in the world, second only to corn starch (see Chapter 3 in Volume 1 of this series). Thailand is by far the largest producer, with 60%–70% of this amount, followed mainly by Asian countries (Indonesia, Vietnam, and India) and Brazil, the only large producer outside Asia. Much of Thai and Vietnamese cassava starch production is exported, mainly to China. This large export volume makes cassava starch the most exported starch in the world.

Although it represents a smaller production than cassava, extensive research has been done with respect to wheat and potato starch, with a very efficient industrial production and starch losses under 5%. Because it is a tropical crop, from regions less developed than Europe (where most wheat and potato starches originate), or North America, researches on cassava starch processing were much smaller, with the technologies already used in the processing of other starches, essentially corn and potato. As these technologies were not developed specifically for cassava, processing is not as efficient as compared to other commercial starches, with higher starch losses and smaller utilization of residues.

Because of its importance, a detailed analysis of the cassava starch production process in the main producing countries is very important. For this analysis, we compare the processes adopted by companies in Thailand, the largest producer of cassava starch and where the largest companies are located (capacity of up to 2000 tons of cassava per day), and Brazil, the only country with large production outside Asia and with companies modern and of medium capacity (from 400 to 600

Starch Industries
DOI: https://doi.org/10.1016/B978-0-323-90842-9.00011-X

tons of cassava per day). We also consider China, which also has some modern companies. It is by far the largest importer and represents a great production potential.

This chapter is based on Vilpoux (2003), with updates from more recent literature and, in the case of Brazil, interviews with companies conducted in 2022 and 2023. Tran et al. (2015) also used field data collected 15–20 years earlier and found that the only significant change was the use of biogas as the main energy source.

2.2 Description of the starch processing operations in the analyzed countries

To have a better understanding of the starch production process, this section includes a detailed approach of all the production steps, from the receipt of raw material to the packaging of the end product.

2.2.1 Raw material

With 1.4 million hectares under cultivation in 2020, cassava is Thailand's fourth largest crop, in area, behind rice, natural rubber, and sugarcane, but ahead of corn. Between the years 2000 and 2020, production increased from 19 to 29 million tons and the average yield from 17.3 to 21.9 t/ha[1], with an increase of more than 25% (FAOSTAT, 2022). Depending on the province and yearly growing conditions, yields can reach 28 t/ha (Arthey et al., 2018).

Despite being the ninth largest crop in Brazil, with 1.2 million hectares in 2020 (FAOSTAT, 2022), cassava and its by-products have a secondary place in the economy and, unlike Thailand, are not export products. Grown almost exclusively by small producers, cassava and its derivatives are essentially destined for the domestic market, even in the case of starch.

In China, cassava can be considered a minor crop, occupying in 2020 only 300,000 ha, or a little more than 0.2% of the total area cultivated in the country (FAOSTAT, 2022). The province of Guangxi, in southern China, produces 70% of national cassava production and 85% of starch, cassava's main derivative (Dou et al., 2021).

Cassava is traditionally a seasonal crop, and the harvest concentrates in the driest period. In Thailand, even if cassava is processed all year round, the main purchasing and processing period is between November and April, when more than 60% of cassava is processed (Arthey et al., 2018). In China, most cassava is processed between November and March. Some companies process Chinese and Thai cassava chips between harvests, thus obtaining starch of different properties. In Brazil, cassava starch companies work all year round, as in Thailand, stopping at most one month for preventive maintenance.

Both in Thailand and in China, most cassava is harvested between 10 and 12 months after planting, even if Arthey et al. (2018) found producers harvesting cassava for less than 8 months or up to 20 months. In Brazil, the situation varies depending on the region. In some regions, cassava is harvested between 10 and 14 months after planting, whereas in others, it is harvested between 14 to 24 months, but in all cases, the harvest age is strongly influenced by cassava prices. In this country, cassava is also grown for direct consumption and flour production, with ages and yields that vary. These products will be discussed in Volume 4 of this series.

In China, cassava yields remained stable in the last 20 years, as they were 15.8 t/ha in 2002, compared with an average of 16.5 between 2015 and 2020 (Table 2.1). However, they are quite different depending on the location, with regions that exceed the average of 20 tons. It is difficult to compare cassava yields

[1]To reduce the effect of seasonal variations, the yield of 2000 considers the average from 1998 to 2002 and that of 2020 the average from 2015 to 2020, last available year.

TABLE 2.1 Cassava average yields in the main starch milling regions in Brazil, Thailand, and China between 2015 and 2021.

	10–14 months
Paraná region of Brazil[a]	25.6
Thailand[b]	21.9
China[b]	16.5

[a]*Average 2015–2021 (IPARDES, 2022).*
[b]*Average 2015–2020 (FAOSTAT, 2022).*
Data from IPARDES. (2022). Instituto Paranaense de Desenvolvimento Econômico e Social Base de dados do estado. *Available at: http://www.ipardes.gov.br/imp/index.php for Brazil, and FOSTAT (2022) for Thailand and China.*

in China with that in other countries since the farming production system in this country comprises very small portions of land, approximately 0.5 ha per family, and insufficient technical aid.

In the last 20 years, average yields in Thailand increased from 17.1 to 21.9 t/ha. As in China, the average cultivated area per farmer is very small.

Thailand has a limited number of varieties for industrial purposes compared to Brazil. This difference can be explained by the fact that Brazil is the center of origin of cassava (for more information, see Chapters 10 and 11 in Volume 2 of this series). This wide availability of varieties can explain the higher productivity observed in this country (Table 2.1). Another possible explanation is the greater use of technology by producers, such as mechanized planting, fertilization, and herbicides. Brazilian producers of industrial cassava for starch production are larger than in Thailand and China, being frequent producers with more than 50 ha or even more than 100 ha. In Thailand, most of the production is carried out by small farmers on farms of less than 4 ha (Arthey et al., 2018). In China, Fu et al. (2018) found an average surface of farm with cassava between 0.3 and 2.4 ha, with a cassava planting area between 0.1 and 1.0 ha.

Sarmento (1997) analyzes starch content in four varieties of cassava selected by IAC (Agricultural Institute of Campinas) for industrial purposes, aged from 10 to 24 months, and found 33%–35% starch content on a wet basis (Fig. 2.1). Samples of three varieties of Thai cassava roots harvested in September 2000, at the end of the rainy season, and two varieties harvested in June, at the beginning of the rainy season, show average starch contents of 28%–36% on a wet basis (Vilpoux, 2003). Although old, these percentages can only have been improved with time (see Chapter 4 for rapid assessment with hydrostatic balance. For more information about dry matter and starch content of cassava, see Chapter 8 in Volume 1 of this series). They give a good indication of the companies' extraction potential. Thus, considering starch contents of 32%–35% of dry matter in industrial varieties of cassava in Brazil, with extraction rates of 25%–29%, extraction allows recovering only 65%–70% of total starch[2]. This efficiency is very low when compared to the extraction processes of the other commercial starches (corn, wheat, and potato). Thus, Bergthaller et al. (1999) and BeMiller and Whistler (2009) mention a total starch recovery of 97%–98% for potato.

2.2.2 Cassava starch industry

There are approximately 87 processing factories in Thailand, the majority with a capacity of 400–800 tons of starch per day and three over 800 (Arthey et al., 2018)

Sriroth et al. (2000) classified starch industries into large (>200 t. starch/day), medium (100 –200 t. starch/day), and small (<100 t. starch/day). In Brazil, most industries are

[2]For the calculations, 3% impurities and 12% moisture were removed from the extracted starch. These percentages can be higher, which further decreases the percentage of extracted starch.

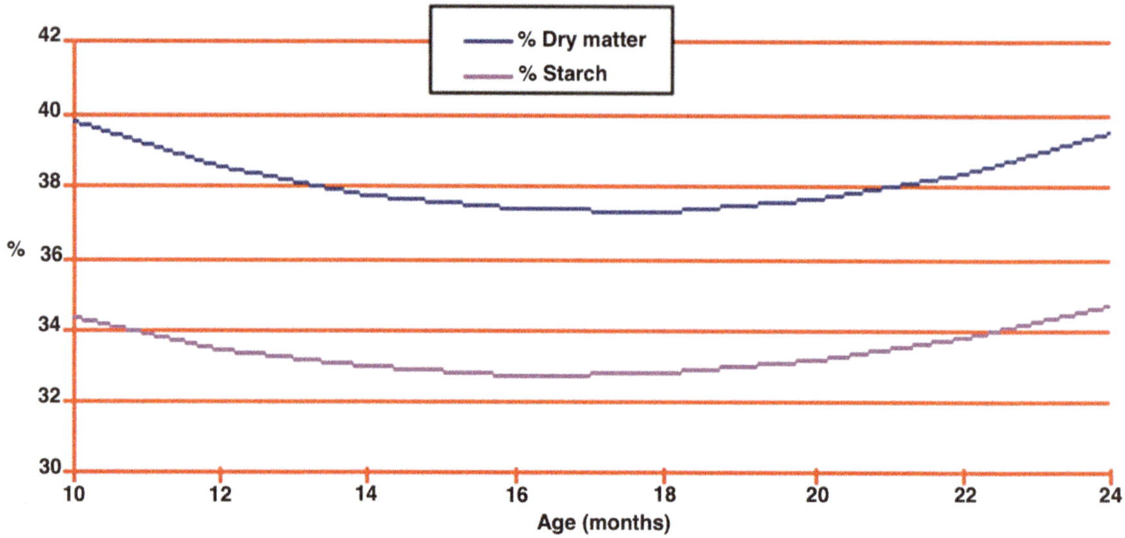

FIGURE 2.1 Average content of starch and dry matter in four varieties of cassava recommended for industrial purposes in Southeast of Brazil. *From Vilpoux, O. F. (2003). Cap. 7. Cassava starch production processes Brazil, Thailand, and China. In: M. P. Cereda & O. F. Vilpoux (Eds.),* Tecnologia, usos e potencialidades de tuberosas amiláceas Latino Americanas, *first ed., 3. Fundação Cargill, São Paulo.*

classified in the small or medium categories. China has approximately 10 large plants with a daily capacity of approximately or over 1000 tons of cassava, or around 250 tons of starch (Huidan Research, 2012; INews, 2020). Even if the top 10 companies represented 89.83% of national production in 2019 (INews, 2020), there are also dozens of starch plants with a daily capacity of 50–100 tons of starch.

As Thai companies largely depend on exports, they are quite concerned with higher traceability and quality standards. Some starch plants in this country have been awarded ISO 9000 certifications and work with GMP (Good Manufacturing Practices) and HACCP (Hazard Analysis and Critical Control Points) systems (Sriroth et al., 2000). Notwithstanding these facts, Thai starch analyzed, with samples collected from a company having ISO 9000 certification, is comparable to that produced in Brazil (Table 2.2).

Thai and Brazilian starch quality are very similar, while the Chinese starch analyzed indicates a lower quality, with a larger volume of impurities.

2.3 Production process

Suvittawat et al. (2014) found an extraction rate of 22.7% in a big cassava starch industry they analyzed. Vilpoux (2003) estimated average extraction in Thailand at 24%, the same percentage as found by Chavalparit and Ongwandee (2009), while Usubharatana and Phungrassami (2015) found extreme values between 22% and 30%. In Brazil, extraction rates are higher, with an annual average over 26%, and maximum rates of up to 32% in the top companies for the winter period[3]. Average extraction rates in China are around 25% (Dou et al., 2021).

[3] The top Brazilian starch plants obtain an average extraction rate of 28%–29%, based on a yearly process of 11–12 months.

TABLE 2.2 Comparison of composition of a few starch samples from Thailand, Brazil, and China.

	Thailand				
	1	2	3	Brazil[a]	China
H_2O (%)	12.04	11.45	11.64	12.28	12.48
% MS					
Starch	92.58	93.48	93.60	96.41	91.80
Fibers	0.50	0.36	0.39	0.17	0.88
Fat	0.15	0.11	0.05	0.10	0.22
Reducing sugars	0.08	0.13	0.09	–	0.91
Ashes	0.13	0.16	0.17	0.12	0.14
Protein	0.00	–	0.01	0.18	0.30
PH	4.90	6.09	6.11	5.43	7.03
Acidity	1.12	1.09	0.75	1.24	1.10

[a]*Average figures obtained from the analysis of production of 34 starch plants;*
-: No information.
From Vilpoux, O. F. (2003). Cap. 7. Cassava starch production processes Brazil, Thailand, and China. In: M. P. Cereda & O. F. Vilpoux (Eds.), Tecnologia, usos e potencialidades de tuberosas amiláceas Latino Americanas, *first ed., 3. Fundação Cargill, São Paulo.*

The mass balance is similar for Thailand and Brazil (Fig. 2.2), but water and energy consumption can vary greatly between processing units. Thus, for Thailand, Jakrawatana et al. (2016) found a variability of ± 50% for water consumption and of ± 30% for energy. These variations can be explained by the proportion of recycled water used in the system and by the power of the equipment, especially the grinding equipment.

In Thailand, water from the dewatering process is recycled to the extraction process, water from the concentration process is recycled to the peeling and grinding processes, and water from screw-pressing of cassava pulp discharged from the extractor is recycled to the grinder. Water from the root washer (peeling process) is discharged to the wastewater treatment and gas recovering (Pingmuanglek et al., 2017). In Brazil, water from the powder washer and from dewatering are recycled to the concentration process, water from extraction is recycled to the grinding process and water from concentration is return to the peeling process. As for Thailand, water from the root washer is discharged to the wastewater treatment to produce biogas.

Quantities of bran are similar, and so is the loss during the process. Starch contents in bran are around 45% in most starch plants of both countries, but in Brazil, it can reach 75% in those with the poorest extraction rates.

In Thailand, sulfur dioxide is used in the centrifugal or extraction stage of starch processing. Chavalparit and Ongwandee (2009) mentioned a quantity of 70 kg per ton of starch, or 16.6 kg per ton of root. Sriroth et al. (1998) examined the effect of sulfur dioxide addition during commercial-scale processing and revealed that, in addition to an obvious bleaching effect, inclusion levels also altered functional properties of starch. They identified an increased stability of the granules, expressed by an increase in gelatinization temperature of about 2°C, and a decrease in swelling at lower temperatures. They also described a lower degree of enzyme or acid hydrolysis. They conclude that entry and possibly exit of material into the granule, such as enzyme, was influenced by the presence of sulfur dioxide. Acid and water were also restricted in their access to the granule. In the presence of sulfur dioxide, enzyme hydrolysis was both lower and of a pattern that indicated only surface activity, whereas in granules that did not contain sulfur dioxide, enzyme activity occurred from within the granule, resulting in breakdown, fragmentation, and increased hydrolysis from 44.8% to 53.5%.

In consequences, Sriroth et al. (1998) recommended not to use sulfur dioxide. In Brazil, this product has never been used.

2.3.1 Reception of raw material

Depending on the country of origin, starch plants have different systems. Cassava starch

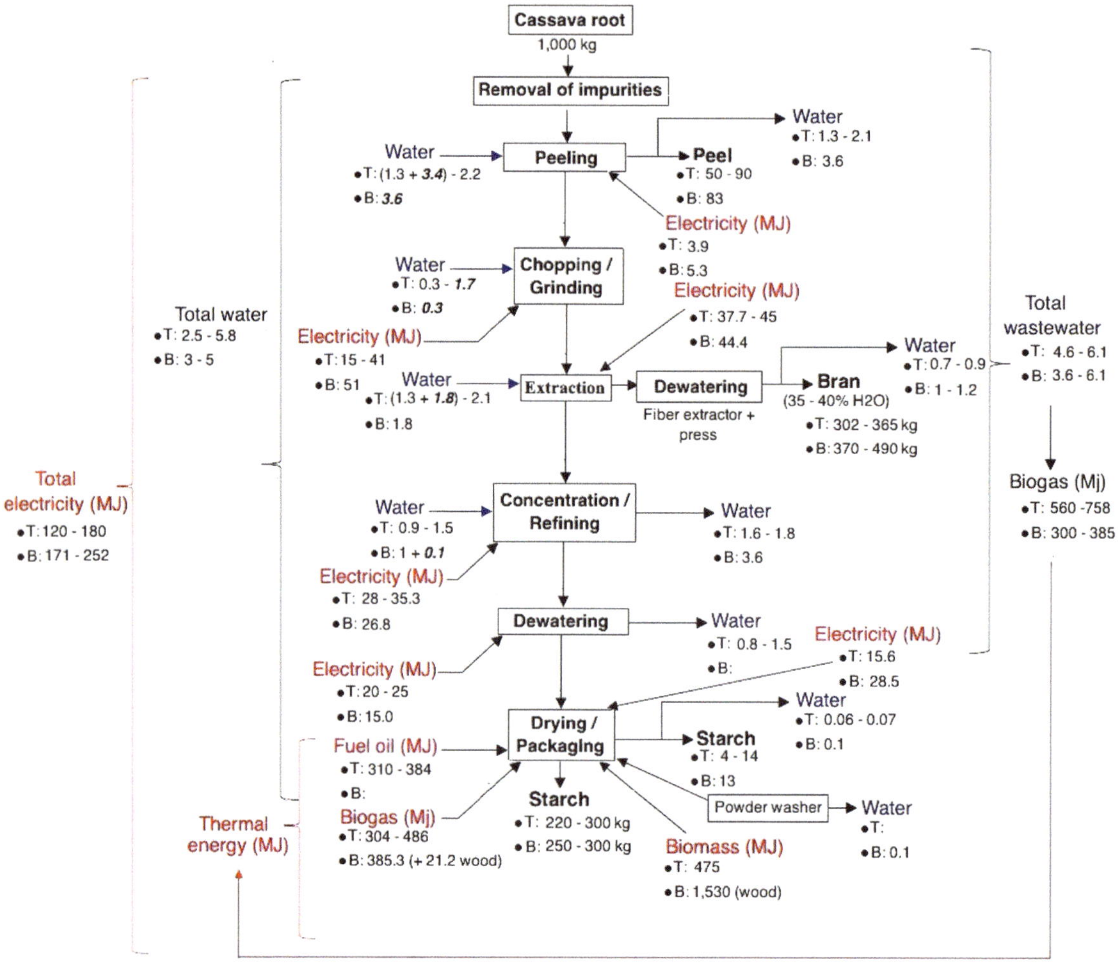

FIGURE 2.2 Mass balance of starch plants in Thailand (T) and Brazil (B). *Data from Sriroth et al. (2000), Vilpoux (2003), Chavalparit and Ongwandee (2009), Tran et al. (2015), Usubharatana and Phungrassami (2015), Jakrawatana et al. (2016), Pingmuanglek et al. (2017), Sánchez et al. (2017), Assawamartbunlue and Luknongbu (2020).*

plants in Thailand and China store the raw material on a concrete floor until the processing. As producers in those countries are small, cassava is delivered in dump trucks in the case of large volumes, by cart, and in the case of certain small producers, mainly in China, by small tractors (Fig. 2.3).

In Brazil, most of the starch plants use an unloading system based on a tipping platform. In the last 10 years, starch factories have changed from side tipping to rear tipping (Fig. 2.4A), which is less harmful to the truck's structure. After being unloaded, the roots go through a rotating sieve where the soil is removed and returned to the truck, to be discounted from the weight of raw material delivered (Fig. 2.4A and B).

A system with forks is sometimes used in China (Fig. 2.5). Cassava is taken out of the truck by means of a fork pushed by several

FIGURE 2.3 Cassava is deposited on concrete floors by means of different unloading systems in Thailand (A) and China (B and C).

FIGURE 2.4 Cassava unloading systems in Brazil: rear tipping platform (A) and screening and earth return system (A and B).

FIGURE 2.5 Cassava unloading system used in China, by means of forks and winches: (A) unloading with forks; (B and C) winches.

people (Fig. 2.5A) and pulled by means of a winch system (Fig. 2.5B).

Because of a higher automation, the Brazilian system seems more adequate. Storing the cassava in silos (Fig. 2.6) allows for the separation of cassava in order of arrival because the removal is done from the bottom of the silo by an endless screw, where the oldest raw material is (Fig. 2.6A). Storing cassava in silos does not adversely affect quality, which is mainly dependent on an effective management of the raw material supply system.

The main problem in the Brazilian automated system lies on the limited unloading capacity, on account of which long queues of trucks can wait to unload. Thus, the trucks make fewer shipments per day, which increases transport costs. The system used in Thailand and China does not result in truck queues.

A possible solution for starch plants in Brazil would be the use of mobile containers. When the truck arrives, the full container is left in the yard and replaced by an empty

FIGURE 2.6 Cassava storing silo in Brazil with the screw on the bottom of the silo (A) and conveyor belt to take the cassava from the silo to the washer/peeler (B).

FIGURE 2.7 Loader tractor to put cassava roots in the production line (A) and silo that stores the roots sent to the peeler (B).

container, thereby releasing the truck to get another shipment. The problem with this system is the need to strongly invest in containers. In this sense, the system used in Asia seems more economical and effective.

Both in Thailand and in China, the cassava is placed in the production line with a loader tractor (Fig. 2.7A), which places the roots inside a small silo (Fig. 2.7B), from where it is carried to the peeler by means of a conveyor belt.

Once in the production line, the cassava roots are sent to the cleaning and peeling operations.

2.3.2 Cleaning and peeling operations

The peeling process may be carried out directly as the process' first cycle, as it is done in the Brazilian starch plants. Chinese and Thai companies prefer starting by a dry-cleaning cycle, where the soil and part of the peel is removed. The cleaning process is carried out by means of a rotary drum (Fig. 2.8), with split openings for discharging soil and part of the peel. This operation enables to easily recover the soil and part of the peel without them being mixed with water, as it happens in a wet peeling process.

In Brazil, the removal of the soil is done before to put the cassava in the silos and the soil is returned to the producers, as explained in Section 2.3.1.

In Brazil, Thailand, and China, peelers are similar: semicylindrical, with rotary paddles, and spray of water recycled from the process itself (Fig. 2.9). As shown in (A)–(C), the shape, size, and coating of these blades may vary.

In Brazil, a cylindrical rotary peeler could be found until the beginning of the 2000s, but is no longer used (Fig. 2.10). It was similar to those used in China and Thailand for the first cleaning cycle.

Once peeled, the roots are passed along a conveyor belt, where they are controlled, and the remaining ligneous parts and stones are removed (Fig. 2.11A and B). In Brazil, most starch plants have improved field control and eliminated this inspection stage. In this case, at the end of the peeler, the cassava roots enter directly into a shredder or cutting machine, before being sent to a reservoir, from where they enter the main grinder (Fig. 2.11C).

After the inspection was carried out to remove any foreign matter, the grinding, or grating, process starts.

2.3.3 Cassava roots grating

Before the grating process, cassava roots are cut into 3–5 cm pieces using shredders to ease the process. In Thailand, the shredders rotate at a speed of 250–600 rpm and have 10–20 HP rotors (Sriroth et al., 2000). In Brazil, shredders with 20–30 HP motors can be found in starch plants, with a rotation of 500 rpm.

Grinding is one of the main steps in the production of starch. On this operation depend the starch extraction rate and the quantity of starch in the bran. It should be noted that there are two types of starch in the final bran: the

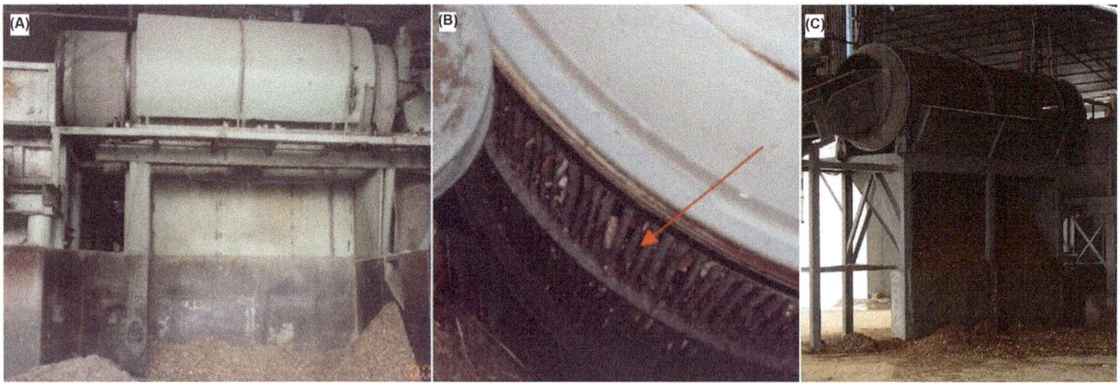

FIGURE 2.8 Cleaning system as adopted in Thailand (A) and China (C), with waste discharge splits as shown by arrow (B).

FIGURE 2.9 Semicylindrical peelers in Brazil (A), Thailand (B), and China (C).

one found in the plant cells that have not been grated and is more difficult to extract and the one found outside the cells, known as linked starch. The recovery of both types of starch depends on an effective grinding process.

In Thailand, the grinding process is carried out through a 1450–2000 rpm grater, measuring 77.5 cm in diameter, and fitted with 80–84 30-cm-long blades with 210 teeth each (Vilpoux, 2003). In Brazil, traditional graters are 80 cm in diameter, with 32-cm-long blades and 180 teeth each, with a rotation speed of 1200–1900 rpm.

FIGURE 2.10 Cylindrical rotary peeler used in Brazil.

Modern grinders in Thailand are made of stainless steel to prevent contamination and darkening (in the presence of iron the cassava mass turns blue) and have a reversible saw system, which doubles their useful life. The saw slots in the drums are shaped like arrow wings, which enable to quickly change the blades (Fig. 2.12B). In Brazil, the space between the teeth is made of wood and the blades of carbon steel (Fig. 2.12A).

Unlike the graters in Thailand and Brazil, made of drums with toothed blades, the type of graters used in China is the hammer mill (Fig. 2.12C). Some companies in Brazil are starting to use a hammer mill in a first grinding, followed by a traditional grinder, while other companies use only a hammer mill, in two grindings.

The disadvantage of hammer mills is the increased heating of the grinded cassava paste, which results in the need to add more water than that used in a grater with toothed blades. On the other side, these grinders have a simpler maintenance.

The most relevant elements for an effective grinding process are the rotating speed (from 50 to 80 m/s in Brazil and from 70 to 90 m/s in China) and the number of teeth per inch

FIGURE 2.11 Control of cassava roots before the grinding process in Brazil (A) and in Thailand (B); pre-grinding at the end of the peeler without inspection (C).

FIGURE 2.12 Different types of graters: drums with toothed blades in Brazil (A) and in Thailand (B) and hammer mill in China (C).

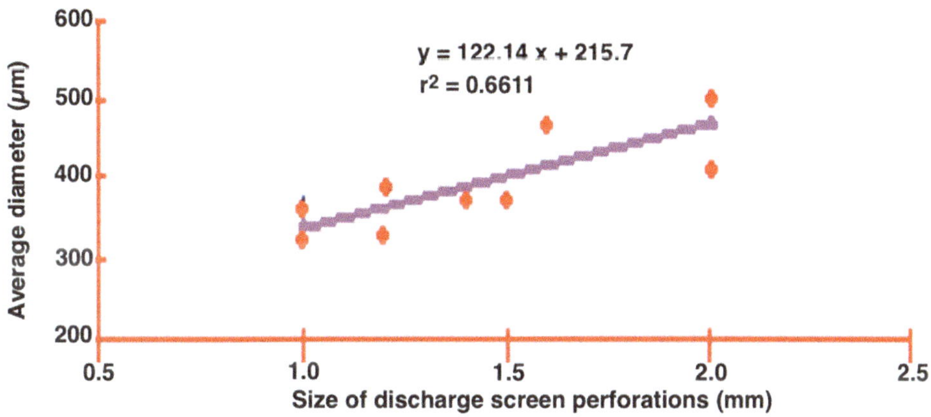

FIGURE 2.13 Average pulp diameter (μm) in relation to the size of discharge screen perforations, in the case of graters fitted with 20 teeth blades per inch. *Based on Vilpoux, O. F. (2003). Cap. 7. Cassava starch production processes Brazil, Thailand, and China. In: M.P. Cereda & O.F. Vilpoux (Eds.),* Tecnologia, usos e potencialidades de tuberosas amiláceas Latino Americanas, *first ed., 3. Fundação Cargill, São Paulo.*

(from 12–20 in Brazil and 17.5 in the graters used in Thailand). The graters used in China have a smaller diameter than in Brazil, measuring 45–60 cm, though the rotating speed is slightly faster.

In addition to the rotating speed and the number of teeth per inch in the blades, the spacing between blades is also important, which on average is 8.8 mm in Brazil. The diameter of the discharge screen perforations in the grater also affects the size of the grated particles. According to Jacquet (1999), the smaller the diameter of those perforations, the more effective the grinding will be (Fig. 2.13). In China, this diameter is 1.7 mm and from 1 to 2 mm in Brazil.

Based on research carried out in Brazil, the effectiveness increases with the use of a second grinding. However, although finer grinding improves extraction, it limits the removal of fibers and results in a more difficult purification process. On the other hand, using a more effective grinding makes the use of a second grinding unnecessary.

2.3.4 Starch extraction and bran removal

In addition to grinding, starch recovery also depends on the extraction, which is usually carried out by means of centrifugal screens (Fig. 2.14). In Thailand, the first extraction cycle (coarse extraction) is done through 35-to-40-mesh screens (355–425 μm), whereas in the second extraction cycle (fine extraction, or refining), 100-to-120-mesh (125–150 μm) or 140-to-200-mesh screens are used. The companies operate with 4–6 sets of extractors, while in Brazil, they operate with four, two for the coarse extraction, one for refining, and one for the disposal of bran. In this country, the extraction cycle is done through 100-mesh screens, refining with 220-mesh screens and bran disposal with 90-mesh screens. The water that comes out of this stage of the process is reused in the grinding.

There are two types of centrifugal screens: the horizontal type, which is used in China and some companies in Thailand (Fig. 2.14C and D), and the vertical type, which is used in most starch plants in Thailand and in Brazil (Fig. 2.14A and B).

FIGURE 2.14 Centrifugal screens used in starch extraction: vertical screens in Brazil (A) and Thailand (B) and horizontal screens in Thailand (C) and China (D).

In the case of Thailand, horizontal centrifugal screens rotate at a speed of 600–800 rpm, with 7.5 HP rotors and measuring $1.14 \times 1.36 \times 2.10$ m. The rotating speed of vertical centrifugal is higher, 1200 rpm, and it uses twice as much power but is more effective (Sriroth et al., 2000). Screen aperture size of the industrial scale extractor is from 125 to 250 μm in diameter, rotational speed is from 600 to 1500 rpm, and basket radius is from 0.40 to 0.65 m (Saengchan et al., 2014). In Brazil, rotating speed of vertical centrifugal is around 800 rpm, with diameter from 800 to 1200 mm.

In Thailand, the screen can be covered with a nylon fabric when used in the last extraction step to remove finer fibers from the starch milk (Fig. 2.15A), whereas in Brazil, only stainless-steel screens are used. In the past, vibrating screens were used in the different countries (Fig. 2.15B), but this sort of equipment has fallen out of use almost completely in Brazil. In China, fixed concave screens, or vertical screen extractor, with no rotors are used (Fig. 2.15C). This equipment is simple, uses little power supply, and allows for an effective removal of finer fibers.

In Thailand, the use of vertical screen extractor in the fine extraction, instead of centrifugal screens, allows for a reduction of starch loss of up to 4%–5%, of 2 m^3 of water and 5 kWh per ton of starch (Chavalparit & Ongwandee, 2009; Jakrawatana et al., 2016; Pingmuanglek et al., 2017). With this equipment, productivity increases 0.3 ton of starch for each 100 tons of fresh roots (Jakrawatana et al., 2016).

Once the starch is removed, in Thailand, the bran is passed through worm presses (Fig. 2.16A) or belt presses, or a series of supplementary

FIGURE 2.15 Systems used to remove finer fiber from starch milk: (A) nylon fabric used in centrifugal screens in Thailand, (B) vibrating screens, and (C) vertical screen extractor in China.

FIGURE 2.16 Water removal equipment for cassava bran: (A) worm presses used in China; (B) series of centrifugal screens used in China, Thailand, and Brazil; (C) filter press used in Brazil, and (D) drying yard used to dry the bran in Thailand.

centrifuge screens, as in Brazil and China (Fig. 2.16B). Belt presses are most effective to decrease the moisture content in the final bran (Thai Tapioca Trade Associations, 2003). There are few starch plants having this type of press in Brazil. Some of them use Brazilian cassava flour presses and other companies uses filter presses (Fig. 2.16C), which is the most effective system for cassava bran pressing, as it reduces bran moisture content from 90% to 40%.

Few starch plants in Brazil and none of them in China dry the bran. In Thailand, this waste can be dried in the same concrete yards where the cassava is received, or in the chip drying yards, where it is used in the production of pellets (Fig. 2.16D). There are some starch plants in Brazil where the bran is dried in continuous ovens also used in some modern Brazilian flour mills (Fig. 2.17). In this case, the dried bran is used in higher value markets such as mineral flocculation or human food. For more information about Brazilian cassava flour production, see the upcoming Volume 4 in this series.

Once extracted, the starch still contains many impurities, mainly microfibers. Purification is therefore necessary, followed by starch concentration to reduce the water content.

2.3.5 Starch milk concentration and refining

Before passing through the centrifugal separators during the concentration and refining processes, the starch milk may be passed through filters, which consist of hydrocyclones

FIGURE 2.17 Continuous oven used for drying cassava bran.

FIGURE 2.18 Sand filters used in Thailand (A) and China (B).

(Fig. 2.18), to remove the sand content. These filters are common in Asia but are not used in Brazil. Centrifugal separators are complex pieces of equipment, whose adjustment may be affected if large elements are passed through them.

In the early 2000s, starch concentration and refining were essentially done using disk centrifuges (Fig. 2.19A and B), but some starch plants in Thailand and China already used hydrocyclones (Fig. 2.19D). In this case, Dou et al. (2021) indicated the use of a 12-stage hydrocyclones.

Today, extraction in these two countries is essentially done by hydrocyclones, while in Brazil, many companies use a disk centrifuge for concentration (Fig. 2.19A) and a series of three to four hydrocyclones for refining (Fig. 2.19C), and a solution is also adopted in some companies in China. Hydrocyclones in Brazil work with motors from 20 to 30 HP.

The first disk centrifuge concentrates the starch milk and is filled with clean water. The hydrocyclones (or the second centrifuge) remove the water added after the concentration cycle, whereby the refining cycle of starch is completed. The residual water removed from the concentration process is used in the washing and peeling process.

In Thailand, the starch milk shows a concentration of 18–20°Be (Sriroth et al., 2000) after the concentration and refining steps. In Brazil, when entering the first centrifugal separator, starch milk concentration is 3–4°Be and 15–18°Be once this step is completed. Immediately afterward, more water is added for the refining process, whereby the concentration decreases to 5–6°Be before

FIGURE 2.19 Disk centrifuges used in Brazil (A) and Thailand (B) and hydrocyclones used in Brazil (C) and China (D).

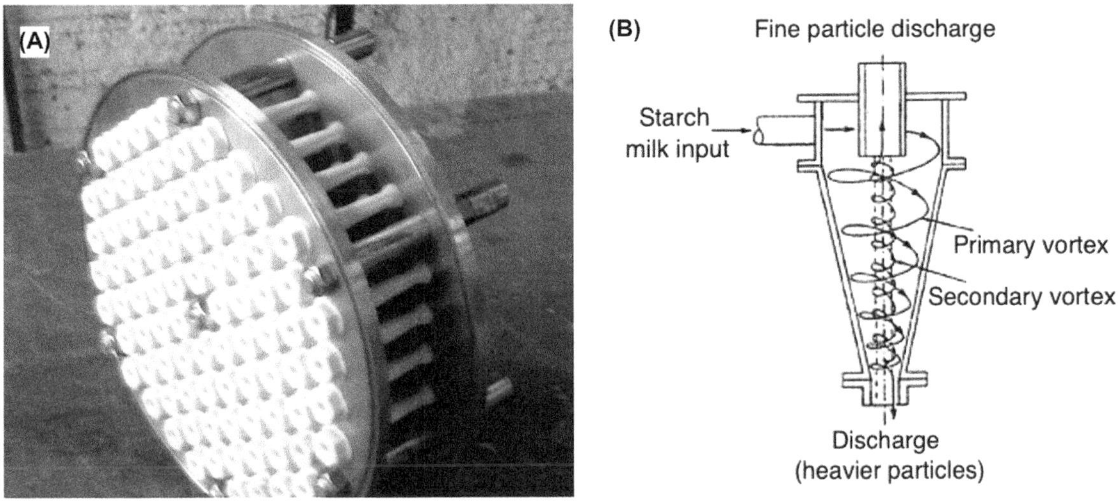

FIGURE 2.20 Internal part of a hydrocyclone (A) and hydrocyclone operation (B).

entering the hydrocyclones. Once the process is completed, after passing through the refining process, the concentration is between 20 and 24°Be.

The advantage of hydrocyclones over disk centrifuges is that they need less water, are simpler to use and to produce, and have a reduced risk of breaking down. In Europe, the use of hydrocyclones in the extraction of potato starch is common.

When using a hydrocyclone, starch milk enters the cyclone at a high speed and forms a vortex (Fig. 2.20B). As it descends, the flow speed increases, and the larger particles are displaced toward the sides of the cyclone before being removed through the bottom discharge. Meanwhile, the liquid containing the smaller particles in the center of the cyclone does not flow out, creates a countercurrent, or secondary vortex, and leaves through the upper end of the cyclone.

Based on the type of particles removed, the starch milk leaves through the upper or the lower ends of the cyclone. The use of several series of hydrocyclones allows for an adequate cleaning of the starch.

2.3.6 Dewatering and drying

There are two main systems for starch milk dewatering: vacuum filter (Fig. 2.21A), adopted by most Brazilian starch plants until the beginning of the 2000s[4], and a basket centrifuge, or peeler centrifuge, used for many years in Thailand and China and now present in all companies in Brazil (Fig. 2.21B and C).

The advantage of basket centrifuges is that they allow to obtain a product with 35%–40% of moisture, whereas with a vacuum filter, a product with approximately 45% of moisture is obtained. The decrease in moisture enables to reduce power consumption and temperature during the drying cycle, which consequently decreases starch gelatinization risks. However, basket centrifuges are more expensive and consume more power.

[4]The vacuum filter is still used in the production of modified starches because it has less starch loss than the basket centrifuge and does not require a return.

FIGURE 2.21 Vacuum filter used in Brazil (A), and basket centrifuges, or peelers, used in Thailand (B) and in Brazil (C), for starch milk dewatering.

The problem with the basket centrifuge is the high loss in starch, which must be returned into the system in the second centrifuge for starch refining. In Brazil, some companies use a decanter centrifuge to concentrate the starch returning from the basket centrifuge (Fig. 2.22).

For drying, the equipment used in the companies in the analyzed countries consist of flash dryers (Fig. 2.23). In Brazil, operating temperature of the equipment ranges from a minimum of 120°C (in the case of companies using basket centrifuges) to 140°C–150°C (in the case of companies using vacuum filter). In Thailand, companies operate using air heated at 170°C–200°C, with a capacity of 8–10 tons of dry starch per hour (Sriroth et al., 2000).

The dryers traditionally used in Brazil comprise only one drying cycle (Fig. 2.23A), whereas dryers in China (Fig. 2.23C) and in Thailand comprise a starch cooling cycle, following the drying cycle. Some starch plants in Brazil use a powder washer after the drying cyclone (Fig. 2.23B), which enables to recover the starch discharged in the air of the cyclone.

The different flash dryers used in starch plants are shown in Fig. 2.24.

In Brazil, many starch plants use the drying systems shown in Fig. 2.24A and B. When using peelers, starch enters the drying process with a lower moisture content, which makes it unnecessary to mix moist and dry starch (Fig. 2.24A). When using vacuum filters, the high moisture content in starch requires lower temperatures to prevent starch gelatinization, with longer drying periods. To reduce the drying period with higher temperatures, it becomes necessary to mix humid starch with a drier product (Fig. 2.24B).

The cooling cycle used in China and Thailand and described in Fig. 2.24C comprises the installation of a second cyclone at the end of the drying process. Starch plant equipment in Asia is operated at high drying temperatures and during reduced periods due to the use of peelers. Storing a product with a very high temperature and residual moisture enables the moisture of starch to condensate in the storing silo, where the starch gelatinizes and adherents plaques are formed. To minimize this problem, the companies carry out a second drying cycle that enables to store a cooled product.

Most Brazilian starch plants use a positive pressure ventilation system (the fan is located at the beginning of the dryer). Starch plants in China and Thailand use a negative pressure, with the fan placed at the end of the dryer (Fig. 2.24C). This system usually allows for a reduction in starch loss and power during the drying process. In the system used in Brazil,

FIGURE 2.22 Decanter centrifuge used in Brazil to concentrate the starch that returns from the basket centrifuge.

FIGURE 2.23 Flash dryers used in Brazil (A and B) and in China (C).

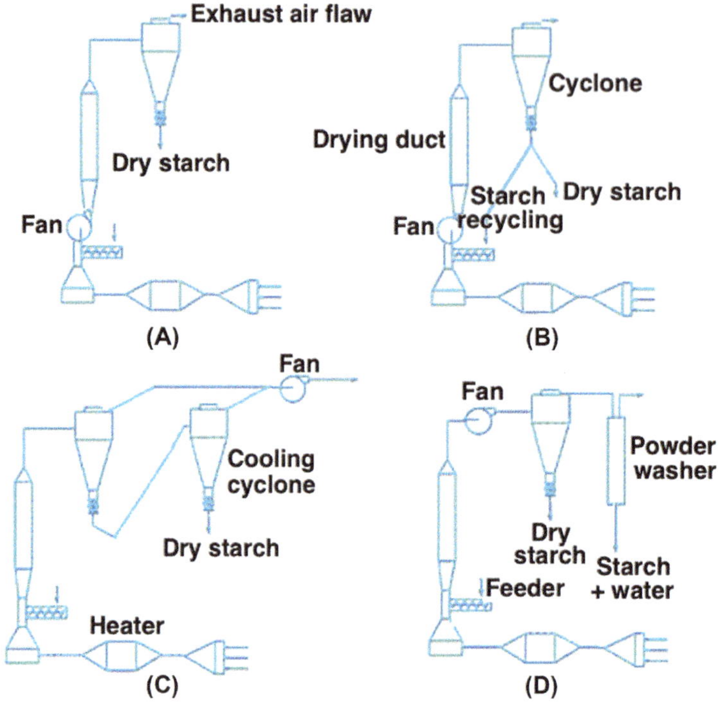

FIGURE 2.24 Drying systems found in starch plants. (A) Direct drying under positive pressure; (B) direct drying under positive pressure and dry starch recycling; (C) direct drying under negative pressure and cooling cycle; and (D) direct drying under mixed pressure (negative and positive) and powder washer. *From Vilpoux, O. F. (2003). Cap. 7. Cassava starch production processes Brazil, Thailand, and China. In: M. P. Cereda & O. F. Vilpoux (Eds.),* Tecnologia, usos e potencialidades de tuberosas amiláceas Latino Americanas, *first ed., 3. Fundação Cargill, São Paulo.*

ventilation is located after the starch feeding location, which means that the starch must pass through the paddles and thus the starch granules may be crushed, or the starch may stick to the paddles, which may affect the equipment performance.

The system illustrated in Fig. 2.24D is another system increasingly common in Brazil, which has two differences with the system described in Fig. 2.24A. This is a mixed-pressure dryer, where the fan is located after the drying duct and immediately before the cyclone and is fitted with a powder washer at the end of the drying cycle. The powder washer allows cleaning the cyclone exhaust airflow and recycling the starch discharged from the system. Installing the fan in the middle of the dryer enables to use only one piece of equipment, but it may affect the efficiency of the system. The best choice would be to locate two fans, one before the feeder and the other at the end of the dryer.

According to Sriroth et al. (2000), with the implementation of systems to reduce losses in the cyclones in Thailand, 37.5% of the starch loss occurring over the drying may be recovered. Fig. 2.2 shows the loss suffered during the drying process in Brazil, which amounts to approximately 5% of the starch produced. In this case, based on a recuperation of 37.5% of the 5% loss, the powder washer may allow recovering 280 tons of starch per year, with a daily production of 50 tons of starch over a period of 300 days a year.

In Brazil, with the powder washer, most of the starch lost in the drying process is recovered and returned to the hydrocyclones.

In addition to the differences in the flash dryers used, there are differences as regards to the fuel used in each country. In Thailand, most industries use biogas produced from wastewater treatment. In Brazil, the use of biogas is also increasingly common, but in some companies with indirect burning, wood is still used as a complement to biogas. In the most modern units, with direct burning in the flash-dry, biogas generated in the treatment is sufficient for drying the entire production and producing electricity for all the process, a situation also found in Thailand.

2.3.7 Packaging

Once the drying cycle is completed, the starch plants use screens to remove the largest particles, which are returned in the process. The presence of gelatinized starch particles is a result of drying starch with high moisture content at high temperatures, which favors gelatinization. The screens used in Thailand, China, and Brazil are shown in Fig. 2.25.

In Brazil, mainly 25 kg bags are used in packaging. Some of the companies also use big bags of 1000–1500 kg or 50 kg bags. In Thailand, a wider variety of packaging is used: 20, 30, and 50 kg bags or 500, 850, and 1000 kg big bags (Fig. 2.26).

Packaging systems are still primitive, and even in Thailand, which exports most of its production, the loss during packaging is high. To minimize losses and prevent explosions when producing "super dry" starch (<10% moisture), it is highly recommended to use ring systems to eliminate static electricity in the silo exhaust or at the bagger feeder.

2.3.8 Use and treatment of wastes

Wastes may be classified into two categories: solid, which includes bran and cassava peel, and liquid, known as cassava wastewater. The use of cassava starch processing residues will be treated in more detail in the upcoming Volume 5 of this series and will

FIGURE 2.25 Dry starch screens used in Thailand and China (A) and Brazil (B).

FIGURE 2.26 Cassava starch packaged in bags (A) and big bags (B) in Thailand and packaging equipment in Brazil (C).

not be detailed here. Peel may be used as fertilizer or animal feed (mixed with other cassava by-products), with free distribution by the companies. There are no commercial alternatives for the use of peel in Brazil. In Thailand, the peel is sometimes used for mushroom cultivation. The research done in Brazil shows the satisfactory results obtained in Shitake cultivations using mixtures of bran and peel (Felinto, 1999).

The bran obtained by companies in Brazil, with 85 to 90% moisture content, is undervalued by most companies. Some of the starch plants dry and sell it to be used in flotation in the mining industry or as animal feed. One company is starting to commercialize cassava bran for food consumption, as a fiber product. In Thailand, bran can be dried in concrete yards like those used to dry chips and used mixed with chips in the production of pellets, or it may be used by some companies as a raw material to produce citric acid.

With regard to cassava wastewater, both in Thailand and in Brazil, most of the companies use decantation ponds. The major change that has occurred in the last 20 years in the cassava starch companies, both in Brazil and in Thailand, is the use of this wastewater for biogas production (Fig. 2.27). Despite the improvements, in Brazil, there are no controls on the efficiency of the treatment and gas production.

Based on information collected in Fig. 2.2 and already mentioned by Vilpoux (2003), biogas can supply 75% of the power needed to operate the flash-dryer in the case of indirect starch drying process and 100% of the power needed in the case of direct drying process (Fig. 2.28A), with surplus for production of the electric power needed for processing (Fig. 2.28B). Besides considerably reducing energy costs, the use of biogas can increase

FIGURE 2.27 Decantation pond covered with tarpaulin for biogas retention in a Brazilian production unit.

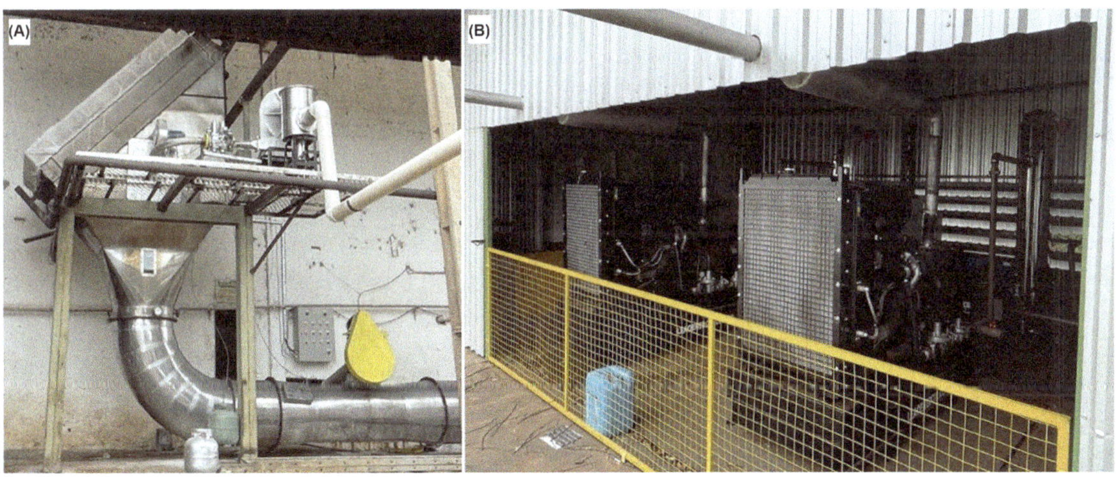

FIGURE 2.28 Direct burning heating system from biogas in flash-dryer (A) and electric energy generators from biogas in a Brazilian cassava starch production company (B).

production in the Brazilian units by 20%. Today, companies stop production for 4 hours a day, during peak hours of energy consumption, when energy costs can be more than 3 times higher. In the case of their own electricity production, they will no longer need to stop and will be able to work 24 hours a day.

In China, in the exception of the modern companies, the cassava wastewater is discharged into unprepared ponds. The problem in this country is even more serious since a great volume of water is used in the process, which may reach up to 20 cubic meters per ton of starch (Dou et al., 2021), which is more than twice the volumes used in Brazil or Thailand.

2.4 Final considerations

Starch processing technologies are similar in Thailand, Brazil, and China, for the most moderns, and show equivalent effectiveness and production costs. Compared with Brazilian starch plants, the larger size of Thai companies and some of the Chinese starch plants does not mean gains in efficiency and quality. The average productivity is higher in the Brazilian companies and the quality of the starch is equivalent.

Even if multinational companies have an important presence in industrial production, especially in Thailand, cassava starch production remains behind the extraction processes of other types of starch. Starch losses of almost one-third in cassava are unthinkable regarding potato, corn, and wheat starch extraction. Recovering part of this starch may provide the companies with a considerably increased competitiveness in relation to other cassava starch plants operating with traditional production systems and to other commercial starch types.

At the same time, enhanced use of byproducts, as for other commercial starches, may further improve cassava competitiveness and strengthen its position as the second most produced starch in the world. The use of biogas was an important step, but others are possible, such as the use of wastewater after biogas production and the use of bran and peel.

Acknowledgments

We thank the cassava starch companies *Podium Industry* and *Pantanal* and the equipment company *Paranavai Maquinas*, for the information provided and for the authorization to take pictures of their equipment.

References

Arthey, T., Srisompun, O., & Zimmer, Y. (2018). *Cassava Production and Processing in Thailand*. Agri Benchmark. Braunschweig, Germany. [Online] <http://www.agribenchmark.org/fileadmin/Dateiablage/B-Cash-Crop/Reports/CassavaReportFinal-181030.pdf>.

Assawamartbunlue, K., & Luknongbu, W. (2020). Specific energy consumption of native starch industry in Thailand. *Energy Reports*, *6*, 299–303.

BeMiller, J., & Whistler, R. (2009). *Starch chemistry and technology* (3rd ed. Elsevier ed.). Burlington: Academic Press.

Bergthaller, W., Witt, W., & Goldau, H. P. (1999). Potato starch technology. *Starch-Stärke*, *51*(7), 235–242.

Chavalparit, O., & Ongwandee, M. (2009). Clean technology for the tapioca starch industry in Thailand. *Journal of Cleaner Production*, *17*(2), 105–110.

Dou, G., Wang, X., Zhao, B., Yuan, X., Pan, C., Tran, T., Zellweger, H., Zhu, K., Guo, Y., Wu, H., & Yin, J. (2021). The transformation and outcome of traditional cassava starch processing in Guangxi, China. *Environmental Technology*, *42*(21), 3278–3287.

FAOSTAT. (2022). Rome, Italy. <http://www.fao.org/faostat/en/#data> (accessed 10.2.22).

Felinto, A.S. (1999). *Cultivo de cogumelos comestíveis do gênero Pleurotus spp em resíduos agroindustriais*. Master Thesis (Food Science and Technology) – Agricultural Science School Luiz de Queiroz - Universidade de São Paulo, Piracicaba.

Fu, H., Qu, Y., & Pan, Y. (2018). Efficiency of cassava production in China: Empirical analysis of field surveys from six provinces. *Applied Sciences*, *8*(8), 1356.

Huidan Research (2012). *Survey and Forecast of China Tapioca Starch Market 2012–2016*. <https://www.prnewswire.com/news-releases/huidianresearchcom-survey-and-forecast-of-china-tapioca-starch-market-2012-2016-170312106.html>.

INews (2020). *China's cassava starch output is declining year by year*. https://inf.news/en/economy/41bfa2c-d8674241a1b6ca17075c32500.html.

IPARDES. (2022) *Instituto Paranaense de Desenvolvimento Econômico e Social Base de dados do estado*. <http://www.ipardes.gov.br/imp/index.php>.

Jacquet, S. (1999). *Otimização do processo de extração nas fecularias de mandioca* (p. 24) Botucatu: Botucatu: Tropical Roots and Starch Center– Universidade Estadual Paulista.

Jakrawatana, N., Pingmuangleka, P., & Gheewala, S. H. (2016). Material flow management and cleaner production of cassava processing for future food, feed and fuel in Thailand. *Journal of Cleaner Production*, *134*, 633–641.

Pingmuanglek, P., Jakrawatana, N., & Gheewala, S. H. (2017). Supply chain analysis for cassava starch production: cleaner production opportunities and benefits. *Journal of Cleaner Production*, *162*, 1075–1084.

Saengchan, K., Nopharatana, M., & Songkasiri, W. (2014). Recovery of tapioca starch from pulp in a conical basket centrifuge–Effects of rotational speed and liquid to solid (L/S) ratio on cake formation and starch–pulp separation efficiency. *Separation and Purification Technology*, *127*, 192–201.

Sánchez, A. S., Silva, Y. L., Kalid, R. A., Cohim, E., & Torres, E. A. (2017). Waste bio-refineries for the cassava starch industry: New trends and review of alternatives. *Renewable and Sustainable Energy Reviews*, *73*, 1265–1275.

Sarmento, S.B. S. (1997). *Caracterização da fécula de mandioca (Manihot esculenta C.) no período de colheita de cultivares de uso industrial*. p. 162. Doctoral Dissertation (Food Science) - Pharmaceutical Science School - Universidade de São Paulo, São Paulo.

Sriroth, K., Wanlapatit, S., Piyachomkwan, K., & Oates, C. G. (1998). Improved cassava starch granule stability in the presence of sulphur dioxide. *Starch-Stärke*, *50*(11-12), 466–473.

Sriroth, K., Piyachomkwan, K., Wanlapatit, S., & Oates, C. G. (2000). Cassava starch technology: The Thai experience. *Starch/Stärke, Weinhein*, *52*, 439–449.

Suvittawat, A., Soni, P., Igel, B., & Khang, D. B. (2014). Cassava root supply chain in Nakhorn Ratchasima province: Challenges and opportunities for local starch mills. *Journal of Food, Agriculture & Environment*, *12*(1), 332–338.

Tran, T., Da, G., Moreno-Santander, M. A., Vélez-Hernández, G. A., Giraldo-Toro, A., Piyachomkwan, K., Sriroth, K., & Dufour, D. (2015). A comparison of energy use, water use and carbon footprint of cassava starch production in Thailand, Vietnam and Colombia. *Resources, Conservation and Recycling*, *100*, 31–40.

Usubharatana, P., & Phungrassami, H. (2015). Carbon footprint of cassava starch production in North-Eastern Thailand. *Procedia CIRP*, *29*, 462–467.

Vilpoux, O. F. (2003). Cap. 7. Cassava starch production processes Brazil, Thailand, and China. In (first ed). M. P. Cereda, & O. F. Vilpoux (Eds.), *Tecnologia, usos e potencialidades de tuberosas amiláceas Latino Americanas* (3). São Paulo: Fundação Cargill.

CHAPTER

3

Extraction of starch from small industrial units: examples with sour cassava starch and starches from canna (*Canna edulis*), sweet potatoes (*Ipomea batatas*), and arrowroot (*Maranta arundinacea*)

Olivier François Vilpoux[1,2] *and Marney Pascoli Cereda*[2,3]

[1]Federal University of Rondonopolis, Post-graduate Program in Environmental Management and Technology - PPGTA, Rondonopolis, Mato Grosso, Brazil [2]Tecsapp Solutions, Campo Grande, Mato Grosso do Sul, Brazil [3]Retired from São Paulo State University, Botucatu, São Paulo, Brazil

3.1 Introduction

Commercial production of starch is traditionally obtained from four raw materials, that is, corn, representing most of the world's production, potato, wheat, and cassava (see Chapter 3 of Volume 1 in this series). The processing of these raw materials is usually done on a large scale with modern equipment. In the case of cassava, the processing was described in Chapter 2 and will not be covered here.

However, in addition to the referred raw materials, it is possible to find alternative regional productions of starches, the main ones being rice, sweet potato (*Ipomea batatas*), arrowroot (*Maranta arundinacea*), and canna (*Canna edulis*, or *C. indica*)[1]. Even if there are some modern production units for processing these raw materials, with facilities like those used for cassava starch (see Chapter 2), most of the production is carried out in small units with different processes and adapted equipment, often locally manufactured. Small units were considered those with a processing capacity between 1 and 10 tons of raw material per day and with an essentially mechanized process.

Similar solutions can also be found in the production of sour starch from cassava, a traditional

[1]Agricultural production of those plants is described in the second volume of this series.

DOI: https://doi.org/10.1016/B978-0-323-90842-9.00008-X

product found in Brazil ("polvilho azedo") and in several Spanish-speaking Latin American countries ("almidone agrio"). This starch has expansion properties without the need for extrusion or leavening agents, but for this, it needs to be submitted to a fermentation and sun drying process (see Chapter 7).

The largest part of starches from sweet potato, arrowroot, and canna is extracted for commercial purposes in Asia, where China is one of the most diversified producers. In Latin America, Colombia and Venezuela also produce starch from arrowroot and canna for local markets, where it is called *achira*. In the past, Brazil used to produce arrowroot starch, but due to the competitiveness of corn and cassava starches, which are cheaper, most of processing units are no longer operating; artisanal production is only remaining. In Brazil, satisfactory results were obtained from the tests made with sweet potato in traditional cassava starch plants, in terms both of extraction and quality. Sweet potato starch production can also be found in Japan. In the case of sour cassava starch, production is limited to a few Latin American countries, such as Brazil, Colombia, Venezuela, and Paraguay.

The benefit of producing starches from alternative sources is obtaining differential products in relation to traditional ones, with market niches and at potentially higher prices. Most of these types of raw materials may be processed in the same way as cassava starch. In addition to sweet potato, tests with arrowroot have been made with satisfactory results in cassava starch units in Brazil. For canna, Vilpoux (2003) describes an industrial unit in the province of Guizhou, China, with a daily grinding capacity of 200 tons of rhizomes. The equipment was like the one used in the traditional cassava starch plants of China (see Chapter 2), except for the hammer mill, which is not efficient for this type of raw material and was replaced by a drum grinder, as in cassava starch processing companies in Thailand and Brazil. Also, in the province of Guangxi, in the South of China, Vilpoux (2003) described an arrowroot starch plant with a daily grinding capacity of 200 tons. This starch plant was also the same as those described for the extraction of cassava starch.

If large-scale starch production processes are well known, those used in small units are little covered in the literature. Vilpoux (2003) described small units of canna starch in the province of Guizhou, China, similar to those existing in the Brazilian State of Minas Gerais, Brazil, producing fermented cassava starch, or sour starch (about fermented cassava starch, see Chapter 7 of this volume). The processing methods in small starch companies are different from the big industries and are described in this chapter. Very small units, with a manual extraction process, are not covered in this chapter, but are addressed in Chapter 9.

3.2 Starch production process

To illustrate starch extraction in small units, this item takes as a basis starch extraction units from canna in the province of Guizhou, China, and sour starch processing in Brazil. Canna is easy to cultivate and has a high productivity, which amounts to 80 t/ha after 8–9 months in nitrogen-rich soil, and 40 t/ha in poorer soils (Hermann et al., 1998). This cultivation is described in Chapter 7 in Volume 2 of this series.

Canna processing to obtain starch in Vietnam is described by Hermann et al. (1998) and is very similar to the processing method used in Southern China. Fig. 3.1A shows a canna plantation in China, where several variety tests are being carried out. Certain varieties of canna are ornamental, with a poor rhizome production, and other specific varieties are grown for the purpose of starch extraction.

The first cycle in the industrial process is the transportation of rhizomes. In the small canna starch processing units, transportation is

FIGURE 3.1 Canna varieties cultivated in an experimental field in China (A) and cultivated in Brazil, 4 months after planting (B).

FIGURE 3.2 Delivery of raw material in China for canna processing.

directly made by the farmers, without using mechanical transportation (Fig. 3.2).

Once they are delivered, the canna rhizomes are stored on the unit's ground. It is not necessary to peel the rhizomes, but only to remove the earth from them. For this purpose, they are washed with water in a fiber-cement tank (Fig. 3.3A). To automate the washing process and increase the plants' capacity, the rhizomes could be sent to a washing and peeling drum,

FIGURE 3.3 Washing of canna rhizomes in a water tank in China (A) and washing and peeling drum used in small cassava units in Brazil (B and C).

like those used in Asia for the initial cleaning cycle of cassava (see Figure 2.8 in Chapter 2) or in some units in Brazil (Figure 2.10 in Chapter 2). The only difference would be the addition of water during this cycle, since in Chinese cassava starch plants, the cleaning process is dry. Doing Holdings, an equipment manufacturer, describes such an equipment for processing sweet potato starch (Doing, 2019).

In Brazil, small production units of sour starch and cassava flour use a similar washing and peeling system but in batches and with the addition of water (Fig. 3.3B and C). In small units, this system could also be used for washing canna, arrowroot, and sweet potatoes. In the case of cassava, it also allows for the removal of the periderm, or bark.

Estrada Rivera (2020) described the same type of equipment for canna starch extraction in Colombia, but with a metal drum, fewer slots, and side filling.

Once washed, canna rhizomes are ground in a drum grinder. This grater is similar to the one used in Brazilian cassava starch or flour companies but has a smaller diameter. For larger volumes of raw material, the same grinder used in starch factories may be used (Vilpoux, 2003). Grinding is carried out under water (Fig. 3.4).

Canna rhizomes have a large quantity of flakes, similar to ginger and arrowroot. These flakes hinder the grinding process and make the use of a hammer mill difficult since the cellulose flakes are not grated and get clogged at

FIGURE 3.4 Canna rhizomes grounded in a drum grater in China.

the output screen of the grater. For cassava and sweet potatoes, a hammer mill can be used. Doing Holdings indicates that for sweet potatoes, milling is essentially done by a hammer mill (Doing, 2017). Despite the difficulties of using a hammer mill for arrowroot and canna, Branco and Cereda, as shown in Chapter 4, used this type of mill successfully.

Once the grinding process is completed, the grated paste passes through a vibrating screen to extract the starch (Fig. 3.5A). These screens are very similar to those that were used in Brazil, until the early 2000s, in small sour starch companies to remove fine fibers from starch suspension, after passing through centrifugal screen (Fig. 3.5B). The grated paste passes through the screen, where it shakes due to the vibrations. A water jet above the screen goes through the paste and draws the starch, whereas the washed fibers are removed at the final section of the equipment.

There is also another sieve system, a rotatory extractor with a hexagonal or round shape (Fig. 3.5C). This type of extractor was described by Estrada Rivera (2020) for canna extraction in small units in Colombia and was tested in Brazil by Branco et al. (2020) to compare starch extraction from arrowroot and cassava. This equipment is described in more detail in Chapter 4.

The vibrating screen is not very effective and results in a great loss of starch. It was used in Brazil only to remove the finest fibers but not as the main extraction equipment. Vibrating screens were used for extraction only in a few small cassava starch plants. The use of centrifugal conical sieve would allow for an enhanced performance. Today small companies can use

FIGURE 3.5 Starch extraction in a vibrating sieve under water jets for canna starch extraction in China (A) and for cassava starch extraction in Brazil (B) and rotatory hexagonal sieve (C).

the same centrifugal sieves as the big ones, but some use smaller ones with nylon sieves, which are cheaper (Fig. 3.6A). Often, to cheapen the process and the initial investment, a smaller number of centrifugal sieves are used than in large cassava starch units, as seen in Chapter 2.

In Asia, a large canna starch extraction company uses a vertical screen extractor (Fig. 3.6B). This type of extractor uses less energy, needs less maintenance, and is cheap. It could be used in smaller companies.

After extraction, the starch milk passes through a decanting channel, as shown in Fig. 3.7. To save space, a small canna starch processing unit used a U-shaped channel, which goes round the processing unit (Fig. 3.7A and B). This system is a variant of a labyrinth separator, used in many small Brazilian cassava starch plants (Fig. 3.7C). The heavier impurities settle at the beginning of the channel, with the starch granules decanting afterward, while the lighter impurities and residual water exit by gravity at the end of the channel.

Canna rhizomes have a high phenolic content, so the starch and the water are black. To clean the starch, a second washing cycle is carried out by using a settling tank (Fig. 3.8A and B). In this tank, the starch is washed and left to decant. The clean starch stays at the bottom, while the dirty liquid with most of the phenols goes to the surface. According to Hermann et al. (1998), this washing process must be repeated several times.

Estrada Rivera (2020) also indicated the use of tiled tanks for canna starch decantation and

FIGURE 3.6 Centrifugal sieves with nylon screens (A) and vertical sieve extractor (B).

FIGURE 3.7 Canna starch extraction unit in China, with a U-shaped decanting channel (A and B) and the decanting channel in the labyrinth used in Brazil (C).

FIGURE 3.8 Decantation tank for canna starch purification (A and B) for sweet potato starch (C) and for cassava sour starch (D).

cleaning, tanks similar to those used in Brazil for cassava starch fermentation. However, unlike the process observed in China, in the process described by Estrada Rivera (2020), the starch goes directly from the extraction process to the tanks, without passing through a sedimentation channel or a labyrinth decanter.

For sweet potato starch, Wang et al. (2020) indicated that the use of decanting in concrete tanks (Fig. 3.8C) allows for a brighter color of starch than the use of more modern concentration and refinement systems, such as those seen in Chapter 2. In this case, extraction can be done with centrifugal sieves. During the decanting of the starch, there is a slight fermentation with increased acidity, which can help to neutralize the phenols' color from the sweet potato and explain the lighter color than in a faster process. The same works for canna starch purification.

The traditional method of sedimentation is widely used in China, even by the biggest producer of sweet potato starch. Wang et al. (2020) described a production unit with this system that produces annually 120,000 tons of starch, 70,000 tons of noodles, and 10,000 tons of vermicelli. Despite the preference for the decantation system in tanks, the Doing Holdings (Doing, 2017, 2019) commercializes large-scale units for sweet potatoes with the use of centrifuge sieves, hydrocyclones, and vacuum filter (for the description of these kind of equipment, see Chapter 2).

In Brazil, sour starch industries also use concrete or tiled tanks (Fig. 3.8D) for the fermentation of cassava starch, always after a labyrinth separator. The permanence of the starch in the tanks increases the acidity, even with a permanence of a few days as in the case of canna and sweet potato starches. In this case, the acidity can cause the tiles to fall in the masonry tanks and the presence of sand in the starch in the case of concrete tanks.

In the decantation tank of canna starch (Fig. 3.8A and B), where a second washing cycle is carried out, the upper layer with phenols is still mixed with the starch. To separate the starch, the upper layer in the tank is sifted and washed using a cloth screen and clean water (Fig. 3.9). Most of the phenols are removed, and the starch retained in the cloth is partially purified.

Processing units in Brazil use a lump breaker, which is a type of mill, to break the starch blocks that come out of the tanks.

After decantation, moisture content in the starch is between 45% and 50%, which is too high to enter the stage of noodles production in the example of canna starch. To reduce moisture content to 35%–40%, a dewatering process is made in traditional fast-rotation basket centrifuges (Fig. 3.10).

After dewatering, most of the starch was sold still humid in noodle processing units, or to a larger starch plant, where it was processed once again for cleaning and drying in a flash dryer (see Chapter 2). Then, at this facility, the canna starch was packed into 25 kg bags (Fig. 3.11) and sold throughout China. In addition to noodles, canna starch is used in the production of candies, pregelatinized starch, and pharmaceutical products. The noodle production in artisanal processing units is presented in Chapter 8.

In sour starch production, in Latin America, the starch is sun-dried on wooden or bamboo frames, in open air to facilitate drying (Fig. 3.12). The starch is placed on white cloths, which cover the wooden or bamboo frames, and turned over by hand on a regular basis.

FIGURE 3.9 Sifting of the settling tank's surface for canna starch.

FIGURE 3.10 Starch basket centrifuge and canna starch dewatering in Southern China.

FIGURE 3.11 Canna starch packing in South of China.

In Colombia, canna starch is sun-dried in plastic-covered greenhouses above 1-meter-high wooden tables (Estrada Rivera, 2020).

In the production of sour starch, the drying cannot be done in a greenhouse covered with plastic, as in the canna drying described by Estrada Rivera (2020). The producers know that if they cover the starch with plastic or cloths, it will not expand as expected. The plastic blocks the UV rays that react with the starch and are important to obtain the characteristic expansion properties of this kind of starch (see Chapter 7 for more explanations).

FIGURE 3.12 Drying systems used in small sour starch industries in Brazil.

After the drying process, a starch with 12%–14% moisture is obtained. For canna starch, moisture content was 48% after decantation, and 12%–13% after drying, with the recovering of 10%–11% of starch, in dry matter. These values are consistent with those obtained by Hermann et al. (1998) in Vietnam, with extraction rates of 10%–15%, most of them between 10% and 11%.

According to Franco et al. (2001), canna rhizomes have 18.5%–23.5% starch content. Based on an average extraction rate of 10.5% in dry starch, total recovery of starch is 45%–60%. The use of a more effective extraction system should result in a significant increase in extraction rates. These results are very low, especially if we consider that canna starch granules are the largest starch granules between the underground starchy crops (Hermann et al., 1998), measuring 35–101 μm in diameter (Leonel & Cereda, 2002). For more information, see also Chapter 10 in Volume 1 of this series. The large size of canna starch granules makes extraction and decanting easier.

3.3 Final considerations

The process of starch production from roots and tubers, irrespective the degree of technology employed, comprises the steps of washing, cell disintegration and release of starch granules, separation of starch from fibers and impurities, starch concentration, and finally drying. Peeling is optional, but the peel may influence the starch quality and purification process, hindering the separation of fibers.

The technologies adopted in small production units for the extraction of starch from canna, arrowroot, sweet potato, or cassava are practically the same all over the world, but with some local specificities. Regional differences include the use, in some places, of small washing and peeling drums, or of vertical centrifugal extraction sieves. This equipment has a lower cost in relation to that used in larger companies and allows for production gains in relation to manual washing and extraction with vibrating screens. Chapter 3 explains the difficulties of developing equipment for the starch extraction industry outside the ready-made machines market.

Starch processing of Canna, in the Guizhou Province, China, provides a livelihood for part of the local community, the farmers, and the employees of the processing plants, which comprise starch and noodle production units. The province of Guizhou is located near the border between China and the Northeast of Vietnam. According to Hermann et al. (1998), in that Vietnamese region, 20,000 hectares of canna were cultivated and used in the extraction of starch for noodle production (for production of noodles from canna starch, see Chapter 8).

Canna's high productivity and a production cycle shorter than that of cassava make it highly interesting for starch production. The combination of a highly productive crop with a modern starch extraction process may increase the interest in this cultivation and make it more competitive as compared with the more traditional commercial types of starch.

In the case of arrowroot, the problem for the development of starch production is the high cost of agricultural production, with low productivity (see Chapter 2 of Volume 2 in this series). The starch is highly valued; the extraction of starch is easier than that of canna or sweet potato, which contain phenolic compounds, but the agricultural production is a difficult hurdle to overcome.

Sweet potato has lower starch yields than cassava and high sugar contents that are lost during extraction. Leonel et al. (1998), in a comparison of starch extraction from cassava and sweet potato in a large Brazilian company with a grinding capacity of 200 tons per day, found extraction percentages of 25.5 for cassava and 18.3 for sweet potato. Another obstacle for the development of sweet potato starch extraction is the high content of phenols, which impairs the final quality of the product, especially its color. Wang et al. (2020) indicated that the use of a settling tank for the purification of the starch allows for a brighter color of starch. However, this process makes large-scale production difficult, since the sedimentation process lasts up to 32 hours. During this time, a slight fermentation occurs, consuming sugar and acidifying the water, which discolors the phenolic compounds.

In the case of the sour starch produced in Latin America, the need for fermentation and drying in the sun, for the product to acquire its expansion capacity, hinders the complete mechanization of the process and makes the product more expensive, besides harming the hygiene of the product with possible contamination with insects and dust. Thus this product is increasingly being substituted by cassava starch, with or without chemical modifications.

References

Branco, F. P., Cereda, M. P., & Naka, M. H. (2020). Comparison of arrowroot (*Maranta arundinacea*) and cassava starch extraction in separation, concentration, and purification using a rotating sieve under water. *Engenharia Agrícola, 40*, 90–95.

Doing. (2017). *How to process sweet potato into starch by sweet potato starch production machine?* <http://www.starch-projectsolution.com/faq/sweet_potato_starch_production_1049.html> (Consulted in February 2023).

Doing. (2019). *Sweet potato starch processing plant.* <https://www.cassavastarchmachine.com/product/Cassava_starch_processing_machine/sweet_potato_starch_processing_plant_151.html> (Consulted in February 2023).

Estrada Rivera, K. D. (2020). *Producción y obtención de almidón de achira Canna edulis con fines comerciales en el municipio de Algeciras Huila.* <https://ciencia.lasalle.edu.co/cgi/viewcontent.cgi?article = 1184&context = ingenieria_agronomica>.

Franco, C. M. L., Daiúto, E., Demiate, I. M., Carvalho, L. J. C. B., Leonel, M., Cereda, M. P., Vilpoux, O., & Sarmento, S. B. S. (2001). Cap. 6. Estrutura dos grânulos de amido. Propriedades gerais do amido. In (first ed.). M. P. Cereda, & O. F. Vilpoux (Eds.), *Propriedades gerais do amido* (Vol. 3). São Paulo: Fundação Cargill.

Hermann, M., Quynh, N. K., & Peters, D. (1998). Reappraisal of edible canna as a high-value starch crop in Vietnam. *CIP Program Report*, *98*, 415–424.

Leonel, M., Jackey, S., & Cereda, M. P. (1998). Processamento industrial de fécula de mandioca e batata doce-um estudo de caso. *Food Science and Technology*, *18*, 343–345.

Leonel, M., & Cereda, M. P. (2002). Caracterização físico-química de algumas tuberosas amiláceas. *Food Science and Technology*, *22*, 65–69.

Vilpoux, O. F. (2003). Cap. 9. Starch obtained from canna (*Canna edulis*) in China. In (first ed.). M. P. Cereda, & O. F. Vilpoux (Eds.), *Tecnologia, usos e potencialidades de tuberosas amiláceas Latino Americanas* (Vol. 3). São Paulo: Fundação Cargill.

Wang, X., Wang, J., Liu, H., Zhao, L., Wang, Y., Wu, X., & Liao, X. (2020). Improving the production efficiency of sweet potato starch using a newly designed sedimentation tank during starch sedimentation process. *Journal of Food Processing and Preservation*, *44*(10), e14811.

CHAPTER

4

Design and evaluation of an equipment for starch release and extraction of underground starchy crops in small processing units

Marney Pascoli Cereda[1,2] *and Fabiano P. Branco*[3]

[1]Retired from São Paulo State University, Botucatu, São Paulo, Brazil [2]Tecsapp Solutions, Campo Grande, Mato Grosso do Sul, Brazil [3]Dom Bosco Catholic University - UCDB and Federal Institute of Mato Grosso do Sul, Campo Grande, Mato Grosso do Sul, Brazil

4.1 Introduction

Starch extraction is a simple process that began with the oldest process recorded, which was from the extraction of starch from wheat, as described in the Treatise "De agri cultura" by Cato the Censor, around 160 BCE (Cato, 1933). The process consisted of soaking the wheat grains in water, changing it twice a day. On the 10th day, after draining the water, the macerated and soft wheat grains were mixed, and the product strained into a new linen cloth before being sun-dried. This whole process was driven by humans or animals.

This simple process has evolved into a strong industrial sector, with the extraction of starches from corn and wheat and the inclusion of other starchy raw materials with more than 60% moisture, such as cassava and potatoes, which therefore do not require maceration and can be simply grated.

This scheme highlights the operations required to produce starch, each of which has the potential to be replaced by equipment (mechanization) or even to be automated. Such equipment already exists on the market and was described in Chapters 2 and 3, which deal with the extraction processes of starch from underground starchy crops, respectively, in large and small units. However, equipment for mechanized starch release and extraction in very small units is more difficult to find.

The starch extraction industry is already consolidated, with equipment found in the international market. However, depending on the raw material to be used and on the size of the company, there may be a need to adapt, improve, or even build new equipment.

Starch Industries
DOI: https://doi.org/10.1016/B978-0-323-90842-9.00010-8

In contrast to the countries of the northern hemisphere, South America has several starchy raw materials that could serve for the starch industry (See Chapter 14 in Volume 1 of this series). However, only cassava is exploited on a large scale, if we exclude potato, which originated in South America but is mainly processed in large processing units in European countries.

All underground starchy crops have this in common: they are harvested with 60% moisture content or more, which requires adaptations with respect to cereal processing, because instead of being ground, they must be grated.

It is also necessary to think about the raw material receiving area. Processing plants as described in Chapter 2 are designed to receive between 200 and 2000 tons of roots or tubers per day. Therefore, when planning a production facility, it is important to know how much raw material is available so that processing is not interrupted. This need is much greater in the case of roots and tubers compared to cereals, due to the perishability of the raw material, which cannot be stored for a long period, and to the high moisture content, which makes transportation more expensive, limiting processing within a small radius of the processing company (Vilpoux, 2011). All this must be established before thinking about the equipment.

South America is not outstanding in mechanical engineering applied to starch extraction projects. Copies of equipment used by European and American companies are more common, which complicates the extraction of starch from underground raw materials because the processes may differ and the equipment used for cereal starch extraction may not be the most adapted.

On the other hand, the construction project of new equipment needs to have a complete survey of the literature because in this area, the search for patents is very important. Among the issues that need to be considered, the first is undoubtedly the user's needs, but keeping an engineering view. In this case, there is a need for safety and ergonomics.

As the underground starchy raw materials have high humidity, they generate a large amount of wastewater, consisting of extraction water added to the raw material water (for the mass balance of a modern processing unit, in Brazil and Thailand, see Chapter 2). This wastewater, as well as the other residues generated, can be used in the field, for fertilization, to generate energy and to produce other products with commercial value, thus ceasing to be a waste and becoming a coproduct. This possibility is just beginning to be perceived in Brazil.

Extraction of starch from underground raw materials requires all the steps mentioned for extracting starch from wheat in Roman times, except the maceration step, since they have sufficient moisture. The first operation is the disintegration of the raw material tissues to release the starch from the cells. Next, it is necessary to separate the fibrous bran from the starch suspension, purify, concentrate, and dry it. In the case of underground starchy raw materials, the starch obtained has few impurities with a low content of lipids and proteins (see Chapters 8 and 10 in Volume 1 of this series). As part of equipment development, it is necessary to evaluate the performance with recognized and standardized methodology and to compare this performance with machines already available in the market.

With this prerequisite, this chapter presents a case study on the design, planning, elaboration, evaluation, and adjustments for a root and tuber disintegrator and starch extractor, to be used by small farmers. Besides cassava, the equipment was tested with arrowroot (*Maranta arundinacea*), a little known but highly valued raw material (See Chapter 2 in Volume 2 of this series).

4.2 Steps and premises for equipment development

First, it is important to consider that it is cheaper and more efficient to buy a machine that is already on the market than to develop a new one. Therefore, it is necessary to think very carefully if it is not possible to adapt an old machine for a new use. Planning a machine takes time, energy, and resources.

Landolsi (2017) reports that machine design is not guided by any rigid rules; however, design is not a random procedure either. There are a number of factors that can aid a machine design process. It is very important to understand the process of mechanical machine design and follow this process to develop designs that deliver the best performance. The author gives a look at the general procedure that helps to design equipment that perform better and are durable and easy to manufacture and to repair.

- Understand the requirements: The first step of engineering a high-performance machine is recognizing the need, that is, understanding the purpose of design and preparing a complete statement of the problem that includes details about the aim or purpose, for which the mechanical design project is proposed.
- Making a written statement: It is necessary to make a written statement of what exactly the problem is for which the machine design must be done. This statement must be very clear and as detailed as possible. It is the list of the aims that are to be achieved from machine design.
- Consider previous experience: It is important to consider any experience and make the necessary changes in the design already available, if any. Further, the designer can also consider their personal judgment to facilitate the production of the machine and machine elements.
- Analyze and evaluate the design mechanism: Analysis of the various possible mechanisms must be done, for them to be considered when designing the machine. From the various options, the best can be selected whenever required.
- *Analysis of forces/constraints/conditions*: It is important to analyze how each component of a machine fits together and interacts, what constraints or conditions act (forces, heat, chemicals, electromagnetic, electrostatic, radiations, etc.), and the energy transmitted by each component. The analysis should, in most cases, return numerical values and/or clear characteristics that can be used as parameters for designing and choosing machine elements.
- *Design/parametrization of elements*: All the components of equipment are subject to stress and other forces/conditions during operation. These stresses affect the strength of the machine, so analyzing these working conditions is important to design equipment that is durable and can endure the stresses of real-world working conditions.
- *Dimensions/properties of the machine elements*: Finally, it is necessary to find out the appropriate dimensions and materials for the machine elements considering the forces and constraints acting, its material, and limits of use. The size of the machine elements should be such that they should not distort or break when loads are applied in working conditions.

When the machine design is ready, Barth et al. (2016) highlighted the steps to be considered for equipment development.

- Selection of the standards parts/components: When the list of components is defined and the characteristic of each part is known, it is recommended to choose what parts can be selected from constructor catalogs that are ready-to-use.

  According to Norton (2013), one or more safety coefficients must be calculated to estimate the probability of failure. There may also be standards that must be adopted. Some of the engineering societies and government, industry, and international

organizations that publish standards and norms of potential interest to mechanical engineering are the American Gear Manufacturers Association (AGMA), American Institute of Steel Construction (AISC), American Iron and Steel Institute (AISI), American National Standards Institute (ANSI), American Society for Metals (ASM International), American Society of Mechanical Engineers (ASME), American Society of Testing and Materials (ASTM), American Welding Society (AWS), Anti-Friction Bearing Manufacturers Association (AFBMA), International Standards Organization (ISO), National Institute of Standards and Technology (NIST), Society of Automotive Engineers (SAE), and the Society of Plastics Engineers (SPE). In South America, it is possible to find equivalent institutions, such as the Brazilian Association of Standards and Techniques (ABNT), which serves all types of machines (forestry, agricultural, civil engineering, and industries of all kinds).
- Make drawings/documentation: After designing the machine and machine elements, it is important to make assembly drawings of the entire machine and detailed drawings of all machine elements. These drawings must clearly specify the dimensions of the assembly and the machine elements, their total number, the material, and method of their production. The designer should also specify the accuracy, surface finish, and other related parameters for the machine elements. It is important to create detailed mechanical drawings of each component separately and of the total machine assembly. Complete specification using CAD software can be of great assistance for the manufacturing process, repairing, and recycling.
- Design iteration for manufacturing: If necessary, it is recommended to make design changes that facilitate the manufacture of the machine and its components so as not to affect the performance of the machine.
- Equipment manual: Finally, once the equipment is finalized and evaluated, it is very important to make a complete manual, which explains to users the components of the equipment, how to turn it on and off, and maintenance and care that the operator should take to ensure his own safety and that of people in the surroundings. It should be encouraged to write down a preventive maintenance schedule and spare parts in case of required maintenance. This care should ensure longer usage time and avoid stops due to problems with the operation of the equipment.
- Other needs: Close to the place where the processing unit will be installed, there is a need for many services that will not be analyzed in this chapter, such as, in the case of a starch processing company, potable and industrial water, place for unloading raw material, energy, and so forth. Other aspects are more related to the evaluation of the company itself or the processes, such as starch extraction efficiency analysis, determination of particle size after milling or grinding, starch recovery, and evaluation of the starch content in the extraction residue and in the extraction water (for more details on the mentioned analyses, it is suggested to refer to Chapter 12 and 13 of Volume 1 in this series).

4.3 Used technologies

For equipment development, a recommended technique is to select and evaluate already existing operations for similar raw materials, that is, in the case of underground starchy crops, with moisture content higher than 60%. These processes are already available on a commercial scale for roots (cassava), tubers (sweet potato),

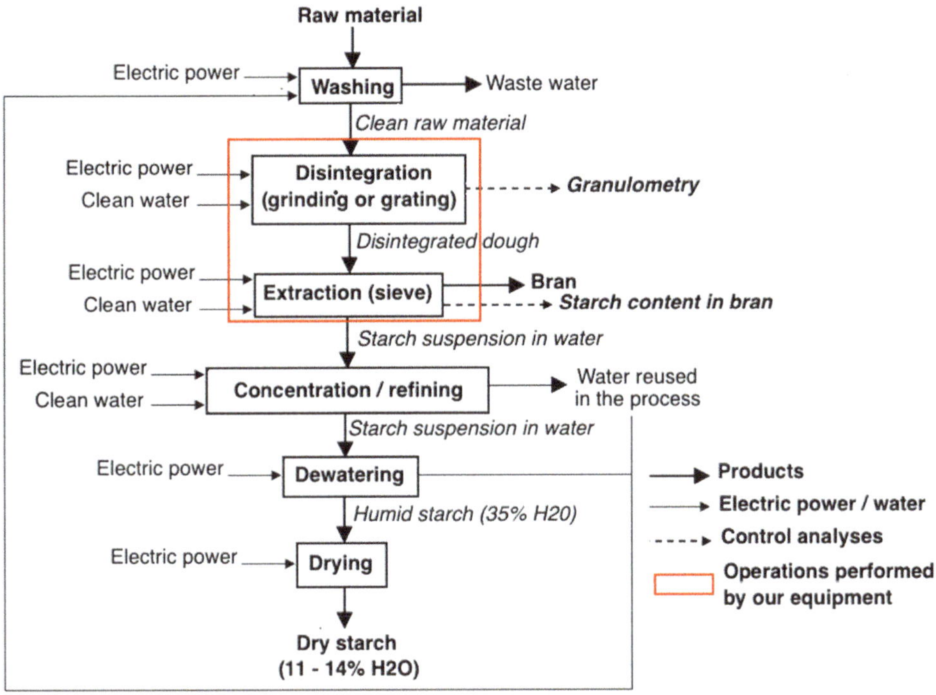

FIGURE 4.1 Operations for extracting starch from underground starchy crops and operations of interest for developing a disintegration-starch extraction equipment. *Adapted from Cereda, M. P., & Vilpoux, O. F. (2003). Processo de produção de fécula de mandioca: comparação Brasil, Tailândia e China. In: M. P. Cereda (ed.),* Agricultura: Tuberosas Amiláceas Latino Americanas. Volume 3: Tecnologias, usos e potencialidades de tuberosas amiláceas latino americanas *(pp. 143–175). 1st ed. São Paulo: Fundação Cargill, see also Chapter 2.*

and even rhizomes (arrowroot), mainly in Asia, where large companies processing these raw materials can be found (see Chapter 2 for starch processing in large units). On a smaller scale, there are many examples in Latin America, such as the production of cassava sour starch in Brazil, Colombia, and Paraguay, or the production of canna (*Canna edulis*) and arrowroot (*Maranta arundinacea*) starches in Colombia and Venezuela, as well as in Asia (See Chapter 3 for starch processing in small units).

For Sajeev et al. (2012), the operations for starch extraction are common to all raw materials and consist of disintegration, extraction, purification, concentration, and drying, but may present small changes depending on the raw material used.

The starch production process consists of performing a series of sequential operations starting with the disintegration of the raw material, followed by separation processes of fiber, water, and starch, and finally starch drying (Saengchan et al.i, 2014).

Each operation uses specific equipment that must compose a harmonic unity as they interact with those of the other operations. Fig. 4.1 shows the operations necessary for starch extraction, indicating the inputs and outputs for each one of them. Basically, the raw material must go through the operations of washing, which in the case of cassava also allows for peeling part of the roots, disintegration, extraction, concentration and refining, dewatering of the starch, and then drying.

Adjustments may occur for each raw material due to its characteristics. As the equipment developed is for disintegration and starch extraction, the analysis should focus only on these two steps.

4.3.1 Evaluation of starch content of roots

Although this item is not directly connected with the development of the equipment, knowing in advance the starch content in the raw material allows for the control of the extraction process. Without dispensing with the official method for determining dry matter (DM) and starch (for more details see Chapter 12 in Volume 1 of this series), the hydrostatic balance is widely used for a quick assessment of DM and for the payment of cassava roots in processing units (Fig. 4.2).

In this method, some care should be taken in the preparation of the roots, which should be previously brushed and washed, but not peeled. The evaluation on a hydrostatic balance assumes that the weight of the fresh raw material includes all its components, including its water content. When it is weighed and immersed in water, the water from the roots will have the same density as the water outside, so only the DM will be weighed. However, the weight under water includes not only the starch but other soluble and insoluble components such as sugars and fibers (Iritani & Weiler, 1976). As it is well established that the starch content in cassava roots represents more than 80% of the DM, there is a strong correlation of this with starch (for more information, see Chapter 8 in Volume 1 of this series).

Despite the lack of precision, estimating the amount of preexisting starch in the raw material is important for determining the extraction yield and paying the suppliers on a quality basis.

In Brazil, to evaluate the DM content in cassava roots with hydrostatic balance, Formula 1, developed by Grossman and Freitas (1950), is usually used.

$$DM = 15.75 + 0.0564 \times P \quad (4.1)$$

where P is the weight of 3.0 kg of roots submerged in water. For evaluation of the starch content, the value of 4.65 is subtracted from the DM value. These formulas are generally used in field research, but the results are quite questionable, as they are influenced by variety, root age at harvest, period of the year, and root size, among others.

Usually, companies use the weight in water of 3 or 5 kg of roots weighed in air, without converting this to DM. Thus, cassava roots are paid according to the value attributed to the gram of cassava weighed in water, which also raises the problem of the reliability of the results.

Stanley and Jewell (1989) point out that for potatoes, density varies significantly between different species grown under similar conditions and even between potatoes of the same species grown under different environmental and nutritional conditions. Another question in the use of the hydrostatic balance is that the value obtained does not represent only the starch. According to Iritani and Weller (1976), during storage of potatoes, a linear relationship between specific weight and accumulated sugar was found, which may also interfere with the determination and results of the hydrostatic balance by increasing the starch content.

Nevertheless, the method of starch evaluation by hydrostatic balance is sufficiently fast and accurate for potatoes. This tuber is harvested at the end of its cycle, but there are doubts whether this method can be applied to other underground starchy crops, which are not harvested at the end of their cycle, such as cassava, and which differ greatly in physiology and morphology (to understand differences in storage organs, we suggest reading Chapter 4 in Volume 1 of this Series).

In the case of cassava, results obtained in Brazil indicate variation of reliability in the use of hydrostatic balance as a function of age at

FIGURE 4.2 Automated hydrostatic balance (A) and traditional manual system (B and C) weighing in air and in water. *Photos provided by Vilpoux O. F.*

harvest. Roots harvested less than 8 months after planting showed low correlation between DM assessed with the balance and DM assessed in laboratory. With increasing age of the plant, the correlation improves. Results of experiments conducted in Brazil show that there was variation in DM with cultivation time, but that the value obtained on the balance always overestimated the DM evaluated in laboratory, with a greater error for lower DM contents. These results agree with those of Maeda and Dip (2000) for potato and cassava and those of Kawano et al. (1987) for cassava between 30% and 40% DM. Pola et al. (2020) observed higher average DM contents with hydrostatic balance in relation to the contents measured in laboratory with 4 cultivars harvested between 10 and 13 months after planting and lower contents when harvested from 14 to 22 months after planting.

Maeda and Dip (2000) checked the relation between specific gravity and water mass percentage in potato and cassava and detailed the care that should be taken to obtain coherent results. As these are cylindrical organs, the wire that holds the container where the roots and tubers are immersed in water must be aligned with the balance axis and attached to the center of gravity of the roots to prevent any part from remaining above the surface of the water, falsifying the results. Considering these precautions, the authors considered that it was possible to obtain a good correlation between density estimated with the hydrostatic balance and DM content for both potato and cassava. However, Carvalho et al. (2007) disagreed with these conclusions for cassava and stated that the hydrostatic balance is not accurate, with great variability of results.

It should be remembered that the main advantage of the hydrostatic balance is its ease of use and speed of response in relation to standard determination. This explains why, in the last 10 years in Brazil, it started to be used by all the large cassava starch processing units for the payment of their raw material.

4.3.2 Disintegration (grating or grinding)

The disintegration, performed by grating, or grinding, is the most important operation (Sajeev et al., 2012) because it destroys the plant cell walls and exposes the starch granules inside (Leonel & Cereda, 2000). The industrial process of grinding, or size reduction, can be done in different equipment, divided into four generic types: compression, impact, friction, and shear (Mckinney, 2006). An ideal disintegrator should consume little energy per unit mass of product and reduce the raw material to a uniform size with adequate size (MacCabe et al., 1993).

The most efficient grater is known by the name of Jahn Rasper (Fig. 4.3A). It is made with serrated blades fixed in a drum and very close together (Fig. 4.3B), similar to the arrangement of a shark's teeth. It is the most commercially used model in large and small industrial scales because of its high efficiency and the fact that the blades can be individually replaced. Each individual blade is made of approximately 1 mm thick steel with variable lengths, which can vary from 10 and 20–30 cm according to the drum where it is installed. In general, the blade is 3 mm high with about 1.5 mm of that being visible in the drum (Cecil, 1992; Grace, 1977).

Cecil (1992) recommended that the blades be fixed by blocks, usually of wood 10 mm thick and 2–3 mm wide. For a diameter of 30 cm, the rotation should be 1500 rpm. If the diameter is 50 cm, the rotation should be 1200 rpm. Above a certain drum speed, however, there is no significant increase in grating efficiency. There is therefore an optimum speed for the grater according to the particle size obtained and the economy of power supply.

This type of grater promotes size reduction through friction and shearing of saw teeth, which therefore must be replaced or sharpened periodically for disintegration efficiency to remain high (Sajeev et al., 2012).

The same system, but much more rustic, was described by Cecil (1992) as a grater with a wooden cylinder covered with a grating surface composed of sheet metal perforated with nail

FIGURE 4.3 Grater used in cassava starch extraction (A) and wooden rotor with carbon steel cutting blades (B). *Photos provided by Vilpoux O. F.*

tips, with a capacity to process up to 500 kg of raw material per day. In the same concept, Sheriff and Balagopalan (1999) described a multipurpose grater they developed for starch extraction, with a capacity of 75–125 kg/h. The authors evaluated the performance of the machine for various tuber crops. Sajeev and Balagopalan (2005) designed a grater for starch extraction from yam, cassava, and potato, with a capacity of 120–200 kg/h.

Despite being used in Asia in large industries, in Latin America, the hammer mill is used little, even though some companies have recently started to use it as the first grind, before the traditional grater, used as the second grind. The hammer mill consists of hammers mounted on a rotating shaft (Fig. 4.4). Under the bottom of the machine are fixed perforated screens, which allow the fragments of the disintegrated material to pass, but retain the larger pieces (Souza et al., 2010). These screens improve the efficiency of the machine, preventing pieces from passing through without being crushed.

Hammer mills are most often used for hard materials (MacCabe et al., 1993). For roots and tubers, hammer mills have the advantage of low maintenance requirements compared to drum and blade teeth mills. When using hammer mills to disintegrate cassava roots, there may be greater heating of the ground material and consequently greater water consumption when compared to the traditional grating process (Cereda & Vilpoux, 2003).

According to Da et al. (2013), the degree of disintegration is a key performance factor in starch extraction because the smaller the particle size obtained after disintegration, the greater the amount of starch extracted due to the destruction of the plant cell in the raw material. The same authors report that the faster drum speeds and sharper blade teethes are decisive in the particle size obtained and therefore improve the extraction efficiency. However, much smaller particles make it difficult to separate the starch from the bran in the subsequent process (Cecil, 1992; Cereda & Vilpoux, 2003).

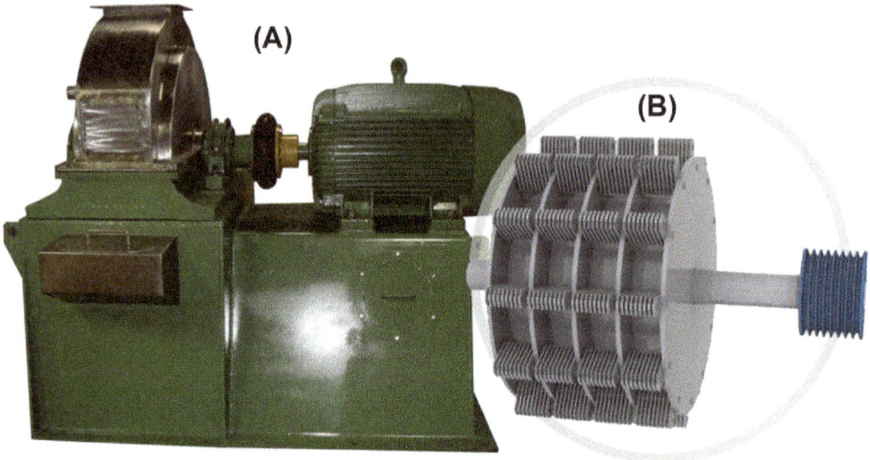

FIGURE 4.4 Hammer mill (A) and rotor for grinding (B). *Photo from Paranavaí Máquinas.*

It is important to note that depending on the efficiency of the grinding process, two types of starch can be found in the bran from the starch extraction, granules inside the plant cells that were not torn during disintegration and granules outside the cells, but which are not removed during the whole processing because they seem to be bound to the cassava fibers (Cereda & Vilpoux, 2003).

The percentage of released starch is evaluated by an index called the rasping effect. Its value after disintegration can vary between 50% and 90% (Grace, 1977). In general, in a typical starch extraction industry, a single grater is used, but the use of a second grater is possible and can recover about 4% of the residual starch. However, a too radical grating can hinder the process of separation of starch from fibers (Cecil, 1992).

Starch retention rates in potato bran in Europe are around 2%, referred to as bound starch (BeMiller & Whistler, 2009). Concerning potato starch extraction, a minimum of 95% recovery is reached in modern potato starch plants, but optimum engineering (rasping, decanting, and sieving) gives recovery rates of 97%–98% (Bergthaller et al., 1999). For cassava, starch losses in the bran are much higher, around 33.5% in Brazilian industries (Leonel & Cereda, 2000) and 41%–62% in small units in Colombia and Vietnam (Da et al., 2013). It is not well known why this difference, but losses in cassava are related to starch retained in plant cells that have not been fully ruptured (Leonel & Cereda, 2000; Saengchan et al., 2015).

The calculation of the motor power to activate the grater or hammer mill is a very important factor to define energy costs and motor wear, but no literature information was found for this calculation. Fig. 4.5 represents the relationship between specific power consumption of cassava, potato, and sweet potato industrial grinders and the raw material grinding capacity, according to information obtained from manufacturers' catalogs (Bosida, 2016; EBS, 2015; Konex, 2016; Larsson, 2016). The results show that the energy consumption increases linearly with the increase in the capacity of the grating machine.

It is normal to increase the motor power as the amount of raw material feeding the grater increases, but the relationship should not be linear. The results obtained indicate that instead of calculating the motor power required for grating, the calculation has been done without technical

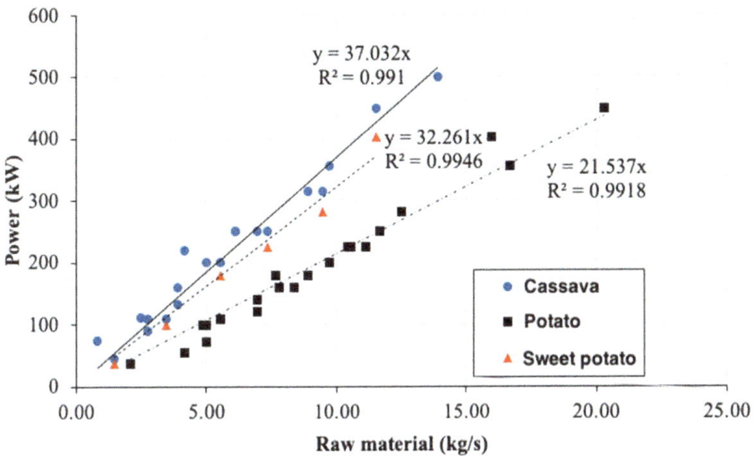

FIGURE 4.5 Power curves as a function of mass flow rate for industrial grinders with cassava, potato, and sweet potato, from manufacturers' catalogs.

analysis, simply with an increase directly proportional to the expected amount of raw material, which causes higher energy costs and nonrational use of motors.

In Fig. 4.5, it is possible to observe that a straight-line equation meets the behavior between the raw material flow and the power of the electric motor used in the equipment, allowing to establish the specific energy consumption in industrial grinders of 37.03 kJ/kg for cassava, 21.54 kJ/kg for potato, and 32.26 kJ/kg for sweet potato.

4.3.3 Starch extraction

After grinding, the disintegrated material must undergo starch extraction, which consists of the mechanical separation of the starch suspended in water from fibers and other impurities. The mechanical separation is based on the difference in size, shape, and density. This operation can be done by sieving, filtration, sedimentation, and centrifugation (MacCabe et al., 1993) since native starch is insoluble in water.

In starch processing units, extraction is performed in separating screens that operate essentially by horizontal or vertical centrifugation (Fig. 4.6A), but can operate by gravity in smaller companies, using vibrating or rotating screens (Fig. 4.6B and C. See also Chapter 3). This operation is decisive in the extraction efficiency because most of the starch losses occur in this step.

Rotating centrifugal sieves are present in large units, used in batteries of 4–6 in sequence. There are two types of centrifugal sieves: vertical (Fig. 4.6A) and horizontal (Fig. 4.7A), mainly found in Asia (see Chapter 2). In Asia, it is also possible to find static concave sieves without motor (Fig. 4.7C), which besides simplicity, consume little energy and allow an efficient elimination of smaller fibers (Cereda & Vilpoux, 2003).

Saengchan et al. (2014) studied the effect of rotation, cone diameter, and water/solids ratio on the performance of centrifugal sieves. The results obtained in a pilot centrifuge showed that increasing the rotation improves the separation up to a limit, as very high rotations reduce the retention time of the material to be separated on the sieve screen. The efficiency improves for larger basket diameters and low water/solids ratios.

4.4 Elements used for the selection of an equipment for starch release and extraction

The starch extraction industry is a good example to discuss equipment development within the necessary recommendations. Cereal

FIGURE 4.6 Vertical rotary centrifugal sieve (A), vibrating sieve (B), and rotary sieve (C). *(A) Photo provided by Vilpoux, O. F.; (B and C) from Borray, G. A. R. (2003).* Concepción de un modelo de agroindustria rural para la elaboración de harina y almidón a partir de raíces y tubérculos promisorios, con énfasis en los casos de achira (Canna edulis), arracacha (Arracacia xanthorriza) y ñame (Dioscorea sp.), *112 p.*

crops in the northern hemisphere provide raw materials that can be stored for planned use throughout the year. These are well known raw materials that can be harvested automatically and are standardized. On the other hand, in the tropical regions of the globe, except for corn, starchy crops are essentially roots and tubers, well adapted to small areas and small

FIGURE 4.7 Open horizontal rotary centrifugal sieve (A) with the basket screen used for extraction (B) and static concave sieve without motor (C). *(A and B): From Saengchan, K., Nopharatana, M., & Songkasiri, W. (2014). Recovery of tapioca starch from pulp in a conical basket centrifuge—Effects of rotational speed and liquid to solid (L/S) ratio on cake formation and starch–pulp separation efficiency.* Separation and Purification Technology, *127, 192–201. https://doi.org/10.1016/j.seppur.2014.02.026; (C): photo provided by Vilpoux O. F.*

producers. At harvest time, the moisture content of these starch-storing organs is above 60%, which reduces storage time, but facilitates tissue breakdown, necessary to extract the starch.

Except for cassava, processed in large scale units, and a few sweet potato (*Ipomea batatas*) processing units in Asia, the other underground starchy crops of South American origin, essentially canna (*Canna indica*) and arrowroot (*Maranta arundinacea*), are processed in small manual or partially mechanized units (see Chapter 3). This is also the case for cassava in most South American countries and in several regions of Brazil. To improve the extraction conditions of these units, it is necessary to develop smaller equipment, with capacity to process between 1 and 10 tons of raw material per day, within the concept of appropriate technology (Dagnino, 2014), as occurs in Valle del Cauca, Colombia (Alarcón & Dufour, 2002; Borray, 2003; Wheatley et al., 1997).

FIGURE 4.8 Original hammer mill, developed for grinding cereals, with two sets of six hammers and cutter disk (A) and modified hammer mill, with six sets and a total of 45 hammers (B).

After consulting the available material, an equipment that includes the crushing and starch extraction operations in the same machine was designed to facilitate operation and simplify the layout. The replacement of the traditional grater by a hammer mill was based on the recommendation of Hoque et al. (2007) who described the hammer mill as more suitable for more fibrous raw materials, such as canna and arrowroot, and that, as reviewed, is also easier to maintain.

A Pegorari Agricultura Têxtil brand hammer mill was used, for being a rustic equipment easily found in the market. It is recommended for grain milling and for this purpose equipped with a rotor-cutter mounted on a milling box with two sets of six hammers (Fig. 4.8A). To adapt the equipment for new use, the hammer mill of six hammers was replaced by a new configuration with 45 rectangular hammers 38.1 mm wide, by 4.76 mm thick, and 80 mm long, arranged in three sets of 7 hammers and other three sets of 8 hammers alternated (Fig. 4.8B). The rotor width was maintained, and the adopted diameter of 280 mm allowed for the use of the original grinding box. The set is driven by a 1.5 kW (2 HP) single-phase electric motor (220 V) with nominal rotation of 1700 rpm. The motor power is transmitted through a pair of 1:1 belt pulleys, which produce the same rotation in the mill shaft.

To approximate what occurs in starch extraction industries, it was necessary to introduce a water feed in the original hammer mill structure. The purpose of the operation under water is to dilute the material during disintegration, also helping to drag it and to clean the hammers. The water feed was kept fixed at 0.75 L per min.

Due to ease of use and construction, a rotary screen system, similar to the one shown in Fig. 4.6C, was selected. This type of extractor was described by Estrada Rivera (2020) in canna extraction in small units in Colombia. This model was preferred to the vibrating sieve, which is more difficult to produce and whose material suffers more wear due to

TABLE 4.1 Characteristics of the evaluated screens with the diameter of the holes (D mm), distance between the holes (Xh mm), and percentage of open area (Oa %).

Tipo	D (mm)	Xh (mm)	Oa (%)
P1	1.5	3.0	22.5
P2	1.8	3.0	32.4

vibrations. It was also preferred to the static concave sieve, which is not used in Brazil, and to the rotary centrifugal sieve, which is more expensive and needs more maintenance.

The hexagonal design of the screen-covered box was selected because it is easy to implement and maintain and has no continuously moving parts, except for the central rotor. The distribution of water in the center of the equipment helps to drag the starch and also to separate the starch suspension in water from the bran.

To adjust the starch extraction, two types of perforated screens were tested (P1 and P2) with the same total area, of 0.056 m^2, but with differences regarding hole diameter (D), distance between holes (Xh), and percentage of open area (Oa) (Table 4.1). The perforated screens are similar to those used in industrial cassava starch extraction equipment (Cereda & Vilpoux, 2003).

The hammer mill was attached to a hexagonal screen-covered cylinder that acts as an extractor, separating the starch from the bran (Fig. 4.9).

FIGURE 4.9 Disintegration and starch extraction equipment developed for small-sized units processing underground starchy crops (Branco et al., 2020). *From Branco, F. P., Cereda, M. P., & Naka, M. H. (2020). Comparison of arrowroot (Maranta arundinacea) and cassava starch extraction in separation, concentration, and purification using a rotating sieve under water.* Engenharia Agrícola, *40(1), 90–95. http://doi.org/10.1590/1809-4430-Eng.Agric.v40n1p90-95/2020.*

The hexagonal-shaped rotary screen was selected for ease of fabrication. It is covered with a 80 mesh (180 μm) nylon screen and rotates around a counterclockwise axis. The hexagonal frame is 2 m long, the diameter on the hexagon is 60 cm, the inclination is 10° from the horizontal plane, and the shaft rotates at 30 rpm, as described by Borray (2003). The rotation of the shaft is done with a 0.5 HP electric motor.

The rotation about the shaft promotes the movement of the disintegrated material within the sieve. Consequently, continuous separation of the starch suspension in water takes place, which passes through the screen openings and is collected in the discharge channel. The fibrous material retained inside the sieve is conducted along the central axis, in a continuous process aided by a water jet and the 10% inclination with which it was designed, which conducts the disintegrated material to its natural discharge at the end of the sieve (Fig. 4.10A). This inclination is planned for this type of extractor, but the design leaves room to adjust the angle further if necessary.

The water jet was constructed with a perforated PVC tube positioned in the center of the hexagonal sieve (Fig. 4.10B) and that allows sprinkling over the disintegrated material at a flow rate of 18 L/min, resulting in a ratio of 6.4 kg of water per kilo of disintegrated material, as recommended by Saengchan et al. (2014) for arrowroot extraction.

The flow of raw material to feed the hammer mill, evaluated from the time required to disintegrate 10 kg of raw material, varied from 1.84 to 2.86 kg/min, depending on the size of the rhizomes and roots used, which represents a processing capacity of 880–1400 kg in 8 hours of processing.

To evaluate the efficiency of disintegration and starch extraction, the machine was tested with arrowroot and cassava. The containers for collecting samples of the starch suspension

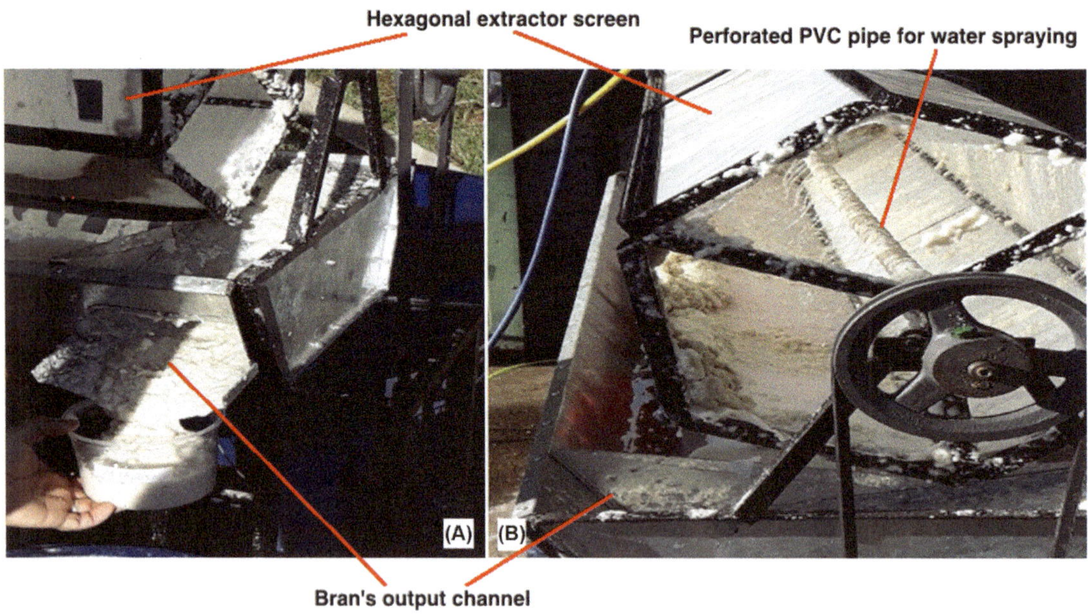

FIGURE 4.10 Detail of the channel to collect the fibrous residue (A) and perforated PVC tube positioned in the center of the sieve to spray water onto the nylon fabric (B).

in water distributed along the whole sieve allowed to verify the points where the extraction occurs.

4.4.1 Energy consumption

Energy consumption, expressed per kilogram of disintegrated raw material, or specific consumption (E_c), was calculated according to Eq. (4.2) (Souza et al., 2010).

$$E_c = \frac{P}{M_{rm}} \quad (4.2)$$

where E_c is specific energy consumption (kJ/kg), P is consumed electric power (kW), and M_{rm} is the raw material mass flow rate (kg/s).

The specific consumption of the adapted hammer mill was compared with commercial industrial grinders operating under similar particle size and moisture conditions (BeMiller & Whistler, 2009). The determination of the specific energy consumption taken as reference was based on the description of commercial grinders available in the manufacturer catalogs Bosida (2016), EBS (2015), Konex (2016), and Larsson (2016). Each commercial grinder was considered in three aspects: raw material, flow rate, and drive power.

When compared to the hammer mill used in the developed equipment, the specific consumption was calculated between 31.47 and 48.91 kJ/kg, a result close to that of the industrial process of disintegration of cassava roots, which was 37.03 kJ/kg, obtained for industrial grinders.

4.4.2 Water consumption

Fig. 4.11 presents the water consumption per kilogram of starch. This figure shows a lower consumption for arrowroot extraction compared to cassava and the material with smaller Sauter mean diameter. Da et al. (2013) indicated an average consumption, for

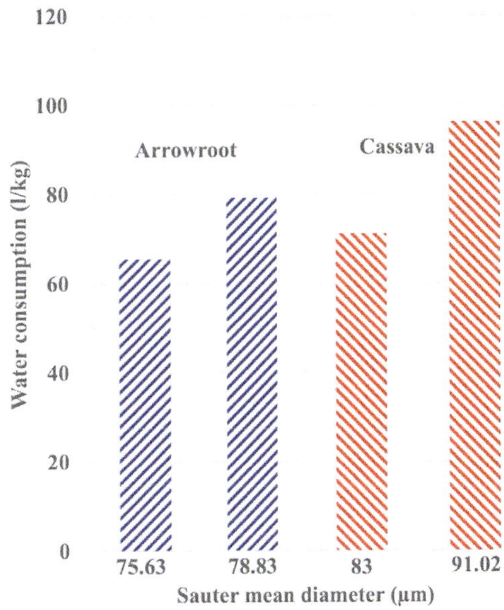

FIGURE 4.11 Evaluation of the volume of water required to extract 1 kg of starch from cassava and arrowroot raw materials and influence of Sauter mean diameter on extraction.

cassava starch extraction in small units, of 17.80–61.10 L/kg, depending on the separation technology and cassava variety, lower results than those indicated in Fig. 4.11.

Tests with higher inclination and faster rotation of the sieve could be performed to verify the improvement of extraction with reduced water use.

The overall water consumption could also be reduced by recycling the extraction water in the washing and peeling of the raw material, as happens in large units.

4.4.3 Starch extraction efficiency

To establish the extraction efficiency, the methodology of Saengchan et al. (2014) and Saengchan et al. (2015) was adapted, with the positioning of container collectors under the extraction sieve, as shown in Fig. 4.9.

With these collectors, samples of the starch suspension in water were taken randomly along the extractor. After shaking, a 50 mL sample was transferred to Falcon tubes previously weighed on an analytical balance. The tubes were then centrifuged for 5 minutes at 2800 rpm. The supernatant water was discarded, and the tube and starch were dried in an oven with air circulation. After drying, the tubes were weighed again, which allowed to quantify the decanted starch at each point of the rotary sieve shaft. The efficiency index as a function of starch concentration was calculated by the starch/water mass ratio in the suspension using Eq. (4.3).

$$C = \frac{M_{starch}}{M_{sol}} \times 100 \tag{4.3}$$

where C is starch concentration in water (%), M_{starch} is dry starch mass (g), and M_{sol} is mass of initial starch-water solution (g).

From the concentration obtained by averaging starch concentrations in water, it was possible to determine the average volume of water per kilogram of starch extracted for each raw material (Eq. 4.4).

$$\frac{w}{S} = \frac{V_{tot}}{W_{ts}} \tag{4.4}$$

where W/S is the ratio of liters of water per kilogram of starch in suspension (l/kg), V_{tot} is total volume of water in the sample (l), and W_{ts} is total mass of starch in the sample (kg).

With the results for each starch collecting container, it was possible to know the efficiency of the extractor sieve along its length for cassava and arrowroot. The results obtained, presented in Fig. 4.12, indicate the concentration of starch in water as a function of position on the sieve. In general, it is observed that as the disintegrated material passes through the rotating sieve, the starch concentration in the suspension decreases.

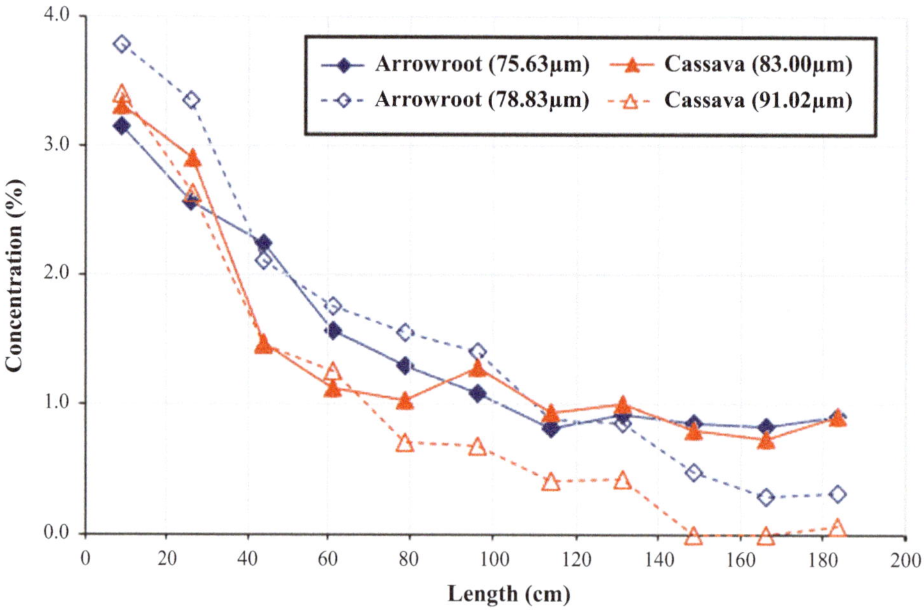

FIGURE 4.12 Concentration of starch in the extraction water along the length of the hexagonal extractor screen, for arrowroot and cassava pulps with different Sauter mean diameters.

The raw materials were previously disintegrated in the hammer mill. Two sieves of different diameters, 1.5 and 1.8 mm, were used, with evaluation of the diameter of the particles from the disintegrated material. For this, the particle size distribution was determined according to the methodology used in Sajeev et al. (2012), in which 100 g samples of disintegrated raw material were submitted to sieving in four sieves of ASTM 20 (850 μm), 40 (425 μm), 80 (180 μm), and 170 (90 μm) mesh, with evaluation in triplicate. The material retained on each sieve was removed, weighed, and dried in an oven with air circulation (Marconi, MA035/1152) regulated at 105°C for 72 hours. Subsequently, the samples were weighed to be expressed in percentage of DM. This value was used to determine the DM retained on each sieve (Xi in Eq. 4.5) and expressed as a percentage.

The comparison between treatments for each sieve used the Sauter mean diameter, which relates the particle diameter, whose surface to volume ratio is the same in all particles and was calculated according to Eq. (4.5) (Cremasco, 2012).

$$\mathrm{SMD} = \frac{1}{\sum \frac{X_i}{D_{pi}}} \qquad (4.5)$$

where X_i is percent of mass retained on each sieve and D_{pi} is the average diameter of each sieve (μm).

The results obtained are shown in Table 4.2.

TABLE 4.2 Sauter mean diameter (SMD) expressed in micrometers (μm) of the disintegrated materials of arrowroot and cassava, with two screen openings, 1.5 and 1.8 mm.

	P1 (1.5 mm)	P2 (1.8 mm)
Cassava	83.00 ± 1.10^{Aa}	91.02 ± 0.88^{Ba}
Arrowroot	75.63 ± 0.81^{Ab}	78.83 ± 0.15^{Bb}

Means followed by the same lower-case letter in the column and the same upper-case letter in the row do not differ significantly from each other by the Tukey test at 5% error probability level.

The arrowroot disintegrated in the hammer mill obtained a smaller particle size than the cassava. For both raw materials, the extraction of a material with larger granulometry occurred only in the first two thirds of the equipment. The water recovered in the last third of the sieve contained little, or no starch in the case of cassava.

The percentage of extraction was not evaluated, which makes comparison between the experiments difficult. However, the fact that the percentages of starch at the end of the extraction sieve were lower with larger particles may mean that more starch was trapped in the nondisintegrated cells, with no possibility of release. In this case, the free starch present in smaller proportion was extracted faster than with smaller particles.

For smaller particle diameters, the starch concentration in the last 75 cm of the sieve remains high, close to 1% for arrowroot and 0.5% for cassava. This result indicates that more starch could be recovered if the rotary sieve had a longer length, which would not change much the production cost of the equipment.

For both raw materials, the concentration in the last sample showed a slight increase over the previous measurement. The reason for this variation was the accumulation of fibrous material at the end of the sieve before flowing out through the outlet channel. Residual starch from the fibrous material returns to the sieve and settles in the last container to increase the starch concentration in the last sample.

In arrowroot starch extraction, the average concentration of starch in water was 1.48% for Sauter mean diameter of 75.63 μm and 1.26% for Sauter mean diameter of 78.83 μm. These results are similar to those obtained by Cecil (1992) for cassava starch extraction in small units, with results of 1%–2%, but are lower than in large Brazilian industries, with concentrations of 2.2%–2.9% (3–4°Be).

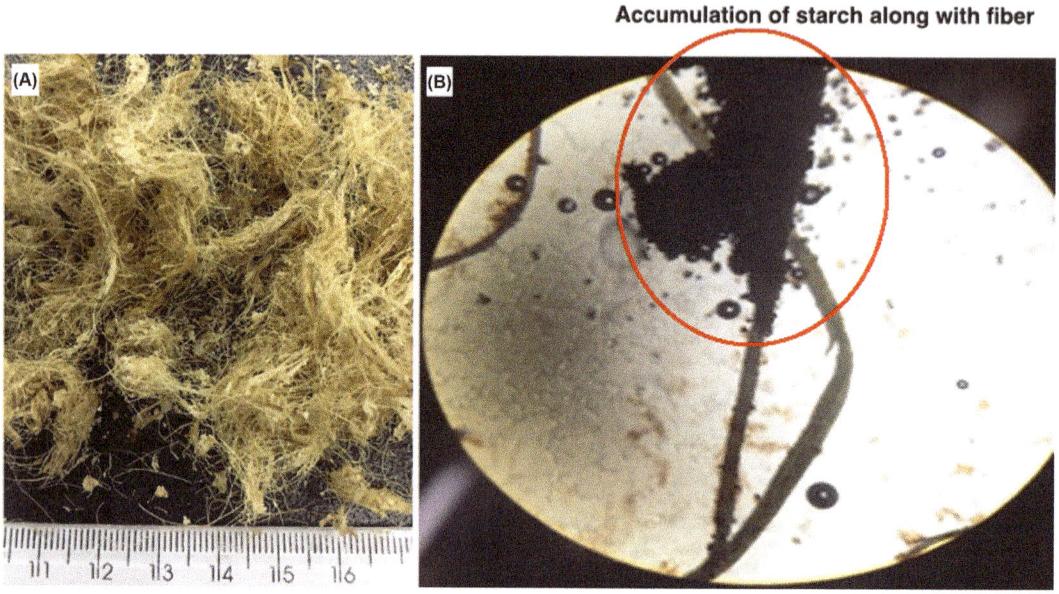

FIGURE 4.13 Arrowroot fibrous residue (A) and microscope view (400x) of arrowroot fibers with accumulation of starch grains (B) (Branco et al., 2019). *From Branco, F. P., Naka, M. H., & Cereda, M. P. (2019). Granulometry and energy consumption as indicators of disintegration efficiency in a hammer mill adapted to extracting arrowroot starch (*Maranta arundinacea*) in comparison to starch extraction from cassava.* Engenharia Agrícola, 39, *341–349. http://doi.org/10.1590/1809-4430-Eng.Agric.v39n3p341-349/2019.*

4.4.4 Difficulties

In some tests with arrowroot, it was not possible to maintain a continuous flow of the output in the hammer mill due to accumulation of fibrous material inside the mill (Fig. 4.13). The arrowroot rhizomes have thicker fibers, which could not be completely broken during the disintegration of the rhizome, with the requirement of stops in the disintegration operation to clean the mill. One could also consider a coarser previous cut of the rhizomes, as occurs with cassava in larger companies (see Chapter 2). With this, it would be possible to obtain a more continuous feed.

The equipment was also evaluated with cassava roots and sweet potato tubers, but arrowroot was the only one that generated fibrous bran as a residue from starch extraction, which should be considered as a characteristic of the rhizome (for more details on the anatomical constitution of underground starchy raw materials, see Chapter 4 of Volume 1 in this Series). On the other hand, it is possible to see in Fig. 4.13B that the arrowroot bran had less mass of disintegrated material that is colored in blue with lugol than is usually the case with cassava starch extraction, whose bran is more viscous and retains more starch granules.

The fiber accumulation occurred between the rotor and the discharge chute of the mill, where there is a perforated steel screen that retains the passage of the disintegrated material until a particle size smaller than the diameter of the holes is reached.

4.5 Final considerations

This chapter presented the analysis of starch extraction as an industrial process from a process engineering perspective. South America,

Africa, and some Asian countries consume large amounts of starch, mainly from local or regional sources of cereals, roots, and tubers. Often, the technologies used for starch extraction are still rudimentary compared to modern units (see Chapters 2 and 3).

Thus, the area of process engineering, mechanical engineering, chemistry, and related areas can contribute significantly to the advancement and development of equipment for the processing and production of starch and derivatives, increasing the efficiency of extraction, quality, and productivity and contributing to the economic development of these regions. In addition, this type of industry can generate energy from the production of biogas (see the upcoming Volume 5 in this Series), byproducts valued for cooking, as well as pharmaceutical potential among others (see the upcoming Volume 4 in this Series).

The tests showed the variations found when processing different raw materials in the same equipment. While for cassava and sweet potato the flow was continuous, the same did not occur for arrowroot rhizomes, due to the accumulation of fibrous material inside the hammer mill. This result indicates the need for prior preparation of the rhizomes to allow for greater fluidity of the disintegration.

Besides allowing for more efficient disintegration, the 1.5 mm perforated screen allows for similar results as industrial equipment with other types of grinding. Tests with a 1.2 mm perforated screen did not show significantly different results, besides making it difficult to pass the material.

The option for a hammer mill proved to be suitable for small-scale processing, due to the easy adaptation of small commercial mills and the reduction of maintenance, with the need for frequent change of saw teethes in traditional graters.

The developed grinding-extraction system presents easy construction and low manufacturing cost, requiring only adjustment of the sieve size. Other project parameters should also be verified, such as the influence of the sieve rotation on the retained starch and mainly the adjustment of the amount of water supplied over the disintegrated material.

Another concern from the engineering point of view is related to working conditions, which in general can contribute to a significant reduction in accidents, work-related illness problems, and ergonomics. Generally, in units that use rudimentary technologies, there is great physical effort, exhausting workdays, and exposure to accident risks.

References

Alarcón, M. F., & Dufour, D. L. (2002). Almidón agrio de yuca en Colombia. In P. B. ; Ospina, H. ; Ceballos, F. ; Alvarez, A. C. ; Bellotti, L. A. ; Calvert, V. B. ; Arias, L. L. F. ; Cadavid, L. B. ; Pineda, R. G. A. ; Llano, & M. I. Cuervo (Eds.), *La yuca en el Tercer Milenio: Sistemas modernos de producción, procesamiento, utilización y comercialización. Centro Internacional de Agricultura Tropical (CIAT); Consorcio Latinoamericano para la Investigación y el Desarrollo de la Yuca* (pp. 470–502). Cali, CO: Proyecto IP-3 de Mejoramiento de Yuca. Available from https://cgspace.cgiar.org/bitstream/handle/10568/55257/capitulo26.pdf?sequence = 1&isAllowed = y, Consultado em: 4/abril/2023.

Barth, M., Renner, J. S., Nunes, M. F., & Sanfelice, G. R. (2016). Características do trabalho na agricultura familiar e sua influência na emigração dos jovens. *Iluminuras, 17*(2006), 256–276.

BeMiller, J., & Whistler, R. (2009). *Starch chemistry and technology* (3rd ed.). Burlington: Academic Press, Elsevier ed.

Bergthaller, W., Witt, W., & Goldau, H. P. (1999). Potato starch technology. *Starch-Stärke, 51*(7), 235–242.

Borray, G.A.R. (2003). *Concepción de un modelo de agroindustria rural para la elaboración de harina y almidón a partir de raíces y tubérculos promisorios, con énfasis en los casos de achira (Canna edulis), arracacha (Arracacia xanthorrhiza) y ñame (Dioscorea sp.)*, 112 p.

BOSIDA. (2016). *RU Series Super Rasper*. http://www.bosidachina.com/English/zlkjpor_view.asp?id = 36.

Branco, F. P., Cereda, M. P., & Naka, M. H. (2020). Comparison of arrowroot (Maranta arundinacea) and cassava starch extraction in separation, concentration, and purification using a rotating sieve under water. *Engenharia Agrícola, 40*(1), 90–95. Available from https://doi.org/10.1590/1809-4430-Eng.Agric.v40n1p90-95/2020.

Branco, F. P., Naka, M. H., & Cereda, M. P. (2019). Granulometry and energy consumption as indicators of disintegration efficiency in a hammer mill adapted to extracting arrowroot starch (Maranta arundinacea) in comparison to starch extraction from cassava. *Engenharia Agrícola, 39*, 341–349. Available from https://doi.org/10.1590/1809-4430-Eng.Agric.v39n3p341-349/2019.

Carvalho, P. R. N., Mezette, T. F., Valle, T. L., Carvalho, C. R. L., & Feltran, J. C. (2007). Avaliação da exatidão, precisão e robustez do método de análise do teor de matéria seca de mandioca (Manihot esculenta, Crantz) por meio da determinação do peso específico (balança hidrostática). *Revista Raízes e Amidos Tropicais, 3*.

Cato, M.P. (1933). *Cato, the Censor, on farming*. <https://soilandhealth.org/wp-content/uploads/01aglibrary/010121cato/catofarmtext.htm> (Consulted in May 2023).

Cecil, J. E. (1992). *Small, medium and large-scale starch processing* (1st ed.). Roma: FAO.

Cereda, M. P., & Vilpoux, O. F. (2003). Processo de produção de fécula de mandioca: comparação Brasil, Tailândia e China. In M. P. Cereda (Ed.), *Agricultura: Tuberosas amiláceas latino americanas. Volume 3: Tecnologias, usos e potencialidades de tuberosas amiláceas latino americanas* (1st ed, pp. 143–175). Fundação Cargill: São Paulo.

Cremasco, M. A. (2012). *Operações unitárias em sistemas particulados e fluidomecânicos*. Blucher: São Paulo.

Da, G., Dufour, D., Giraldo, A., Moreno, M., Tran, T., Velez, G., Sanchez, T., Le-Thanh, M., Marouze, C., & Marechal, P. A. (2013). Cottage level cassava starch processing systems in Colombia and Vietnam. *Food and Bioprocess Technology, 6*, 2213–2222. Available from https://doi.org/10.1007/s11947-012-0810-0.

Dagnino, R. (2014). *Tecnologia social: Contribuições conceituais e metodológicas* (2nd ed.). Campina Grande: EDUEPB.

EBS. (2015). *Cevadeira 300T - Manual de operação e manutenção*. 37 p.

Estrada Rivera, K.D. (2020). *Producción y obtención de almidón de achira Canna edulis con fines comerciales en el municipio de Algeciras Huila*. https://ciencia.lasalle.edu.co/cgi/viewcontent.cgi?article = 1184&context = ingenieria_agronomica.

Grace, M. R. (1977). In M. R. Grace (Ed.), *Cassava processing* (3rd ed.). Roma: FAO.

Grossman, J., & Freitas, A. G. (1950). Determinação do teor de matéria seca pelo método de peso específico em raízes de mandioca. *Revista Agronômica, 14*, 75–80.

Hoque, M., Sokhansanj, S., Naimi, L., Bi, X., Lim, J.,& Womac, A.R. (2007). Review and analysis of performance and productivity of size reduction equipment for fibrous materials. In: ASABE ed. *2007 ASABE Annual International Meeting*. Minneapolis, p. 18. https://doi.org/10.13031/2013.23342.

Iritani, W. M., & Weiler, L. (1976). The influence of early storage (Pre-Holding) temperatures on sugar accumulation in Russet Burbank potatoes. *American Potato Journal, 53*(5). Available from https://doi.org/10.1007/BF02854986.

Kawano, K., Fukuda, W. M. G., & Cenpukdee, U. (1987). Genetic and environmental effects on dry matter content of cassava root 1. *Crop Science, 27*(1), 69–74.

KONEX. (2016). Rasper MODEL DRS/DRD 300, 400, 500, 600. http://www.konexindo.com/index.php/rasper.

Landolsi, M.W. (2017). *Machine design procedure: What are the rules and the factors of the process?* <https://conceptek.net/en/engineering/machine-design-procedure-what-are-the-rules-and-the-factors-of-the-process> (Accessed 29.03.23).

LARSSON. (2016). *LARSSON Rasper GL550/300*. p. 2. Available at: https://www.larssonsweden.com/wp-content/uploads/2016/06/DO0044001-A3-EN-Rasper-GL550.pdf.

Leonel, M., & Cereda, M. P. (2000). Extração da fécula retida no resíduo fibroso do processo de produção de fécula de mandioca. *Ciência e Tecnologia de Alimentos, 20*(1), 122–127. Available from https://doi.org/10.1590/S0101-20612000000100023.

MacCabe, W. L., Smith, J. C., & Peter, H. (1993). *Unit operations of chemical engineering* (5th ed.). McGraw-Hill.

Maeda, M., & Dip, T. M. (2000). Curvas de porcentagem mássica de água versus peso específico em vegetais in natura-otimização de processos industriais pela seleção via teste da matéria-prima. *Food Science and Technology, 20*, 309–313. Available from https://doi.org/10.1590/S0101-20612000000300006.

Mckinney, L.J. (2006). *Grain processing: Particle size reduction methods*. p. 42–45.

Norton, R. L. (2013). *Projeto de máquinas: Uma abordagem integrada* (4th ed.). Porto Alegre: Bookman, Bookman ed.

Pola, A. C., da Costa Nunes, E., & Moreto, A. L. (2020). Matéria seca em raízes de mandioca determinada pelos métodos da balança hidrostática e de secagem em estufa. *Agropecuária Catarinense, 33*(1), 56–60.

Saengchan, K., Nopharatana, M., & Songkasiri, W. (2014). Recovery of tapioca starch from pulp in a conical basket centrifuge—Effects of rotational speed and liquid to solid (L/S) ratio on cake formation and starch–pulp separation efficiency. *Separation and Purification Technology, 127*, 192–201. Available from https://doi.org/10.1016/j.seppur.2014.02.026.

Saengchan, K., Nopharatana, M., Lerdlattaporn, R., & Songkasiri, W. (2015). Enhancement of starch-pulp separation in centrifugal-filtration process: Effects of particle size and variety of cassava root on free starch granule separation. *Food and Bioproducts Processing, 95*, 208–217. Available from http://linkinghub.elsevier.com/retrieve/pii/S0960308515000644.

Sajeev, M. S., & Balagopalan, C. (2005). Performance evaluation of multi-purpose mobile starch extraction plant for small scale processing of tuber crops. *Journal of Root Crops, 2*, 106–110.

Sajeev, M. S., Nanda, S. K., & Sheriff, J. T. (2012). An efficient blade type rasper for cassava starch extraction. *Journal of Root Crops, 38*(2), 151–156.

Sheriff, J. T., & Balagopalan, C. (1999). Evaluation of multi-purpose starch extraction plant. *Tropical Science, 39*, 147–152.

Souza, L. H., de, Dias, G. P., Souza, L. C. V., de, Dias, I. G., & Rodrigues, D. E. (2010). Avaliação da demanda energética de um desintegrador-picador-moedor (DPM) na moagem de milho. *REVENGE Engenharia na Agricultura, 18*(9), 480–487.

Stanley, R., & Jewell, S. (1989). The influence of source and rate of potassium fertilizer on the quality of potatoes for French fry production. *Potato Research, 32*, 439–446.

Vilpoux, O. F. (2011). Desempenho dos arranjos institucionais e minimização dos custos de transação: Transações entre produtores e fecularias de mandioca. *Revista de Economia e Sociologia Rural, 49*, 271–294.

Wheatley, C., Scott, G. J., Best, R., & Wiersema, S. (1997). *Métodos para agregar valor a raíces y tubérculos alimenticios: Manual para el desarrollo de productos* (1st ed.). Cali: CIAT.

CHAPTER

5

Starch hydrolysis: physical, acid, and enzymatic processes

Marney Pascoli Cereda[1,2]

[1]Retired from São Paulo State University, Botucatu, São Paulo, Brazil [2]Tecsapp Solutions, Campo Grande, Mato Grosso do Sul, Brazil

5.1 Introduction

Chemically, starch is composed only of glucose molecules. Li et al. (2023) stated that starch-based sugars are an important group of starch derivatives used in food, medicine, chemistry, and other fields. To obtain sugars, starches are hydrolyzed by acid, heat, and pressure, or enzymatically, or combinations of each of these methods. More recently, processes with cold hydrolysis and ultrasonic pretreatment have been tested. Each process has its advantages and disadvantages, but the products obtained by enzymes are the most valued.

Hydrolyzed products can be marketed in dehydrated form or, more commonly, in the form of syrups, which are easier to transport and store. Several countries in South America have units that process starch to obtain sugars, mostly international companies that use corn starch.

In general, any starch can be used to produce hydrolysates. In the northern hemisphere, starches from cereals (corn and wheat) and potatoes predominate. Rice starch is also used, but to a limited extent. Some countries in Asia, such as Thailand, India, and China, as well as Brazil, produce glucose syrups from cassava starch on an industrial scale, demonstrating the feasibility of the process with this raw material.

In European countries and the United States, more than 50% of the starch production is to produce hydrolysates such as glucose, maltose, dextrin, maltodextrins, and fructose for soft drinks, in most cases concentrated as syrups. These starch derivatives are used by a very wide range of companies and are destined for a variety of markets, the largest being food, pharmaceuticals, and cosmetology (for more on these application areas, read Chapters 11 and 16).

Hydrolysates, for their specific properties, ease of storage and use, are an excellent option for starch valorization. In Brazil, only one cassava starch extracting company produces glucose by the acid process. Syrups can vary in sugar concentration and glucose monomer profile to serve different markets. A large market in Brazil is the maltose syrups for breweries and lately for microbreweries, as a complement

Starch Industries
DOI: https://doi.org/10.1016/B978-0-323-90842-9.00016-9

to malt (vegetable enzymes). The use of syrups also allows the beverages to be standardized.

In general, hydrolysates obtained from starches of underground starchy raw materials have the advantages of requiring lower gelatinization temperature and have lower protein and lipid content than starches from cereals. In this case, other raw materials could be considered in each country or region (for raw material types and products, see Volume 2 of this series). In South America, Brazil is one of the world's leading producers of cassava, with industrial processing of the root into flour and starch (for the types and processes of flour production, please refer to the upcoming Volume 4 in this series).

Two processes for manufacturing starch sugars are most used, acid and enzymatic. The acid process allows glucose and dextrin to be obtained, but with little control of the process and of the monomers generated. Obtaining syrups and sugars by the enzymatic process, on the other hand, allows for the production of a wide range of hydrolysates of a more constant quality in the same line of equipment, but requires a more effective control of the process, more experienced operators, an equipped laboratory, and trained technicians.

In Brazil, Cargill and Ingredion are American multinationals and the largest producers of hydrolysates. The company Indemil, located in Paranavaí, state of Paraná, Brazil, produces glucose syrup via the acid process, both from cassava and corn starch. Companies producing and selling commercially available enzymes disclose their specific characteristics and requirements, such as pH, the most suitable starch concentration, which allows for more controllable operations.

The data used in this chapter are the results of a European Community (EC) project for the valorization of cassava products and were initially published in Portuguese (Surmely et al., 2004) and disseminated by the Cargill Foundation. The results presented are of an increase in scale from laboratory to pilot scale, with the intention of motivating more industries to produce cassava and others underground starchy crops for hydrolysates, highlighting the difficulties, and evaluating the necessities of the equipment and the enzymatic process with the change in scale. Samples of the products obtained were analyzed in laboratory and established the necessities of industrial conditions.

5.2 Reducing sugars and dextrose equivalent

In general, the degree of hydrolysis of starch derivatives is measured by the number of D-glucose monomers generated. They are designated by DE (dextrose equivalents, which means reducing sugar).

Reducing sugars are the carbohydrates that have aldehyde or ketone radicals on their carbon 4, which therefore have reducing terminals. When this terminal carbon is free, it establishes in the glucose molecule an aldehydic function, as occurs in glucose, or a ketonic function, as is the case with fructose, but both with the property of reducing copper. Fig. 5.1 shows the glucose molecule in its planar configuration, with the carbon numbers and the reducing radicals.

Aldehydic and ketonic functional groups of the carbohydrate molecules correspond to an active site for reactions. Depending on the presence or absence of these radicals, carbohydrates are classified as reducing and nonreducing (Fig. 5.2), and by reducing reaction, D-glucose may be transformed into a D-gluconate.

Both the identification and the determination of reducing radicals use copper as an indicator, since it changes color when it is in reduced or oxidized form. The determination of reducing carbohydrates can be done by hot reaction with a reduced copper solution that is blue in color, but when it changes to the oxidized form, it turns brick-orange. This reaction

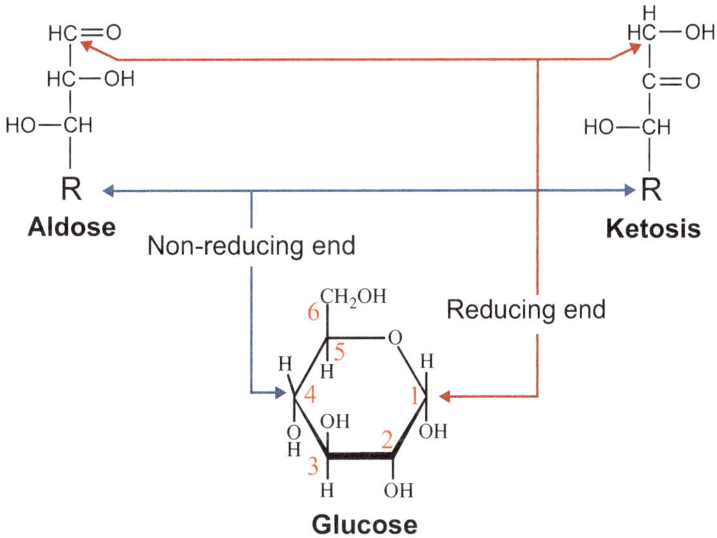

FIGURE 5.1 Reducing and nonreducing ends of a glucose molecule according to functional groups.

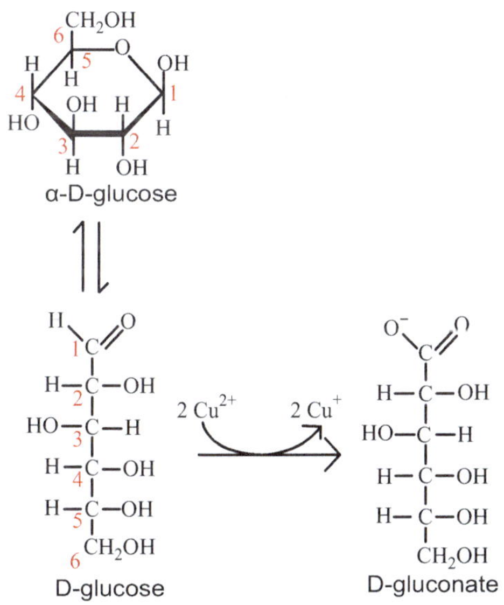

FIGURE 5.2 Glucose molecule in its reduced and oxidized forms, using copper as indicator.

is the basis for methods of analysis such as Layne-Einon or Somogyi-Nelson, among many others (for methods to determine or identify reducing radicals, see Chapter 12 in Volume 1 of this series).

Although glucose is represented in flat form, organic molecules always present curved spatial forms, due to the angles of the glycosidic bonds. In the case of glucose, the angle is 107 degree, so that in its spatial forms, the glucose molecule can be configured as in Fig. 5.3, in concave or convex shapes, but always presenting a reducing and a nonreducing end. The spatial configuration confers greater reactivity between available radicals, since they can be closer or more exposed to the action of enzymes, for example.

The DE expresses the degree of hydrolysis of the starch since each break in the amylose or amylopectin chain creates reactive sites until the basic monomers, glucose and fructose, are reached. Fig. 5.4 represents the glucose molecules that are the basis of the starch molecule. The linkage between two glucose molecules, both reducing, gives a maltose molecule, with only one reducing radical located at C_1, since the reducing radical of the second glucose molecule will be involved in the glycosidic bond (C_1-C_4). By this mechanism occurs the polymerization of the glucose units to synthesize

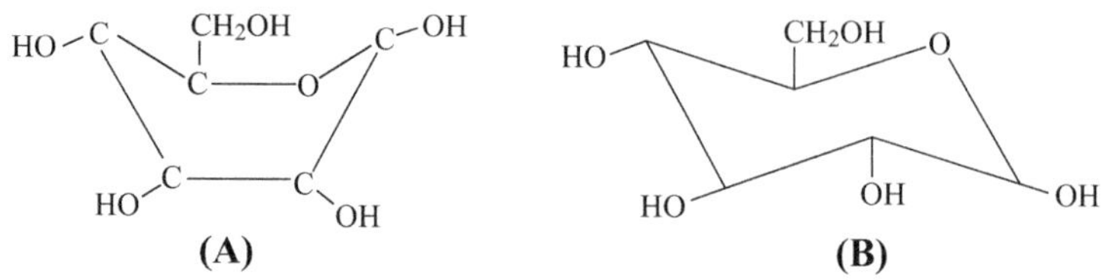

FIGURE 5.3 Spatial shapes of α-D-glucose: concave (boat-shaped) (A) and convex (chair-shaped) (B).

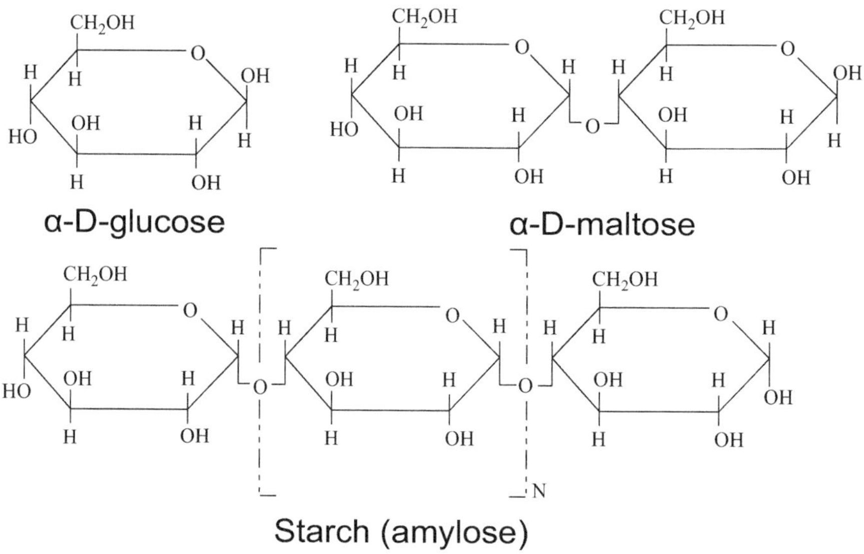

FIGURE 5.4 Linkages between glucose (α-D-glucose) molecules to gives maltose and starch.

starch, while the reducing ends decrease in number and the polysaccharide becomes less and less reducing.

Hydrolysis is the basis for the formation of sugars. When it occurs, the number of reducing ends increases and represents the degree of polymerization of the starch molecule. By convention, the reducing value of glucose is considered to be 100%. Therefore, the degree of hydrolysis is represented and expressed in DE, and the higher it is, the greater the starch depolymerization. The application of this criterion can be expressed as follows:

- DE 100 when the proportion of glucose molecules is 100.
- DE 52.6 when the proportion of maltose molecules is 100%.
- DE close to 0 when the starch is native (not hydrolyzed).

As an example, two hydrolysates prepared under different conditions may have the same DE and have a completely different sugar profile (Fig. 5.5). Consequently, the properties of the two hydrolysates will also be different.

The two starch hydrolysates in Fig. 5.5 have the same DE 42. However, one is glucose syrup, with decreasing amounts of glucose and maltose polymerization unities (DP_1 and DP_2) and products of 3–7 glucose residues. The remainder (DP > 7) is dextrin of varying degrees of polarization. With the same DE, the profile of maltose syrup is quite different, with

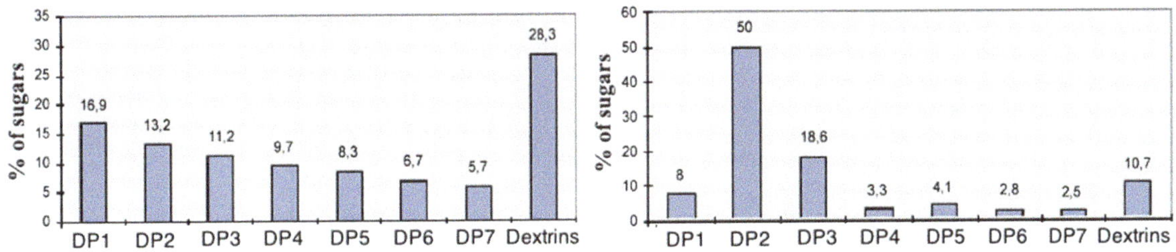

FIGURE 5.5 Carbohydrate profile of two syrups with the same DE 42; the left one is glucose syrup (acid process), and the right is maltose syrup (enzymatic process). *DP*, Degree of polymerization; *DP1*, glucose; *DP2*, maltose; *DP3*, maltotriose; *DP* > 7, dextrin (by HPLC analysis).

predominance of maltose unities, and low contents of glucose, maltotriose (DP_1 and DP_3), and dextrin.

5.3 Starch hydrolysis

If the starch is treated by specific acids or enzymes, the result is the fractionation of the polymer chain with the release of smaller molecules (French, 1984). This process is called hydrolysis, and if it is complete, it will give only glucose. However, this is not what happens in practice. Starch hydrolysis can be accomplished essentially by physical (heat), chemical (with organic or inorganic acids), and enzymatic processes. The use of heat is difficult to control and may produce pyrodextrins, a polymer that is often brownish in color and very uneven, which is used in the preparation of glues (for more details on pyrodextrins, see Chapter 15). Chemical or acid hydrolysis is used when the product does not require quality, or when the market price of the product is low, because it is more difficult to obtain good quality products. Enzymatic modifications are more valued, and their advantages over acid hydrolysis are:

- With acid hydrolysis, the neutralization produces a significant amount of salt, which can alter the taste of the final product, while in enzymatic hydrolysis, the mineral content is very low and the removal of the salts formed is possible with the use of ion exchange resins, producing a lightly conductive syrup.
- In enzymatic hydrolysis, less activated carbon is used to remove colored compounds and those responsible for undesirable odors and tastes.
- The enzymatic hydrolysis allows simplifying the production line, with unitary reactors for liquefaction, saccharification, and decolorization.
- The enzymatic hydrolysis also allows for energy savings. Traditional acid liquefaction, in a batch process, requires pressure cooking at a high temperature. The enzymatic liquefaction of cassava starch can be performed at 85°C, lasting a few minutes.
- Finally, enzymatic hydrolysis enables the manufacture of a full range of hydrolysates with the same line of equipment, only replacing the enzyme used in the process.

Still, both acid and enzymatic hydrolysis produce a mixture of intermediate sugar residues, with more uniformity when enzymatic hydrolysis is used, as will be seen later.

5.3.1 Chemical hydrolysis of starch

The hydrolysis of starch is based on the fact that the glycosidic bond is stable under alkaline conditions but is broken under acidic conditions. Chemical hydrolysis of starch is still being used in Brazil, due to its low investment.

The main commercial product is dextrin, but acid hydrolysis also allows for the production of glucose in syrup form.

5.3.2 Enzymatic hydrolysis of starch

The enzymatic conversion products of starch vary in chain size, starting from glucose (1 molecule) reaching carbon residues of 5–11 glucoses, characterizing the dextrins, of higher molecular weight. Starch, the raw material for hydrolysates, is a polysaccharide consisting of straight and branched chains, where each molecule consists of several hundred glucose units, linked together (for more information on the structure of starch, see Chapter 9 in Volume 1 of this series).

Starch granules are insoluble in cold water. When the water penetrates the amorphous areas of the starch granule, it forms hydrogen bonds with the free hydrophilic groups of the starch molecule. These bonds have weak strength, but are in a very high number, which prevents dissolution of the granule (Swinkels, 1985). Under favorable conditions, starch granules swell slightly in cold water (10%–15% of diameter), in a process reversible by drying. However, when the temperature of the suspension is sufficient, breaking of the hydrogen bonds occurs, the starch granule swells irreversibly, and the phenomenon of gelatinization occurs. With the gelatinization occurs new bond formation between water molecules and hydroxyl groups, released by the ingress of water (Thibault & Colonna, 1988). Starch gelatinization is the first step in the hydrolysis process, because enzymes attack granular starch very slowly (for more information on the relationships of starch with cold and heated water, see Chapter 10 in Volume 1 of this series).

Because hydrolysis is a process that requires heat, it is important to use starches from underground organs that generally present lower gelatinization temperatures than cereals. Although there are pronounced differences between the behavior of the starches, with respect to their functional properties, the hydrolysis products will be very similar, especially the lower molecular weight hydrolysates.

5.4 Enzymes

Enzymes are the most important agents in hydrolysis reactions. If used in the hydrolysis process, they are also called diastases, or amylolytic enzymes. They are proteic compounds that act as biological catalysts in all energetically possible metabolic reactions and accelerate these reactions by specific activation (Chaplin & Bucke, 1990). Enzymes also make possible to rapidly reach the equilibrium state of the reaction by decreasing the activation energy and thus increasing its speed. Enzymes do not undergo changes during the reaction, and at the end of each cycle, they present the same activity again (Lehninger & Kamoun, 1985).

According to Cuvellier (1993), enzymes can be of origin:

- Vegetable: they can be proteases, such as papain from papaya, bromelin from pineapple, and ficin from fig. A vegetal amylotitic complex widely used in the brewing industry is cereal malt.
- Animal: only bovine and pig pepsins are currently of industrial interest.
- Microbial: this is the almost exclusive source of commercial enzymes, especially bacterial enzymes, many of which are thermoresistant. The advantages of this source are numerous, the greatest of which is that it occupies little space and does not use animals. Other advantages are geographical and seasonal independence, use of a cheap substrate, increased production yields by selection of microbial strains, and a production of enzymes with diverse properties and specificities.

- The main needs of enzymes:
  - pH: all enzymes are sensitive to variations in the hydrogen concentration (H^+) of the medium. There is a pH for which the enzymatic activity is maximum, which can be basic or acidic. The optimum pH of an enzyme is not necessarily equal to the pH of its natural medium.
  - Temperature: as for pH, enzymes have a temperature, or a temperature range, at which the activity is maximum. In general, it is preferable to use enzymes that can withstand varying conditions and temperature increases.

5.4.1 Amylolytic enzymes

Amylolytic enzymes belong to the category of hydrolases, enzymes that catalyze hydrolysis reactions, and can be classified by the mechanism and action they exert. Whatever the mechanism or their action, amylolytic enzymes always act at the nonreducing end of the starch chain.

Mechanism of action of amylolytic enzymes:

- Endo enzymes randomly hydrolyze the glycidic bonds in the middle of the molecule. Among them are α-amylase, CGTase, and pullulanase.
- Exo enzymes hydrolyze the molecule from a nonreducing end. Among them are β-amylase and amyloglucosidases.

5.4.1.1 Amylolytic enzymes and their actions

- α-amylase is an endo enzyme that randomly hydrolyzes the α-(1–4) bonds of starch but does not act on the α-(1–6) bonds. This enzyme produces only high molecular weight α-dextrins. The α-amylase is a liquefying enzyme that can reduce the viscosity of gelatinized starch, which is why it is used early in the hydrolysis process.
- β-amylase specifically catalyzes the hydrolysis of the α - (1–4) of starch from a nonreducing end (exo enzyme). When it acts on amylose, it produces only maltose, a sugar composed of two glucose units joined by α - (1–4) bonding, with only one reducing end, unless the number of glucose units in the starch chain is odd, in which case one glucose may be left over. The β-amylase is considered a saccharifying enzyme because it produces sugars from starch.

Not being able to hydrolyze the α-(1–6) bonds, the joint action of α- and β-amylases on the amylopectin produces dextrin-limits, a generic designation for a group of dextrins of different molecular weights that resist the enzymes by their α-(1–6) bonds.

- Glucoamylase or amyl-1,6-glucosidase is both a liquefying and saccharifying enzyme, capable of completely hydrolyzing starch into glucose from a nonreducing end. It is also the only one capable of hydrolyzing both α-(1–4) and α-(1–6) bonds at the same time. The result of enzymatic conversion of starch by glucoamylase is total production of glucose units.
- Deramifying enzymes: these include several enzymes with activity on α-(1–6) bonds, such as R-enzymes, pullulanase, isoamylase, and oligo-1,6-glucosidase. These enzymes are capable of hydrolyzing only the α-(1–6) bonds of starch. They have different characteristics and specialties, but in general, they remove the branches, turning the starch into straight dextrins (amylose).
- Cyclizing or CGTases: these endoenzymes can simultaneously cut dextrins from 7, 8, and 9 glucose molecules and promote their cyclization to form cyclodextrins. Cyclodextrin-like enzyme derivatives will be covered in the upcoming Volume 5 of this series.

Fig. 5.6 summarizes the various possibilities of starch modification by enzymatic route,

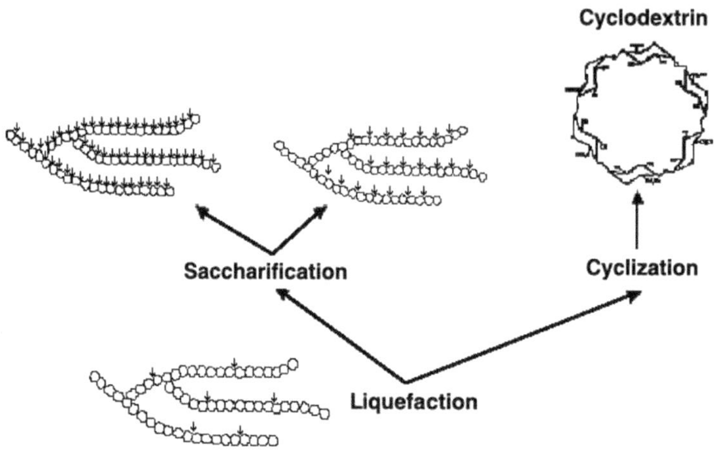

FIGURE 5.6 Possibilities of starch modification by enzymatic route. Legend: the arrows represent the chain break points.

representing enzymes promoting liquefaction, saccharification, and cyclization.

The International Unit of Enzyme Activity is the KATAL (kat), defined as the amount of enzyme required to transform one Mol of substrate per second under standard experimental conditions. It can also be expressed in μMol of substrate per minute. Despite this international standardization, each enzyme manufacturer uses its own units, under experimental conditions that are also particular.

5.4.1.2 *Commercial amylolytic enzymes*

Although it is possible to isolate biological agents capable of producing enzymes of good quality, the ease of purchasing and using commercial enzymes makes them to be preferred for industrial processes. Even so, malt is still the most widely used enzyme preparation in the production of beverages based on starchy raw materials, such as beer and distilled spirits.

There are several companies producing enzymes on an international scale, but to use them well, it is necessary to know their characteristics, which are provided by the manufacturers, along with other technical information.

In the specific case of this 5-volume series, where authors aimed at providing knowledge for better utilization of starches extracted from underground starchy crops, it is necessary to evaluate the processes of hydrolysis of these starches on an industrial scale and establish the performance of some enzymes present in roots and tubers. Surmely et al. (2004) evaluated commercial enzymes available in Brazil in the early 2000s by using commercial cassava starch. The authors remember that although technical information on commercial enzymes is available, it generally refers to use on corn starch and information on the action of commercial enzymes on tropical tuber starches is rare. Two types of α-amylases can be used, thermolabile or thermoresistant, the former with the advantage of being easily inactivated, which translates into less process energy (Teague & Brumm, 1992).

Food grade enzyme preparations are in the form of a nonviscous, brown liquid and are divided into three categories due to the characteristics of the preparations:

- α-thermo-resistant amylase, whose temperature of action is above 90°C;

- α-amylase thermo-sensitive, whose temperature of action is between 70 and 80°C;
- amyloglucosidase and β-amylase saccharification enzymes.

Despite attempts of international standardization, technical sheets and articles cite different temperatures and pH, making it difficult to compare enzyme activities. In addition, the presence of calcium stabilizes the enzymes, especially in relation to heat, but there are enzymes that do not need this correction. Another general observation is that when enzymes are used outside the optimal parameters, the hydrolysis profile can completely change.

5.5 Evaluation of some commercial amylolytic enzymes

Surmely et al. (2004) evaluated the enzymatic activity as the increase in the DE in relation to the dose of all enzymes available at the time in the Brazilian market with doses varying from 1 to 2 or 3 mL of the enzymatic preparation per kg of dry cassava starch, according to each technical sheet. When necessary, the calcium concentration was adjusted in each trial, according to the technical recommendations.

Many commercial enzymes are available today, but only as an example, three evaluations made by the authors were highlighted, with the profile of the hydrolysates obtained by high performance liquid chromatography (HPLC). The enzymes used were provided by the company Novozymes of Brazil; these enzymes can be replaced periodically by preparations more powerful and more suitable to the present needs.

5.5.1 Performance of the thermoresistant enzymes Thermamyl

As a representative of a heat-resistant α-amylase, whose action temperature is above 90°C, Novozymes' Termamyl 120 L was selected (Table 5.1), a product that was used for a long time, but was replaced by Termamyl SC DS. According to the Gluten Free Home Brewing website (2020), the difference comes from the fact that Termamyl SC is more concentrated and requires half the amount for the same effect.

5.5.1.1 *Applications*

- Food industry: sugars, alcohol, beverages, and beer.
- Industrial fermentation to produce vitamins, amino acids, antibiotics, and so forth.
- Textile industry and adhesives.

Termamyl 120 L was used. The performance evaluation is presented in Fig. 5.7, with the evolution of the initial DE as a function of time, for each enzyme concentration, in 35% cassava starch suspension and temperature maintained at 90°C.

Note: If Termamyl 120 L is changed by Termamyl SC DS, the quantities to be used should vary, for similar results.

TABLE 5.1 Characteristics of the thermally resistant α-amylase type enzyme Termamiyl 120 L, produced by *Bacillus amyloliquefaciens* strain.

	Termamyl 120 L
Characteristics	
Appearance	Food grade nonviscous liquid
Color	Dark brown
Density (g/mL)	1.2
Viscosity (cPs)	1–25
Activity (KNU/g)	120
Optimum conditions	
pH	6–8
Temperature (°C)	90–105
Calcium (mg/kg)	30–60

Note: KNU = Kilo New Unit; 1 KNU/g activity means that 1 gram of enzyme hydrolyzes 5.26 g of starch per hour, as evaluated by the Novozymes standard method.

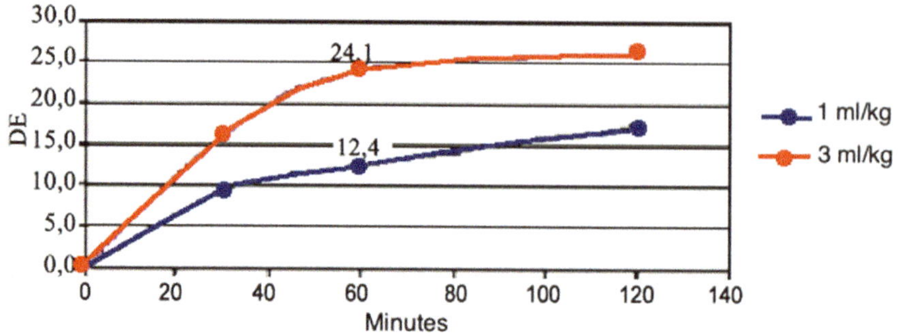

FIGURE 5.7 Evolution of DE with two concentrations of the enzyme Thermamyl 120 L, as a function of reaction time, for 35% w/v cassava starch at 90°C. *DE*, Dextrose equivalents.

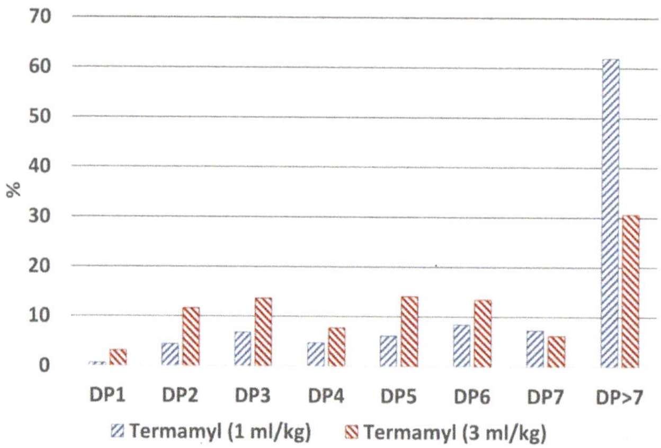

FIGURE 5.8 Hydrolysis profile of cassava starch (35% w/v) with two Thermamyl 120 L concentrations at 90°C for 60 minutes. *DP*, Degree of polymerization; *DP1*, glucose; *DP2*, maltose; *DP3*, maltotriose; *DP > 7*, dextrin (HPLC analysis).

It is possible to observe that the maximum DE reached by Thermamyl 120 L enzyme at a dose of 3 mL enzyme per kg of starch was 24.1 after 60 minutes, with stabilization close to this value. For enzyme concentration of 1 mL/kg, the reaction was slower and did not reach the maximum DE even after 120 minutes of the experiment.

The performance should also consider the profile of the hydrolysates obtained by HPLC, as can be seen in Fig. 5.8.

Fig. 5.8 shows that at a dose of 3 mL/kg, Thermamyl generated 30.6% of Gp > 7 carbohydrates (dextrin), while at a concentration of 1 mL/kg, the concentration of this group increased to about 62%, indicating incomplete activity.

5.5.2 Performance evaluation of the thermosensitive enzyme Ban 120 L

Ban is a bacterial amylase from Novozymes, an α-amylase produced by submerged fermentation of a bacterium of the genus Bacillus. The enzyme hydrolyzes the α-1,4 bonds of amylose and amylopectin, which results in a rapid reduction in the viscosity of gelatinized starch. The products from the hydrolysis are dextrins and soluble oligosaccharides such as maltodextrins. Today, BAN 120 L has been replaced by BAN 480 L, which is more concentrated and has a KNU 4 times higher (Table 5.2). Since 1 KNU/g means that 1 gram of enzyme hydrolyzes 5.26 g of starch per hour, a KNU of 480 means that 1 gram of enzyme hydrolyzes 2524.8 g of starch and a KNU of 120, 631.2 g.

TABLE 5.2 Characteristics of NovoZyme Ban 120 and 480 L thermosensitive α-amylase type enzymes from *Bacillus licheniformis* strain.

	Ban 120 L	Ban 480 L
Characteristics		
Color	Light brown	Light to dark brown
Density (g/mL)	1.21	1.21
Viscosity (cPs)	Not found	10–50
Activity (KNU/g)	120	480
Optimum conditions		
pH	5–7	5–7
Temperature (°C)	70–80	70–80
Calcium (mg/kg)	250	250

Note. KNU = Kilo New Unit; 1 KNU/g activity means that 1 gram of enzyme hydrolyzes 5.26 g of starch per hour, as evaluated by the Novozymes standard method.

5.5.2.1 Applications

- Food industry: sugars, alcohol, beverages, and beer.
- Fermentation industry: vitamins, amino acids, antibiotics, and so forth.
- Textile paper industry and adhesives.

BAN 120 L was used and evaluated from the evolution of the initial DE as a function of time, for each enzyme concentration, in 35% cassava starch suspension and temperature maintained at 80°C (Fig. 5.9).

The increase in DE as a function of time, for a given enzyme concentration, with 35% starch and a reaction temperature of 80°C, is presented in Fig. 5.9, which indicates that the profiles of the two BAN concentrations are similar and continue to increase after 120 minutes, varying only the DE obtained.

The performance should also consider the profile of the hydrolyzed products obtained, as can be seen in Fig. 5.10.

Fig. 5.10 identifies a similar profile for the two Ban contents, but with higher dextrins for the 1 mL/kg.

5.5.3 Performance of the thermosensitive saccharifying enzyme AMG 300 L

AMG is a food-grade amyloglucosidase, produced from a selected strain of *Aspergillus*

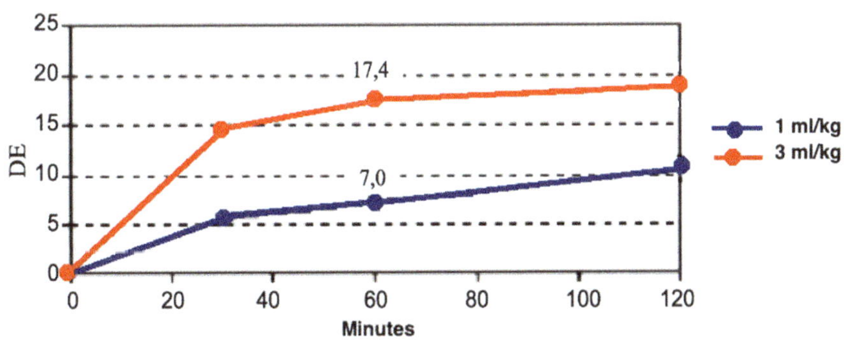

FIGURE 5.9 Evolution of DE with two concentrations of the enzyme BAN 120 L, as a function of reaction time, for cassava starch (35% w/v) at 80°C. Note: If Ban 120 L is changed by Ban 480 L, the quantities to be used should vary, for similar results. *DE*, Dextrose equivalents.

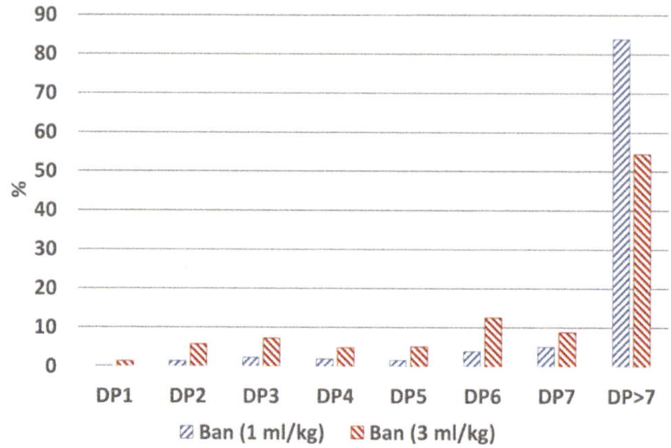

FIGURE 5.10 Hydrolysis profile of starch (35% w/v) with two concentrations of Ban 120 L at 80°C for 60 minutes. *DP*, Degree of polymerization; *DP1*, glucose; *DP2*, maltose; *DP3*, maltotriose; *DP* > *7*, dextrins (HPLC analysis).

niger. The enzyme hydrolyzes the α-1,4 and α-1,6 bonds, yielding a fully liquefied starch. During hydrolysis, the glucose units at the nonreducing end of the saccharide are gradually eliminated. The speed of hydrolysis depends on the type of bond and the length of the chain. AMG is recommended for saccharification of starch in glucose production. The new AMGs have higher efficiency, with AGU of 300, compared to 120 in the evaluated enzyme (Table 5.3).

TABLE 5.3 Characteristics of NovoZyme AMG 300 L thermosensitive β-amylase type enzymes.

	AMG 300 L
Characteristics	
Color	Light brown
Density (g/mL)	1.2
Activity (AGU/g)	120
Optimum conditions	
pH	4.0–4.5
Temperature (°C)	58°C–70°C
Calcium (mg/kg)	not required

Note: 1 AGU: enzyme concentration that hydrolyzes one micromol of maltose per minute, according to Novozymes A/S standard.

5.5.3.1 Applications

- Food industry: sugars, alcohol, beverages, and beer.
- Fermentation industry: vitamins, amino acids, antibiotics, and so forth.
- Fermentation industry.

The performance evaluation of the enzyme AMG 300 L is presented in Fig. 5.11, with the evolution of the initial DE as a function of time, for each enzyme concentration, in 35% cassava starch suspension and temperature maintained at 60°C.

The activity of the enzyme AMG 300 L allows for obtaining DE higher than in the enzymes previously described, which explains its recommendation for glucose production. The evolution of DE is similar between the two doses of enzymes, with a tendency to stabilize after 4 hours.

The performance should also consider the profile of the hydrolyzed products obtained, as can be seen in Fig. 5.12.

Fig. 5.12 presents the carbohydrate profile of the hydrolyzed starch. It is observed that the oligosaccharides DP1 (glucose) and DP2 (maltose)

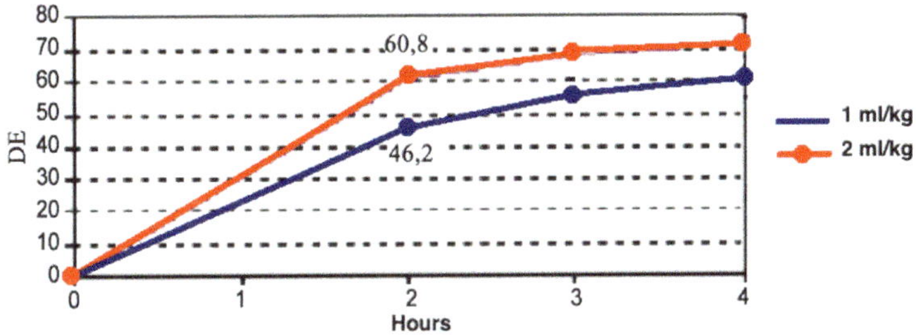

FIGURE 5.11 Evolution of DE with two concentrations of the enzyme AMG 300 L, as a function of reaction time, for 35% w/v cassava starch at 60°C. *DE*, Dextrose equivalents.

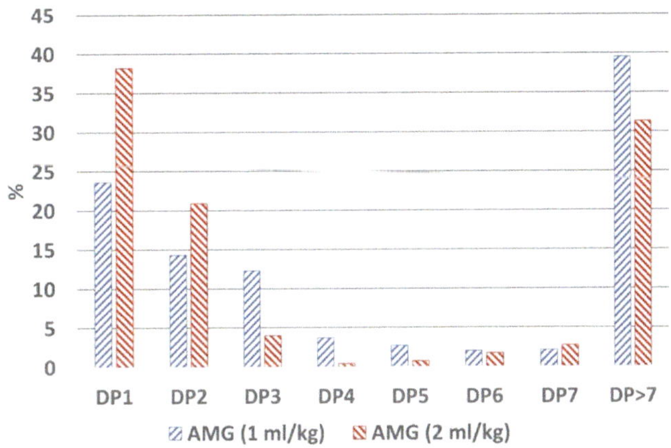

FIGURE 5.12 Profile of starch hydrolysates (35% w/v) with two concentrations of AMG 300 L at 60°C for 120 min. *DP*, Degree of polymerization; *DP1*, glucose; *DP2*, maltose; *DP3*, maltotriose; *DP > 7*, dextrins (HPLC analysis).

are produced in large proportion, being majorities at the concentration of 2 mL/kg. In this case, the doubling of the enzyme concentration changed the profile of the hydrolysate, with predominance of glucose and maltose. The contents of intermediate oligosaccharides (DP3 to DP6) were lower with the higher amount of enzyme.

5.6 Hydrolysis process

Once the performances of some of the commercially available enzymes for starch hydrolysis have been discussed, the main processes that can be carried out on a semiindustrial scale are seen. After commenting the performances of some of the commercially available enzymes for starch hydrolysis, the main processes that can be carried out on a semiindustrial scale are seen.

5.6.1 Free and immobilized enzymes

In general, free enzymes are most used in industrial scale in hydrolyzed starch processing. Immobilized enzymes are still under research or only applied in some areas, where the products are more valuable or the enzymes too expensive.

According to Hebert and Scriban (1973), immobilized enzymes are confined on a mineral or organic support but exhibit the same mode of action as free enzymes. The classification of

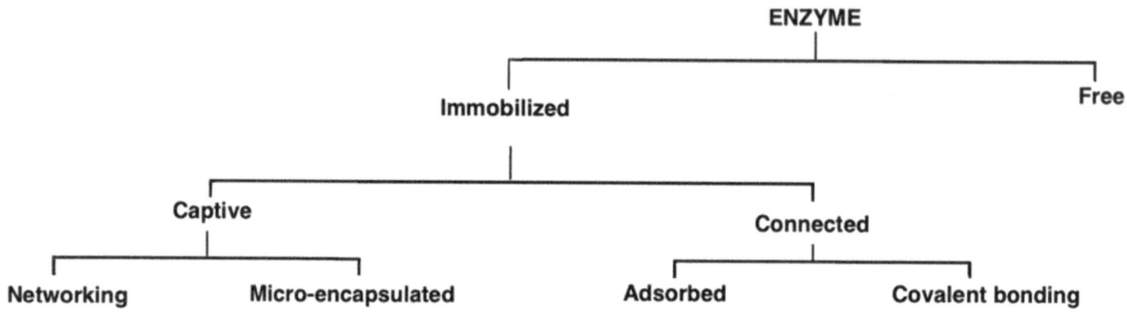

FIGURE 5.13 Classification of immobilized enzymes according to support.

immobilized enzymes established according to the mode of immobilization is presented in Fig. 5.13.

The main advantages of immobilized enzymes are the possibility of continuous processing, increased enzyme stability, enzyme-free end products, improved control of hydrolysis, low process cost compared to free enzymes, and the possibility of using multienzyme systems.

The costs of the support and the stability of these enzymes are the two most important parameters in selecting the best immobilization method (Chaplin & Bucke, 1990). It is worth remembering that the kinetic characteristics of the enzyme are altered with immobilization, generally by modifying the spatial availability. Immobilization also creates tensile forces that modify the tertiary structure of the enzyme.

5.6.2 Starch hydrolysates and production processes

Regarding the hydrolysis agents used, starches can be converted by three processes: acid, acid-enzyme, and enzyme-enzyme.

5.6.2.1 Acid hydrolysis

The simplest process of starch hydrolysis is with acid. In this process, an acid, usually concentrated HCL in a proportion of 0.1%–0.2% of the dry starch weight, is added to a starch paste with a concentration around 50% in dry matter.

TABLE 5.4 Percentages of carbohydrates in acid hydrolysates of corn starch as a function of DE.

DE	Glucose	Maltose	Maltotriose	Maltotetraose	Others
10	2.3	2.8	2.9	3.0	89.0
20	5.5	5.9	5.8	5.8	77.0
30	10.4	9.3	8.6	8.2	63.5
40	16.9	13.2	11.2	9.7	49.0
50	25.9	16.6	12.9	10.0	34.6

Reproduced with permission from Lloyd, N. E., & Nelson, W. J. (1984). Glucose and frutose containing sweetners from starch. In R. Whistler, J. N. Bemiller, E. F. Paschall (Eds.), Starch chemistry and technology *(2nd ed., pp. 611–660). Orlando: Academic Press.*

This suspension is then subjected to heat for a period that depends on the process selected.

Modern plants can employ continuous converters that are more efficient. This equipment better controls the DE and minimizes the formation of undesirable dyes during conversion due to nonenzymatic browning (Maillard reaction). The continuous process also reduces the formation of substances generated during conversion, such as levulinic acid, formic acid, and so forth. Lloyd and Nelson (1984) described one of these processes, where a 40% starch suspension is acidified to pH 1.8–2.0 with HCl and pumped to a converter, which simultaneously injects live steam for heating. The live steam can be provided by jet cooker or by reactors with steam heating.

Table 5.4 shows the carbohydrate composition in corn starch hydrolysate as a function of

DE. For production of syrups with DE greater than 50, the acid process is not employed, because the longer process time generates high concentration of coloring substances and undesirable bitter taste, usually caused by furfural.

When hydrochloric acid is used, neutralization is done with NaOH, resulting in NaCl that is soluble and provides more transparent syrup than hydrolysis with sulfuric acid. This acid is neutralized with calcium carbonate, which generates calcium sulfate, which is insoluble and clouds the syrup. The formation of NaCl, however, has the disadvantage of providing a salty taste to the syrup and requires treatment in an ion exchange resin column for its removal. Hydrolysis with sulfuric acid requires finer filtration, but in compensation does not require ion exchange resin in the purification of the syrup.

5.6.2.2 *Acid-enzyme hydrolysis*

In this process, the starch is first hydrolyzed with acid (liquefaction) and then treated with a specific enzyme. This step is called saccharification, which, depending on the type of enzyme used, produces a higher proportion of glucose, maltose, or dextrins. The acid/enzyme conversion process can produce several types of hydrolysates, but it is not suitable to obtain hydrolysates with more than 90% glucose, due to retrogradation reactions that are catalyzed by the acid, originating β-glycosidic bonds, which are not hydrolyzed by glucoamylases (Engasser, 1988).

5.6.2.3 *Enzyme-enzyme hydrolysis*

The liquefaction of starch in this process is catalyzed by α-amylase. The origin, dosage, temperature, pH, and the time of action of these enzymes influence the final composition of the product after saccharification (Colonna et al., 1988). The liquefied starch should not present residual starch, which consists of part of the granular starch that was not gelatinized, but also part of the gelatinized starch that was not liquefied. This residual starch has high molecular weight molecules, which hinders the filtration of the hydrolysate (Surmely et al., 2004).

It is an expensive technology, with sophisticated equipment, facilities, and labor. However, the difficulty is essentially a matter of resource availability, since the technology is simple and there is plenty of information on processing from the main starches, such as corn, cassava, and wheat.

Setting up a hydrolysate plant for other raw materials is more complicated due to the lack of available technical data.

5.6.3 Processes for obtaining hydrolysates

Processes can be classified into batch and continuous. What characterizes a continuous process is that the input flow rate is equal to the output flow rate. Batch processes are still the most common.

Li et al. (2023) stated that the production of starch sugars involves basically starch liquefaction and saccharification processes. The production cost of starch sugars can be reduced by increasing the initial concentration of starch slurry. However, the usage of a highly concentrated starch slurry is characterized by challenges such as low reaction efficiency and poor product performance during the liquefaction and saccharification processes. However, despite the difficulties, the authors believe that this will be the target of further research in commercial starch hydrolysis.

Continuous processing of starch hydrolysates can be done in columns or in an ultrafiltration reactor. Swinkels (1985) cited that in continuous enzymatic conversion processes, a starch suspension containing thermoresistant enzyme α-amylase is instantaneously gelatinized by steam injection (jet cooker) and continuously discharged into a holding column where the enzyme viscosity reduction reaction occurs. The low-viscosity suspension is then pumped to another column where the enzyme is deactivated by heating with steam (jet cooker) at very high temperature.

Continuous separation of the sludge using ultrafiltration is possible. Separation and concentration processes using synthetic membranes are applied in many areas, such as chemical purification, biotechnology, pharmaceuticals, food processing, water, and wastewater treatment, as well as reactors for enzymatic processes. In this case, ultrafiltration has the advantage of allowing for modular installation. The function of the membrane is to act as a thin, permeable and selective barrier, but in practice, some enzymes always pass through the filter membrane.

Ultrafiltration is a technique that allows, by passing through a selective permeable membrane, to separate solvent and small molecules, called filtrate or solution permeate. The macromolecules, including most of the enzymes, do not pass through the filter and are retained in the solution, which progressively condenses and becomes the concentrate (Dornier, 1992). The separation takes place with the help of the difference in pressure (less than 10 bars) imposed on either side of the membrane. The mixture flows tangentially or parallel to the porous wall, to limit the clogging of the porous wall caused by the accumulation of particulate material. This process is called tangential ultrafiltration (Fig. 5.14).

5.6.3.1 *Starch hydrolysis process yield*

The calculation of the yield of the process is established on the amount of raw material used. Each break in the starch chains leads to the formation of smaller chains and is accompanied by the addition of one molecule of water (hydrolysis) for each broken bond, which leads to an increase in the weight of hydrolyzed starch and consequently an increase in yield. It is common to obtain yields above 100%.

For example, in the theoretical case of total hydrolysis of starch into glucose molecules, 1 g of starch would give 1.1 g of glucose, with a yield of 110%. This yield increase is dependent on the DE, which also corresponds to the percentage of water molecules entering the starch chains. The enzymatic conversion of a starch to glucose of DE 38 corresponds to a yield of 103.8%.

5.6.4 Products obtained from starch hydrolysis

Starch hydrolysates include all the products of the fractionation of its polymeric chains,

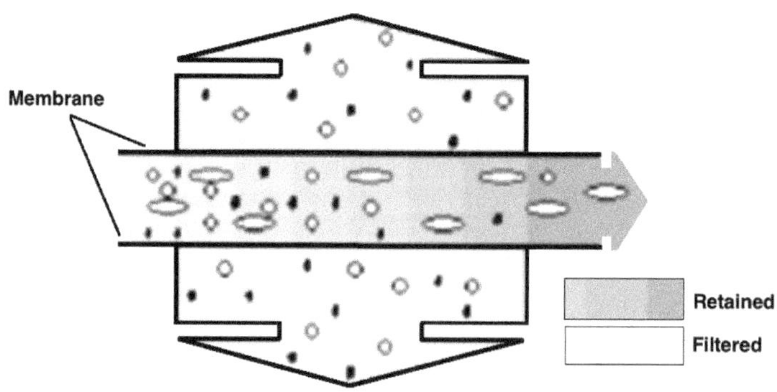

FIGURE 5.14 Tangential ultrafiltration scheme. *Reproduced with permission from Surmely, A., Alvarez, H., Cereda, M. P., & Vilpoux, O. (2004). Starch hydrolysis. In* Cereda & Vilpoux, Technology, Uses and Potentialities of Latin American Starchy Tuberoses *(v. 3, pp. 378–449). São Paulo, Fundação Cargill (In Portuguese).*

regardless of the catalysts used (acids and enzymes) or the degree of fractionation. It brings together many different products such as glucose, maltose, maltotetraose, maltodextrins, fructose, cyclodextrins, dextrins, and so forth. Hydrolysates have physical, functional, energetic, and organoleptic properties that are characteristic of each type of product (Teague & Brumm, 1992).

Three categories of hydrolysates for food use are most common in Brazil. These are glucose, maltose, maltodextrin, and their derivatives, such as sorbitol, an alcohol product derived from glucose. Oligosaccharides, such as iso-maltose, are hydrolyzed derivatives that have been conquering the market, specifically those aimed either at health, such as dietary fiber, or at favoring the growth of lactic bacteria, such as probiotics.

Starch hydrolysates can be marketed in syrup form, which is most common, and in powder form. Maltodextrins and glucose powders are dehydrated in spray-dryers (see Chapter 12). Liquid syrups are cheaper, but their high viscosity complicates transportation and marketing. Transportation must be done in special tank trucks, with a heating system for unloading. The producing companies can install syrup storage tanks at the buyers, on a free-lease basis, due to the cost of their installation. The syrups stored for a long time can darken due to the Maillard reaction, especially those derived from cereals, which may contain residual protein.

The dry product option is an interesting solution for small consumers, as hydrolyzed products in bags are more stable due to lower water activity and are easier to transport and dose. They can be made by dehydrating the syrup by spray-drying or drum drying.

5.6.4.1 Glucose

The term glucose syrup refers to all purified and concentrated aqueous solutions of D-glucose polymers with DE between 20 and 80 (Howling, 1992). Glucose refers to hydrolysates that contain glucose molecules in the range of 5%–95%, provided that they predominate over the other polysaccharides present in the hydrolysate. Pure glucose or dextrose is composed of 100% D-glucose. It can be produced by hydrolysis of starch or sucrose, such as from sugar cane, followed by separation of the D-glucose and the other mono- or polysaccharides. On a commercial scale, the raw material for glucose production is starch. Commercial glucose is a mixture of hydrolysates where glucose predominates over maltose, maltotriose, and other saccharides in varying proportions. The more intense the hydrolysis, the higher the formation of glucose, therefore the higher the DE (Table 5.5).

In general, the properties of glucose syrup are determined by the glycidic spectrum and the average degree of polymerization of the constituents of this syrup (Howling, 1992). These physical and functional properties vary as a function of DE (Table 5.6).

5.6.4.2 Maltose

The product of hydrolysis is considered maltose syrup when it contains at least 30% maltose in a purified and concentrated aqueous solution. In addition to the content, maltose must predominate over the other hydrolysates. According to Okada and

TABLE 5.5 Composition of commercial glucoses and their DE values.

	Dextrose equivalents			
	30	35	42	55
Glucose	10	14	19	31
Maltose	9	12	14	18
Maltotriose	8	10	11	13
Other saccharides	69	64	56	48

Reproduced with permission from Lloyd, N. E., & Nelson, W. J. (1984). Glucose and frutose containing sweetners from starch. In R. Whistler, J. N. Bemiller, E. F. Paschall (Eds.), Starch chemistry and technology *(2nd ed., pp. 611–660). Orlando: Academic Press.*

TABLE 5.6 Physical and functional properties of hydrolysates as a function of DE.

	Hydrolysis intensity
Properties	**Starch → Low DE → High DE → Glucose**
Viscosity	←
Binding and cohesive power	←
Sweet taste	→
Hygroscopic power	→
Water-holding power	←
Anti-crystallizing power	←
Heat dimming	→
Freezing temperature	←
Aroma enhancement	→
Fermentability	→

Nakakuki (1992), on an industrial scale, the maximum concentration of maltose that can be obtained by starch hydrolysis is approximately 90%. In the maltose syrup market, syrups are classified into three categories (Alexander, 1992).

- High conversion syrup: DE 62–63, with 35% glucose and 30%–45% maltose.
- High maltose syrup: DE 48–52, with 5% glucose and 48%–52% maltose.
- Extra-high maltose syrup: DE 50–60, with 5% glucose and 70%–85% maltose.

According to Alipio (2000), the main use of maltose syrup is as an adjunct or even substitute for malt, in the manufacture of beer. In addition to this use, maltose syrup can be used in several other products, as reported by Schenk and Hebeda (1992), who cited powdered milk, liquid foods, caramel, chocolate, cream, marmalade, jam, soda, wine, whiskey, sauce, seasoning, cream, jam, dehydrated seaweed, ice cream, frozen food, soup, instant coffee, spice, powders for soft drinks, and juices.

A fraction of the hydrolysates, classified as malto-oligosaccharides have been valued as food for probiotic lactic bacteria, and as such are considered dietary fibers, which opens a great market for maltose.

5.6.4.3 *Fructose*

Fructose, unlike glucose and maltose, has only five carbons in its structure (Fig. 5.15).

Fructose is naturally present in fruits, honey, and some vegetables. It is generally considered beneficial to human health because it can be consumed even by patients with type 2 diabetes, provided it is consumed in moderate amounts. However, there are many consumers who have symptoms of fructose intolerance. Ali et al. (1998) stated that fructose consumption may develop hypoglycemic and severe abdominal symptoms after taking foods containing fructose and cognate sugars. Continued ingestion of noxious sugars leads to hepatic and renal injury and growth retardation. Parenteral administration of fructose or sorbitol may be fatal.

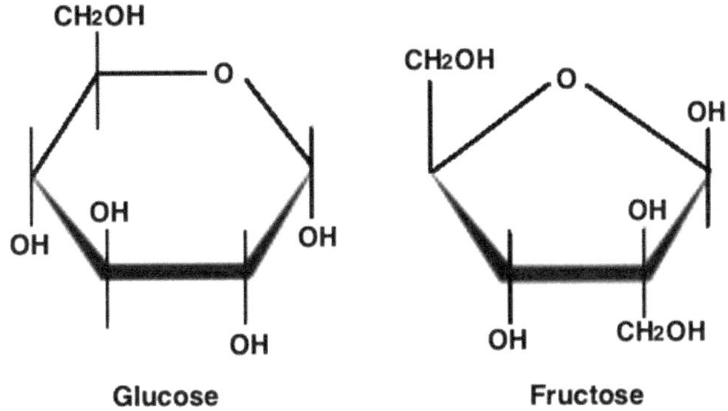

FIGURE 5.15 Structure comparison of glucose and fructose (Qi & Tester, 2019). *Reproduced with permission from Qi, X., & Tester, R. F. (2019). Fructose, galactose and glucose—In health and disease.* Clinical Nutrition ESPEN, 33, *18–28.*

Because it is not feasible to extract fructose from its natural sources, developing fructose from other more readily available carbohydrates has proven to be an attractive market because of its lower sweetness. The largest market is in the United States, where soft drink consumption is very high and being overweight has become a state problem.

As Vuilleumier (1993) mentioned, high-fructose syrups (HFS) are manufactured and used in many countries throughout the world. They are produced from sugarcane and sugar beet pulp and a variety of starch raw materials including corn, rice, cassava, and wheat. Production of HFS is highly dependent on local sucrose and economics of agricultural raw materials. HFS is produced and consumed in the largest quantity in the United States by using corn starch as the raw material. Crystalline fructose is being used in a growing number of food applications, but its production represents a small percentage of total fructose sweeteners.

In South America, it was considered that there was no room for fructose or fructose syrup production using starch as a raw material, due to low cane sugar prices. However, the increasing production of ethanol from corn is changing this perception.

Moorthy (1994) reported the manufacturing process of fructose syrup from cassava starch. The author points out that due to the increased demand for sweeteners, HFS have several applications in the pharmaceutical and food industries. Starch hydrolysis can be performed with acids such as hydrochloric and sulfuric acids or with enzymes. The product hydrolyzed with acids, after concentration, is neutralized with alkali such as sodium hydroxide and is isomerized with cyclohexamine at a temperature of 60°C–65°C. The isomerization increases with time, but the product becomes increasingly colored, so that the best reaction time was set at 6 hours. The syrup produced under these conditions has 22% fructose, evaluated as total sugar content. The syrup is column-filtered, first with strongly acidic cation exchange resin, followed by weakly anionic resin. Concentration results in a colorless syrup with about 85% solids. When kept at the same temperature or cooled, white crystals are deposited. These crystals, when separated and dissolved in adequate amounts of water, provide very sweet and transparent syrup. The enzymatic process with glucose-isomerase can be used to improve isomerization. After hydrolysis, the syrup is decolorized with activated carbon, the concentration is increased to 40%–42%, and the pH is adjusted to 8.2.

The results of Moorthy (1994) were later confirmed by Surmely et al. (2004). Isomerization was carried out with 12 g Sweetzyme IT Extra from

Novozymes and 0.5 g magnesium sulfate to 400 mL of syrup. The medium was kept stirred for 6 hours at 61°C. The isomerized syrup was decolorized by activated carbon that can be replaced by ion exchange resin column. The DE and fructose content were determined, and both the isomerized syrup and the syrup from acid and enzymatic hydrolysis had a final fructose content of 40%.

5.6.4.4 *Other hydrolysates obtained from starch*

Several other hydrolysates can be generated and occupy specific markets.

5.6.4.4.1 Maltodextrins

Maltodextrin is a purified and concentrated aqueous solution, liquid or dry, containing a mixture of glucose, maltose, maltotriose, and other saccharides, whose DE is less than 20 because it has undergone a lower degree of hydrolysis than glucose and maltose syrups (Alexander, 1995).

The properties of maltodextrins as a function of DE follow the same variations as those reported for glucose syrup (Table 5.6).

5.6.4.4.2 Dextrins

Dextrins have many industrial and food applications. Their most common use is as a base for adhesives for cartons, envelopes, and stamps (for more information on starch-based adhesives, see Chapter 15).

Dextrins are also used to polish cereals and processed fruits, and in the decoration of ceramics, in addition to their use in the textile and pharmaceutical industries. In the food area, dextrins enter food preparation as a thickening agent and have applications in brewing, baking, juice and cocoa drinks, distilled liquors, confectionery products, and so forth.

The physicochemical process to obtain dextrin involves the use of heat (110°C–120°C) on starch moistened by diluted acid solution (1000 parts starch, 250 parts water, 2 parts HCl). The product is subsequently dried, ground, graded, and packaged. The process is inexpensive, but yields less and produces lower quality starch. The enzymatic process circumvents these problems. The enzyme can be of vegetable or microbial origin, defining the technological process to be employed. A starch suspension with 35% of its weight in water is heated until gelatinization. Then it is cooled to 65°C–75°C and added about 2% enzyme, depending on the source. At the end of dextrinization, the material is boiled to inactivate the enzymes and then filtered to form the dextrin syrup, which may or may not be concentrated.

5.6.4.4.3 Isomalto-oligosaccharides

Isomalto-oligosaccharide (IMO) syrup is used in an industry that has shown great growth, that of health products, to counteract sweeteners and dietetics. IMOs are glyco-oligosaccharides that have α-1,4 and α-1,6 linkages. They are produced in two steps, the transformation of liquefied starch into iso-maltose, with hydrolysis via β-amylase, and then the transformation of this iso-maltose into panose, isomalto-triose, branched oligosaccharides, and other sugars, using a transglycoside. Thus, IMOs contain a wide variation of sugars, including the limiting dextrins resulting from the action of the α- and β-amylases.

IMO syrup is almost half as sweet as sucrose and does not mask other delicate flavors. It has a viscosity similar to sucrose solution and lower than that of a maltose syrup. The humectance is higher due to the great moisture retaining power of its branched oligosaccharide components, which can prevent the hardening of starchy foods. At the same time, the water activity is lower than that of maltose syrup and the same as that of sucrose solutions, which allows for the growth of microorganisms in foods to be controlled. IMO is stable over a pH range of 3–6 and upon heating at 100°C for 1 hour.

The great advantage and what differentiates it in the market of healthy products is that IMO is not metabolized by microorganisms, which allows this syrup to present reduced possibility of deterioration by growth of microorganisms, while maintaining the characteristics of thickening and sweetening other fermented foods such as bread, drinks containing lactic acid, spices and sauces obtained by fermentation such as soy-based "shoyu", Japanese sake, and so forth. Another advantage is that the intake of branched oligosaccharides improves stool consistency and relieves constipation (fiber effect) since they are little altered by the action of digestive enzymes in the small intestine. Added to this beneficial factor is the favoring of the proliferation of bifidobacteria, since they serve as food for this group of bacteria. It is worth pointing out that *E. coli*, Clostridium perfringens, and other harmful intestinal bacteria, since they cannot metabolize it, do not proliferate, which improves the intestinal flora.

Another important characteristic is the low cariogenicity. Branched oligosaccharides such as iso-maltoses, iso-maltotrioses, panoses, and so forth are not substrates for the glycosyltransferase of *Streptococcus mutans* to synthesize the water-insoluble glucans that are responsible for the bacteria sticking to the teeth. They are also used very little by the microorganism to form the lactic acid that demineralizes tooth enamel. The caloric power of IMO is 2.7–3.3 kcal/g, which characterizes it as a low caloric product (Alipio, 2000).

5.7 Processes and products obtained by hydrolysis of starch extracted from underground raw materials

Starches of any origin can be used as raw material in the manufacture of hydrolysates. In most countries, including South America, corn starch is most frequently used, because it is commercially available and because there is already a well-developed and established technology for its use. The second most common starch is cassava starch, a product produced mainly in Asia and South America.

Because of the lack of technical information for tropical starches, pH, temperature, and enzyme dosage were adjusted. The analysis of the results obtained is presented below.

5.7.1 Syrups from cassava starch obtained in laboratory scale

As presented in Section 5.5, enzymatically obtained syrups are the main hydrolyzed products marketed globally, but generally from corn starch. In the following examples at laboratory scale and later at pilot scale, the batches all used the same cassava starch starter, the same enzymes, and the analyses were performed in the same equipment and with the same methodology. Both the laboratory and pilot trials were performed with 35% starch concentration. The laboratory batches were of 200 mL and the pilot scale batches of 200 L. It is noteworthy that the products obtained were different types of syrups.

- **Liquefaction**: Under stirring, the starch in water was added until the established suspension of 35% in dry matter was obtained. In this step, two phenomena are combined simultaneously. The first is the gelatinization of the starch to allow the attack by the enzyme BAN. The strong bonds between amylose and amylopectin chains need to be broken, and therefore heating of the suspension is required. With heating, the BAN enzyme reduces the starch chains to smaller ones (dextrins) and prepares the medium for the saccharification step. The amount of BAN 120 L enzyme added to the solution was established for each hydrolysate and expressed as mL/kg dry matter of starch. The starch suspension was heated at a rate of 1.5°C per minute until it reached 80°C, a temperature that was maintained for 45 minutes and then cooled at a rate of 1°C per minute until it reached the selected temperature for each saccharification enzyme. The enzyme should be previously diluted and added under stirring in a water bath.

- **Saccharification**: The function of the saccharification step is to break the smaller chains (dextrins) from the liquefaction to give the desired hydrolysate in its unpurified form. Besides the type of enzyme, its concentration, the saccharification time, and temperature are very important to obtain a particular hydrolysate. Any modification of one of these parameters leads to changes in the sugar profile of the hydrolysate and, consequently, the type of hydrolysate produced. For saccharification of the liquefied substrate, enzyme was added at a minimum concentration of 1 mL/kg dry matter of starch, under stirring. The saccharifying enzyme suspension was previously diluted and the selected temperature was maintained for the time set for each selected enzyme to provide the desired DE.
- **Deactivation**: Deactivation stops the starch hydrolysis reaction by inhibiting the liquefaction and saccharification enzymes, which is achieved by increasing the temperature. The saccharification medium was heated with a temperature increase of 1.5°C per minute until it reached 95°C, which was maintained for 10 minutes under stirring. Then the suspension was cooled down to 75°C at the rate of 1°C per minute.
- **Decolorization**: After hydrolysis was used activated carbon in the sugar solution to adsorb the compounds responsible for bad odor and color. The retention of these fractions depends on the concentration of activated carbon used and the contact time. The temperature adopted in this step results from an adjustment between two antagonistic effects. At high temperatures, the formation of the protein/amino acid complex is higher, the solution viscosity is low, and consequently, the adsorption by the carbon is high. For hydrolysates from starch, the formation of the complex is reduced by the low protein content and the optimum temperature is 75°C. During processing, 0.2% activated carbon should be added to the suspension under stirring, calculated in relation to the dry matter content of the hydrolysate, and kept at 75°C for 30 minutes.
- **Centrifugation/filtration**: Centrifugation followed by filtration removes residual starch (gelatinized nonhydrolyzed starch), fat, and activated carbon with adsorbed substances. Starch from underground raw materials is characterized by low contents of lipids and proteins (to learn more about the composition of raw materials and their starches, Chapter 8 in Volume 1 of this series is suggested). Centrifugation has the function of prefiltration and allows for the reduction of filter clogging and to obtain the necessary filtration surface.
- **Demineralization**: The demineralization of the solution occurs by passing the generated liquid through ion exchange resins (anionic and cationic). The function of these resins is to adsorb mineral salts, but they can also completely decolorize the solution and favor the final deactivation of enzymes.
- **Evaporation**: The function of evaporation is to reduce the water activity and to increase the osmotic pressure of the hydrolyzed syrup to ensure its preservation and prevent the development of microorganisms. To avoid deterioration, the syrups need to be concentrated to a minimum Brix of 80°, which was obtained by using a rotary evaporator with the water bath set at 70°C.
- **Drying**: After evaporation, the function of drying is to further reduce the water activity

and increase the osmotic pressure of the hydrolyzed syrup to ensure its preservation in storage. For drying in spray-dryers, generally used in industries, the hydrolysate is first concentrated to at least 20° Brix.

5.7.1.1 *Glucose syrup*

To produce a glucose syrup of DE 38, the enzyme BAN 120 L at a concentration of 3 mL/kg starch was used. Then the suspension was heated at the rate of 1.5°C per minute until it reached 80°C, at which temperature it remained for 45 minutes. Next, the saccharification step was carried out at 60°C with the enzyme AMG together with Dextrozyme. The suspension was cooled at a rate of 1°C per minute from 80°C to 60°C. After saccharification, the deactivation of the enzymes was done by increasing the temperature at 1.5°C per minute until the temperature reached 95°C, holding this temperature for 10 minutes and then reducing it to 1°C per minute. The finalization in the industry is done by hot filtration due to the high viscosity, but in the laboratory scale, it was done by filtration. The evolution of the DE as a function of time and enzyme concentrations is presented in Fig. 5.16.

Dextrozyme is an enzyme that combines the properties of AMG and Pululanase (de-ramifying), which favors the total conversion of starch into glucose and is therefore particularly suitable for starch due to the amylopectin content. In the case of partially hydrolyzed syrup (Fig. 5.16), the difference in action between AMG and Pululanase was minimal. In this case, it was recommended to use only AMG. Generally, the working range of the adjusted pH for AMG is between 4.0 and 4.5. Since the pH of the liquefied syrup was naturally between 5.5 and 6.0 at the 1 mL/kg concentration of this enzyme, no adjustments to the values recommended by the data sheets were necessary. Fig. 5.17 shows the profile of the syrups obtained with the enzyme-enzyme process, for two levels of DE.

Fig. 5.18 shows the profile of a commercial DE 38 glucose syrup, made with the acid process, a cheaper product and with large application in the Brazilian market. The comparison of the profiles shows that the enzymatic process generates a larger amount of DP1, DP2, and DP3 than the acid process, with similar amounts of dextrins. The acid process is characterized, for DE 38 syrup, by a gradual production of oligosaccharides from DP10 to DP1, being glucose the main product.

5.7.1.2 *Maltose syrup*

For this type of syrup, the enzyme BAN 120 L was added at a concentration of 2.5 mL/kg of dry starch. The suspension was heated at 1.5°C per minute until it reached 80°C and remained at this temperature for 45 minutes before cooling at the rate of 1°C per

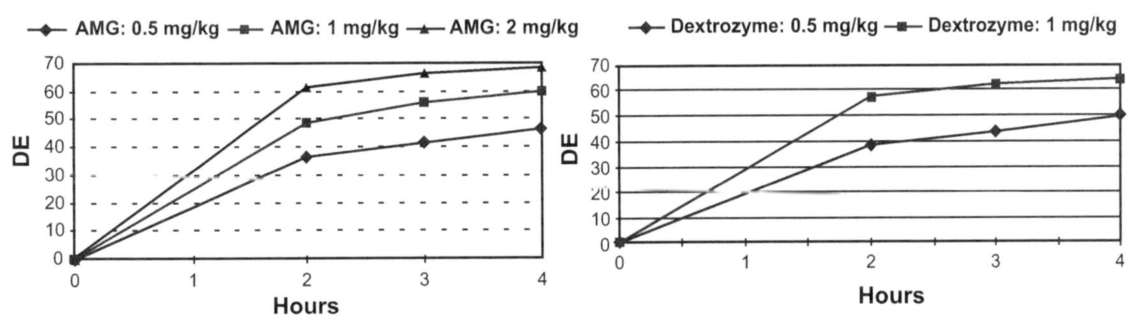

FIGURE 5.16 DE kinetics for glucose syrups as a function of AMG and Dextrozyme doses. *DE*, Dextrose equivalents.

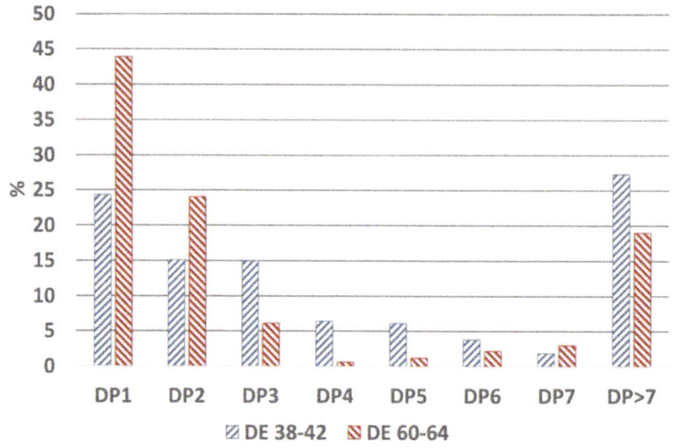

FIGURE 5.17 Profile in syrup hydrolysates of glycoses of DE 38–42 and DE 60–64. *DP*, Degree of polymerization; *DP1*, glucose; *DP2*, maltose; *DP3*, maltotriose; *DP* > *7*, dextrins (HPLC analysis).

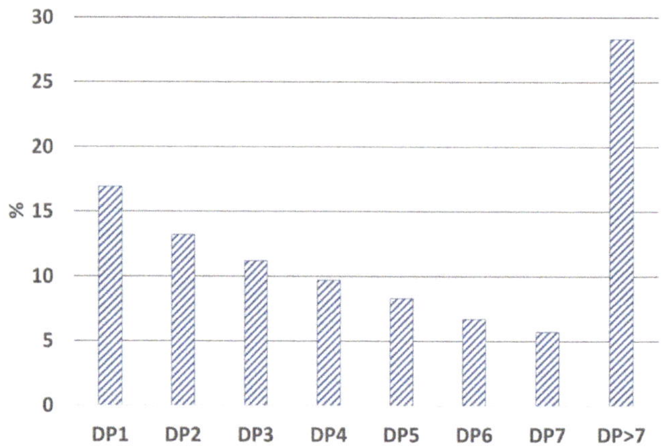

FIGURE 5.18 Profile of commercial glucose syrup DE 38, obtained from cassava starch by acid process. *DP*, Degree of polymerization; *DP1*, glucose; *DP2*, maltose; *DP3*, maltotriose; *DP* > *7*, dextrins (HPLC analysis).

minute. After this step, Fungamyl 800 L enzyme was added at a concentration of 1 mL/kg of dry starch. The suspension was heated to 60°C and kept at this temperature for 150 minutes. The deactivation was done by increasing the temperature at 1.5°C per minute, up to 95°C and kept at this temperature for 10 minutes and then reducing it to 1°C per minute.

The syrup with the profile shown in Fig. 5.19 is a maltose syrup for use in breweries, which means that it has a sugar profile similar to that of a wort obtained from malt and is classified as a malt adjunct.

5.7.1.3 Maltodextrin syrup DE 10

DE 10 maltodextrin syrup can be obtained with the same enzyme BAN 120 L at a concentration of 1 mL/kg dry starch. The suspension is heated at 1.5°C per minute until it reaches 80°C and held for 45 minutes at this temperature. The enzyme is inactivated by increasing the temperature at 1.5°C per minute, until it

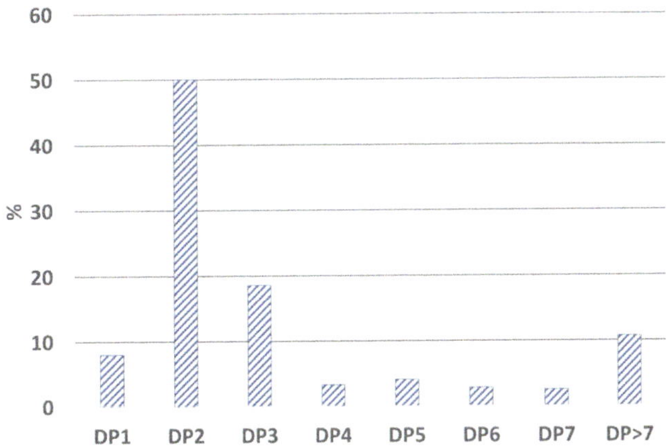

FIGURE 5.19 Profile of maltose syrup hydrolysates obtained from cassava starch, for use in brewing. *DP*, Degree of polymerization; *DP1*, glucose; *DP2*, maltose; *DP3*, maltotriose; *DP* > *7*, dextrins (HPLC analysis).

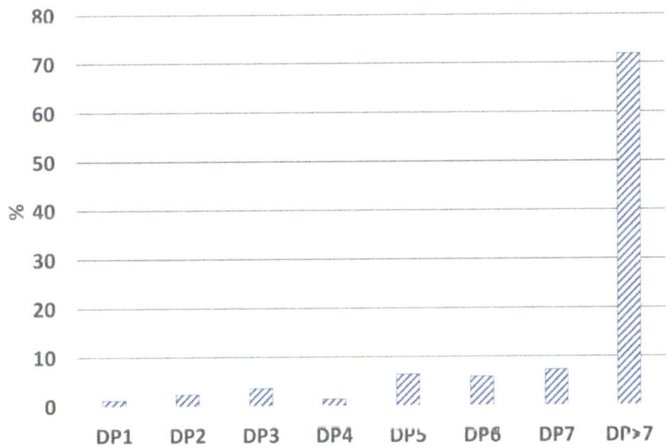

FIGURE 5.20 Profile of maltodextrin DE 10 obtained from cassava starch with BAN 120 L. *DP*, Degree of polymerization; *DP1*, glucose; *DP2*, maltose; *DP3*, maltotriose; *DP* > *7*, dextrins (HPLC analysis).

reaches 95°C, and held for 10 minutes. Then the suspension is cooled down at a rate of 1°C per minute. The profile of the maltodextrin syrup hydrolysate is shown in Fig. 5.20.

The most common maltodextrins range is from DE 2–4 to DE 9–12 and can be as high as 20. Fig. 5.21 shows two profiles of maltodextrins DE 18–20 and DE 3–4, with very different profiles.

The profile of the hydrolysates can be tailored to meet a customer's specific demand. This is possible by dosing the enzymes and the reaction time, but also by blending syrups. Blends can also be made with other sugars (sucrose, invert sugar, malt extract, etc.) to obtain desired physical, functional, energetic, and organoleptic characteristics. The purification and coloration of these hydrolysates can also be customized by suppressing, modifying, or adding steps in the manufacturing process, such as caramelizing the sugars.

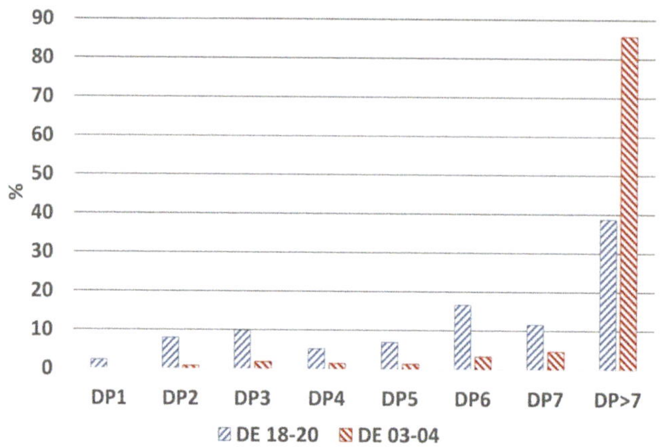

FIGURE 5.21 Syrup profiles of DE 18–20 and DE 3–4 maltodextrins obtained from cassava starch. *DP*, Degree of polymerization; *DP1*, glucose; *DP2*, maltose; *DP3*, maltotriose; *DP* > 7, dextrins (HPLC analysis).

Once the laboratory testing adjustments were completed, the next phase was to manufacture the products on a pilot, or semiindustrial, scale.

5.7.2 Syrups derived from cassava starch obtained in semiindustrial tests

Pilot scale testing is of utmost importance when it comes to the use of little-known materials and for which little literature exists, as is the case with starches extracted from underground raw materials of South American origin. Of the starchy underground raw materials, cassava is the one with the most commercial use, followed by potato, and very recently, sweet potato (for a comparison of these potentialities, please refer to Chapter 3 in Volume 1 of this series). The other underground starchy crops of South American origin have little production of starch and derivatives, with local production of starch from arrowroot (*Maranta arundinacea*) and canna (*Canna indica*) standing out (see Chapter 3).

While it is possible to find well-established scripts for cereal starches, this is not the case for tropical ones. For this reason, the availability of roadmaps at the pilot scale level, in addition to facilitating the commercial use of tropical starches, also allows for avoiding the common error of considering the behavior of cassava starch as equal to that of corn. This confusion has caused great losses for companies unaccustomed to operating with tropical starches, mostly by not considering the lower gelatinization temperature and the virtually zero presence of impurities, lipids, and protein.

The tests in semiindustrial scale had as objective the following:

- Establish scripts for industrial production, adjusted for starches extracted from starchy underground raw materials originating from South America.
- Evaluate the extension of laboratory tests to pilot scale, so that they are valid for production of various types of hydrolysates.
- To deepen and evaluate the behavior of the equipment and process in response to scale-up, in comparison with laboratory tests.

In order not to repeat the process description, only the important points of each step are highlighted, and the necessary modifications are summarized in flow charts. A single starter of commercial cassava starch was adopted for all processes. The enzymes were the same as those used in the laboratory step (Item 5.6.1). The steps in the production of hydrolysates from cassava starch were liquefaction, saccharification, denaturation, and decolorization and were performed in a single equipment, a stirred reactor of INOXIL brand, with a capacity of 100 L, equipped with a half-cane coil, decentralized mixer with rotation of 400 rpm and 2 blades, in a position of 1/3 and 2/3 of the bottom of the tank, single centralized deflector, outlet at the bottom of the tank, and temperature transmitter located in the middle of the deflector.

- **Liquefaction**: Because it has a lower gelatinization temperature than cereal starches, the liquefaction of cassava starch can be performed with a thermolabile α-amylase, which according to Teague and Brumm (1992) leads to energy savings in the deactivation of the enzyme at the end of the process. The thermolabile α-amylase is added to a starch suspension with pH adjusted to 6.5 (Nebesny, 1990). The mixture is then heated to deactivate the enzymes and stop hydrolysis at the appropriate DE. According to Swanson et al. (1986), deactivation of the thermolabile α-amylase can be done between 90 and 115°C, or at 85°C for 30 minutes. When the desired DE is reached, the pH is reduced to 5.0 and the temperature is raised to the boiling point for 10 minutes and allowed to cool to add the saccharification enzyme.

  Alternatively, it is possible to immediately add a saccharification enzyme, without deactivating the liquefying enzyme, performing a concomitant process. Chung and Chang (1985) suggested for liquefied starch an optimum DE between 8 and 12, suitable for any of the hydrolysates. Liquefied starch, consisting mostly of dextrins (oligo- and polysaccharides), is transformed into hydrolysates containing mainly low molecular weight sugars. As reported by Reeve (1992), this step is performed with the help of specific enzymes, depending on the type of hydrolysate desired.

  The most important parameter in liquefaction is the amount of residual starch, which should be as small as possible. For this, the action of the liquefaction enzymes on the gelatinized starch must be optimized. This parameter is influenced by the amount of enzyme and the agitation. The agitation not only is defined by the rotation speed of the paddles, but also depends on several other parameters, such as the number of paddles, the surface and shape of the paddles, their positions on the shaft, and the position of the shaft in the reactor (centralized, decentralized, tilt angle, etc.). The number and positions of the baffles and other factors also influence this. Finally, the agitation is a function of the turbulence of the medium. The higher the agitation, the lower the amount of residual starch and the lower the amount of enzyme required; in this case, the rotation of the agitation shaft was set at 400 rpm. The explanation for the higher conversion efficiency is the increased contact surface between substrate and enzymes.

  The second important point is the control of the temperature increase, which should be gradual. On a laboratory scale, the temperature rise was controlled by a water bath. On an industrial scale, this increase is a function of the amount of steam injected and the energy consumption of the liquefaction medium. In the range of 65°C–70°C, an increase in steam injection is needed to maintain a temperature rise of 1.5°C per minute. In this temperature range, the phenomenon of gelatinization occurs for most starches from roots and tubers of tropical origin, a phenomenon that is accompanied by an increase in viscosity and energy absorption. On an industrial scale for high viscosity substrate the setting of a 400 rpm, agitation is difficult.

- **Saccharification**: In the change of scale between laboratory and semiindustrial process, it was necessary to increase the concentration of saccharification enzymes. To achieve the same sugar profile, that is, the same hydrolysate, it was necessary to add 20% more Fungamyl or AMG enzymes. On the other hand, at this stage of production, agitation is not as important as for liquefaction, serving only to homogenize the medium and ensure adequate contact between the enzymes and the liquefied starch substrate. In this case, the rotation of the stirring shaft was set at 80 rpm. Microbial contamination can occur at this stage, when simpler sugars appear. Depending on the type of hydrolysate produced and consequently on the saccharification enzyme used, the temperature for this step varies between 55°C and 65°C, which is favorable for the development of microorganisms and particularly of yeasts.
- **Denaturation**: In this stage, the liquefaction and saccharification enzymes need to be inactivated by heat to fix the sugar profile of the hydrolysates. The reaction of denaturing or inhibition of the enzymes is kinetic and depends on temperature and application time. The temperature was set at 95°C for 10 minutes.
- **Decolorization and filtration**: These two steps were kept as described for laboratory scale due to the lack of pilot scale equipment.
- **Continuous centrifugation**: The efficiency of the centrifugation was evaluated by the percentage of residual starch removed. There are only two parameters for setting the continuous centrifugation, the speed of rotation, and the residence time of the liquid, that is, the time of centrifugation of a fixed volume of liquid. The parameters for achieving maximum efficiency in this step are the longest possible residence time and the highest rotation speed. An Alfa Laval centrifuge, model LAPX 202, made of AISI 304 L stainless steel with a rotation of 3500–10,000 rpm and a minimum flow rate of 60 L/h was used. The best parameters were obtained with 60 L/h of centrifuged liquid with rotation at 10,000 rpm. Under these conditions, an 80% removal of residual starch and a reduction of 1/3 in filtration time was observed compared to that evaluated in the laboratory.
- **Filtration**: On a pilot scale, an Aquafil filter press was used, with PVC filter plates, AISI 304 L stainless steel structure, filter surface of 1.2 m^2 and working pressure of 0.6 bars. This type of equipment does not allow for the formation of an auxiliary precoat filtration, as occurs with diatomaceous earth filters or a filtration under pressure (5 bars), conditions required for a filtration of this type of fluid. For all the hydrolysates, filter clogging problems were observed, accompanied by a corresponding decrease in filtrate flow. The inadequacy of this filter made the total removal of residual starch and discoloration impossible because the filter could not retain all the activated carbon. For this type of filtration, a vertical plate precoat filter with a pressure of 5–6 bars, as well as a vacuum filter system with a diatomaceous earth precoat is more suitable.
- **Evaporation**: This step was performed with the help of an OROSM stainless steel AISI 304 L single effect evaporator, cylindrical with round bottom, double envelope jacket heating and outlet at the bottom of the tank. The useful volume of the evaporator was 25 L, with 80 rpm rotation and vacuum up to 0.1 bar. The working temperature was 65°C and the pressure 0.1 bar (absolute pressure). The rotation for this type of equipment allows to increase the evaporation surface. The hydrolysates are evaporated until the desired Brix is reached, which was set at 82° for the concentrated syrups. The most important point in the evaporation step is to establish the number of effects, and thus which type of evaporator

should be used. The choice between them is a matter of amortizing the cost of the equipment, that is, a compromise between investment and energy consumption. The greater the number of effects, the less deleterious effect on the final product.

- **Spray-dryer drying**: All types of hydrolysates were dried in spray-dryer, with inlet Brix hydrolysate at 35 for glucose and maltose and 25 for maltodextrin. The spray-dryer used was from ALFA LAVAL, in stainless steel AISI 304 L, with conical bottom, scraped surface, air heating by steam with a maximum inlet temperature of 200°C. The evaporation capacity of the equipment was 25 kg of water per hour at a temperature of 190°C. The air temperature at the spray-dryer inlet was 170°C, and the outlet temperature after the cyclone was 90°C. Temperatures lower than these values of inlet and outlet induce a higher humidity in the spray-dryer and, consequently, an adherence of hydrolyzed particles on the wall of the equipment. An exit temperature higher than 100°C impairs the quality of the final product. The maximum Brix of entry in the spray is 54; above this value, the hydrolyzed syrup is very viscous and even hot; the dryer produces threads of molten sugar.
- **Drum-dryer**: Only maltodextrins DE 9 and 20 were dried in a drum-Dryer. The equipment used was NORDON stainless steel AISI 304 L, with a dryer cylinder length of 40 cm and diameter of 45 cm. The diameter of the secondary cylinders was 12 cm, with a thickness between the dryer cylinder and the secondary cylinders of 0.15 mm and maximum rotation of the dryer cylinder of 1 rpm. The drying was performed with a concentrated hydrolysate of 82°Brix, with a drum-dryer cylinder temperature of 135°C. The contact time for maltodextrin DE 9 was 4 seconds and for maltodextrin DE 20 28 seconds. The great advantage of this dryer, compared to the spray-dryer, besides being more compact, is that it is fed with concentrated liquid hydrolysates of 82°Brix, with reduced energy consumption.

To move from pilot or semiindustrial scale production to industrial scale, there is the need for a new change of scale and the use of more auxiliary equipment such as deaerators and dust filters. In industrial scale, the equipment is not a problem, because all of them are available in the market, but there is the need for adaptations in the enzymatic preparations and some temperature and agitation controls.

We conclude that the laboratory results are important, since the change to semiindustrial scale did not represent major changes, only an increase in the amount of enzyme in the saccharification phase. The production process of hydrolysates in the laboratory allowed for a good approximation for the first scale-up, which suggests a good extrapolation from semiindustrial to industrial scale.

The amount of enzyme used varied from 1 to 3 mL/kg of dry starch. The temperature was programmed to increase by 1°C per minute. For the liquefaction at 90°C, the reaction time varied from 30, 60, and 120 minutes. The deactivation of the enzymes was done by adjusting the pH to 3.5 with diluted acid and heating at 95°C for 15 minutes. For each enzyme and use, minor adjustments were made.

- Liquefaction enzyme: BAN 120 L from Novozymes.
- Saccharification enzymes: FUNGAMYL 800 L and AMG 200 L from Novozymes.

It is important to remember, as seen in item 5.5, that more modern enzymes have a higher activity and, consequently, it is possible to estimate that the quantities needed today with the enzymes available on the market would be smaller. The literature describes the processes always based on starch, starting from the dry commercial product. The physicochemical composition of the used commercial cassava starch is presented in Table 5.7.

However, other raw materials derived from cassava can be evaluated for comparison purposes, although the most usual is the dried starch suspension in water. Thus, the starch milk coming out of the second centrifuge, or the hydro cyclones, with a starch content of 35% could be used. The mass of grated root, at the exit of the grater before the starch extraction operation, without adding water, and the final extraction residue, after the last passage through the rotary sieves (GL), called bran, were also evaluated. The high starch content in bran can be explained by the fact that the tests were conducted in the early 2000s, with a less efficient extraction. Today, this content should be around 45% (see Chapter 2).

Each product, even if belonging to the same category, such as glucose syrups of DE 38, 64, and so forth, requires specific parameters and experimental conditions. In this topic, the production of three types of hydrolysates is described: glucose syrup DE 38, maltose syrup, and maltodextrin syrup DE 10, in concentrated and dehydrated liquid form. The amount of starch used was the same as in the laboratory trials, 35% dry weight in water.

5.7.2.1 *Starch suspension in water*

The pilot tests were adjusted to the conditions of the available equipment. The materials, commercial cassava starch, and enzymes were the same as in the laboratory trials (Fig. 5.22). The production yield for the starch suspension was 99.7%. The only loss with this feedstock is residual starch, but this loss is limited and can be corrected with the use of industrial scale equipment. In addition, it is compensated by yield greater than 100% in the transformation of starch to glucose DE 38. The protein content was 0.03%, lipids 0.07%, and fiber 0.01%, which can be considered low compared to the same products made with corn starch. The low levels of minerals in the hydrolyzed starch suspension would not require the passage through the ion exchange resin column or use of activated carbon, which represents cost reduction, besides the operation at lower temperature and use of thermosensitive enzyme.

This topic presented the script for hydrolysate production from starch suspension in water. The following topics analyze the production of hydrolysates from nontraditional products, such as grated cassava root and cassava starch extraction residue or bran.

5.7.2.2 *Grated cassava root*

In the case of using corn as a raw material for the production of enzymatic hydrolysates, it is not possible to directly use the grain, first because it requires increased humidity, since it is stored at 12% moisture, but also because when transformed into paste, it would present

TABLE 5.7 Average physical-chemical composition of different cassava raw materials.

		% on dry basis					
Raw material	**% Dry matter**	**Starch**	**Sugars**	**Fat**	**Protein**	**Fibers**	**Ash**
Dry starch	12.65	95.00	-*-	0.10	0.15	0.35	0.20
Starch milk	35	95	-*-	0.1	0.15	0.35	0.2
Grated pulp	38	81	2.2	1.0	3.10	4.20	2.5
Bran	14	78	-*-	1.0	1.30	17.00	0.8

-*-: without results.

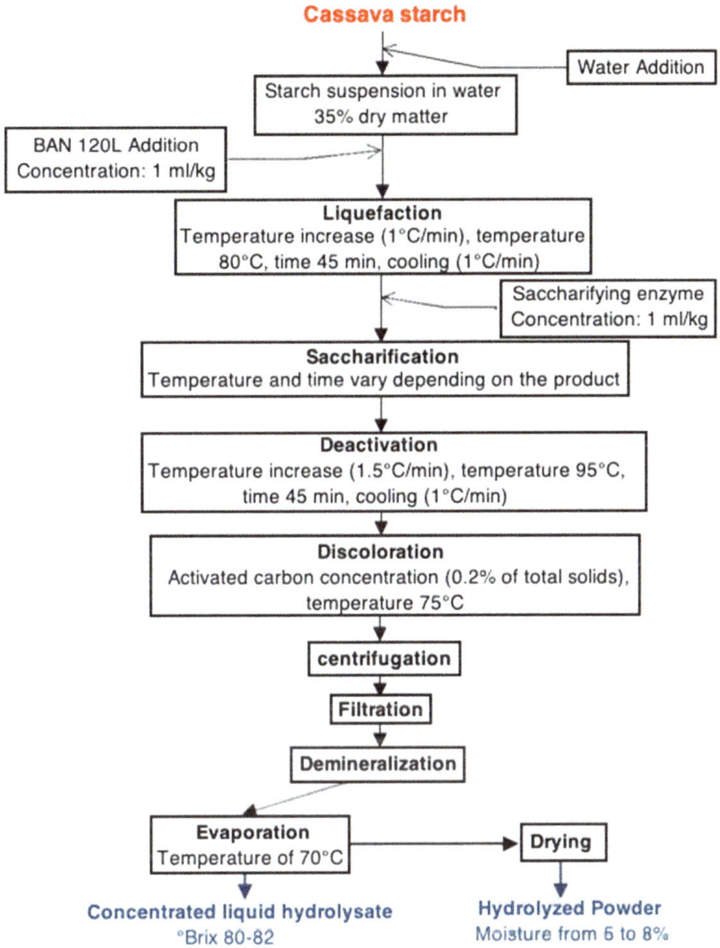

FIGURE 5.22 Script for production of liquid, concentrated, and powder hydrolysates from cassava starch suspension in water.

a high content of fibers, lipids and proteins, which would have to be removed in order not to harm the quality of the hydrolysate. However, this is not the case with cassava root, which furthermore presents around 60% moisture, sufficient to be hydrolyzed (Table 5.7).

It can be seen in Table 5.7 that before hydrolysis, there are already sugars in the cassava grated pulp. The DE at the zero point was 2.2, which corresponds to the amount of natural sugars in the cassava root. These sugars were identified and, even without the addition of enzymes, were dosed as glucose (55%) and fructose (45%), probably corresponding to the natural hydrolysis of sucrose, detected also in the manipueira (water from pressing the grated pulp).

The production flow from grated cassava roots is shown in Fig. 5.23.

When pressing the hydrolyzed grated pulp suspension, there is a loss of almost 15% of the hydrolysate, which is explained by the fact that

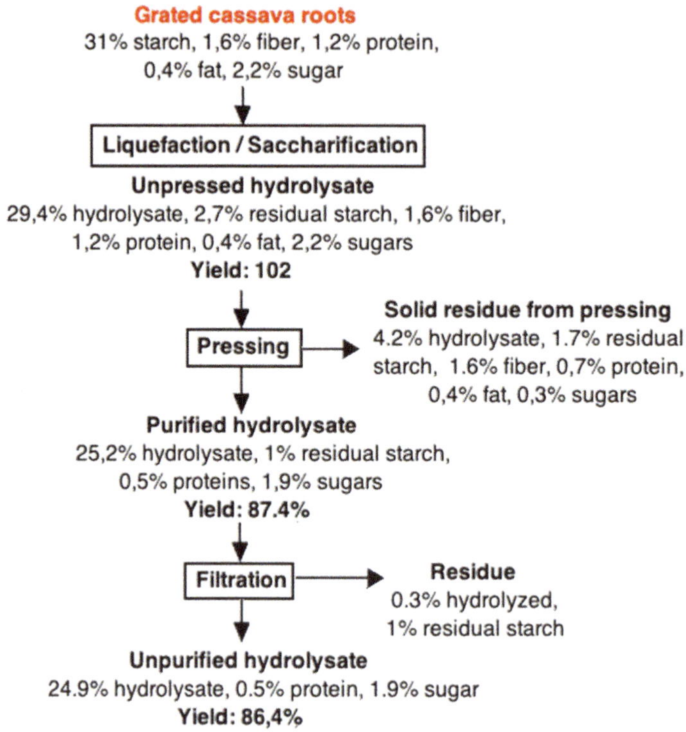

FIGURE 5.23 Mass balance in the production of glucose syrup DE 38, with grated cassava pulp.

fibers retain a lot of liquid. An increase of the pressing pressure slightly reduces this loss but does not allow for a full recovery of the hydrolysate. However, this loss is less than the starch loss during extraction, which has been estimated around 30%, with extraction yields around 28% (see Fig. 2.2 in Chapter 2).

The greatest difficulty in using grated roots as raw material is the material viscosity at the grater outlet. The viscosity interferes in the homogenization of the pulp in the liquefaction and saccharification phases, but also in the pumping of the saccharified material to the press. One solution to this problem is the joint use of pectinase as reported by Leonel and Cereda (2000), where it was used for the hydrolysis of dried cassava bran, with good results in viscosity reduction and pressing yield. Recovering the hydrolysate retained in the fibers after pressing, by washing and new pressing is possible, but it makes the process more expensive in an additional step and reduces the Brix degree of the syrup.

5.7.2.3 Glucose syrup DE 38 from cassava bran

The yield of glucose syrup production from cassava bran was 58.2%, as presented in Fig. 5.24. This yield seems low compared to that of starch suspension in water, but it should be remembered that this is an often undervalued residue.

When the conversion is done from the bran, the viscosity is high because of the large amount of water bound to the fiber, and the enzymes have difficulty to act on the gelatinized starch. Another possibility is that fewer granules gelatinize due to the lack of more direct contact with heat. Finally, after pressing, the fiber retains almost 30% of hydrolyzed liquid, similar to what

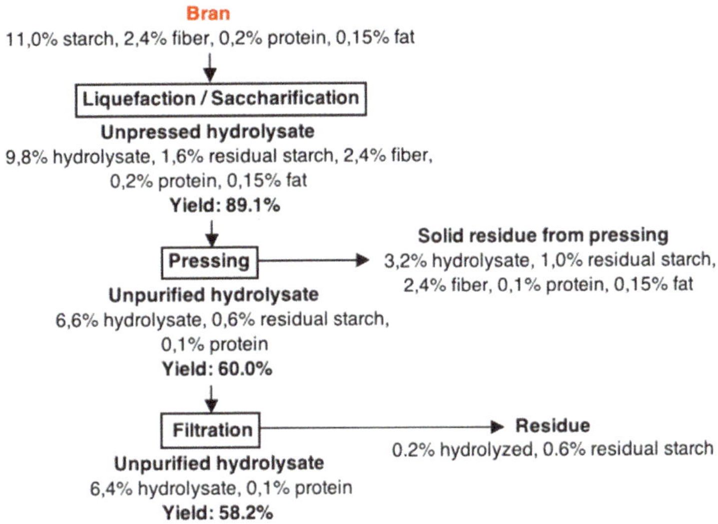

FIGURE 5.24 Mass balance in DE 38 glucose syrup production with cassava bran.

occurs with grated cassava roots. The yield could be increased by using pectinase as an auxiliary enzyme, as reported by Leonel and Cereda (2000). Thus, with the addition of 8 kg of pectinase per ton of fiber, the authors reduced the retention of sugars in the bran from 31.41% to 9.33%, which is equivalent to a yield superior to 85%, compared to the 58.2 indicated in Fig. 5.24.

Recovering the hydrolysate retained in the fibers after pressing by washing with water and pressing again presents the same problem already pointed out for the grated root pulp. A suspension of 11% starch in the bran requires 25%–30% more enzymes than a suspension of 35% starch without fibers, to produce the same hydrolyzed profile. This can be explained by a greater contact between enzyme and substrate in the fiber-free case, due to a better spatial proximity.

5.8 Hydrolysate of other underground starchy raw materials

From the results obtained with cassava starch, the same script can be used for other underground starchy raw materials, with the same equipment, varying only the type and content of enzymes, reaction time, temperature, and pH to obtain dextrins, glucose, maltose, maltodextrin, and others. Based on this premise, Surmely et al. (2004) conducted a series of experiments, always with raw materials grown in the same place and with the same methodology.

5.8.1 Hydrolysate of grated cassava roots with sweet potato

As reported earlier, commercial enzymes are most used because they are more available, are more standardized, and have more predictable results. However, it is sometimes possible to use indigenous enzymes.

A long discussed, but not perfectly elucidated question is the presence of amylolytic enzymes in cassava roots. In every plant that accumulates starch, there must be a mechanism for hydrolysis to release energy that will be used by the plant when needed. Veiga et al. (2002) cited the isolation and characterization of a thermoresistant alpha-amylase obtained from cassava roots of the variety EMBRAPA - IM 158. The authors report that this autochthonous enzyme was distinguished from others also isolated from cassava roots by its high amylolytic activity. The authors also report that the maximum activity on a 1% suspension of soluble potato starch occurred at 60°C and pH 6.2. The enzyme was characterized and showed inhibition at pH less than 5.0 and in the presence of Cu^{2+}, but had its activity increased when the medium was supplemented with Ca^{2+} and Mg^{2+}. The kinetic values were Km 9.24 and Vmax 0.45 mg/min. The enzyme was stable when the extract was heated at 70°C for 10 minutes.

Although potentially interesting, this amylase showed lower temperature resistance than BAN (70°C–80°C). A better knowledge of native cassava amylases could establish a new target for breeders in the selection of varieties with higher enzyme content, indicated for hydrolysate processing.

Nowadays, natural and organic products have been more valued. In hydrolysis processes with commercial enzymes, it is almost impossible to achieve traceability and certification of organic products. The use of vegetable enzymes may be an easier way to get a certification. Certified organic maltodextrins have been imported from Europe for specific markets. Walter and Purcell (1973) reported the presence of thermoresistant α- and β-amylases in sweet potato. For this reason, an experiment with a mixture of cassava and sweet potato pulp was realized.

Cassava root has on average 30%–35% starch and 60%–65% water, a concentration suitable for the hydrolysis process. For the experiment, 180 g of cassava root was mixed with 10 g of sweet potato and 10 mL of water. The process adopted was a batch process with a volume of 200 mL and agitation of 100–150 rpm. After beaten in an industrial blender, the mixture remained for 45 minutes at 80°C for the liquefaction phase. The saccharification was done by adding another 10 g of grated sweet potato and leaving it to stand for 2 hours at 50°C. At the end of the process, the bran was separated, and the liquid was centrifuged at 12,000 rpm for 10 minutes. When the hydrolysis was finished, the liquid was filtered through filter paper covered with a 5 mm layer of diatomaceous earth in a Buchner funnel. The bran was pressed in a manual press with a 0.08 mm opening screen. Concentration of the hydrolysate was done in a rotary evaporator with 500 mL flask, temperature 20°C–95°C, and vacuum 1.0–0.1 bar absolute and in a static water bath. Fig. 5.25 shows the profile of the hydrolysate obtained.

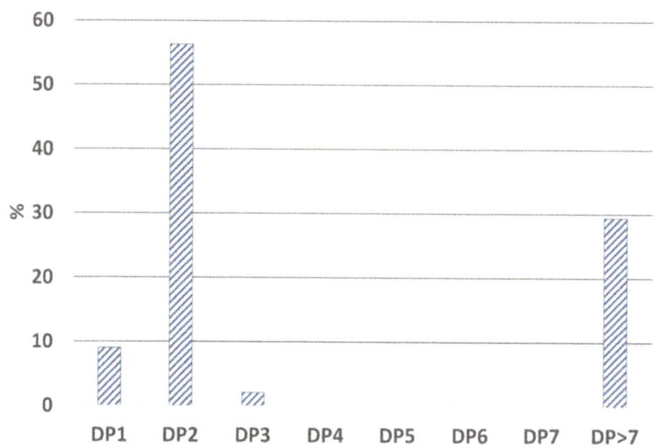

FIGURE 5.25 Profile of the hydrolysate obtained from the mashing of grated cassava and sweet potato pulp. *DP*, Degree of polymerization; *DP1*, glucose; *DP2*, maltose; *DP3*, maltotriose; *DP > 7*, dextrins (HPLC analysis).

The profile shows a predominance of maltose, followed by glucose, but unlike maltose syrup for brewing use, a higher proportion of dextrins and low content of oligosaccharides of degree of polymerization 4–7 were detected. However, residual starch was found in the decanted material that came out of the centrifugation of the hydrolysate. Due to the high production of maltose, the autochthonous enzymatic activity of sweet potato shows good potential.

The same assay was done with yam (Dioscorea sp.), also cited in the literature as presenting high content of amylases, but the enzymes present were not sufficient to obtain hydrolysis of cassava starch, as occurred with sweet potato. Therefore, to date, only sweet potato has shown enzyme content with sufficient activity to fully hydrolyze cassava starch, in addition to its own starch.

5.8.2 Enzymatic production of hydrolysates from starch of other Latin American starchy tubers

Considering the little reliable information to establish the enzymatic hydrolysis process for starches extracted from underground starchy crops, four raw materials produced in the same location where the experiments were performed (cassava, canna, sweet potato, and arrowroot) were selected. Table 5.8 highlights temperatures on average higher for corn starch relative to the other starches, but with large variations from literature data.

Among starches extracted from underground crops of South American origin, besides potato and cassava starches, the most produced is from sweet potato, but with local production of arrowroot and canna starches (see Chapter 3).

5.8.2.1 Preparation of glucose syrup from cassava, canna, arrowroot, and sweet potato

The script established for grated cassava root was used for canna, arrowroot, and sweet potato starches. The starch suspension in water was kept at 35%, and the liquefaction was done with BAN enzyme from Novozymes at a proportion of 3 mL/kg at 80°C for 45 minutes. The saccharification was done with the enzyme AMG from Novozymes at a proportion of 2 mL/kg at a temperature of 60°C for 4 hours. Activated carbon, Carbomafra Brand, and diatomaceous earth were used as auxiliary

TABLE 5.8 Gelatinization and paste temperatures of various starches extracted from underground organs and comparison with corn starch.

		Temperatures by DSC				
Botanic name	**P_t (°C)**	**T_O**	**Tp**	**T_C**	**T_C-T_O**	**ΔH (J/g)**
Corn *(Zea mays)*	75.8	61.6–76.9	66.7–76.9	74.1–88.7	10.5–15.8	11.1–17.3
Potato *(Solanum tuberosum)*	63.2–70.2	58.2–62.9	60.8–70.0	67.4–83.3	6.6–21.3	12.7–19.7
Cassava *(Manihot esculenta)*	67.4–74.3	56.2–65.5	62.2–71.0	69.2–80.6	8.4–15.5	6.7–18.1
Sweet potato *(Ipomoea batatas)*	70.7–87.7	60.2–80.5	65.0–80.6	75.8–82.6	2.0–13.0	15.1–19.1
Arrowroot *(Maranta arundinacea)*	71.7	62.6–67.8	69.8–73.6	75.7–81.4	13.1–13.6	10.4 11.3
Canna *(Canna edulis)*	68.0–90.0	56.8–66.8	61.2–70.1	65.7–74.8	2.5–13.1	4.5–15.7

Symbols: Pt: Pasting temperature; T_O: onset temperature; T_P: peak temperature; T_C: conclusion temperature; Tc–To: gelatinization temperature range; ΔH: gelatinization enthalpy; (-) data not found in the literature.
Data from Chapter 1 in the 1st Volume of this series.

products in the filtration. The results are shown in Figs. 5.26 and 5.27.

The hydrolysis profile of sweet potato did not highlight native α-amylase activity because the starch extraction process eliminates all enzymes originally present in the raw material.

The evolution of the DE of cassava and canna hydrolysis were quite similar, while that of sweet potato and arrowroot starches followed the same pattern, both for liquefaction and saccharification, remembering that, as a characteristic of the α-amylase action of BAN, liquefaction produces mainly oligosaccharides of polymerization degree 3–6.

At the end of the process, the canna hydrolysate contained 50% glucose, while for cassava, it was just under 40%. The percentage of dextrins from Canna, considered as oligosaccharides of Gp > 7, was much lower than that obtained with cassava starch, which demonstrated that the hydrolysis of canna starch was more intense (Fig. 5.27).

The arrowroot and sweet potato hydrolysates are characterized by a lower DE than the

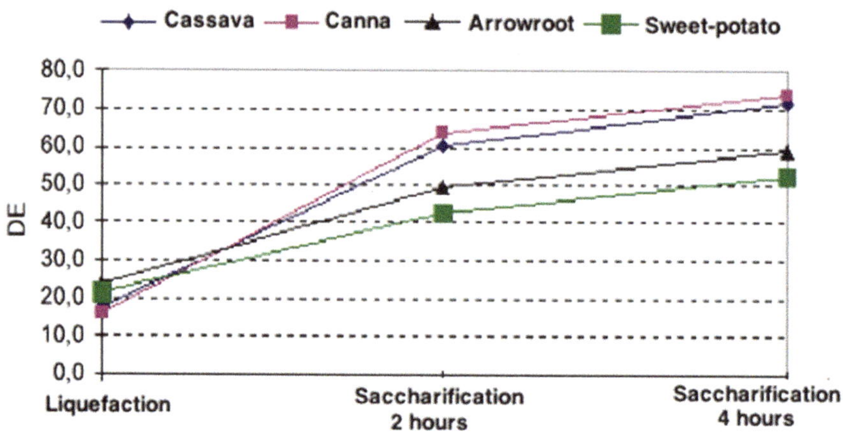

FIGURE 5.26 Profile of DE variation during liquefaction and saccharification of four starches from underground crops.

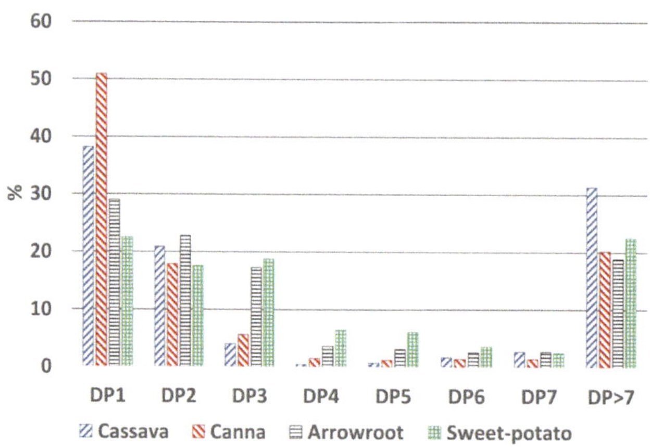

FIGURE 5.27 Hydrolysates from cassava, canna, arrowroot, and sweet potato after 2 h of saccharification at 60°C.

canna and cassava hydrolysates (Fig. 5.26), but in contrast have profiles with higher percentages of oligosaccharides from Gp 3 to 7 (Fig. 5.27). The paste and gelatinization temperatures (Table 5.8) are not sufficient to explain these differences.

Chung and Chang (1985) suggested that the DE of the hydrolysate should not exceed a value of 15 to optimize saccharification. The results presented in Fig. 5.26 show that after liquefaction, the sweet potato and arrowroot solutions reached DE 23, which may explain the differences in behavior during saccharification. In contrast, this result suggests that the starch from these roots is more accessible to the liquefaction enzymes, which may be interesting for dextrin or maltodextrin production.

These results provide some specific behavioral characteristics of the starches evaluated, with canna starch characteristically more susceptible to the saccharifying enzyme, while sweet potato and arrowroot starch were characterized by greater susceptibility to α-amylase.

Sarian et al. (2023) proposed the utilization of Indonesian root and tuber starches for glucose production. They justify themselves by pointing out that in most tropical countries, carbohydrate-based agricultural products occur in large quantities. For the authors, the wider utilization of local starch crops will offer various economic and ecological advantages and local starch crops can be a catalyst for rural industrial development and eventually open new markets. Among the raw materials suggested by the authors, sweet potato, canna, and arrowroot are crops native to South America. As raw materials to be valued, the authors also mention "sago" starch, a starch extracted from a palm tree, but they do not mention cassava starch, which is widely produced in Asia.

Sarian et al. (2023) cited the route used to obtain glucose syrup on a laboratory scale. However, the process adopted differs from the one described in this chapter because although it uses two commercial enzymes, hydrolysis is performed by the cold-starch hydrolyzing process that uses the enzyme preparation Stargen 002, a commercial enzyme mixture marketed by DuPont Industrial Bioscience (Palo Alto, USA). It is a blend from Genencor of *Aspergillus kawachi* glucoamylase and *Trichoderma reesei* α-amylase (Hargono et al., 2018). This enzymatic preparation takes at least 24 hours to finish the hydrolysis process, with a slightly higher concentration than the one adopted in the chapter, of 40% (w/v) starch/water. The results of DE after 24 hours reached 88.4 and 86.3 g/L, respectively, for sago and sweet potato starches compared to corn starch (89.6 g/L), while arrowroot and edible canna gave much lower amounts, 53.3 and 39.7 g/L, respectively. The authors cite previous articles, in which it was proved that the action of cold-starch hydrolyzing enzymes is greater for cereal starches than for starches extracted from underground organs (Sarian et al., 2023).

Although the use of cold process has an obvious energy saving advantage, the authors do not comment on how to keep glucose syrup free of microbial contamination in a warm climate country.

5.9 Final considerations

As can be seen, the process of starch hydrolysis has made more advances in enzyme availability than in the process itself, which has continued without noticeable changes for dozens of years with the use of the same equipment.

Microbial enzymes are the most widely used in commercial preparations, due to their rapid growth and low cost, with emphasis on thermoresistant ones. Miłek and Lamkiewicz (2022) highlighted the importance of these agents and the need for constant search for α-amylase producing strains of Bacillus spp with better performance.

On the other hand, the use of heat is mandatory in hydrolysis, because of the need to gelatinize the starch for hydrolysis to occur. It is one of the most expensive inputs for hydrolysate production. Since heat is one of the relevant process expenses, it is important to investigate other sources. Montalbo-Lomboy et al. (2010) evaluated the use of ultrasound on corn slurry saccharification yield and particle size distribution in both batch and continuous-flow at concentration of 28% w/v prepared and sonicated in batches at various amplitudes (192–320 μm $_{\text{peak-to-peak (p-p)}}$) for 20 or 40 seconds, using a catenoidal horn with the ultrasonic systems operating at a frequency of 20 kHz.

The use of ultrasound allows the process to be conducted with continuous flow. Experiments were conducted by pumping corn slurry at various flow rates (10–28 L/min) through an ultrasonic reactor at constant amplitude of 12 μm_{p-p}. The reactor was equipped with a donut-shaped horn. After ultrasonic treatment, commercial glucoamylases (STARGEN 001) were added to the samples, and liquefaction and saccharification proceeded for 3 h. The authors concluded that the sonicated samples were found to yield 2–3 times more reducing sugars than unsonicated controls. Although the continuous flow treatments released less reducing sugar compared to the batch systems, the continuous flow process was more energy efficient. Overall, the study suggests that both batch and continuous ultrasonication enhanced saccharification yields and reduced the particle size of corn slurry. However, due to the large volume involve in full scale processes, an ultrasonic continuous system is recommended.

The comparison with starches extracted from underground crops is necessary since some of them gelatinize at lower temperatures. This observation leads to the fact that starch sources have evolved little, with corn being the main raw material used and researched.

Acknowledgments

The authors would like to thank engineers Hugo Alvarez and Rodolphe Surmely for taking an active part in the project that provided the results and information selected for this chapter.

The authors thank LNF Latino Americana on behalf of Eduardo Bihre, Food Engineer, Business Manager, for the support in obtaining enzymes and technical information.

The authors would also like to thank the Cargill Foundation, in the person of Cesar Renato Néris, for kindly waiving copyrights.

References

Alexander, R. J. (1992). Maltodextrins: Production, properties and applications. In F. W. Schenk, & R. E. Hebeda (Eds.), *Starch hydrolysis products* (pp. 233–275). New York: VCH Publishers.

Alexander, R. J. (1995). Potato starch: New prospects for an old product. *Cereal foods world (USA)*, *40*, 763–764.

Ali, M., Rellos, P., & Cox, T. M. (1998). Hereditary fructose intolerance. *Journal of Medical Genetics*, *35*(5), 353–365. Available from https://doi.org/10.1136/jmg.35.5.353.

Alipio, R. (2000). Xarope de Isomalto-oligossacarídeo. Food *Ingredients*, n. 9, 24–25. São Paulo.

Chaplin, M. F., & Bucke, C. (1990). *Enzyme technology*. CUP Archive.

Chung, D. H., & Chang, H. K. (1985). *Food analysis* (p. 171) Seoul: Ginro Yeanku Co.

Colonna, P., Buléon, A., & Mercier, C. (1988). Starch and enzymes: Innovations in the products, processes and uses. *Chimica Oggi*, *10*, 9–14.

Cuvellier, G. F. (1993). Réacteurs enzymatiques à enzymes libres et à enzymes immobilisées. Cas des industries de l'amidon. In R. Scriban (Ed.), *Biotechnologie* (pp. 422–430). Paris: Lavoisier.

Dornier, M. (1992). Approche des conditions optimales de filtration tangentielle intégrant la consommation énergétique. In *AGRENA-CEE, 1992, Rennes. Comptes Rendus du Colloque* (pp. 71–76). Rennes.

Engasser, J. M. (1988). Bioreactor engineering: The design and optimization of reactors with living cells. *Chemical Engineering Science*, *43*(8), 1739–1748.

French, D. (1984). Organization of starch granules. *Starch: Chemistry and technology* (pp. 183–247). Academic Press.

Gluten Free Home Brewing. (2020). Termamyl SC DS Thermo-Stable Amylase Enzyme. Available in <https://glutenfreehomebrewing.com/STOREProduct/1052/Termamyl-SC-DS-Thermo-Stable-Amylase-Enzyme-4-OZ.html>.

Hebert, J.P., & Scriban, R. (1973). Les enzymes immobilisées. Leurs utilisations, rôle possible en brasserie. *Bios*, n. 12, 574–590. Douai.

Hargono, H., Jos, B., & Kumoro, A. C. (2018). Kinetics of the hydrolysis of cassava starch by glucoamylase and a granular starch hydrolyzing enzyme. *Scientific Study & Research: Chemistry & Chemical Engineering, Biotechnology, Food Industry, 19*(4), 443.

Howling, D. (1992). Glucose syrup: Production, properties and applications. In F. W. Schenk, & J. N. Hebeda (Eds.), *Starch hydrolysis products* (pp. 277–317). New York: VCH Publishers.

Lehninger, A. L., & Kamoun, P. (1985). *Principes de biochimie* (pp. 207–221). Flammarion Médecine-Sciences.

Leonel, M., & Cereda, M. P. (2000). Avaliação da concentração de pectinase no processo de hidrólise-sacarificação do farelo de mandioca para obtenção de etanol. *Food Science and Technology, 20*, 220–227.

Li, Z., Kong, H., Li, Z., Gu, Z., Ban, X., Hong, Y., Cheng, L., & Li, C. (2023). Designing liquefaction and saccharification processes of highly concentrated starch slurry: Challenges and recent advances. *Comprehensive Reviews in Food Science and Food Safety, 22*, 1597–1612. Available from https://doi.org/10.1111/1541-4337.13122.

Lloyd, N. E., & Nelson, W. J. (1984). Glucose and frutose containing sweetners from starch. In R. Whistler, J. N. Bemiller, & E. F. Paschall (Eds.), *Starch chemistry and technology* (2nd ed., pp. 611–660). Orlando: Academic Press.

Miłek, J., & Lamkiewicz, J. (2022). The starch hydrolysis by α-amylase Bacillus spp.: An estimation of the optimum temperatures, the activation and deactivation energies. *Journal of Thermal Analysis and Calorimetry, 147*, 14459. Available from https://doi.org/10.1007/s10973-022-11738-1.

Montalbo-Lomboy, M., Khanal, S. K., van Leeuwen, J. H., Raman, D. R., Dunn, L., Jr, & Grewell, D. (2010). Ultrasonic pretreatment of corn slurry for saccharification: A comparison of batch and continuous systems. *Ultrasonics Sonochemistry, 17*(5), 939–946. Available from https://doi.org/10.1016/j.ultsonch.2010.01.013.

Moorthy, S. N. (1994). *Tuber crop starches* (pp. 18–40). Thiruvananthapuram: Central Tuber Crops Research Institute. (Technical Bulletin Series).

Nebesny, E. (1990). Carbohydrate compositions and molecular structure of dextrins in enzymatic high maltose syrups. *Starch-Stärke, 42*(11), 437–444.

Okada, M., & Nakakuki, T. (1992). Oligosaccharides: Production, properties and applications. In F. W. Schenk, & R. E. Hebeda (Eds.), Starch hydrolysis product: *Worldwide technology, production, and applications* (pp. 335–366). New York: VCH Publishers.

Qi, X., & Tester, R. F. (2019). Fructose, galactose and glucose–In health and disease. *Clinical Nutrition ESPEN, 33*, 18–28.

Reeve, A. (1992). Starch hydrolisis: Processes and equipment. In F. W. Schenk, & R. E. Hebeda (Eds.), *Starch hydrolysis product: Worldwide technology, production, and applications* (pp. 79–120). New York: VCH Publishers.

Sarian, F. D., Leemhuis, H., Nurachman, Z., van der Maarel, M. J., & Dessy, N. (2023). Utilization of Indonesian root and tuber starches for glucose production by cold enzymatic hydrolysis. *Biologia (Lahore, Pakistan), 78*, 1889–1897.

Schenk, F. W., & Hebeda, R. E. (1992). *Starch hydrolysis product: Worldwide technology, production, and applications.* New York: VCH Publishers, 650p.

Surmely, A., Alvarez, H., Cereda, M.P., & Vilpoux, O. (2004). Starch hydrolysis. In *Cereda & Vilpoux, Technology, Uses and Potentialities of Latin American Starchy Tuberoses* (v. 3, pp. 378–449). São Paulo, Fundação Cargill (In Portuguese).

Swanson, T. R., Carroll, J. O., Britto, R. A., & Duhart, D. J. (1986). Development and field confirmation of a mathematical model for amyloglucosidase/pullulanase saccharification. *Starch-Stärke, 38*(11), 382–387.

Swinkels, J. J. M. (1985). Differences between commercial native starches. *Starch Conversion Technology, 37*, 15–46.

Teague, W. M., & Brumm, P. J. (1992). Commercial enzymes for starch hydrolysis products. In F. W. Schenk, & R. E. Hebeda (Eds.), *Starch hydrolysis products* (pp. 45–77). New York: VCH Publishers.

Thibault, J.F., & Colonna, P. (1988). *Proprietes fonctionnelles: Pectines et amidon. Les cahiers de l'ENSBANA* (pp. 157–171).

Veiga, E.M., Pires, T.C. R., Magalhaes, C.P., & Finardi Filho, F. (2002). Characterization of a heat stable alpha-amylase from Amazonian viriety of cassava roots (Manihot esculenta Crantz). In *IFT 2002—Annual Meeting, 2002, Anaheim. Procedings of IFT 2002—Annual Meeting*. Chicago, IL: IFT.

Vuilleumier, S. (1993). Worldwide production of high-fructose syrup and crystalline fructose. *The American Journal of Clinical Nutrition, 58*(5), 733S–736S. Available from https://doi.org/10.1093/ajcn/58.5.733S.

Walter, W. M., Jr, & Purcell, A. E. (1973). Alpha-amylase in sweet potatoes. A comparison between the amyloclastic and chromogenic starch methods of analysis. *Journal of Food Science, 38*(3), 548–549.

CHAPTER

6

Commercial starches modified by physical, chemical, enzymatic and combined methods: processes and applications

Ivo Mottin Demiate[1], Camila Delinski Bet[1,2] and Vivian Cristina Ito[2,3]

[1]Food Engineering, State University of Ponta Grossa, Ponta Grossa, Parana, Brazil [2]UniCesumar, Ponta Grossa, Parana, Brazil [3]Food Science and Technology Graduate Program, State University of Ponta Grossa (UEPG), Ponta Grossa, Parana, Brazil

6.1 Introduction

Carbohydrates play a central role in carbon and energy conversion of the primary metabolism of all living organisms, both autotrophic and heterotrophic. The most abundant polymers on our planet are the polysaccharides, with starch in the first position as storage carbohydrate (Apriyanto et al., 2022).

Starch represents the bulkiest source of energy in human nutrition due to its availability, relative low cost, and multiple possibilities of processing and ingestion. From cheap and easy to produce crops to the most elaborated processed foods, starch performs in accordance with several needs and gives texture, consistency, smooth mouthfeel, softness, stability to emulsions, gel formation, film formation, binding, adhesion, moisture retention, and other properties (Kaur et al., 2012). In different foods, it can also produce crispness and expanded snacks or biscuits that are very tasteful and popular worldwide (Oliveira, Demogalski, et al., 2021, Oliveira, Macnaughtan, et al., 2021). For more information on starch expanded products, or sour cassava starch, see Chapter 7 in this volume and the upcoming Volume 4.

Besides food uses, starch also is used in many industrial processes, including paper and cellulose, textile, chemical, packaging, refinement and separation of metals, and oil industry, among others. Special properties are then necessary to fulfill all those distinct

Starch Industries
DOI: https://doi.org/10.1016/B978-0-323-90842-9.00013-3

demands. Resistance to temperature as well as to shear force, for example, is relevant and searched to accomplish the industrial expectation. The scientific research has provided a vast amount of materials that include the characterization of dozens of starches from tubers, corms, stems, stalks, seeds, cereals, and fruits. The commercially relevant starches, however, are restricted to a few crops, that is, corn, cassava, wheat, potato, sweet potato (in China and Japan), and rice (for more information on starch production, see Chapter 3 in Volume 1 of this series).

Natural or native starch, extracted and purified from the abovementioned vegetable sources, presents well described behaviors and performs in a predictable way that is vastly reported in the scientific literature. As expected, due to its role as an energy reserve in biological systems, starch exhibits sensitivity to both chemical and enzymatic hydrolysis, as well as physical stresses. Due to the multiple purposes and possibilities to benefit industrial processes from starch, many researchers have focused on changing technological properties by exposing starch to controlled conditions, trying to overcome limitations and improving its industrial performance. Chemical, physical/mechanical, enzymatic/biological, and combined methods were studied, and many commercial products were developed and made available for distinct applications both in food and non-food industries.

Modification of starch macromolecules involves derivatization (etherification, esterification, cross-linking, and grafting), decomposition (acid or enzymatic hydrolysis and oxidation), or physical treatment (annealing, heat-moisture treatment [HMT], irradiation, ultrasonication, high hydrostatic pressure, among others). Marked changes in physicochemical properties of starch are achieved with such modifications (Pandiselvam et al., 2019; Singh et al., 2007).

When considering food applications, from the nutritional point of view, one of the most relevant aspects of starch to be considered is its digestibility. Due to its chemical composition, a homopolymer of glucose, starch digestion can impact drastically the human body if ingested in high amounts and short time. Specifically, in the case of people suffering from type 2 diabetes, which is a growing and endemic global disease, the concept of resistant starch (RS) has enormous importance. Modified starches can positively impact the digestibility profile of the corresponding native counterparts with formation of RS classified as RS4. RS is not digested by human intestine amylases but reaches the colon and is fermented to short fatty acids by the microbiota from that final part of the human digestive tract, representing a potential benefit to the health (Ai et al., 2013; Shaikh et al., 2020); see also Chapter 15 in Volume 1 of this series.

The world starch market involves several billions of dollars annually and grows at high rates, with corn as the most prominent source and North America as the leading global player (Adewale et al., 2022); see also Chapter 3 in Volume 1 of this series.

The studies on chemical modification of starch have become popular to the scientific community in the early 1940s. Acetylation was introduced in the mid-1950s and produced improved starches for use in salad dressing and pies, while in the 1970s, hydroxypropylation of starch represented a significant improvement, responsible for greater stability of puddings and frozen foods (Mhaske et al., 2022). The most evident issues with native starch were related with its insolubility in cold water, easiness to retrograde (Haq et al., 2019), and the gelling tendency of pastes that easily result in syneresis (Kaur et al., 2012), as well as the tendency to lose viscosity and thickening power on continuous cooking mainly in acidic conditions or under high temperatures (pressure cooking) (Wurzburg, 1995).

There was relevant development of chemically modified starches before the 1980s, and from that point on, food legislation has been updated and improved to make foods safer and with cleaner labels, that is, with the very essential ingredients and, if possible, without chemical additives (artificial colorants, preservatives, flavorings, stabilizers, emulsifiers, thickeners, and others). This explains why the starch industry has been searching continuously for new natural starch sources as well as for alternative methods for starch tailoring to food applications (physical, biological/enzymatic methods) (Wurzburg, 1995); see also Chapter 14 in Volume 1 of this series. In the last two decades, concerning chemically modified starches, most research and development was directed to non-food applications.

In the present chapter, commercial modified starches' process and applications will be reported and discussed in a systematic approach to make it easier for the reader to have a broad vision of the scenario. Physical, chemical, enzymatic, and mixed processes will be considered to cover the subject with a general vision of the modified starch potential and applications.

6.2 Physical modifications of starch

In recent years, great efforts have been made to broaden starch applicability, by achieving specific enhanced properties for specific types of applications. Physical modification techniques are increasingly popular because they are low cost and have no toxic effects compared to chemical processes (Schafranski et al., 2021). They present variations in parameters as pH, temperature, and pressure (Fonseca et al., 2021).

Practices to physically modify starch include thermal (annealing and HMT) and nonthermal (irradiation, ultrasonication, high hydrostatic pressure, pulsed electric field, and micronization) treatments. Nonthermal physical modification can be environmentally friendly and energy-saving and satisfy the needs of consumers for safe and minimally processed foods (Grgić et al., 2019).

Table 6.1 compares the effects of physical modifications (thermal and nonthermal) on starch properties and applications, as well as their advantages and disadvantages.

According to Wu et al. (2022), nonthermal physical modification can damage the starch granule morphology, depolymerize molecular chains, and decrease crystallinity. These structural changes lead to swelling power and increased solubility, decreased viscosity and gelatinization temperatures, and an increase or decrease in in vitro digestibility, which highly depend on starch type as well as treatment condition.

Thermal treatment is a physical method that alters functional properties from starch by giving a minimal effect without gelatinization (Kim et al., 2023). Hydrothermal starch modifications, such as annealing and HMT are the most used methods effective in changing starch properties [water absorption capacity, relative crystallinity (RC), and pasting properties], maintaining the molecular integrity of starch. These treatments also influence starch digestibility because they are mainly promoted by the disruption in starch structure, increasing accessibility of starch molecular chains to the amylolytic enzymes, altering its crystalline structure, and enhancing starch digestibility (Fonseca et al., 2021).

HMT and annealing strongly affect starch digestibility, differently and to different extents depending on the botanical source of the starch.

6.2.1 Heat-moisture treatment

HMT consists of heat-treating starch with high temperature ($\sim$84°C–120°C) and reduced moisture ($\sim$10%–30%), insufficient to gelatinize starch, for a period that can vary from 15 min to 16 h, to maintain granular integrity (Fonseca et al., 2021; Schafranski et al., 2021). The efficacy of the method depends on the parameters—moisture content, temperature,

TABLE 6.1 Summary of comparison of effects of physical modifications on properties and applications, advantages, and disadvantages of starch.

Modification	Effect on starch properties	Application	Advantages	Disadvantages
Heat-moisture treatment[1]	• Solubility and swelling power reduced; • Gel hardness and digestibility increased; • Retrogradation and breakdown viscosity reduced; • Thermal stability increased.	• Biodegradable films: higher tensile strength, stretch resistance, greater rigidity, and extensibility. • Edible films: higher thickness, tensile strength and elongation; lower solubility. • Nanocrystals: higher yield, shorter period of production time, and higher relative crystallinity. • Food products: increased cooking time and volumetric and gravimetric yield in parboiled rice; firmer texture, improved taste and distinct flavor for noodles.	• Effectiveness; • Flexibility in relation to heat sources; • Low cost; • Nongeneration of chemical residues.	• Modification depends on the process parameters (moisture content, temperature, and heating time)
Annealing[1]	• Solubility and swelling power reduced; • Gel hardness and digestibility increased; • Retrogradation and breakdown viscosity reduced; • Thermal stability increased.	• Food packages: biodegradable films, edible films. • Nanocrystals: increased production yields and thermal stability, and decreased crystallinity. • Food products: improved pasting properties; increased content of slowly digestible starch; increased sensory evaluation scores, cooking qualities, and texture properties in noodles.	• Effectiveness; • Flexibility in relation to heat sources; • Low cost; • Nongeneration of chemical residues	• Modification depends on the process parameters (temperature, and heating time)
Gamma radiation[2]	• Granules retained original shapes and integrity, despite some starches with fractured and cleaved granules; • Crystalline type unaffected and relative crystallinity reduced; • Molecular chains depolymerized; • Amylose content reduced; • Solubility and digestibility increased.	• Preservation and sterilization of food: gamma radiation damages the DNA of microorganisms. • Food industry: this technology inactivates pathogenic and spoilage microorganisms, reduces enzymatic activity in the final products without compromising food quality, and prolongs the product's shelf life. • Manufacture of feed suitable for livestock and poultry: efficiency of decontamination of aflatoxin residues.	• Low temperature; • High efficiency; • Easy to operate; • No chemical residues.	• It is possible to negatively influence the physicochemical and sensorial properties of starch-based foods.

Ultrasonication[2]	• Pores and cracks on the surface; • Granules deformed; • Molecular chains degraded with amylose content increased; • Crystalline type unchanged and relative crystallinity reduced; • Solubility and swelling power increased; • Starch pastes more elastic characteristic; • T_o, T_p, T_c, ΔH, and viscosities decreased;	• Development of starch-based edible film: films can have better transparency, moisture resistance and joint structure as ultrasound (US) improves the compatibility between a hydrophobic material and a hydrophilic polymer. • Bioethanol production: to hydrolyze the starch to increase enzyme susceptibility during fermentation. • Emulsification: acoustic cavitation intensifies the breaking up of the oil droplets, forming emulsions with smaller droplets compared to mechanical agitation. • Pretreatment to reduce oil absorption in frying: US can disrupt the structure of the granules causing uneven distribution of the oil.	• Simple; • Efficient; • No residues; • Higher product yields.	• Lack of standardization and reproducibility; • Temperature increases due to heat generated.
High hydrostatic pressure (HHP)[2]	• Granule morphology and crystalline structure unaffected with pressure below 500 MPa and destroyed at 600 MPa; • Amylose content increased; • Solubility and swelling power increased at low temperatures and reduced at high temperatures; • Gelatinization peak disappeared at 600 MPa; • T_o, T_p, T_c, ΔH, and viscosities decreased.	• Fat replacer: HHP allows to achieve better spreadability, consistency, elasticity and yield stress, emulsion stability and smaller granule size. • Production of edible films: HHP promotes higher enthalpy, improves thermal stability of films and water resistance, and can be used to promote cold gelatinization. • Encapsulation: HHP can force molecular permeation of the component from amorphous regions to intercrystalline amorphous regions and crystalline regions. • Fried foods	• Preserve product quality; • Short processing time; • No toxicity.	• High equipment cost.

(Continued)

TABLE 6.1 (Continued)

Modification	Effect on starch properties	Application	Advantages	Disadvantages
Pulsed electric field[2]	• Molecular chains degraded; • Granule surface roughened and displayed cavities and fissures; • T_o, T_p, T_c, ΔH, and paste viscosity decreased. • Loss of relative crystallinity • RDS increase and SDS reduction	• Inactivation of microorganisms and enzymes: lysis of microorganisms occurs due to irreversible structural changes in the membrane, with the formation of pores and destruction of the semipermeable barrier of the membrane. • Modification of starchy foods can improve starch extraction and alter starch properties by decreasing relative crystallinity, gelatinization temperatures and enthalpies, viscosity, and pasting temperature. • 3D printing: reduction of the apparent viscosity peak, firmer and more resistant gels, which is associated with higher printability.	• No residues; • Energy-saving; • Short time.	• Not suitable for solid products.
Micronization[2]	• Molecular chains degraded; • Granule surface rough and granules aggregated; • Relative crystallinity and short-ranged order decreased; • Solubility, swelling power, and digestibility increased; • T_o, T_p, T_c, ΔH decreased; • Starches displayed weakened shear-thinning behavior.	• Fat replacer; texture improver and gel agent: reduced size, larger surface area, better water absorption, viscosity, stability, less lump formation, and less perception of granularity in the mouth.	• Preserve product quality.	• Some methods have low yields and are costly; • Temperature increase.
Cold plasm	• Chain depolymerization; • Reduction of amylose content; • Reduction of relative crystallinity; • Changes in paste properties; • Onset temperature reduction.	• Starch films: increases hydrophobic characteristics in the starch film and elasticity.	• Green technology; • Low power and pressure.	• High cost.

ΔH, Gelatinization enthalpy change; T_c, conclusion gelatinization temperature; T_o, onset gelatinization temperature; T_p, peak gelatinization temperature.
Based on: [1]Fonseca et al. (2021); [2]Wu et al. (2022).

and heating time—as well as botanical source and starch characteristics, such as structure, organization, and amylose/amylopectin ratio (Fonseca et al., 2021).

Punia (2020) reported that modifications in starch caused by HMT include effects on thermal stability, crystallinity, pasting properties, retrogradation, swelling capacity, morphology, gelatinization, and digestibility. For example, HMT promotes increase in the thermal stability of starches by a shift in the onset, peak, and final gelatinization temperatures to higher values. These changes in thermal properties can be attributed to the formation of a stable configuration, due to a realignment of polymer chains with the noncrystalline regions of starch after hydrothermal treatment (Fonseca et al., 2021).

This physical treatment facilitates the intertwining of amylose and amylopectin and stimulates the diffusion of starch molecules. In this way, starches altered by HMT have a higher temperature demand to reach swelling, which in turn destabilizes the crystalline domain of the starch, leading to an increase in transition temperatures. In the gelatinization, pressure is exerted on the crystalline region. Thus, the melting temperatures of starch crystallites are controlled by the amorphous region that surrounds them (Ambigaipalan et al., 2014).

In addition, HMT promotes an increase in the exchanges of the starch chains, causing the crystalline structure to rupture and separate the double helical structure. The broken crystals subsequently rearrange themselves. Due to this disruption close to the surface of starch granules, their interiors are more susceptible to the action of α-amylases (Gunaratne & Hoover, 2002; Schafranski et al., 2021).

HMT also influences in increased mobility of starch chains and helical structures, resulting in significative structural changes; morphological variations to granules such as size, surface cracking, decreased birefringence, and agglomeration of granules or partial gelatinization; as well as molecular degradation of starch chains (BeMiller & Huber, 2015).

Wattananapakasem et al. (2021) evaluated the HMT of geminated-black rice, with moisture contents of 25% and 30%, at 90°C for 1 and 2 h. Regarding the morphology of the granules, the authors reported the presence of agglomerations and pores after applying this modification. In addition, there was also an improved pasting property, with higher pasting temperature when compared to untreated starch.

Starches can be classified into RS, slowly digestible starch, and rapidly digestible starch. RS helps to control blood sugar and to fight type 2 Diabetes and obesity, while rapidly digestible starch has the opposite effect. Regarding starch content, when starch digestibility increases, the fractions of RS and slowly digestible starch decrease (for more information on RSs, refer to Chapter 15 of Volume 1 in this series).

According to Iuga and Mironeasa (2020), HMT promotes alterations in starch digestibility, enhancing the nutritional value of starch-based products by increasing the slowly digestible starch and RS contents. This change of starch digestibility is important for the health of the population, particularly for individuals with diabetes (Khatun et al., 2019). It is worth mentioning that HMT parameters (temperature and moisture content), starch properties (granule size, crystallinity, amylose, amylopectin content, interactions, and organization between those two macromolecules), as well as the starch botanical source, promote different effects on starch digestibility (Pratiwi et al., 2018).

Furthermore, the RS content is related to various starch properties such as crystalline structure, amylose and amylopectin ratio, gelatinization, and retrogradation and also depends on the processing and storage conditions of food, mainly temperature. These aspects are crucial for the enzymatic susceptibility of starch (Liu et al., 2016).

Brahma and Sit (2020) modified potato starch using HMT, with moisture contents of

30% and 35% at 100 and 120°C for 2 h. The results for the RS values were higher with 35% moisture content, increasing by 95% in the treatment at 100°C and by 88% in the treatment at 120°C. This behavior proposes the formation of a rigid structure that reduces the accessibility of the enzymes in the analysis to disrupt starch molecules (Fonseca et al., 2021). Liu et al. (2016) used the HMT in corn starch at 110°C for 16 h using 20%, 25%, 30%, and 35% of moisture content. Os autores observaram que neste caso houve increased of slowly digestible starch and RS contents. At the highest moisture content (35%), esse efeito acentuou para RS que had their highest value. Moreover, the degree of hydrolysis decreased with the increase in the moisture content used in this hydrothermal treatment. Another study was conducted by Wang et al. (2016) using HMT, with regular corn starch and high-amylose corn starch (moisture contents of 20%, 25%, and 30%, at 120°C for 2 h). nestas essas condições the authors reported um efeito contrário with the increase in moisture content used. The slowly digestible starch content increased, and but the RS also increased considerably when compared with native starches.

Organic amaranth starch was modified by HMT at 10%, 15%, and 20% moisture content in hermetically sealed flasks, which after moisture equilibration were autoclaved for 15, 30, or 60 min at 120°C. The samples modified under higher moisture content, and longer exposure time showed increased thermal stability, possibly due to new bonds induced by higher molecular mobility. With 10% moisture, there was a reduction in the onset and peak temperature, but they were increased in the higher moisture conditions (20%). A pregelatinization of the starch may occur under the conditions studied, which favors the increase in gelatinization parameters. No changes were observed in the morphology of the granules, although agglomerates were formed. As for the diffraction pattern, there were no changes, but peaks related to the amylose-lipid complexes were observed after modification and reduction of RC with increasing moisture content (Bet et al., 2018).

6.2.2 Annealing

Annealing is a hydrothermal process that subjects starch with intermediate or excessive water content, ranging from 40% to 70% (m/v), with temperatures above the glass transition temperature and below the gelatinization temperature of starch, under a period that varies from minutes to days. As with HMT, the process parameters and starch characteristics define the effect of annealing treatment. Nevertheless, in annealing, the moisture content does not vary, since the process is held in excess of water (Punia, 2020).

Studies indicate that effects of annealing on starch properties include increase in starch granular size, gelatinization temperature, crystallinity, thermal stability, and molecular mobility because an increase in the mobility of the amorphous regions to a crystalline state occurs (Fonseca et al., 2021).

Alcázar-Alay and Meireles (2015) described that annealing on starch promotes the reorganization of molecular chains due to the hydration of the granule, improving the arrangement of double helices, occurring a reversible swelling of the granule and rendering an ordered structure with higher granule stability.

Regarding the effect of annealing on starch pasting properties, Hu et al. (2020) modified sweet potato starch using annealing at different time periods of treatment at 50°C for 1, 3, and 5 days. As the time increased in annealing, there was an increase in the pasting temperature and a decrease in breakdown, peak, final viscosities, and in setback, when compared with native starch. Thereby, as annealing time elapsed, these changes in pasting properties were more prominent. Annealing promoted more resistance to high temperatures and weakened retrogradation, improving the pasting properties.

It is possible to say that changes occur with associations among the chains within the amorphous region of the granule as well as changes in crystallinity, promoted during hydrothermal treatment (Fonseca et al., 2021).

Samarakoon et al. (2020) evaluated the effect of annealing on the properties of waxy starches from different botanical sources, such as potato, rice, corn, and barley. The treatment changed the properties and the structure of starches from different botanical sources differently. The RC for corn starch increased by 9% after annealing; nevertheless, it was not changed for barley, rice, and potato starch.

Song et al. (2013) modified potato and sweet potato starches. For digestibility parameters, after annealing, the RS content decreased, and rapidly digestible starch and slowly digestible starch contents increased. Fonseca et al. (2021) reported that increase in rapidly digestible starch can be attributed to the easy and rapid attack of the enzymes of digestion on the amorphous region and melted crystalline defects. Thus, loss of the α-helical structure upon partial gelatinization of annealed starches justifies the increase in slowly digestible starch and the decrease in RS.

Liu et al. (2016) determined the digestibility of corn starch after annealing process (at 50°C for 24 h). There was an increase of slowly digestible starch and RS contents and decreased rapidly digestible starch, when compared to native starch. It is important to mention that annealing does not gelatinize starch. Thereby, increase in RS is due to structural changes such as increased crystallinity and molecular reorganization.

6.2.3 Gamma radiation

Starch in the industry can be modified accordingly by several treatments, such as chemical, enzymatical, and thermal methods using heat and physical methods using radiations for better applications in the food and other segments. Gamma radiation is the most common physical process used for modification of starch in the food industry, as it is fast, is independent of a catalyst, penetrates deeper, is less toxic, requires minimal sample preparation, is cost-effective, and is environmentally friendly (Ito et al., 2018; Sunder et al., 2022).

In recent decades, food irradiation has been widely researched and its effects are well-established to preserve, reduce microbial load or sterilization, increase the shelf-life of products, and above all to maintain the quality of food (Ito et al., 2019). Its efficacy and safety are proven by official agencies including the Food and Agricultural Organization (FAO), International Atomic Energy Agency (IAEA), and the World Health Organization (WHO).

Commonly, a low dose of gamma radiation, less than 2 kGy, can be applied for improving the quality of food; a medium dose, between 2 and 10 kGy, can be applied to increase the shelf-life of food products; and higher dose, more than 10 kGy, can be applied for sterilization of food products (FDA, 2023).

Sunder et al. (2022) reported that starch irradiated with gamma rays can produce free radicals, generate sugars due to cleavage of amylopectin branches, and exhibit variation in digestibility, morphology, crystallinity, thermal property, amylose content, and chemical composition. However, it is important to mention that these behaviors will depend on the botanical source of starch and irradiation conditions (Chung & Liu, 2009).

Ionizing radiation generates free radicals that can induce molecular changes and fragmentation of starch due to glycosidic bonds breakage, promoting the decomposition of amylose and amylopectin into smaller molecules (Bhat & Karim, 2009; Chung & Liu, 2009).

This treatment has been applied to increase the pasting properties of corn and potato starches in various industries (Sunder et al., 2022). Moreover, it has been used to modify gelatinization temperature, decrease moisture content, and alter the structural feature of rice starch (Kumar et al., 2017; Polesi et al., 2018).

The irradiation can cause increased starch digestibility due to the loss of the granular structure and to the molecular fragmentation of the starch, which facilitates the access of the amylolytic enzymes (Bhat & Karim, 2009).

Studies using gamma radiation up to 50 kGy in starch of corn (A-type polymorphs) (Chung & Liu, 2009), potato (B-type polymorphs), and bean (C-type polymorphs) (Chung & Liu, 2010) reported that irradiation increases the RS content, irrespective of the polymorph. In contrast, Polesi et al. (2018) investigated the effect of gamma radiation on the digestibility of rice starch (A-type polymorph), with 1, 2, and 5 kGy doses, and results presented no increase in RS with increasing radiation doses.

Several researchers have observed reduction in the starch digestibility due to the formation of β-bonds, carboxyl groups, and structural modifications (Atrous et al., 2015; Punia et al., 2020). Rombo et al. (2004) mentioned that β-bonds reduce digestibility because they are partially digested by the amylolytic enzymes, whereas the carboxyl groups of the irradiated starch inhibit the enzymatic attack, leading to an increase in RS content, when compared to the native starch.

6.2.4 Ultrasonication

Ultrasound (US) has been used for the physical modification of starches since 1933, employing clean technology, reducing energy and solvent consumption, optimizing time, and avoiding the generation of residues in the obtained extract, which would have to go through a later separation system (Chemat et al., 2017; Hu et al., 2013; Zheng et al., 2013).

It comprises waves with a frequency above the audible range (> 15–20 kHz) and requires a medium for propagation. It can be divided into three frequency ranges, including power ultrasound (16–100 kHz), high-frequency ultrasound (100 kHz–1 MHz), and diagnostic ultrasound (1–10 MHz). According to the ultrasonic region of the spectrum, two groups are distinguished. The first, called low-intensity US, uses low-power (<1 W/cm^2) and high-frequency (100 kHz–10 MHz) sonication for process controls and nondestructive testing. The second group (high intensity or power US) comprises high power (10–1000 W/cm^2) and low frequency (16–100 kHz) sonication, used to cause changes in the product on which it is applied (Zhu, 2015).

There are two types of US devices, differentiated by the means of propagation of the waves, which can occur with the aid of a probe, or by immersion of the food in ultrasonic baths (Adekunte et al., 2010). Different results can be registered using the two devices, with a higher probability of clumping when higher powers are used (Jambrak et al., 2010).

US baths consist of a stainless-steel vessel and a set of transducers placed at the bottom of a metal base, vibrating in phase, transmitting ultrasonic energy into the liquid. They operate at a frequency of 35 kHz, while in probe-type systems, the transducer is attached to the probe, which transmits the vibration to the medium. They operate at a frequency of 20 kHz, and these low frequencies can create bubbles with larger diameters and stronger pressure peaks than frequencies of around 35 kHz (Ozuna et al., 2015).

The frequency is important for analyzing the duration of rarefaction, the phase in which cavitation bubbles grow, which is inversely proportional to the ultrasonic frequency; therefore, at high frequencies, larger amplitudes and intensities are required to generate cavitation. This means that this vibrational frequency directly influences the acoustic effect. When the vibration frequency is lower, the generated bubbles have more time to reach their peak, resulting in higher energy released during implosion and a higher rarefaction cycle. Under higher frequencies, there is less time for bubble growth, and with that, lower shock waves and rarefaction cycles. Temperature and viscosity are also factors that influence this phenomenon. Thus, the effect of modification may be more intense or not, according to the frequency of the equipment (Ashokkumar, 2015; Chemat et al., 2017).

Analyzing these factors, the use of a probe is more efficient in the modification of starches, because the ultrasonic intensity is delivered on a small surface (by the tip of the probe) when compared to the bath. In addition, it is immersed directly in the reaction flask, so that there is less attenuation of the ultrasonic field, especially for small sample volumes, not neglecting the temperature rise (Chemat et al., 2011). Therefore, it is extremely important to pay attention to the distance between the tip of the probe and the treated sample (Cárcel et al., 2012).

When the sound wave passes through the fluid, it causes a longitudinal displacement of the particles and a pressure gradient is generated in the medium, between the pressure of the acoustic wave produced and the vapor pressure of the solvent used (Cárcel et al., 2012).

During the reduced pressure cycle, the intensity of the acoustic wave is high enough and can overcome the intermolecular forces (surface tension) of the liquid, generating vacuum cavities into which gasses and vapors present in the liquid can diffuse, causing the bubbles to expand. When the pressure is positive, the flow is reversed, so the gasses flow from the bubble into the liquid, although less is lost, causing the bubble to increase in size with each cycle. When these bubbles reach a certain volume at which they can no longer absorb energy, they implode abruptly (high pressure/compression cycle), releasing a large amount of energy, increasing the temperature of the system (Ozuna et al., 2015). According to Haaj et al. (2013), cavitation bubbles can reach a diameter between 10 and 20 μm at a frequency of 24 kHz.

The imploded bubble also produces high shear forces and jets of liquid into the solvent that can have enough energy to physically damage the cell wall or cell membrane (Chemat et al., 2011). Therefore, during sonication, an increase in the velocity of liquid layers occurs and waves with high shear energy are formed (Jambrak et al., 2010; Kadam et al., 2015; Pingret et al., 2012). When the bubbles collapse on the surface of a solid material, the high pressure and temperature generate microjets directed to the solid surface and can destroy the cell walls of the plant matrix or cause changes in its structure (Chemat et al., 2011).

This combination of factors can disturb the plant structure by implosions, turbulence at the microscopic level, and collisions between particles and shear, in parallel to the diffusion of water that breaks the crystalline structure, in the case of starches. This set of effects promoted by ultrasound and responsible for the changes caused is called acoustic cavitation. It is possible to promote changes in viscosity, reduction of molecular weight, and inactivation of enzymes and bacteria by the intensity of cavitation (Jambrak et al., 2010; Kadam et al., 2015; Pingret et al., 2012).

Yang et al. (2019) performed US modification of a probe (diameter: 10 mm) with frequency factor 22 kHz, on rice starch with different power levels (150, 300, 450, and 600 W) at 25°C for 20 min in pulsed mode (5 s on and 5 s off). A volume of 50 mL was used, and the probe was penetrated at 2 cm depth of the pastes. They found cracks and pores on the granule surface, with minimal effects on granule size, reduction of the amorphous region, improved particle size uniformity, increased pasting properties, and reduced gelatinization parameters (temperature and enthalpy).

Starch suspensions from arrowhead tubers (*Sagittaria sagitifolia* L.) (10%, w/w) were modified by a tri-frequency ultrasonic bath (20/40/60 kHz) with power density of 300, 600, and 900 W for 15 and 30 min, with pulses of 10 s on and 4 s off, to avoid temperature rise. The results show an increase in swelling power (due to cavitation that induced an increase in the temperature of the internal matrices of the modified starches), disintegration of the structure accompanied by gradual loss of crystallinity (under higher intensity), increased exposure of hydrogen bonds, increased solubility (due to hydrolysis of the granules), increased water and oil holding capacity (induction of empty fractions in the starch molecule due to cavitation), destruction of agglomerates, and increased gelatinization temperature with reduced gelatinization enthalpy. Under higher intensity, surface cracking and roughness of starch granules were observed (Raza et al., 2021).

Starch from purple taro (*Colocasia esculenta* B. Tini) was modified by probe ultrasound (Vibra-Cell 500 W, Sonics and Material Inc, USA, 20 kHz), in 10% (w/v) suspensions with vibration amplitude ranging from 40%, 50%, and 60%, for 30 min. After modification, a reduction in brightness, reduced thermal stability (possibly by weakening bonds), slight reduction in gelatinization temperature and RC, less tendency to retrograde, and increased peak viscosity were observed. Small cracks and deformations were visualized in the starch granules (Martins et al., 2020).

Corn and cassava starch were submitted to modification by high power US at 40 kHz frequency with short periods of time (10 and 20 min) at 26°C. There was a decrease in the enthalpy of gelatinization (being lower mainly after 10 min of sonication) without affecting the temperatures of this endothermic process. The RC suffered a decrease and small cracks were detected on the surface of the granules (Rahaman et al., 2021).

Therefore, US modification promotes changes in the native characteristics of starch, such as the appearance of damage (pores, channels, and cracks) on the granule surface and even reduction in molecular weight (Wang et al., 2023; Yu et al., 2013). Decrease in viscosity and increase in solubility and clarity due to the increase of free macromolecules rather than the breakdown of starch molecules were also reported by some authors (Zuo et al., 2009).

6.2.5 High hydrostatic pressure

High hydrostatic pressure (HHP) is classified as a physical, nonthermal, low-energy modification to optimize the functional properties of starches. It is capable of breaking or changing noncovalent interactions in starch molecules (hydrogen bonds or Van der Waals force), although it does not change covalent bonds (Zhi-guang et al., 2020).

This modification is based on the isostatic principle, where pressure acts in all directions instantaneously and homogeneously, regardless of the size and geometry of the material; and on Le Chatelier's principle, where with the reduction of volume, there is an increase in pressure, shifting the system to a new state of equilibrium (Balasubramaniam et al., 2015).

To perform this modification, starch suspensions are prepared (<30%), which are treated in a HHP container filled with water at the desired pressure (usually 100–600 MPa) under given time (≤30 min) and room temperature (20°C–30°C). After treatment at the desired pressure, the pressure is reduced to atmospheric pressure (0.1 MPa) (in which the noncovalent bonds change) in a short time, so the pressurized suspension is dried or frozen and stored at room temperature (Castro et al., 2021).

Under high pressure and in excess water at room temperature, starch gelatinization can be induced, and the pressure range is specific for each type of starch, noting that type B starches are more resistant to pressure than diffraction patterns A and C. Therefore, in this modification, parameters such as water content, pressure, type of starch, temperature, and treatment time should be controlled (Bauer & Knorr, 2005).

Depending on the intensity of treatment, the modification by HHP can alter the gelatinization of starch, as well as its physicochemical properties, leaving the granules intact or partially damaged, causing this starch to have a gelatinization characterized by low leaching of amylose and lower rate of enzymatic reactivity. The type of starch itself, the amount of water, pressure levels, temperature range, and treatment time are important variables to be considered (Liu et al., 2018; Oliveira, Demogalski, et al., 2021, Oliveira, Macnaughtan, et al., 2021).

Therefore, high pressures can weaken the starch structure, and as the HHP level is increased, the morphology of the granules may exhibit rough surfaces, deformed structures, and aggregate formation, or even total collapse of the structure. In general, crystallinity tends to be reduced, with disruption of the crystalline region and a change in the diffraction pattern from C to B. Under high pressure conditions, HHP can reduce the proportion of double helices and the degree of order of the starch. Regarding water solubility and swelling power, these properties when measured at high temperatures tend to decrease after treatment at high HHP levels, and the opposite behavior is observed when low HHP levels are employed (Dominguez-Ayala et al., 2022).

Increased gelatinization temperatures may occur by the interaction between double helices in the crystalline region, causing better ordering of the helices in cereal starches, but the enthalpy is reduced. The paste properties are also decreased, by the lower interactions between starch chains, as hydrogen bonds are decreased. The water holding capacity can be increased with the higher intensity in the HHP modification, while the oil holding capacity is reduced. Regarding digestibility, starches with amylose content between 10% and 40% show reduced rapidly digestible starch and increased slowly digestible starch and RS with higher pressure intensity. On the other hand, for waxy starches, RS levels can be decreased with increased pressure, possibly due to the destruction of starch granules and the breakdown of double helices (Dominguez-Ayala et al., 2022).

Wheat, potato, and cassava starch were subjected to modification by HHP in suspensions at 5% (m/v). Suspensions (1.5 mL) were stirred to avoid sedimentation and already pressurized in a high pressure multivessel with temperature control at different pressures, times, and temperatures. Wheat starch was the most sensitive to pressure, due to the type A diffraction pattern, and potato starch (type B) was the most resistant. Cassava starch showed phase transition to gelatinization at higher temperatures and pressures than wheat and lower than potato starch. Keeping temperature and treatment time constant, increasing pressure causes an increase in the degree of gelatinization. As the temperature increases, the pressure required for complete gelatinization decreases. Keeping temperature and pressure constant, the longer the treatment time, the higher was the degree of gelatinization achieved, although above 60 min, the degree of gelatinization of potato starch suspensions no longer changed (Bauer & Knorr, 2005).

Suspensions at 15% (m/v) pea starch were subjected to HHP treatment (S-IL-100-350-09-W; Stansted Fluid Power Ltd., UK) in vacuum-packed polypropylene bags. The samples were subjected to pressures of 150, 300, 450, and 600 MPa at 30°C for 25 min, with a pressurization rate set at 1/MPas. They were then vacuum-filtered, oven-dried at 40°C for 24 h, and passed through a 100-mesh sieve. The modification resulted in cracking under pressures below 450 MPa. However, at 600 MPa, the starch granules collapsed completely. There was a change in the diffraction pattern, with a reduction in the intensity of peaks in the diffractogram, with the destruction of the crystalline structure under high pressures. The water absorption index, water solubility, and swelling power of pea starch at 30°C–70°C were higher for the samples treated at 600 MPa (possibly due to aggregation of amylose under pressure, interfering with lipid-starch bonds, increasing water retention), but at higher temperatures, these properties showed reduction (rearrangement of starch molecules, which inhibited hydration and swelling or due to disintegration of granules inhibiting amylose solubilization) (Liu et al., 2018).

Regarding paste properties, pea starch samples submitted to pressures up to 450 MPa exhibited an increase in viscosity, but at 600 MPa, they had a decrease. HHP may have induced the formation of amylose-lipid complexes, loss of molecular ordering, and the interlacing of amylopectin, which limited the dispersion and swelling of the granules. With this, more time was needed for swelling, as molecular interactions become stronger. The gelatinization temperatures decreased, as well as the gelatinization enthalpy, for the treatments up to 450 MPa, evaluated by differential exploratory calorimetry, since at 600 MPa, no gelatinization peak was detected. Finally, starches modified by HHP showed a reduction in the degree of hydrolysis, as well as in the amount of rapidly digestible starch and slowly digestible starch, while the levels of RS increased (Liu et al., 2018).

Therefore, HHP-modified starches can be used to produce edible films due to higher water solubility and development of porous starches used to encapsulate and control release of food ingredients, as fat replacer or due to low oil retention in the development of healthier fried foods (Dominguez-Ayala et al., 2022).

6.2.6 Pulsed electric field

Pulsed electric field (PEF) is a physical and emerging method of modifying starches, which can affect conformation, microstructure, granule size, paste properties, solubility, swelling power, as well as thermal stability, in vitro digestibility, and structural transition. For this, low energy consumption and a short period of time are required (Maniglia et al., 2021).

According to Sing et al. (2022) and Castro et al. (2021), it consists of applying an electric field directly to foods that have low viscosity and electrical conductivity, such as juices, milk, soy milk, pea soup, and even liquid egg. In this way, the products flow through the electrodes, and a short time (nano to milliseconds) is employed to prevent damage from temperature rise. The pulses generated can be bipolar, square wave, exponential, or oscillatory in shape, but the most energy efficient are bipolar square pulses (applied voltage is held constant at the maximum value for a certain time, called pulse width, after which the voltage is decreased by up to 37%).

The modification is based on a PEF system consisting of a pulse generator and a treatment chamber (where the sample is placed). The generator will convert the high voltage alternating current into direct high voltage energy (propagated in one direction only), to be used by the system. This energy is stored in a capacitor, and in the treatment chamber between the two electrodes, electrical pulses are produced by rapid discharges of electrical energy, controlled by a switch, causing an electrical potential differential, thus generating the electrical pulse fields (Castro et al., 2021).

The secret of PEF modification is in the electropulsation (exposure of cells to electric pulses) that promotes alteration in the cell membrane, increasing its permeability and conductivity, depending on the treatment time and electric field intensity. The changes occur because the cell senses the difference between the electrical voltage of the intra- and extracellular environment (basal transmembrane potential) and may cause the formation of hydrophilic pores and increased permeability, when a minimum transmembrane potential is reached (Kotnik et al., 2019).

The modification of starches by PEF presents the advantage of being a simple, fast process, not requiring chemical solvents, as well as not producing residues, thus generating clean-label products (Conde et al., 2022).

To perform the modification, a room temperature starch suspension has its conductivity adjusted between 50 and 200 μS with potassium chloride solution, which will be pumped into the PEF chamber and subjected to the treatment conditions. The electric field strength can range from 2.86 to 50 kV/cm, with duration between 214 and 1272 μs and frequency between 600 and 1008 Kz. Pulses can occur between 6 and 40 μs, mainly between 20.16 and 21.37 μs. The samples should be cooled to room temperature, since due to the Joule effect, heat generation occurs with the passage of electric current in a conductive material, proportional to the square of the electric current intensity. Thus, cooling prevents the starch from gelatinizing. However, it is worth noting that increasing the temperature during treatment improves conductivity and should be kept between 50 and 200 μS/cm, optimizing the transmembrane potential for the modification to occur. They are then vacuum-filtered and dried at 40°C (Castro et al., 2021).

The changes caused by PEF lie in changes in granule morphology, disappearance of the Maltese cross under application of high electric fields, mainly for type A and B starches, reduction in RC, decrease in lamellar repeat distance, reduction in gelatinization temperatures, enthalpy, and paste properties. An increase in fast digesting starch content and decrease in slow digesting starch are reported, while maintaining the RS content, due to changes in the starch chains (Duque et al., 2022; Park & Kim, 2021).

Potato starch (8%) with conductivity set to 200 μS/cm by potassium chloride was subjected to PEF at 30, 40, and 50 kV/cm. The operation was continuous, at a flow rate of 60 mL/min, with square wave, unipolar frequency of 1008 Hz and pulses of 40 μs duration. Granule damage, aggregation, reduction in gelatinization parameters and RC, also associated with amylopectin chain scission, and loss of helical order were observed (Han et al., 2009). The presence of phosphate esters in potato starch may favor deeper penetration of electrical pulses into the crystalline lamellae.

In waxy rice starch, the application of square wave pulses at a controlled flow rate and conductivity promoted increased rapidly digestible starch and reduced gelatinization parameters and RC (Zeng et al., 2016).

Maniglia et al. (2021) subjected 8% suspensions of wheat and cassava starch with conductivity set to 1 Ms/cm at 25°C, while maintaining a constant flow rate of 2 L/h, to a bench-scale continuous-flow PEF unit. Treatment chambers were connected to a high-voltage pulsed power generator (20 kV–100 A) that delivered square monopolar pulses (1–10 μs, 1–1000 Hz). The pulse length was fixed at 5 μs, already; the electric field intensity was adjusted according to the variation of the applied voltage; and the total power input was linked to the pulse repetition frequency. Three combinations of field intensity and energy input were tested (T1: 15 kV/cm–25 kJ/kg; T2: 25 kV/cm–25 kJ/kg; and T3: 25 kV/cm–50 kJ/kg). During treatment, the temperature remained 25°C ± 2°C, while the maximum temperature at the chamber exit was 10°C.

The internal microstructure of cassava starch granules was affected, with the loss of birefringence being visualized. There were no morphological changes in cassava starch granules, but in wheat starch, especially under higher field intensity, there was damage and cracking on the surface (Maniglia et al., 2021).

In general, modification with PFE can damage the internal structure of the granules, facilitating water absorption and swelling, loss of RC, reduced enthalpy and gelatinization temperatures, and increased rapidly digestible starch (Park & Kim, 2021).

6.2.7 Cold plasma

Plasma is considered the fourth state of matter, composed of electrons, ions, free radicals, and also of free and excited atoms, which originate from the application of energy, whether electric, magnetic, thermal, microwave, or radiofrequency, to a gas. The composition of the gasses that are fed into the plasma unit, for example, carbon dioxide and argon plasma, influences the results and can lead to the introduction of hydroxyls, ketones, esters, and free radicals. Besides this variable, the humidity, power, and time of treatment also interfere in the results obtained when using cold plasma for modification of starches (Okyere et al., 2019).

Cold plasma is generated in a gaseous phase by electrical discharges that can be generated in two ways: (1) between electrodes with low frequency discharges or high voltage direct current discharges, which require low pressure (10^{-1} to 10^{-3} Torr) in the reactor; (2) electrode discharges generated by absorbing electric field energy at resonance frequency (microwave or radio frequency). However, it is also possible to generate the stable glow plasma at atmospheric pressure with a 50 Hz source, which can cause dextrinization of starch (Lii et al., 2002).

It can come in two types: high temperature (thermal, hot plasma, which requires high pressure 10^5 Pa and power 50 MW) and low temperature (melt discharges, which operate at low pressure or atmospheric pressure). Starch modification has been reported for cross-linking, depolymerization, and oxidation, mainly because it is a clean technology that requires low power and low pressure. It can cause cross-linking, dehydroxylation, dehydrogenation of starch chains, decrease in paste properties, increase in gelatinization enthalpy, and increase in RS content (Singh et al., 2022).

According to Carvalho et al. (2021), the changes caused by cold plasma are dependent on the botanical source and treatment conditions. Effects related to increased solubility, water binding capacity, and reduced paste temperature may be useful in the food industry. In addition, decrease of molecular weight, cooking time of rice, reduction of enthalpy, and increase of gelatinization temperatures have also been observed (Okyere et al., 2019).

The modification using low-pressure direct current generation of stable low energy glow plasma, in air atmosphere at 10^{-2} Torr, was tested, avoiding to promote the total destruction of the granules due to electrons, ions, generated radicals, and temperature increase, or by oxidation by ozone and nitrogen oxide. Native starches from cassava, corn, corn with high amylose content, oats, potato, rice, waxy corn, and wheat were submitted to the treatment. To do so, 1 g of each sample was placed in a reactor with 10 cm internal diameter coupled to stainless steel electrodes. After pressure adjustment, a potential gradient was applied from one of two sources (high voltage direct current power supply operating at 1–2 kV, with a constant current of 1 mA; or by inductor at 15 kV and 27.2 MHz). Pulses of 1 min on and 1 min off were used to avoid heating the samples, and the experiment was conducted for 10, 20, and 30 min at 1, 10^{-1}, and 10^{-2} Torr. It was visualized that starches with low moisture content (<2.5%) had significant hydrolysis, and amylose was more sensitive to plasma treatment compared to amylopectin, as high amylose oat and corn starch were resistant to treatment. There was also pH reduction, resulting in oxidation and degradation, due to oxidation of the surface on contact with plasma, leading to the formation of carboxylic acids. No pores were visualized on the surface of the potato starch granule, although in native and waxy corn starch, as well as rice, pores were characteristic.

Modification of 20 g of Caeté starch (*Goeppertia allouia*), a starchy raw material from the Amazon region, was conducted with dielectric barrier plasma at 25°C and atmospheric air with different voltages for plasma generation (7, 10, 14, and 20 kV). The average pressure was 101 kPa, with a distance between the aluminum electrodes of 15 mm, with a plasma excitation frequency of 200 Hz and a duration of 15 min. The samples submitted to a voltage of 14 kV showed greater depolymerization of the chains, with reduced amylose content, less tendency to retrogradation, and surface oxidation of the granules, favoring the appearance of new carbonyl and carboxyl groups, with reduced pH. This influenced the observed increase in paste viscosity. At higher voltages (20 kV), there was formation of oxidized complexes with cross-linking of the chain. There were no changes in the RS content, but the digestibility increased (Carvalho et al., 2021). For more information on Caeté, see Chapter 16 in Volume 2 of this series.

In contrast to these results, Okyere et al. (2019) reported reduced viscosities and onset gelatinization temperature of waxy starches from corn, rice, and potato, as well as an increase in RS for corn and potato. Possibly there was also formation of single V-type helices in waxy starches of corn and rice. The enthalpy increased for all three samples, and the RC reduced only for potato waxy starch. A plasma unit with a radio frequency power supply (13.56 MHz), CO_2 (25 cm^3/min), and argon gas for plasma generation (15 cm^3/min) was used for the modification. Power of 120 W was applied to the sample to increase the dissociation of CO_2 to carbon monoxide and oxygen. Samples of 15 g were subjected to the plasma chamber for 30 min, shaken, and recoiled for another 30 min.

6.2.8 Micronization

Micronization is a process relying on mechanical force to reduce particle size to the micron-scale range, that is, a reduction in the size of the particle to less than 10 microns, achieved through diverse mechanisms. Although this modification is known to be used in the pharmaceutical industries, in recent decades, several studies have revealed its potential in the food industries (Dhiman & Prabhakar, 2020).

Micronization can be applied to both solid materials and liquid materials. For solid materials, colloid milling, jet milling, and ball milling are used, while the liquid materials involve microfluidization and high-pressure homogenization, the ball milling being the most common process (Xia et al., 2017).

It is noteworthy that this modification technique is a mechanical process, but some authors reported that during the micronization, the mechanical forces, like shear, collision, and friction, increase the temperature at the point of impact with increasing time; thus starch granules are broken down into smaller particles. In this way, the changes of starch caused by micronization might be from both mechanical and thermal effects (BeMiller & Huber, 2015; Suki et al., 2016; Wang, Li, et al., 2020; Wang, Liu, et al., 2020).

Regarding the effect of micronization on granule morphology, several studies showed that alterations occur, such as granule deformation with increasing roughness and irregularity (He et al., 2020; Wang, Li, et al., 2020; Wang, Liu, et al., 2020), appearance of hollows in the granule center (Wang et al., 2018), loss of birefringence (Li et al., 2014), and aggregation of starch granules (He et al., 2020; Wang, Li, et al., 2020; Wang, Liu, et al., 2020). These behaviors positively correspond with the treatment intensity (time, speed, energy, pressure, etc.) (Wu et al., 2022).

Research conducted by He et al. (2020), Wang et al. (2018), Wang, Liu, et al. (2020), and Wang, Li, et al. (2020) evidenced a reduction in starch crystallinity, reflecting a transformation of the crystalline structure into amorphous starch (Soe et al., 2019). In addition, Wu et al. (2022) reported that ball milling techniques exhibit a stronger effect on starch crystalline structure, when compared with jet milling (0.6–0.8 MPa) and microfluidization (30–180 MPa).

For thermal properties, micronization caused decrease in the gelatinization temperatures (T_o, T_p, and T_c) and ΔH of potato (Wang, Li, et al., 2020; Wang, Liu, et al., 2020), cassava

(Moraes et al., 2013), and rice starches (González et al., 2018). Wu et al. (2022) reported that this treatment causes a lower apparent viscosity and starch paste displays a weakened shear-thinning gel-like behavior with elevated elasticity.

Some researchers have observed an increased in the starch digestibility after micronization, which was observed in wheat starch (McAllister & Sultana, 2011), potato and maize starch (Sushil et al., 2011), and barley starch with amylose of different contents (Emami et al., 2010).

6.3 Chemical modifications of starch

Chemical modification opened new horizons for starch uses throughout the industrial processes that could benefit from it. Certainly, that technology was responsible for the large and fast growth of the starch market due to the new perspectives for employing the polymer in processes in which native starch properties did not allow the needed performance that time. The market of sweeteners and starch derivatives is larger than that of native starch alone, showing the importance of adding value to this polysaccharide (Adewale et al., 2022).

Besides the industrial development of many processes and products for modifying starch, the scientific literature brought a huge volume of papers, books, and other technical communications to the community. There are classic works in which most information is concentrated, mainly consistent books and review papers that focus on the chemical modification of starches. One of the most known books on modified starches was edited by Wurzburg (1986).

The chemical modification implies changing in some way the starch macromolecules (amylose and amylopectin), which can be partially degraded (hydrolyzed or oxidized) or have some hydroxyl groups replaced with different functional groups (carboxyl, ester, ether, and amino groups) (Haq et al., 2019). Chemical modification of starches improves, in general, paste behavior (consistency, smoothness, and clarity) imparting freeze-thaw and cold storage stabilities. The starch source, reaction conditions (pH, temperature, time, presence and type of catalysts, and reagent concentration), type and extent of substituent (degree of substitution or molar substitution), as well as the substituent distribution throughout the starch macromolecules must be considered (Singh et al., 2007).

Some important concerns, however, arise from the chemically modified starches, including the availability and price of raw materials, restrictions to the use of chemically modified starch (Kaur et al., 2012), and environmental issues due to elevated amounts of water used and wastewater/pollutants generated upon starch processing (Adewale et al., 2022). For more information on starch restrictions in international legislations, see Chapter 6 in Volume 1 of this series.

Cassava starch is one of the most valued in the global market due to unique physicochemical and technological properties (nonallergenic, mild flavor, high paste viscosity and transparency/clarity, low tendency to retrograde, and freeze-thaw stability) (Wang et al., 2022). Some examples of uses of chemically modified starches are shown in the Annex 1.

6.3.1 Acid hydrolysis of starch

Acid-modified, acid-thinned, or thin-boiling starches are very popular and have broad applications in the food industry as gelling agents, used in gum candies (starch jellies), for example. These modified starches also find uses in the pharmaceutical, textile, paper, and many other industries. Roughly, acid hydrolysis involves inorganic acids (hydrochloric acid and sulfuric acid, mainly), which promote breakage of glycosidic linkages through nucleophilic substitution reaction and addition of water molecules.

Among the main properties reported to acid-modified starches, increased solubility, decreased swelling power, and hot paste viscosity are well documented. The increase in RC has been reported in some studies; however, there are others that did not detect changes in RC values. Lowering of amylose is also described as a behavior of acid-modified starches (Haq et al., 2019; Ulbrich & Flöter, 2019b), a fact that can be associated with the greater susceptibility of amorphous fraction of the granules to the acid attack when compared to the crystalline region.

All these changes in starch depend on the reaction conditions (acid concentration, reaction temperature, and time), as well as the starch source (Ulbrich et al., 2016). The reaction conditions include temperature that is below the gelatinization temperature to avoid excessive digestion of the starch granules (Adewale et al., 2022). Controlled process conditions (acid concentration, temperature, and reaction time) allow for the production of thin-boiling starch with the desired properties for specific industrial applications (Ulbrich & Flöter, 2019a, 2019b).

Known by forming low viscosity (high fluidity) hot pastes even at high solid concentrations but with high tendency to retrograde, resulting in relative rigid gels during cooling, acid-modified starches show superior behavior when used in the production of biodegradable films since higher solid concentrations can be considered with clear benefit to the process (faster drying of the films and production of thicker film layers) when compared with the respective native starches (Wurzburg, 1986).

Acid-modified starches, when used in food applications, may present superior nutritional quality due to their high tendency to retrograde. RS can be formed, mainly RS type 3 with potential improvement of functional properties of different processed foods (Koksel et al., 2008). Additionally, these acid-thinned starches are potential substitutes to the comparatively expensive pectin and to gelatin, in this case representing a key ingredient for vegan products (Ulbrich & Flöter, 2019b).

Depending on the reaction conditions, the acid modification can produce starch nanocrystals (SNCs). These biodegradable and renewable nanoparticles have potential as reinforcing agents in polymer matrices, with improvement of mechanical properties. Additionally, due to the very large surface with high amounts of hydroxyl groups, SNCs are easily modified to be used in different applications (emulsion stabilizers, drug carriers, and many others) (Dai et al., 2019).

The production of SNCs from cassava, beans, and corn starches was described as well as their chemical modification (oxidation) to produce Pickering emulsions. Sulfuric acid (3.16 mol/L) was used for partial hydrolysis of the starch granules with production of the SNCs, which were oxidized by sodium hypochlorite solution (4.0% v/v) at pH 9.5. The results have shown a good performance of the oxidized SNCs as stabilizers of the emulsions, with those from cassava starch being superior to the ones produced from corn and bean starches (Daniel et al., 2020).

6.3.2 Oxidation of starch

Carbonyl and carboxyl groups replace hydroxyl groups of C-2, C-3, and C-6 positions of the D-glucopyranosyl unit of the starch macromolecules. The reaction is promoted mainly by strong oxidants (sodium hypochlorite, sodium periodate, hydrogen peroxide, ozone, among others) under controlled pH and temperature. The functional properties of oxidized starches are deeply altered when compared to their native counterparts and include high paste clarity (transparency), lower paste viscosity at high solid concentration, viscosity stability, and improved adhesion capacity as well as enhanced film formation. The most popular uses of oxidized starches are in the textile and in the paper industry, as sizing and coating agents. Other uses are in the health sector (controlled release of drugs and wound healing), packaging (films with controlled release of bioactive compounds) (Adewale et al., 2022), and in the food industry (Daniel et al., 2020; Demiate et al., 2000; Wojeicchowski et al., 2018).

When considering commercial oxidized starches, sodium or calcium hypochlorite are the most reported oxidizing agents. The pH range of the reaction is critical to the process, with favorable formation of carboxyl groups at high pH values (mild to moderate alkalinity). As carboxyl groups are formed, part of the alkalinity will be neutralized and the reaction course can change, unless action is taken to restore the favorable pH value for the desired modification. Due to the alkaline conditions of the starch oxidation by hypochlorite, reaction becomes complex as oxygen from air can promote degradation of starch besides oxidation. Weakening of glycosidic linkages through oxidation of secondary hydroxyls can be associated with degradation of starch (scission of glycosidic bonds). In most commercial starches produced by hypochlorite treatment, carboxyl contents up to 1.1% are found by analysis (Smith, 1967). Carboxyl groups play a major role in stabilizing amylose molecules and minimizing retrogradation, one of the most important properties of this type of modified starch. Differently from native and acid-modified starches, after pasting in hot water and cooling, rigid gels will not form and will remain clear, if compared to the same sources. This stability of the pastes is one of the main advantages of the oxidized starches in applications where this property is needed (Wurzburg, 1986).

Ozone is being studied as an efficient oxidant for starch modification. With the advantage of being considered safe without formation of chemical residues or producing toxic waste, it fits as a green technology, and food use of oxidized starches with use of ozone may benefit from green labeling. Like in conventional chemical oxidized starch, ozonized starch presents carbonyl and carboxyl radicals as well as partial degradation of the polymeric structure. The resulting starch has a shear thinning behavior and low retrogradation tendency, with greater cooking stability (Pandiselvam et al., 2019).

6.3.3 Cross-linking of starch

Cross-linking means that covalent chemical bonds will be introduced inside the starch granule (intra- and intermolecular), reinforcing its structure that will become more suitable for applications that involve high shear, high acidity, as well as high temperatures of processing. Pastes will remain viscous even after extended cooking time or retort temperature, making this type of starch very useful for acidic food and for processes that include intensive cooking (Singh et al., 2007).

Reagents used as cross-linking agents include phosphoryl chloride ($POCl_3$), sodium tripolyphosphate (STPP), sodium trimetaphosphate (STMP), epichlorohydrin (EPI), and a mixture of adipic acid, acetic anhydride, and vinyl chloride. Toxicity of some of these reagents such as EPI is always a concern, and for that reason, STMP has been a better agent, nontoxic, and effective as a cross-linking reagent for starch (Bakouri & Guemra, 2019). Frequently, cross-linked starches contain about one cross-link for every 100–3000 anhydroglucose units (AGUs). The literature is abundant of studies and patents covering the use of bi- or polyfunctional reagents in the cross-linking of starch (Wurzburg, 1986).

One important fact that has impacted the chemical modification of starches was the discovery of natural phosphorylation (monostarch phosphate) in potato (*Solanum tuberosum*) starch at the beginning of the 20th century. Phosphorylation of starch involves two possibilities, that is, monostarch phosphate (MSP) and distarch phosphate (DSP) (cross-linked). Esterification of one phosphate to one hydroxyl group of the glucose unities produces MSP. In the case of DSP, two glucose unities are linked to the same phosphate group by cross-linking MSP. During phosphorylation, the pH will influence the production of MSP or DSP. Briefly, MSP associates with increased swelling power, viscosity, paste clarity, and freeze-thaw stability, whereas DSP helps to keep the granular structure tougher, with improved resistance to high temperature, low pH, and retrogradation (Ramadan & Sitohy, 2020).

Cross-linking is a very used and a useful way of modifying both native and all types of modified starches, toughening the granular structure that will maintain its integrity even on swelling. They can be used in high viscosity thickeners, with resistance to viscosity breakdown (stirring cooking, high temperatures/retort, and at low pH values), and to mechanical shear (Wurzburg, 1986).

Organic acids, mainly citric and malic, are commonly employed in food processing for esterification reactions that will be followed by cross-linking. The use of such organic acids is safe, as they are not harmful to health and can also decrease the starch digestibility, that is, increase the contents of slowly digestible starch and RS (Fan et al., 2021; Fashi et al., 2023).

6.3.4 Esterification of starch

In nature, phosphorylation is the only covalent modification of starch that was well described. It deeply alters physicochemical properties of starch, that is, swelling power, viscosity/pasting, surface charge, crystallinity, and thermal stability. Potato (*Solanum tuberosum*) starch stands out with the highest level of phosphorylation among the conventional/commercial starches (Apriyanto et al., 2022). For more information on starches with high phosphorus content, Chapter 8 of Volume 1 in this series is suggested.

Inorganic and organic starch esters can be produced by several reagents and reaction conditions. Phosphorylated starch and acetylated starch are examples of such starch esters. Ester derivatives of phosphoric (starch phosphates) are those in which only one starch hydroxyl is involved in the starch phosphate linkage and the resulting product is a monoester (Wurzburg, 1986).

Acetylated starches are known and produced for more than 100 years from low substitution (DS range 0.01–0.20) to very high substitution (DS range 2.0–3.0). One of the main properties of these acetylated starches is its cold paste stability, that is, low tendency to retrograde and syneresis due to the presence of acetyl groups (Wurzburg, 1986).

Succinic anhydride reacts with starch and forms a half-ester; starch succinates tend to swell in cold water, depending on the level of modification, and have lower gelatinization temperatures when compared with the corresponding native starch. The inclusion of a bulky radical (hydrophilic succinate group) results in increased viscosity of the starch, with improved stability and clarity of the paste as well as freeze-thaw stability (Wurzburg, 1986).

Starch alkenyl-succinates have special and desired properties for food use, mainly in stabilizing emulsions. In this kind of modified starch, a hydrophobic region is incorporated to the hydrophilic starch backbone. Starch octenyl-succinate (OS-starch) is very useful for stabilizing oil-in water emulsions in the food, pharmaceutical and industrial areas. One of the best examples of successful application of high-viscosity OS-starch is as a stabilizer for high-viscosity high-oil systems such as salad dressings (Wurzburg, 1986).

Besides being used as a stabilizer in emulsions, the amphiphilic properties of OS-starch also favor its use in encapsulation and as a fat replacer (Ai et al., 2013).

OS-starches (DS 0.019–0.058) from normal and from high-amylose corn (34.3% and 68.4% of amylose, respectively) were produced and studied considering their digestibility (in vitro and in vivo). The results showed that there were changes on the RS contents, related with the gelatinization temperature in the case of high-amylose starch that had a decrease in this value. In general, the RS contents increased, and other interesting conclusions were possible for the in vivo digestion, which showed that about 20%–30% of cooked OS-starch (high-amylose starch) was not digested and found in the feces, which was a much higher amount when compared with that from the native starch counterpart test (Ai et al., 2013).

Organic acids (citric, lactic, malic, and succinic) may react with starch macromolecules under specific reaction conditions to produce hydrolysis, esterification, and cross-linking, altering technological properties (swelling power, solubility, paste viscosity, retrogradation, digestibility, and others). Besides, fatty acids (stearic and myristic) can also complex with amylose inside the starch granule and promote relevant changes in physicochemical properties, mainly by introducing amphiphilic behavior (Karma et al., 2022).

6.3.5 Etherification of starch

Viscosity stability was always searched for starch applications, and this is one of the advantages of hydroxyethylated starches. Ether bond is stable for hydrolysis even at high pH. Hydroxypropyl starch is produced by reacting propylene oxide with starch under alkaline reaction. This modified starch has valuable properties for food and non-food uses. Its production cost, however, is a limitation for non-food applications.

Hydroxypropyl starch is a high-performance modified starch mainly for the food industry, presenting improved cold storage stability, clarity, as well as textural properties of its paste (short texture). Generally, the desired DS is 0.1, that is, 1 hydroxypropyl group per 10 glucose unit (AGU) (Wurzburg, 1986).

6.4 Enzymatic modifications of starch

Amylolytic enzymes (α-amylase, β-amylase, glucoamylase, and cyclodextrin-glycosyltransferase) are tools for modifying starches in several ways. Producing porous starch is one example, which makes it possible to facilitate the access of small molecules inside the granules (flavor carrier, drug carrier, heavy metal absorber, etc.). An important fact related with porous starches is that most studies were carried out for cereal starches, despite the importance of other sources (tuber and roots, legumes, and palm stem) (Latip et al., 2021).

The formation of holes and channels throughout the starch granules by amylases can represent an aid for increasing chemical modification efficiency due to the deeper access of the reagents, mainly esterification with organic acids (Fashi et al., 2023).

Debranching enzymes (isoamylase and pullulanase), branching enzymes, and transferases can be used to modify starch and alter its digestibility. Slowing down the digestibility is always important to avoid the undesirable high glycemic index that represents a serious issue in modern life. Together with increasing slowly digestible starch, the food industry is searching for RS because together they can benefit human health, by keeping blood glucose stability, regulating postprandial blood glucose levels, and acting as prebiotic in the colon (Fan et al., 2021, see also Chapter 15 in Volume 1 of this series).

Cassava starch production is growing, and this starch has great potential for developing new food and non-food uses due to its technological properties (tendency toward soft gel production, freeze-thaw stability, and transparency). It is a differentiated raw material for producing modified starches that have innovative functional properties. Among these properties, enzyme-modified starches are reported as improved from the nutritional point of view, considering the slow or more restricted digestion (increased slowly digestible and RSs levels) (Mhaske et al., 2022). For more information about hydrolyzed starches see Chapter 5. Some examples of uses of enzymatically modified starches are shown in the Annex 1.

6.5 Multiple modifications of starch

As can be seen in the previous sections, various techniques have been used for starch modification with the intention of improving the performance and expanding the practical application of starch. Thereby, these treatments (physical, chemical, and enzymatic) can also be performed combined with other modifications. The combinations or multiple modifications of starch are an asset due to their improvement of properties, opening new applications and new perspectives for science and technologies of starch modification in different fields.

Maior et al. (2020) evaluated modification of organic acids (lactic, citric, and acetic) with HMT. The authors found no changes in the diffraction pattern and a slight reduction in RC for all samples. The samples treated with HMT and more citric acid and lactic acid did not show gelatinization curves. Samples combined with HMT and acetic acid showed a reduction of 15% in gelatinization enthalpy, and samples treated only with HMT showed no difference in this property. For digestibility, HMT increased the RS content; the treatments with lactic and citric acid produced the highest values, with a fivefold increase in relation to the native maize starch. The results indicate a possible application for acid-heat treatment starches in functional foods with reduced viscosity and to produce low-carbohydrate foods for consumers with chronic diseases.

The effects of combined methods with infrared HMT treatment were determined by Ismailoglu and Basman (2015) in corn starch and by Ismailoglu and Basman (2015) in wheat starch. The authors used microwave powers (550 and 730 W), exposure times (30, 60, and 90 min), and moisture content (20% and 30%) in both studies. They reported that gelatinization temperature values increased significantly as the infrared power increased from 550 to 730 W. The pasting properties were affected by a decrease in viscosity only for the samples treated using 550 and 730 W and 30% of moisture content.

Chi et al. (2019) employed heating combined with annealing treatment to synergistically modulate structure and digestibility of corn starch and potato starch. Dry heating decreased starch molecular weight and created small molecular fractions with suitable chain length, resulting in starch better rearrangement during annealing treatment. Regarding digestibility, slowly digestible starch and RS decreased after combined modification for corn starch and increased for potato starch. This behavior can be attributed to the different molecular structure of starches, shown in their diffraction patterns (A-type for corn and B-type for potato).

Nemţanu and Braşoveanu (2021) investigated the effects of corona electrical discharge treatment with electron beam irradiation on starch physicochemical properties. The results showed that the combined treatments can produce more pronounced changes in starch properties such as acidity, paste clarity, gel consistency, and viscosity. These techniques can modify the technological properties of starch, depending on the application sequence of corona electrical discharge because of the degradation phenomenon. In addition, the use of corona electrical discharge as a pretreatment to electron beam irradiation of starch can reduce the minimum irradiation dose required for certain industrial applications by influencing more the sensitivity of starch to radiation.

Ghalambor et al. (2022a, 2022b) studied the characteristics of sago starch doubly modified by acid hydrolysis and hydroxypropylation. There was no change in granule sizes, but the treated starches had reduced paste properties and glass transition temperature. With the higher proportion of propylene oxide and shorter hydrolysis time, there was a higher degree of substitution and swelling capacity and less syneresis. However, with longer hydrolysis time, both swelling and syneresis increased, besides increasing gelatinization temperatures and decreasing retrogradation temperatures. After the double modification, sago starch had an increase in water solubility. With this, the authors pointed out the use of doubly modified sago starch for products such as frozen dough or frozen bakery products.

6.6 Food uses of modified starch

Modifications are performed to confer the starch properties necessary for its industrial applicability, making the process more profitable, easy, and with the desirable quality. Thus, some of the various applications of modified starches for food use will be exposed below (for more details on the standards for the use of modified starches in foods, see Chapter 6 of Volume 1 in this series).

Cold plasma-modified corn starch was applied for use in mayonnaise by Laricheh et al. (2022), improving its physicochemical characteristics. The modified starch exhibited improved solubility, oil absorption, and gel stiffness, and the mayonnaise produced resulted in a more stable emulsion, with less syneresis and less sensitivity to oxidation, and appeared lighter than mayonnaise containing unmodified starch. Even after 1 month, the mayonnaise formulated with cold plasma-modified starch showed better odor, flavor, texture, and general acceptance than the control sample, conferring to this modification the possibility of use as a stabilizer of food emulsions.

Dielectric barrier discharge plasma modification applied to potato starch can be used to produce films with lower water vapor permeability and higher tensile strength (Guo et al., 2022).

Ultrasonic modification of yam starch was indicated for use in bakery products, extruded products, and as a thickening agent, since it increased water and oil absorption capacity and viscosity (Singh & Sharanagat, 2020).

Photochemical modification of cassava starch, using lactic acid (20 g/kg) and UV-C irradiation (75 min), showed good results in the production of cheese bread, with higher specific volume and lower viscosity, ensuring product quality (Santos et al., 2021). For more information about modification on cassava starch using lactic acid and UV-C, see Chapter 7.

In the food industry, some substances are sensitive to high temperatures, light, oxygen, and pH, such as anthocyanins, pigments that are widely used to confer color or for their antioxidant properties. Therefore, encapsulation with modified starches is an alternative to increase the stability of these compounds. Among the modifications, the chemical ones are preferred because they improve the interactions between the starch and the anthocyanin, improving its stability, and favoring the encapsulation. Thus, esterification (acetylation and succinylation), etherification (carboxymethylation and hydroxypropylation), oxidation, and size reduction (hydrolysis and nanoprecipitation) are the most used treatments. It is worth noting that in both pharmaceutical and food applications, the Food and Drug Administration allows an average degree of substitution of up to 0.2 for acetylated starches and 0.05 for succinylated ones (Schlindweinn et al., 2022).

Another compound that needs encapsulation is resveratrol, a photosensitive molecule with low solubility in water and common oils. Thus, Matos et al. (2018) prepared an oil-in-water emulsion stabilized by quinoa starch particles modified with octenyl succinic anhydride (OSA) to carry resveratrol. Pickering emulsions stabilized by quinoa starch showed promise for resveratrol transport in functional foods showing to be more stable versus creaming phenomena.

Esterified starches are used as encapsulating agents for their higher viscosity, freeze-thaw stability, lower gelatinization temperature, and reduced hydrogen bonds between amylose and amylopectin (Olagunju et al., 2020).

Added to this, the succinylated starches, with the introduction of OSA groups, induce the production of esters that contain hydrophilic and hydrophobic groups, giving the polysaccharide formed an amphiphilic power, being able to be used in the encapsulation of bioactive compounds, carbonated drinks, juices, energy drinks, and salad dressings (Bhosale & Singhal, 2007). The low gelatinization temperature and enthalpy, lower digestibility, higher swelling power, higher viscosity, and paste clarity found in these starches may contribute to the quality of these products (Altuna et al., 2018). For more information about the use of starch as an encapsulation agent, see Chapter 12.

Corn starch esterified with OSA groups was used as salt encapsulating agent in the internal water phase of a water-in-oil emulsion to increase the perception of salinity. The higher the degree of substitution, the lower the perception of salinity in the mouth, by decreasing the digestibility of starch by salivary amylase. Thus, 1.5%–2% substitution can already assist in reducing the salt content of emulsion-based foods (Chiu et al., 2017).

By reducing the molecular weight of carboxymethylated starches, it is possible to form a layer on the bacterial cell surface and favor cell penetration to destroy it, making this starch a potential antimicrobial agent for the food industry (Chakka & Zhou, 2020).

Hydroxypropylated starches show high water absorption and swelling power due to the larger volume of amorphous regions and lower tendency to retrograde, being widely used by the food industry, forming stable pastes and gels at room temperature (Ju et al., 2012).

Sago starch hydroxypropylated or with the combination of acid hydrolysis and hydroxypropylation was incorporated into hamburger buns improving moisture retention during storage. They are recommended to maintain quality and delay retrogradation of buns (Ghalambor et al., 2022a, 2022b).

Starches are widely used as thickeners in the food industry. Thus, corn and pearl millet starch were submitted to hydroxypropylation and succinylation to be used in white sauce. Even after 14 days of storage, the white sauces showed no syneresis, improving the stability to cold storage. The results were better for pearl millet, which was indicated for gluten-free sauce production use (Shaikh et al., 2020).

Cassava starch modified by acid hydrolysis showed good potential for application in yogurt, with less tendency to retrogradation, as pointed out by Awolu and Olofinlae (2016).

The materials used for food packaging are usually produced with nonbiodegradable petroleum-based polymers. With the appeal of sustainability, replacing these materials with biopolymers has been of interest. Thus, pine nut starch hydrolyzed with different concentrations of HCl was used to produce biofilms with glycerol and xanthan gum, showing higher water solubility and lower water vapor permeability, without changing the glass transition temperature, and with good resistance to high temperatures (Luchese et al., 2015). For more information about starch in films and biodegradable plastics, see Chapters 13 and 14.

The use of modified starches as fat replacement has been a very interesting strategy in healthier diets, because maintaining the quality and acceptability of products that had only the fat content removed is still a challenge. Starch-based fat replacers can mimic the physicochemical characteristics of fat, creating a mouthfeel similar to fat, making the product sensory pleasing, and reducing the calorie intake of the food. To this end, chemical and enzymatic modifications (maltodextrins) are preferred. With chemical modification, the starch granules act like fat globules in modulating the structure and sensory characteristics. Maltodextrins can form thermoreversible gels due to their strong water-holding capacity, with a size (1−3 μm) similar to fat globules, resembling their texture (Chen et al., 2019).

Cross-linked starch (introduction of chemical blocking groups, mono- or difunctional) is also used in the food industry, as it exhibits good water retention, gel formation, and viscous flow and reduces the caloric value of food due to lower digestibility (Peng & Yao, 2017).

It is worth noting maltodextrins with low dextrose equivalent (DE < 20), which can improve viscosity, lubricity, and softness in foods. Commercially, there are already options such as Paselli SA2 from potato starch, Maltrin M040 from corn starch, N-oil from tapioca starch, and Oatrim from oat starch. It is important to be aware of the limitations that can occur with low freeze-thaw and acid and heat stability (Chavan et al., 2016). For more information about maltodextrins, it is suggested to consult Chapter 5.

Replacing 20% fat in mayonnaise with cross-linked rice starch gels showed superior sensory quality compared to those containing only fat, in addition to enabling greater freezing stability (Puligundla et al., 2015). Sweet potato starch modified with citric acid also showed good sensory acceptance without affecting the microbiological stability of ice cream, although the freezing ability could still be optimized (Chen et al., 2019).

Pregelatinized esterified starches can also be used in bread dough as a fat substitute but also as a 10% flour substitute (Hadnadev et al., 2014), conferring higher specific volume to the breads, although the water holding capacity was decreased. Hadnadev et al. (2015) also produced cookies with 50% vegetable fat substitution with emulsions using esterified starch, with acceptance similar to whole grain cookies.

Other uses can be cited as in confectionery fillings with maltodextrins (Hadnadev et al., 2014), cheeses (Basiony & Hassabo, 2021), yogurts with cross-linked starch (Lobato-Calleros et al., 2014), and sausages with extruded and phosphorylated starch (Limberger et al., 2011).

Phosphate cross-linked starch by extrusion can be used as encapsulating agents for *Bifidobacterium breve* bacteria at a concentration of 10% (w/w), increasing probiotic survival when subjected to stressful conditions such as long storage time (Murúa-Pagola et al., 2020).

Enzymatically modified starches also have advantages for the food industry, such as pullulanase-modified cassava starch that with higher viscosity and lower syneresis had optimum acceptability in fermented milk drinks (Lopez-Ochoa et al., 2022).

6.7 Non-food uses of modified starch

Although the food industry is the main market for starch, other industries also benefit from the properties of this polysaccharide, especially after modification. The paper industry ranks second, using starch as a sizing agent (incorporation of protective filler materials) and binder. The modifications help to impart a smooth finish to the paper, improving water resistance (with hydrophobic groups facing outward). Among the most widely used starches are cationic starches, with the introduction of ammonium, iminium, phosphonium, and sulfonium groups (Adewale et al., 2022).

For the textile industry, starch is important to reduce breakage and fraying in looms. It is applied as an adhesive coating on the cold surface, increasing abrasion resistance. Modifications such as cationization, double etherification, and grafting can be used (Shen et al., 2021).

Starch can be used in solid pharmaceutical formulations in the form of powders, granules, capsules, and tablets acting, depending on its concentration, as a diluent, binder, disintegrant, antiadherent, or lubricant. However, to correct some limitations of native starch, such as low disintegration time, difficulty in compressibility, and poor agglutination, starch modifications have been encouraged, improving the drug delivery system. Among them, chemical modification (etherified, esterified, oxidized, cross-linked, or hydrolyzed) is performed with the substitution of hydroxyl at C6, C2, and/or C3 position, thus changing the structure of amylose and amylopectin, conferring higher added value, biodegradability, and better physicochemical properties. Other applications could be related to unconventional and nanostructured drug delivery systems. In the pharmaceutical industry, the main modification is the pregelatinization, since it improves the fluidity, disintegration, and hardness of starches, aimed at the formulation of tablets, requiring a considerably smaller amount than conventional starch. An example of this is the pregelatinized corn starch that is applied in capsules (diluent) or tablets (disintegrator and binder) (Garcia et al., 2020). For more information about starch for pharmaceutical use, see Chapter 16.

Enzymatic modification to produce cyclodextrins is also of broad importance in the pharmaceutical industry, due to the ability to form inclusion complexes with hydrophobic substances. This is essential for molecular encapsulation techniques. Also in the pharmaceutical industry, sodium starch glycolate is used as an excipient in the superdisintegrating function for its flow properties. Hydroxypropylated starch increases shelf life, freeze-thaw stability, cold storage, clarity, and paste texture, making it important for controlled drug release. Acetylates can be used in tablets for their stability, resistance to retrogradation, good water absorption capacity, swelling power and solubility. As an encapsulating agent, octenyl succinate anhydride starch is highly desirable. Oxidi-zed starches are used in the sustained and controlled release of drugs (Garcia et al., 2020).

With their amphiphilic property, as well as gelatinization at low temperatures and higher swelling power, succinylated starches can be used in paints, adhesives, and biodegradable films (Bhosale & Singhal, 2007).

Etherification-derived carboxymethylated starch exhibits higher water absorption and, due to lower retrogradation, has been used in the textile, paper, plastics, pharmaceutical, and biomedical industries and, furthermore, has been recognized as a safe and inexpensive chemical fertilizer (Kan et al., 2017).

Acid hydrolyzed starches are widely used in the paper and textile industry (Wang et al., 2017), or even in the production of glues and gums (Doué et al., 2014).

Wheat starch subjected to dry heat treatment can be used to produce hydrogels applied as inks in three-dimensional printers. After this modification, the granules increased in size without changing their morphology and had slight depolymerization and decreased RC. The inks exhibited lower viscosity, higher resting structure resistance, as well as resistance to external stresses, higher firmness, and lower syneresis. Thus, better printability and reproducibility can be achieved (Maniglia et al., 2020).

To increase the use of renewable energy, energy storage devices have been explored. Among them, biodegradable polymer electrolyte systems are economical, environmentally friendly, and energy-efficient. Thus, ion-liquid plasticized starches are suitable for use in solid biopolymer electrolytes and ionically conductive polymers can be used in fuel cells, lithium batteries, and dye-synthesized solar cell (converts solar energy into electrical energy) (Adewale et al., 2022).

Nanotechnology is also an area where researchers are investigating the use of nano-scale starch, such as nanocrystals (crystalline platelets derived from the breakdown of the semicrystalline structure) and nanoparticles (amorphous), for filling biocomposites, mainly conferring strength. There are already two commercial nanoparticles that have been produced from starch called Mater-Bi (Novamont) and Eco-Sphere (Ecosynthetix). The first comes from the complexation of unstructured and modified corn starch with vegetable oil, and its application is geared toward agricultural use, transport bags, and even food packaging. Eco-Sphere, on the other hand, is a substitute for bio latex, used in adhesives and paper coating (Adewale et al., 2022). For more information about composites, see Chapters 13 and 14.

Some very promising uses of modified starch are related to regenerative medicine, aimed at developing materials that structurally and functionally replace tissue and improving its healing. For this, the starch should have high purity, with no residual proteins or fibers. A commercial example to be mentioned is Caxomer iodine, which consists of three-dimensional hydrophilic starch cross-linked with 0.9% (w/w) EPI as iodine physically immobilized in the microsphere matrix. When placed under the wound, it is able to absorb exudate and surface particulate material. As the dressing is moistened, the iodine present is released for bactericidal action against *Staphylococcus aureus* (Woo et al., 2021).

From these examples, it is understood that the trend in the use of modified starches is growing and always finds new market opportunities, increasing the generation of revenue for food industries. This is because the modification allows for the improvement of quality and yield, making industrial operations more viable and economically profitable.

6.8 Trends in the modified starch market

Trends in the modified starch market include studying new starch sources, beyond the conventional starch, with an emphasis on improved properties, industrial scale-up, and applications in different sectors. For example, application of starch in food packaging is an increasing trend, exploring the production of biodegradable films, edible films, nanocrystals, encapsulation of compounds (e.g., ethylene gas), as well as active and intelligent packaging.

Starch digestibility is always an issue when considering the prevalence of obesity and some nontransmissible chronic diseases, especially type 2 diabetes. Therefore, changing the digestibility profile of commercial starches is an ongoing challenge. Several studies were made, and many others are being carried out looking for alternatives that reduce the rapid digestive starch fraction and increase both slowly digestible starch and RS. New starch modifications and strategies for slowing down the activity of amylases are in focus. Among several strategies, the reaction of starch with phenolic compounds is one of them and can be effective even after starch gelatinization (Xu et al., 2021). For more information about interaction between starch and phenolic compounds, see Chapter 8 in Volume 1 of this series and Chapter 10 in this volume.

Starch market is continuously growing due to new demands from the consumers. Modified starch market value is 2.7 times greater than that of the native one. The next generation of uses will benefit from emerging technologies like 3D printing and nanotechnology, and these will enable better performance in packaging, energy storage materials, and regenerative medicine fields. Technological advances are always underway and the exciting 4D technology is one of such advances. In this case, modification of starch can make the polymer responsive to specific triggers (temperature, pH, humidity, mechanical stress, or sensitivity to certain chemical compounds). The printed objects could suffer shape changes or modification on permeability, allowing for controlled release of entrapped substances (bioactive compounds, drugs, agrochemicals, etc.) (Adewale et al., 2022).

Some factors also are expected to contribute to driving the starch market in some way. The increasing global demand for vegetable protein from pulses (pea, lentil, beans, and chickpea) is increasing the availability of pulse starches, which presently do not find applications that could displace corn starch. The complete utilization of raw materials is a very present concern from the global society, and the industry will have to search for the solution of many gaps in its processes. The circular economy concept is very important and up-to-date, essential for the environmental, social, and governance factors achievement and will have great influence in the starch processing chain. Circular economy will be covered in the upcoming Volume 5 in this series.

6.9 Final considerations

Starch chemistry and technology have been exploited for more than a century, and the evolution since then is notable. There is a vast range of industrial fields that benefit from polymers in their processes, and polysaccharides, including starch, are gaining more importance due to biodegradability, renewability, and sustainability. Technologi-cal properties of starch and its derivatives deliver viscosity, lubricating action, moisture retention, texture, and crispness, among other relevant effects in processes and products. The relative low cost and availability makes starch a very important option to be adopted when compatible polymeric properties are necessary.

In general, as starch is the most important energy reserve for plants and the main energy source in human nutrition, it is expected that the food uses are very relevant on the worldwide starch market. The use of isolated starch from different sources (mostly corn, but also cassava, wheat, potato, and rice) is well established and can be considered an economic and social development indicator of countries and regions. The developed countries from North America, Western Europe, and Asia dominate both the starch technology and market. A vast number of chemical, physical, and biological strategies have been studied, and commercial methods were developed for modifying critical starch properties aiming to improve its performance in different processes and products.

Chemical modification was rapidly developed as soon as new industrial processes demanded improved properties. In non-food uses, that approach is still very useful and makes it possible to have relatively cheap tailor-made solutions. In food uses, on the other hand, alternative technologies for modifying starch properties are underway because of undesired chemical exposition pointed out by the market, with potential health risks, as well as environmental issues. The desired clean label of processed foods is pushing the starch processors to search for new possibilities.

References

Adekunte, A. O., Tiwari, B. K., Culle, P. J., Scannell, A. G. M., & Donell, C. P. O. (2010). Effect of sonication on colour, ascorbic acid and yeast inactivation in tomato juice. *Food Chemistry*, *122*(3), 500–507. Available from https://doi.org/10.1016/j.foodchem.2010.01.026.

Adewale, P., Yancheshmeh, M. S., & Lam, E. (2022). Starch modification for non-food, industrial applications:

Market intelligence and critical review. *Carbohydrate Polymers, 291*, 119590. Available from https://doi.org/10.1016/J.CARBPOL.2022.119590.

Ai, Y., Nelson, B., Birt, D. F., & Jane, J. L. (2013). In vitro and in vivo digestion of octenyl succinic starch. *Carbohydrate Polymers, 98*(2), 1266–1271. Available from https://doi.org/10.1016/J.CARBPOL.2013.07.057.

Akinterinwa, A., Oladele, E., Adebayo, A., Gurgur, E., Iyanu, O. O., & Ajayi, O. (2020). Cross-linked-substituted (esterified/etherified) starch derivatives as aqueous heavy metal ion adsorbent: A review. *Water Science and Technology, 82*(1), 1–26. Available from https://doi.org/10.2166/wst.2020.332.

Akinterinwa, A., Reuben, U., Atiku, J. U., & Adamu, M. (2022). Focus on the removal of lead and cadmium ions from aqueous solutions using starch derivatives: A review. *Carbohydrate Polymers, 290*, 119463. Available from https://doi.org/10.1016/J.CARBPOL.2022.119463.

Alcázar-Alay, S. C., & Meireles, M. A. A. (2015). Physicochemical properties, modifications and applications of starches from different botanical sources. *Food Science and Technology, 35*, 215–236.

Altuna, L., Herrera, M. L., & Foresti, M. L. (2018). Synthesis and characterization of octenyl succinic anhydride modified starches for food applications. A review of recent literature. *Food Hydrocolloids, 80*, 97–110. Available from https://doi.org/10.1016/j.foodhyd.2018.01.032.

Ambigaipalan, P., Hoover, R., Donner, E., & Liu, Q. (2014). Starch chain interactions within the amorphous and crystalline domains of pulse starches during heat moisture treatment at different temperatures and their impact on physicochemical properties. *Food Chemistry, 528*(143), 175–184.

Apriyanto, A., Compart, J., & Fettke, J. (2022). A review of starch, a unique biopolymer – Structure, metabolism and in planta modifications. *Plant Science, 318*, 111223. Available from https://doi.org/10.1016/j.plantsci.2022.111223.

Ashokkumar, M. (2015). Applications of ultrasound in food and bioprocessing. *Ultrasonics Sonochemistry, 25*, 17–23. Available from https://doi.org/10.1016/j.ultsonch.2014.08.012.

Atrous, H., Benbettaieb, N., Hosni, F., Danthine, S., Blecker, C., Attia, H., & Ghorbel, D. (2015). Effect of γ-radiation on free radicals formation, structural changes and functional properties of wheat starch. *International Journal of Biological Macromolecules, 80*, 64–76. Available from https://doi.org/10.1016/j.ijbiomac.2015.06.014.

Awolu, O. O., & Olofinlae, S. J. (2016). Physico-chemical, functional and pasting properties of native and chemically modified water yam (*Dioscorea alata*) starch and production of water yam starch-based yoghurt. *Starch, 68*(7–8), 719–726. Available from https://doi.org/10.1002/star.201500302.

Bakouri, H., & Guemra, K. (2019). Etherification and cross-linking effect on physicochemical properties of *Zea mays* starch executed at different sequences in 1-butyl-3-methylimidazolium chloride [BMIM]Cl ionic liquid media. *International Journal of Biological Macromolecules, 125*, 1118–1127. Available from https://doi.org/10.1016/J.IJBIOMAC.2018.12.140.

Balasubramaniam, V. M., Martínez-Monteagudo, S. I., & Gupta, R. (2015). Principles and application of high pressure-based technologies in the food industry. *Annual Review of Food Science and Technology, 6*(1), 435–462. Available from https://www.doi.org/10.1146/annurev-food-022814-015539.

Basiony, M., & Hassabo, R. (2021). Composition and quality of low-fat halloumi cheese made using modified starch as a fat replacer. *Starch, 74*(3–4), 2100211.

Bauer, B. A., & Knorr, D. (2005). The impact of pressure, temperature and treatment time on starches: Pressure-induced starch gelatinisation as pressure time temperature indicator for high hydrostatic pressure processing. *Journal of Food Engineering, 68*(3), 329–344. Available from https://doi.org/10.1016/j.jfoodeng.2004.06.007.

BeMiller, J. N., & Huber, K. C. (2015). Physical modification of food starch functionalities. *Annual Review of Food Science and Technology, 6*(1), 19–69. Available from https://doi.org/10.1146/annurev-food-022814-015552.

Bet, C. D., Oliveira, C. S. D., Colman, T. A. D., Marinho, M. T., Lacerda, L. G., Pumacahua-Ramos, A., & Schnitzler, E. (2018). Organic amaranth starch: A study of its technological properties after heat-moisture treatment. *Food Chemistry, 264*, 435–442. Available from https://doi.org/10.1016/j.foodchem.2018.05.021.

Bhat, R., & Karim, A. A. (2009). Impact of radiation processing on starch. *Comprehensive Reviews in Food Science and Food Safety, 8*(2), 44–58.

Bhosale, R., & Singhal, R. (2007). Effect of octenylsuccinylation on physicochemical and functional properties of waxy maize and amaranth starches. *Carbohydrate Polymers, 68*, 447–456. Available from https://doi-org.ez82.periodicos.capes.gov.br/10.1016/j.carbpol.2006.11.011.

Bismark, S., Zhifeng, Z., & Benjamin, T. (2018). Effects of differential degree of chemical modification on the properties of modified starches: Sizing. *The Journal of Adhesion, 94*(2), 97–123. Available from https://doi.org/10.1080/00218464.2016.1250629.

Boetje, L., Lan, X., Silvianti, F., van Dijken, J., Polhuis, M., & Loos, K. (2022). A more efficient synthesis and properties of saturated and unsaturated starch esters.

Carbohydrate Polymers, 292, 119649. Available from https://doi.org/10.1016/j.carbpol.2022.119649.

Brahma, B., & Sit, N. (2020). Physicochemical properties and digestibility of heat moisture–treated potato starches for different treatment conditions. *Potato Research, 63,* 367–383.

Cárcel, J. A., García-Pérez, J. V., Benedito, J., & Mulet, A. (2012). Food process innovation through new technologies: Use of ultrasound. *Journal of Food Engineering, 110* (2), 200–207. Available from https://doi.org/10.1016/j.jfoodeng.2011.05.038.

Carvalho, A. P. M. G., Barros, D. R., Silva, L. S. D., Sanches, E. A., Pinto, C. D. C., Souza, S. M. D., Clerici, M. T. P. S., Rodrigues, S., Fernandes, F. A. N., & Campelo, P. H. (2021). Dielectric barrier atmospheric cold plasma applied to the modification of Ariá (*Goeppertia allouia*) starch: Effect of plasma generation voltage. *International Journal of Biological Macromolecules, 182,* 1618–1627. Available from https://doi.org/10.1016/j.ijbiomac.2021.05.165.

Castro, L. M. G., Alexandre, E. M. C., Saraiva, J. A., & Pintado, M. (2021). Starch Extraction and modification by pulsed electric fields. *Food Reviews International, 22,* 8755–9129. Available from https://doi.org/10.1080/87559129.2021.1945620.

Chakka, V. P., & Zhou, T. (2020). Carboxymethylation of polysaccharides: Synthesis and bioactivities. *International Journal of Biological Macromolecules, 165,* 2425–2431. Available from https://doi-org.ez82.periodicos.capes.gov.br/10.1016/j.ijbiomac.2020.10.178.

Chakraborty, I., Pooja, N., Mal, S. S., Paul, U. C., Rahman., & Mazumder, N. (2022). An insight into the gelatinization properties influencing the modified starches used in food industry: A review. *Food and Bioprocess Technology, 15*(6), 1195–1223. Available from https://doi.org/10.1007/s11947-022-02761-z.

Chavan, R. S., Khedkar, C. D., & Bhatt, S. (2016). Fat replacer. In B. Caballero, P. Finglas, & F. Toldrá (Eds.), *The encyclopedia of food and health* (2, pp. 589–595). Oxford: Academic Press, https://doi.org/10.1002/2Ffsn3.1303.

Chemat, F., Huma, Z. H., & Khan, M. K. (2011). Applications of ultrasound in food technology: Processing, preservation and extraction. *Ultrasonics Sonochemistry, 18,* 813–835. Available from https://doi.org/10.1016/j.ultsonch.2010.11.023.

Chemat, F., Rombaut, N., Sicaire, A. G., Meullemiestre, A., Fabiano-Tixier, A. S., & Albert-Vian, M. (2017). Ultrasound assisted extraction of food and natural products. Mechanisms, techniques, combinations, protocols and applications. A review. *Ultrasonics Sonochemistry, 34,* 540–560. Available from https://doi.org/10.1016/j.ultsonch.2016.06.035.

Chen, Y., She, Y., Zhang, R., Wang, J., Zhang, X., & Gou, X. (2019). Use of starch-based fat replacers in foods as a strategy to reduce dietary intake of fat and risk of metabolic diseases. *Food Science & Nutrition, 8*(1), 16–22. Available from https://doi.org/10.1002/fsn3.1303.

Chi, C., Li, X., Lu, P., Miao, S., Zhang, Y., & Chen, L. (2019). Dry heating and annealing treatment synergistically modulate starch structure and digestibility. *International Journal of Biological Macromolecules, 137,* 554–561. Available from https://doi.org/10.1016/j.ijbiomac.2019.06.137.

Chittapun, S., Jangyubol, K., Charoenrat, T., Piyapittayanun, C., & Kasemwong, K. (2020). Cationic cassava starch and its composite as flocculants for microalgal biomass separation. *International Journal of Biological Macromolecules, 161,* 917–926. Available from https://doi.org/10.1016/j.ijbiomac.2020.06.116.

Chiu, N., Tarrega, A., Parmenter, C., Hewson, L., Wolf, B., & Fisk, I. D. (2017). Optimisation of octinyl succinic anhydride starch stabilised w1/o/w2 emulsions for oral destabilisation of encapsulated salt and enhanced saltiness. *Food Hydrocolloids, 69,* 450–458. Available from https://doi-org.ez82.periodicos.capes.gov.br/10.1016/j.foodhyd.2017.03.002.

Christensen, S. J., Madsen, M. S., Zinck, S. S., Hedberg, C., Sørensen, O. B., Svensson, B., & Meyer, A. S. (2023). Enzymatic potato starch modification and structure-function analysis of six diverse GH77 4-alpha-glucanotransferases. *International Journal of Biological Macromolecules, 224,* 105–114. Available from https://doi.org/10.1016/J.IJBIOMAC.2022.10.107.

Chung, H. J., & Liu, Q. (2009). Effect of gamma irradiation on molecular structure and physicochemical properties of corn starch. *Journal of Food Science, 74*(5), C353–C361.

Chung, H. J., & Liu, Q. (2010). Molecular structure and physicochemical properties of potato and bean starches as affected by gamma-irradiation. *International Journal of Biological Macromolecules, 47*(2), 214–222.

Conde, L. A., Kebede, B., Leong, S. Y., & Oey, I. (2022). Changes in starch in vitro digestibility and properties of cassava flour due to pulsed electric field processing. *Foods, 11,* 3714. Available from https://doi.org/10.3390/foods11223714.

Dai, L., Zhang, J., & Cheng, F. (2019). Succeeded starch nanocrystals preparation combining heat-moisture treatment with acid hydrolysis. *Food Chemistry, 278,* 350–356. Available from https://doi.org/10.1016/J.FOODCHEM.2018.11.018.

Daniel, T. H. G., Bedin, A. C., Souza, É. C. F., Lacerda, L. G., & Demiate, I. M. (2020). Pickering emulsions produced with starch nanocrystals from cassava (*Manihot esculenta* Crantz), beans (*Phaseolus vulgaris* L.), and corn (*Zea mays* L.). *Starch - Stärke, 72*(11–12), 1900326. Available from https://doi.org/10.1002/star.201900326.

Demiate, I. M., Dupuy, N., Huvenne, J. P., Cereda, M. P., & Wosiacki, G. (2000). Relationship between baking behavior of modified cassava starches and starch chemical structure determined by FTIR spectroscopy. *Carbohydrate Polymers, 42*(2), 149–158. Available from https://doi.org/10.1016/S0144-8617(99)00152-6.

Dhiman, A., & Prabhakar, P. K. (2020). Micronization in food processing: A comprehensive review of mechanistic approach, physicochemical, functional properties and self-stability of micronized food materials. *Journal of Food Engineering*, 110248. Available from https://doi.org/10.1016/j.jfoodeng.2020.11024.

Dominguez-Ayala, J. E., Soler, A., Mendez-Montealvo, G., & Velazquez, G. (2022). Supramolecular structure and technofunctional properties of starch modified by high hydrostatic pressure (HHP): A review. *Carbohydrate Polymers, 291*, 119609. Available from https://doi.org/10.1016/j.carbpol.2022.119609.

Doué, G., Bédikou, M., Koua, G., Mégananou, R.-M., & Niamké, S. (2014). Enzymatic and acid conversion of new starches from improved orphan crops: Prospects for renewable materials uses in food and non-food industries. *SpringerPlus, 3*, 498. Available from https://doi-org.ez82.periodicos.capes.gov.br/10.1186/2193-1801-3-498.

Duque, S. M. M., Leong, S. Y., Agyei, D., Singh, J., Larsen, N., Sutton, K., & Oey, I. (2022). Understanding the mechanism of how pulsed electric fields treatment affects the digestibility and characteristics of starch in oat flour. *Applied Science, 12*, 10293. Available from https://doi.org/10.3390/app122010293.

Emami, S., Meda, V., Pickard, M. D., & Tyler, R. T. (2010). Impact of micronization on rapidly digestible, slowly digestible, and resistant starch concentrations in normal, high-amylose, and waxy barley. *Journal of Agricultural and Food Chemistry, 58*(17), 9793–9799. Available from https://doi.org/10.1021/jf101702e.

Fan, L., Ye, Q., Lu, W., Chen, D., Zhang, C., Xiao, L., Meng, X., Lee, Y.-C., Wang, H.-M. D., & Xiao, C. (2021). The properties and preparation of functional starch: A review. *Food Reviews International*, 1–25. Available from https://doi.org/10.1080/87559129.2021.2015375.

Fashi, A., Fallah Delavar, A., Zamani, A., & Noshiranzadeh, N. (2023). Solid state malic acid esterification on fungal α-amylase treated corn starch: Design of a green dual treatment. *Food Chemistry, 410*, 135439. Available from https://doi.org/10.1016/j.foodchem.2023.135439.

FDA. (2023). Food and Agricultural Organization, Available from http://www.fda.gov//radiation-emitting-products (Acessed on 23 January 2023).

Fonseca, L. M., Halal, S. L. M. E., Dias, A. R. G., & Zavareze, E. D. R. (2021). Physical modification of starch by heat-moisture treatment and annealing and their applications: A review. *Carbohydrate Polymers, 274*, 118665.

Garcia, M. A. V., Garcia, C. F., & Faraco, A. A. G. (2020). Pharmaceutical and biomedical applications of native and modified starch: A review. *Starch, 72*(7–8), 1900270.

Ghalambor, P., Asadi, G., Nafchi, A. M., & Ardebili, S. M. S. (2022a). Investigation of dual modification on physicochemical, morphological, thermal, pasting, and retrogradation characteristics of sago starch. *Food Science & Nutrition, 10*(7), 2285–2299. Available from https://doi-org.ez82.periodicos.capes.gov.br/10.1002/fsn3.2837.

Ghalambor, P., Asadi, G., Nafchi, A. M., & Ardebili, S. M. S. (2022b). The effects of dually modified (hydrolyzed-hydroxypropylated) sago starch on the quality characteristics of frozen dough and bread. *Starch, 75* (3–4), 2200183. Available from https://doi-org.ez82.periodicos.capes.gov.br/10.1002/star.202200183.

González, L. C., Loubes, M. A., & Tolaba, M. P. (2018). Incidence of milling energy on dry-milling attributes of rice starch modified by planetary ball milling. *Food Hydrocolloids, 82*, 155–163. Available from https://doi.org/10.1016/j.foodhyd.2018.03.051, S0268005X17321641.

Grgić, I., Ačkar, Đ., Barišić, V., Vlainić, M., Knežević, N., & Medverec Knežević, Z. (2019). Nonthermal methods for starch modification-A review. *Journal of Food Processing and Preservation, 43*(12), e14242. Available from https://doi.org/10.1111/jfpp.14242.

Gui, Y., Wei, X., Yang, N., Guo, L., Cui, B., Zou, F., Lu, L., Liu, P., & Fang, Y. (2022). Comparison of structural and functional properties of maize starch produced with commercial or endogenous enzymes. *International Journal of Biological Macromolecules, 209*, 2213–2225. Available from https://doi.org/10.1016/J.IJBIOMAC.2022.04.202.

Guida, C., Aguiar, A. C., & Cunha, R. L. (2021). Green techniques for starch modification to stabilize Pickering emulsions: A current review and future perspectives. *Current Opinion in Food Science, 38*, 52–61. Available from https://doi.org/10.1016/j.cofs.2020.10.017.

Gunaratne, A., & Hoover, R. (2002). Effect of heat-moisture treatment on the structure and physicochemical properties of tuber and root starches. *Carbohydrate Polymers, 49* (4), 425–437. Available from https://doi.org/10.1016/S0144-8617(01)00354-X.

Guo, Z., Gou, Q., Yang, L., Yu, Q.-L., & Han, L. (2022). Dielectric barrier discharge plasma: A green method to change structure of potato starch and improve physicochemical properties of potato starch films. *Food Chemistry, 370*, 130992. Available from https://doi.org/10.1016/j.foodchem.2021.130992.

Haaj, S. B., Magninb, A., Petrier, C., & Boufia, S. (2013). Starch nanoparticles formation via high power ultrasonication. *Carbohydrate Polymers*, *92*, 1625–1632. Available from https://doi.org/10.1016/j.carbpol.2012.11.022.

Hadnadev, M., Hadnadev, T., Dokic, L., Pajin, B., Torbica, A., Saric, L., & Ikonic, P. (2014). Physical and sensory aspects of maltodextrin gel addition used as fat replacers in confectionery filling systems. *Food Science and Technology*, *59*, 495–503. Available from https://doi-org.ez82.periodicos.capes.gov.br/10.1016/j.lwt.2014.04.044.

Hadnadev, T. D., Hadnadev, M., Pokić, M., Rakita, S. & Krstonosić. (2015). Functionality of OSA starch stabilized emulsions as fat replacers in cookies. *Journal of Food Engineering*, 167, 133-138. https://doi.org/10.1016/j.jfoodeng.2015.02.002.

Han, Z., Zeng, X. A., Yu, S. J., Zhang, B. S., & Chen, X. D. (2009). Effects of pulsed electric fields (PEF) treatment on physicochemical properties of potato starch. *Innovative Food Science & Emerging Technologies*, *10*(4), 481–485. Available from https://doi.org/10.1016/j.ifset.2009.07.003.

Haq, F., Yu, H., Wang, L., Teng, L., Haroon, M., Khan, R. U., Mehmood, S., Bilal-Ul-Amin., Ullah, R. S., Khan, A., & Nazir, A. (2019). Advances in chemical modifications of starches and their applications. *Carbohydrate Research*, *476*, 12–35. Available from https://doi.org/10.1016/J.CARRES.2019.02.007.

He, X.-H., Luo, S.-J., Chen, M.-S., Xia, W., Chen, J., & Liu, C.-M. (2020). Effect of industry-scale microfluidization on structsural and physicochemical properties of potato starch. *Innovative Food Science & Emerging Technologies*, *60*, 102278. Available from https://doi.org/10.1016/j.ifset.2019.102278.

Hu, A., Lu, J., Zheng, J., Sun, J., Yang, L., Zhang, X., Zhang, Y., & Lin, Q. (2013). Ultrasonically aided enzymatical effects on the properties and structure of mung bean starch. *Innovative Food Science and Emerging Technologies*, *20*, 146–151. Available from https://doi.org/10.1016/j.ifset.2013.08.005.

Hu, A., Wang, X., Li, L., Xu, T., & Zheng, J. (2020). Effects of annealing time on structure and properties of sweet potato starch. *Cereal Chemistry*, *97*(3), 573–580.

Ismailoglu, S. O., & Basman, A. (2015). Effects of infrared heat-moisture treatment on physicochemical properties of corn starch. *Starch - Stärke*, *67*(5–6), 528–539. Available from https://doi.org/10.1002/star.201400266.

Ismailoglu, S. O., & Basman, A. (2015). Physicochemical properties of infrared heat-moisture treated wheat starch. *Starch - Stärke*, *68*(1–2), 67–75. Available from https://doi.org/10.1002/star.201500160.

Ito, V. C., Bet, C. D., Wojeicchowski, J. P., Demiate, I. M., Spoto, M. H. F., Schnitzler, E., & Lacerda, L. G. (2018). Effects of gamma radiation on the thermoanalytical, structural and pasting properties of black rice (*Oryza sativa* L.) flour. *Journal of Thermal Analysis and Calorimetry*, *133*(1), 529, 7.

Ito, V. C., Zielinski, A. A. F., Demiate, I. M., Spoto, M. H. F., Nogueira, A., & Lacerda, L. G. (2019). Effects of gamma radiation on the stability and degradation kinetics of phenolic compounds and antioxidant activity during storage of (*Oryza sativa* L.) black rice flour. *Brazilian Archives of Biology and Technology*, *62*, e19180470. Available from https://doi.org/10.1590/1678-4324-2019180470.

Iuga, M., & Mironeasa, S. (2020). A review of the hydrothermal treatments impact on starch based systems properties. *Critical Reviews in Food Science and Nutrition*, *60*, 3890–3915.

Jambrak, A. R., Herceg, Z., Subaric, D., Brncic, M., Brncico, S. R., Bosiljkov, T., Cvek, D., Tripalo, B., & Gelo, J. (2010). Ultrasound effect on physical properties of corn starch. *Carbohydrate Polymers*, *79*(1), 91–100. Available from https://doi.org/10.1016/j.carbpol.2009.07.051.

Jiang, M., Hong, Y., Gu, Z., Cheng, L., Li, Z., & Li, C. (2018). Effects of acid hydrolysis intensity on the properties of starch/xanthan mixtures. *International Journal of Biological Macromolecules*, *106*, 320–329. Available from https://doi.org/10.1016/j.ijbiomac.2017.08.028.

Ju, B., Yan, D., & Zhang, S. (2012). Micelles self-assembled from thermoresponsive 2-hydroxy-3-butoxypropyl starches for drug delivery. *Carbohydrate Polymers*, *87*, 1404–1409. Available from https://doi-org.ez82.periodicos.capes.gov.br/10.1016/j.carbpol.2011.09.028.

Kadam, S. U., Tiwari, B. K., Álavrez, C., & O'Donnel, C. P. (2015). Ultrasound applications for the extraction, identification and delivery of food proteins and bioactive peptides. *Trends in Food Science and Technology*, *46*, 60–67. Available from https://doi.org/10.1016/j.tifs.2015.07.012.

Kan, L., Zhao, Q., Hu, J., Wu, Y., & Ouyang, J. (2017). Synthesis and physicochemical properties of carboxymethyl chestnut starch. *Journal of Food Processing and Preservation*, *41*, e13229. Available from https://doi-org.ez82.periodicos.capes.gov.br/10.1111/jfpp.13229.

Karma, V., Gupta, A. D., Yadav, D. K., Singh, A. A., Verma, M., & Singh, H. (2022). Recent developments in starch modification by organic acids: A review. *Starch - Stärke*, *74*(9–10), 2200025. Available from https://doi.org/10.1002/star.202200025.

Kaur, B., Ariffin, F., Bhat, R., & Karim, A. A. (2012). Progress in starch modification in the last decade. *Food Hydrocolloids*, *26*(2), 398–404. Available from https://doi.org/10.1016/J.FOODHYD.2011.02.016.

Khatun, A., Waters, D. L. E., & Liu, L. (2019). A review of rice starch digestibility: Effect of composition and heat-moisture processing. *Starch/Staerke*, *71*, 1–14.

Kim, H.-Y., Ye, S.-J., & Baik, M.-Y. (2023). Physicochemical properties of pressure moisture treated (PMT) and heat moisture treated (HMT) starches. *Innovative Food Science*

& Emerging Technologies, 87, 103392. Available from https://doi.org/10.1016/j.ifset.2023.103392.

Koksel, H., Ozturk, S., Kahraman, K., Basman, A., Ozbas, O. O., & Ryu, G. H. (2008). Evaluation of molecular weight distribution, pasting and functional properties, and enzyme resistant starch content of acid-modified corn starches. *Food Science and Biotechnology*, *17*(4), 755–760.

Kotnik, T., Rems, L., Tarek, M., & Miklavčič, D. (2019). Membrane electroporation and electropermeabilization: Mechanisms and models. *Annual Review of Biophysics*, *48*, 63–91. Available from https://doi.org/10.1146/annurev-biophys-052118-115451.

Kumar, P., Prakash, K. S., Jan, K., Swer, T. L., Jan, S., Verma, R., Deepika, K., Dar, M. Z., Verma, K., & Bashir, K. (2017). Effects of gamma irradiation on starch granule structure and physicochemical properties of brown rice starch. *Journal of Cereal Science*, *77*, 194–200. Available from https://doi.org/10.1016/j.jcs.2017.08.017.

Laricheh, R., Fazel, M., & Goli, M. (2022). Corn starch structurally modified with atmospheric cold-plasma and its use in mayonnaise formulation. *Food Measure*, *16*, 1859–1872. Available from https://doi-org.ez82.periodicos.capes.gov.br/10.1007/s11694-022-01296-3.

Latip, D. N. H., Samsudin, H., Utra, U., & Alias, A. K. (2021). Modification methods toward the production of porous starch: A review. *Critical Reviews in Food Science and Nutrition*, *61*(17), 2841–2862. Available from https://doi.org/10.1080/10408398.2020.1789064.

Li, W., Cao, F., Fan, J., Ouyang, S., Luo, Q., Zheng, J., & Zhang, G. (2014). Physically modified common buckwheat starch and their physicochemical and structural properties. *Food Hydrocolloids*, *40*, 237–244. Available from https://doi.org/10.1016/j.foodhyd.2014.03.012.

Lii, C.-Y., Liao, C.-D., Stobinski, L., & Tomasik, P. (2002). Behaviour of granular starches in low-pressure glow plasma. *Carbohydrate Polymers*, *49*(4), 499–507.

Limberger, V. M., Brum, F. B., Patias, L. D., Daniel, A. P., Comarela, C. G., Emanuelli, T., & da Silva, L. P. (2011). Modified broken rice starch as fat substitute in sausages. *Ciência e Tecnologia. de Alimentos, Campinas*, *31*(3), 789–792.

Liu, H., Lv, M., Wang, L., Li, Y., Fan, H., & Wang, M. (2016). Comparative study: How annealing and heat-moisture treatment affect the digestibility, textural, and physicochemical properties of maize starch. *Starch/Stärke*, *68*, 1158–1168.

Liu, M., Wu., Na-Na, Yu, G.-P., Zhau, X.-T., Chen, X., Zhang, M., Tian, X.-H., Liu, Y.-X., Wang, L.-P., & Bin, T. (2018). Physicochemical properties, structural properties, and in vitro digestibility of pea starch treated with high hydrostatic pressure: Properties of high hydrostatic pressure treated pea starch. *Dis Stärke*, *70*(1–2), 1700082.

Lobato-Calleros, C., Ramirez-Santiago, C., Vernon-Carter, E., & Alvarez-Ramirez, J. (2014). Impact of native and chemically modified starches addition as fat replacers in the viscoelasticity of reduced-fat stirred yogurt. *Journal of Food Engineering*, *131*, 110–115. Available from https://doi.org/10.1016/j.jfoodeng.2014.01.019.

Lopez-Ochoa, J. D., Cadena-Chamorro, E., Ciro-Velasquez, H., & Rodríguez-Sandoval, E. (2022). Enzymatically modified cassava starch as a stabilizer for fermented dairy beverages. *Starch/Stärke*, *74*(7–8), 2100242.

Luchese, C. L., Frick, J. M., Patzer, V. L., Spada, J. C., & Tessaro, I. C. (2015). Synthesis and characterization of biofilms using native and modified *pinhão* starch. *Food Hydrocolloids*, *45*, 203–210. Available from https://doi.org/10.1016/j.foodhyd.2014.11.015.

Maior, L., de, O., de Almeida, V. S., Barretti, B. R. V., Ito, V. C., Beninca, C., Demiate, I. M., Schinizler, E., Filho, M. A. S. C., & Lacerda, L. G. (2020). Combination of organic acid and heat–moisture treatment: Impact on the thermal, structural, pasting properties and digestibility of maize starch. *Journal of Thermal Analysis and Calorimetry*, *143*, 265–273. Available from https://doi.org/10.1007/s10973-019-09241-1.

Maniglia, B. C., Lima, D. C., Júnior, M. D. M., Oge, A., Le-Bail, P., Augusto, P. E. D., & Le-Bail, A. (2020). Dry heating treatment: A potential tool to improve the wheat starch properties for 3D food printing application. *Food Research International*, *137*, 109731.

Maniglia, B. C., Pataro, G., Ferrari, G., Augusto, P. E. D., Le-Bail, P., & Le-Bail, A. (2021). Pulsed electric fields (PEF) treatment to enhance starch 3D printing application: Effect on structure, properties, and functionality of wheat and cassava starches. *Innovative Food Science & Emerging Technologies*, *68*, 102602. Available from https://doi.org/10.1016/j.ifset.2021.102602.

Martins, A., Beninca, C., Bet, C. D., Bisinella, R. Z. ,B., de Oliveira, C. S., Hornung, P. S., & Schnitzler, E. (2020). Ultrasonic modification of purple taro starch (*Colocasia esculenta* B. Tini): Structural, psychochemical and thermal properties. *Journal of Thermal Analysis and Calorimetry*, *142*, 819–828. Available from https://doi-org.ez82.periodicos.capes.gov.br/10.1007/s10973-020-09298-3.

Matos, M., Laca, A., Rea, F., Iglesias, O., Rayner, M., & Gutiérrez, F. (2018). O/W emulsions stabilized by OSA-modified starch granules versus non-ionic surfactant: Stability, rheological behaviour and resveratrol encapsulation. *Journal of Food Engineering*, *222*, 207–217. Available from https://doi.org/10.1016/j.jfoodeng.2017.11.009.

McAllister, T. A., & Sultana, H. (2011). Effects of micronization on the in situ and in vitro digestion of cereal grains. *Asian-Australas Journal of Animal Science*, *24*(7), 929–939.

Mhaske, P., Wang, Z., Farahnaky, A., Kasapis, S., & Majzoobi, M. (2022). Green and clean modification of cassava starch – effects on composition, structure, properties and digestibility. *Critical Reviews in Food Science and Nutrition*, *62*(28), 7801–7826. Available from https://doi.org/10.1080/10408398.2021.1919050.

Moraes, J., Alves, F. S., & Franco, C. M. L. (2013). Effect of ball milling on structural and physicochemical characteristics of cassava and Peruvian carrot starches. *Starch/Stärke*, *65*(3–4), 200–209. Available from https://doi.org/10.1002/star.201200059.

Murúa-Pagola, B., Castro-Becerra, A. L., Abadía-García, L., Castaño-Tostado, E., & Amaya-Llano, S. L. (2020). Protective effect of a cross-linked starch by extrusion on the survival of *Bifidobacterium breve* ATCC 15700 in yogurt. *Journal of Food Processing and Preservation*, *45*(1), e15097. Available from https://doi.org/10.1111/jfpp.15097.

Nemţanu, M. R., & Braşoveanu, M. (2021). Exposure of starch to combined physical treatments based on corona electrical discharges and ionizing radiation. Impact on physicochemical Properties. *Radiation Physics and Chemistry*, *184*, 109480. Available from https://doi.org/10.1016/j.radphyschem.2021.109480.

Nhouchi, Z., Watuzola, R., & Pense-Lheritier, A. M. (2022). A review on octenyl succinic anhydride modified starch-based Pickering-emulsion: Instabilities and ingredients interactions. *Journal of Texture Studies*, *53*(5), 581–600. Available from https://doi.org/10.1111/jtxs.12663.

Obadi, M., & Xu, B. (2021). Review on the physicochemical properties, modifications, and applications of starches and its common modified forms used in noodle products. *Food Hydrocolloids*, *112*, 106286. Available from https://doi.org/10.1016/J.FOODHYD.2020.106286.

Okyere, A. Y., Bertoft, E., & Amponsah, G. A. (2019). Modification of cereal and tuber waxy starches with radio frequency cold plasma and its effects on waxy starch properties. *Carbohydrate Polymers*, *223*, 115075. Available from https://doi.org/10.1016/j.carbpol.2019.115075.

Olagunju, A. I., Omoba, O. S., Enujiugha, V. N., Wiens, R. A., Gough, K. M., & Aluko, R. E. (2020). Influence of acetylation on physicochemical and morphological characteristics of pigeon pea starch. *Food Hydrocolloids*, *100*, 105424. Available from https://doi-org.ez82.periodicos.capes.gov.br/10.1016/j.foodhyd.2019.105424.

Oliveira, D. I. D., Demogalski, L., Dias, A. H., Pereira, L. A. A., Alberti, A., Los, P. R., & Demiate, I. M. (2021). Traditional sour cassava starch obtained with alterations in the solar drying stage. *Food Science and Technology*, *41*(1), 319–327. Available from https://doi.org/10.1590/fst.16120.

Oliveira, L. C., Macnaughtan, B., Gouseti, O., Villas-Boas, F., Clerici, M. T. P. S., Bakalis, S., Muttakin, S., & Cristianini, M. (2021). Extending the functionality of arrowroot starch by thermally assisted high hydrostatic pressure. *Journal of Food Processing and Preservation*, *45*(9), 15756. Available from https://doi.org/10.1111/JFPP.15756.

Otache, M. A., Duru, R. U., Achugasim, O., & Abayeh, O. J. (2021). Advances in the modification of starch via esterification for enhanced properties. *Journal of Polymers and the Environment*, *29*(5), 1365–1379. Available from https://doi.org/10.1007/s10924-020-02006-0.

Ozuna, C., Paniagua-Martínez, I., Castaño-Tostado, E., Ozimek, L., & Amaya-Llano, S. L. (2015). Innovative applications of high-intensity ultrasound in the development of functional food ingredients: Production of protein hydrolysates and bioactive peptides. *Food Research International*, *77*(4), 685–696. Available from https://doi.org/10.1016/j.foodres.2015.10.015.

Pandiselvam, R., Manikantan, M. R., Divya, V., Ashokkumar, C., Kaavya, R., Kothakota, A., & Ramesh, S. V. (2019). Ozone: An advanced oxidation technology for starch modification. *Ozone: Science & Engineering*, *41*(6), 491–507. Available from https://doi.org/10.1080/01919512.2019.1577128.

Park, S., & Kim, Y. R. (2021). Clean label starch: Production, physicochemical characteristics, and industrial applications. *Food Science and Biotechnology*, *30*, 1–17. Available from https://doi-org.ez82.periodicos.capes.gov.br/10.1007/s10068-020-00834-3.

Peng, X., & Yao, Y. (2017). Carbohydrates as fat replacers. *Annual Review of Food Science and Technology*, *8*, 331–351. Available from https://doi-org.ez82.periodicos.capes.gov.br/10.1146/annurev-food-030216-030034.

Pingret, D., Fabiano-Tixier, A. S., Bourvellec, C. L., Renard, C. M. G. C., & Chemat, F. (2012). Lab and pilot-scale ultrasound-assisted water extraction of polyphenols from apple pomace. *Journal of Food Engineering*, *111*, 73–81. Available from https://doi.org/10.1016/j.jfoodeng.2012.01.026.

Polesi, L. F., Sarmento, S. B. S., & Canniatti-Brazaca, S. G. (2018). Starch digestibility and functional properties of rice starch subjected to gamma radiation. *Rice Science*, *25*(1), 42–51. Available from https://doi.org/10.1016/j.rsci.2017.08.003.

Prachayawarakorn, J., & Kansanthia, P. (2022). Characterization and properties of singly and dually modified hydrogen peroxide oxidized and glutaraldehyde crosslinked biodegradable starch films. *International Journal of Biological Macromolecules*, *194*, 331–337. Available from https://doi.org/10.1016/J.IJBIOMAC.2021.11.150.

Pratiwi, M., Faridah, D. N., & Lioe, H. N. (2018). Structural changes to starch after acid hydrolysis, debranching, autoclaving-cooling cycles, and heat moisture treatment (HMT): A review. *Starch/Stärke*, *70*, 1–13.

Puligundla, P., Cho, Y., & Lee, Y. (2015). Physicochemical and sensory properties of reduced-fat mayonnaise formulations prepared with rice starch and starch-gum mixtures. *European Journal of Food and Agriculture*, *27*(8), 463–468. Available from https://doi-org.ez82.periodicos.capes.gov.br/10.9755/ejfa.2015.04.081.

Punia, S. (2020). Barley starch modifications: Physical, chemical and enzymatic - A review. *International Journal of Biological Macromolecules*, *144*, 578–585.

Punia, S., Dhull, S. B., Kunner, P., & Rohilla, S. (2020). Effect of γ-radiation on physico-chemical, morphological and thermal characteristics of lotus seed (*Nelumbo nucifera*) starch. *International Journal of Biological Macromolecules*, *157*, 584–590. Available from https://doi.org/10.1016/j.ijbiomac.2020.04.181, https://doi-org.ez82.periodicos.capes.gov.br/10.1007/s10924-022-02474-6.

Rahaman, A., Kumari, A., Zeng, X. A., Muhammad, A. F., Siddique, R., Khalifa, I., Siddeeg, A., Ali, M., & Manzoor, M. F. (2021). Ultrasound based modification and structural-functional analysis of corn and cassava starch. *Ultrasonics Sonochemistry*, *80*, 105795. Available from https://doi.org/10.1016/j.ultsonch.2021.105795.

Ramadan, M. F., & Sitohy, M. Z. (2020). Phosphorylated starches: Preparation, properties, functionality, and techno-applications. *Starch/Stärke*, *72*(5–6), 1900302. Available from https://doi.org/10.1002/star.201900302.

Raza, H., Ameer, L., Ma, H., Liang, Q., & Ren, X. (2021). Structural and physicochemical characterization of modified starch from arrowhead tuber (*Sagittaria sagittifolia* L.) using tri-frequency power ultrasound. *Ultrasonic Sonochemistry*, *80*, 105826. Available from https://doi.org/10.1016/j.ultsonch.2021.105826.

Rombo, G. O., Taylor, J. R. N., & Minnaar, A. (2004). Irradiation of maize and bean flours: Effects on starch physicochemical properties. *Journal of Science and Food Agriculture*, *84*(4), 350–356.

Samarakoon, E. R. J., Waduge, R., Liu, Q., Shahidi, F., & Banoub, J. H. (2020). Impact of annealing on the hierarchical structure and physicochemical properties of waxy starches of different botanical origins. *Food Chemistry*, *303*, 125344.

Sanchez de la Concha, B. B., Agama-Acevedo, E., Nuñez-Santiago, M. C., Bello-Perez, L. A., Garcia, H. S., & Alvarez-Ramirez, J. (2018). Acid hydrolysis of waxy starches with different granule size for nanocrystal production. *Journal of Cereal Science*, *79*, 193–200. Available from https://doi.org/10.1016/j.jcs.2017.10.018.

Santos, T. P. R., Leonel, M., Mischan, M. M., & Cabello, C. (2021). Study and application of photo-modified cassava starch with lactic acid and UV-C irradiation. *LWT*, *139*, 110504. Available from https://doi.org/10.1016/j.lwt.2020.110504.

Schafranski, K., Ito, V. C., & Lacerda, L. G. (2021). Impacts and potential applications: A review of the modification of starches by heat-moisture treatment (HMT. *Food Hydrocolloids*, *117*, 106690. Available from https://doi.org/10.1016/j.foodhyd.2021.106690.

Schlindweinn, E. B., Chacon, W. D. C., Koop, B. L., Fonseca, J. M., Monteiro, A. R., & Valencia, G. A. (2022). Starch-based materials encapsulating anthocyanins: A review. *Journal of Polymers and the Environment*, *30*, 3547–3565.

Shaikh, M., Ali, T. M., & Hasnain, A. (2020). Comparative study on the application of chemically modified corn and pearl millet starches in white sauce. *Journal of Food Processing and Preservation*, *44*(4), e14393. Available from https://doi-org.ez82.periodicos.capes.gov.br/10.1111/jfpp.14393.

Shen, Y., Yao, Y., Wang, Z., & Wu, H. (2021). Hydroxypropylation reduces gelatinization temperature of corn starch for textile sizing. *Cellulose*, *28*(8), 5123–5134. Available from https://link.springer.com/article/10.1007/s10570-021-03852-4.

Singh, H., Blennow, A., Gupta, A. D., Kaur, P., Dhillon, B., Sodhi, N. S., & Dubey, P. K. (2022). Pulsed light, pulsed electric field and cold plasma modification of starches: Technological advancements & effects on functional properties. *Journal of Food Measurement and Characterization*, *16*, 4092–4109. Available from https://doi-org.ez82.periodicos.capes.gov.br/10.1007/s11694-022-01487-y.

Singh, J., Kaur, L., & McCarthy, O. J. (2007). Factors influencing the physico-chemical, morphological, thermal and rheological properties of some chemically modified starches for food applications—A review. *Food Hydrocolloids*, *21*(1), 1–22. Available from https://doi.org/10.1016/J.FOODHYD.2006.02.006.

Singh, R., & Sharanagat, V. S. (2020). Physico-functional and structural characterization of ultrasonic-assisted chemically modified elephant foot yam starch. *International Journal of Biological Macromolecules*, *164*, 1061–1069. Available from https://doi.org/10.1016/j.ijbiomac.2020.07.185.

Smith, R. J. (1967). Characterization and analysis of starches. In R. L. Whistler, & E. F. Paschall (Eds.), *Starch: Chemistry and technology* (2, pp. 569–635). New York: Academic Press.

Soe, M. T., Chitropas, P., Pongjanyakul, T., Limpongsa, E., & Jaipakdee, N. (2019). Thai glutinous rice starch modified by ball milling and its application as a mucoadhesive polymer. *Carbohydrate Polymers*, *232*, 115812. Available from https://doi.org/10.1016/j.carbpol.2019.115812.

Song, H. Y., Lee, S. Y., Choi, S. J., Kim, K. M., Kim, J. S., Han, G. J., & Moon, T. W. (2013). Digestibility and physicochemical properties of granular sweet potato starch as affected by annealing. *Food Science and Biotechnology*, *23*, 23–31.

Sorndech, W., Tongta, S., & Blennow, A. (2018). Slowly digestible- and non-digestible α-glucans: An enzymatic approach to starch modification and nutritional effects. *Starch – Stärke*, *70*, 1700145. Available from https://doi.org/10.1002/star.201700145.

Suki, F. M. M., Azura, A. R., & Azahari, B. (2016). Effect of ball milled and ultrasonic sago starch dispersion on Sago Starch Filled Natural Rubber Latex (SSNRL) films. *Procedia Chemistry*, *19*, 782–787. Available from https://doi.org/10.1016/j.proche.2016.03.085.

Sunder, M., Mumbrekar, K. D., & Mazumder, N. (2022). Gamma radiation as a modifier of starch – Physicochemical perspective. *Current Research in Food Science*, *5*, 141–149. Available from https://doi.org/10.1016/j.crfs.2022.01.001.

Sushil, D., Ashok, K. S., Bernadine, M. F., Jovin, H., & Michael, J. G. (2011). Cryo-milling of starch granules leads to differential effects on molecular size and conformation. *Carbohydrate Polymers*, *84*(3), 1133–1140. Available from 10.1016/j.carbpol.2011.01.002.

Ulbrich, M., Beresnewa-Seekamp, T., Walther, W., & Flöter, E. (2016). Acid-thinned corn starch-impact of modification parameters on molecular characteristics and functional properties. *Starch/Stärke*, *68*(5–6), 399–409. Available from https://doi.org/10.1002/star.201500210.

Ulbrich, M., & Flöter, E. (2019a). Impact of process parameters on the acid modification of potato starch. *Starch/Stärke*, *71*(1–2), 1800111. Available from https://doi.org/10.1002/star.201800111.

Ulbrich, M., & Flöter, E. (2019b). Functional properties of acid thinned potato starch: Impact of modification, molecular starch characteristics, and solution preparation. *Starch/Stärke*, *71*(11–12), 1900176. Available from https://doi.org/10.1002/star.201900176.

Wang, B., Zhong, Z., Wang, Y., Yuan, S., Jiang, Y., Li, Z., Li, Y., Yan, Z., Meng, L., & Qiu, L. (2023). Recent progress of starch modification assisted by ultrasonic wave. *Food Science and Technology*, *43*, e107522. Available from https://doi.org/10.1590/fst.107522.

Wang, D., Ma, X., Yan, L., Chantapakul, T., Wang, W., Ding, T., Ye, X., & Liu, D. (2017). Ultrasound assisted enzymatic hydrolysis of starch catalyzed by glucoamylase: Investigation on starch properties and degradation kinetics. *Carbohydrate Polymers*, *175*, 47–54. Available from https://doi-org.ez82.periodicos.capes.gov.br/10.1016/j.carbpol.2017.06.093.

Wang, H., Zhang, B., Chen, L., & Li, X. (2016). Understanding the structure and digestibility of heat-moisture treated starch. *International Journal of Biological Macromolecules*, *88*, 1–8.

Wang, L., Liu, S., Hou, Y., Lang, S., Wang, C., & Zhang, D. (2020). Changes in particle size, structure, and physicochemical properties of potato starch after jet-milling treatments. *Journal of Food Processing and Preservation*, *44*, e14569. Available from https://doi.org/10.1111/jfpp.14569.

van der Maarel, M. J. E. C., & Leemhuis, H. (2013). Starch modification with microbial alpha-glucanotransferase enzymes. *Carbohydrate Polymers*, *93*(1), 116–121. Available from https://doi.org/10.1016/J.CARBPOL.2012.01.065.

Wang, Q., Li, L., & Zheng, X. (2020). A review of milling damaged starch: Generation, measurement, functionality and its effect on starch-based food systems. *Food Chemistry*, *315*, 126267. Available from https://doi.org/10.1016/j.foodchem.2020.126267.

Wang, X. M., Zhu, X. M., Zhang, N. H., Tu, Z. C., Wang, H., Liu, G. X., & Ye, Y. H. (2018). Morphological and structural characteristics of rice amylose by dynamic high-pressure microfluidization modification. *Journal of Food Processing and Preservation*, *42*(10), e13764. Available from https://doi.org/10.1111/jfpp.13764.

Wang, Z., Mhaske, P., Farahnaky, A., Kasapis, S., & Majzoobi, M. (2022). Cassava starch: Chemical modification and its impact on functional properties and digestibility, a review. *Food Hydrocolloids*, *129*, 107542. Available from https://doi.org/10.1016/J.FOODHYD.2022.107542.

Watcharakitti, J., Win, E. E., Nimnuan, J., & Smith, S. M. (2022). Modified starch-based adhesives: A review. *Polymers*, *14*(10), 2023. Available from https://doi.org/10.3390/polym14102023.

Wattananapakasem, I., Penjumras, P., Malaithong, W., Nawong, S., Poomanee, W., & Kinoshita, H. (2021). Effect of heat–moisture treatment of germinated black rice on the physicochemical properties and its utilization by lactic acid bacteria. *Journal of Food Science and Technology*, *58*, 4636–4645. Available from https://doi.org/10.1007/s13197-020-04951-z.

Wojeicchowski, J. P., Siqueira, G. L. D. A. D., Lacerda, L. G., Schnitzler, E., & Demiate, I. M. (2018). Physicochemical, structural and thermal properties of oxidized, acetylated and dual-modified common bean (*Phaseolus vulgaris* L.) starch. *Food Science and Technology*, *38*(2), 318–327. Available from https://doi.org/10.1590/1678-457x.04117.

Woo, K., Dowsett, C., Costa, B., Ebohon, S., Woodmansey, E. J., & Malone, M. (2021). Efficacy of topical cadexomer iodine treatment in chronic wounds: Systematic review and meta-analysis of comparative clinical trials. *International Wound Journal*, *18*(5), 586–597. Available from https://doi.org/10.1111/iwj.13560.

Wu, Z., Qiao, D., Zhao, S., Lin, Q., Zhang, B., & Xie, F. (2022). Nonthermal physical modification of starch: An overview of recent research into structure and property alterations. *International Journal of Biological Macromolecules*, *203*, 153–175.

Wurzburg, O. B. (1986). In (1st ed.). O. B. Wurzburg (Ed.), *Modified starches: Properties and uses* (Vol. 1). CRC Press.

Wurzburg, O. B. (1995). Modified Starches. In A. M. Stephen (Ed.), *Food polysaccharides and their applications* (pp. 67–98). New York: Marcel Dekker.

Xia, W., Li, J.-H., Wei, X.-Y., Wang, F., & Lin, Y.-Y. (2017). Effect of micronization on the solubility, viscosity and structural properties of tapioca starch. *IOP Conference Series: Materials Science and Engineering*, *231*, 012128. Available from https://doi.org/10.1088/1757-899x/231/1/012128.

Xu, T., Li, X., Ji, S., Zhong, Y., Simal-Gandara, J., Capanoglu, E., Xiao, J., & Lu, B. (2021). Starch modification with phenolics: Methods, physicochemical property alteration, and mechanisms of glycaemic control. *Trends in Food Science & Technology*, *111*, 12–26. Available from https://doi.org/10.1016/J.TIFS.2021.02.023.

Yang, W., Kong, X., Zheng, Y., Sun, W., Chen, S., Liu, D., Zhang, H., Fang, H., Tian, J., & Ye, X. (2019). Controlled ultrasound treatments modify the morphology and physical properties of rice starch rather than the fine structure. *Ultrasonics Sonochemistry*, *59*, 104709. Available from https://doi.org/10.1016/j.ultsonch.2019.104709.

Yu, S., Zhang, Y., Ge, Y., Zhang, Y., Sun, T., Jiao, Y., & Zheng, X. Q. (2013). Effects of ultrasound processing on the thermal and retrogradation properties of nonwaxy rice starch. *Journal of Food Process Engineering*, *36*(6), 793–802. Available from https://doi.org/10.1111/jfpe.12048, 2013.

Zeng, F., Gao, Q.-Y., Han, Z., Zeng, X. A., & Yu, S.-J. (2016). Structural properties and digestibility of pulsed electric field treated waxy rice starch. *Food Chemistry*, *194*, 1313–1319. Available from https://doi.org/10.1016/j.foodchem.2015.08.104.

Zheng, J., Li, Q., Hu, A., Yang, L., Lu, J., Zhang, X., & Lin, Q. (2013). Dual-frequency ultrasound effect on structure and properties of sweet potato starch. *Starch/Stärke*, *65* (7–8), 621–627. Available from https://doi.org/10.1002/star.201200197.

Zhi-guang, C., Jun-rong, H., Hua-yin, P., Qi, Y., & Chen-lu, F. (2020). The effects of HHP (high hydrostatic pressure) on the interchain interaction and the conformation of amylopectin and double-amylose molecules. *International Journal of Biological Macromolecules*, *155*, 91–102. Available from https://doi.org/10.1016/J.IJBIOMAC.2020.03.190.

Zhu, F. (2015). Impact of ultrasound on structure, physicochemical properties, modifications, and applications of starch. *Trends in Food Science and Technology*, *43*, 1–17. Available from https://doi.org/10.1016/j.tifs.2014.12.008.

Zuo, J. Y., Knoerzer, K., Mawson, R., Kentish, S., & Ashokkumar, M. (2009). The pasting properties of sonicated waxy rice starch suspensions. *Ultrasonics Sonochemistry*, *16*(4), 462–468. Available from https://doi.org/10.1016/j.ultsonch.2009.01.002.

Annex 1. Chemically and enzymatically modified starches and their applications

Class	Modification	Description	Applications	References
Partial degradation/ decomposition	Acid modification	HCl, H_2SO_4, and H_3PO_4 at temperatures lower than gelatinization; lower molecular weight, decreased viscosity, increased syneresis, and gelling property	Wood adhesives, paper, textile, food, and beverage; 3D additive manufacturing; porous starch	Wang et al. (2017), Jiang et al. (2018), Singh et al. (2007), Latip et al. (2021), Koksel et al. (2008), Kaur et al. (2012), Ulbrich and Flöter (2019b) Ulbrich et al. (2016) Ulbrich and Flöter (2019a), Chakraborty et al. (2022)
		H_2SO_4 at temperatures lower than gelatinization	Nanocrystals	Sanchez de la Concha et al. (2018), Daniel et al. (2020)
	Oxidation	Granular starch is treated with oxidant chemicals (NaOCl, H_2O_2, and ozone) at alkaline conditions; decreased viscosity/ shear thinning; freeze-thaw stability; increased solubility of starch granules (higher molecular breakdown and increased electronegativity)	Laundry finishing; textile, drilling; binding material industries; food industry (noodles, battered/breaded coatings to provide adhesion in the coating of foods); biodegradable packaging; adhesives	(Obadi & Xu, 2021), Pandiselvam et al. (2019) (Prachayawarakorn & Kansanthia, 2022) Wojeicchowski et al. (2018), Watcharakitti et al. (2022), Chen et al. (2019)
Substitution/ Derivatization	Esterified	Reaction involving an acid anhydride and alcohol, acid chloride and alcohol, carboxylic acid and alcohol; organic and inorganic esters; substitution of hydroxyl groups by bulky ester groups (hydroxypropyl, acetyl, succinate-octenylsuccinate, free fatty acids)	Food, pharmaceutical, textile, paper, industries; adhesives, biodegradable package materials, encapsulation of bioactive compounds; Pickering emulsion	(Otache et al., 2021), (Watcharakitti et al., 2022), Altuna et al. (2018), (Boetje et al., 2022), (Chakraborty et al., 2022), Chen et al. (2019), (Guida et al., 2021), (Nhouchi et al., 2022)
	Cross-linked	Reinforced starch granules by covalent linkages between macromolecules in the presence of multivalent reagents (sodium trimetaphosphate - STMP, monosodium phosphate, sodium tripolyphosphate - STP, epichlorohydrin, phosphoryl chloride)	Resistance to high cooking Time and temperature (pressure cooking), increased acidity, severe shear; food, textile size, and pharmaceutical industries; adsorption of heavy metals	(Bismark et al., 2018), (Akinterinwa et al., 2020), (Akinterinwa et al., 2022), Chen et al. (2019)

Class	Modification	Description	Applications	References
	Etherified	Substitution of some hydroxyl groups with ether groups, suppressing the formation of hydrogen bonding; hydroxypropyl, hydroxyethyl, and carboxymethyl groups are hydrophilic functional groups that can be bonded to starch molecules through ether linkage; substituents can be nonionic or alkyl ether (hydroxypropyl or hydroxyethyl substituents), anionic (carboxymethyl or sulfonic substituents), cationic (ammonium, amino, imino or phosphonium substituents) and amphoteric (both anionic and cationic substituents)	Starch ethers can become soluble in cold water and be used as thickeners, dissolvable films and coatings, absorbents, and hydrogels; films, foams, and packaging materials; coating, flocculants, drug delivery, additives, paper making, and water-oil emulsion	Bakouri and Guemra (2019), Xu et al. (2021), Akinterinwa et al. (2020), Chen et al. (2019), Chittapun et al. (2020)
Enzymatic modification	Hydrolases, transferases	Branching Enzymes, amylomaltase, α-transglucosidase, debranching enzymes	Modification of starch digestibility; Increased SDS and RS; development of "healthy starches"; conversion of starch into malto-oligosaccharides, which are functional ingredients, in several foods and beverages. Environment-friendly methods for starch modification, clean label status; improved functional and technological properties; porous starch granules	(Sorndech et al., 2018), van der Maarel and Leemhuis (2013), Punia et al. (2020), Chen et al. (2019) Christensen et al. (2023), Gui et al. (2022), He et al. (2020)

CHAPTER

7

Sour starch, a lactic fermented starch with heat expansion properties: innovations and patents

Marney Pascoli Cereda[1,2]

Retired from São Paulo State University, Botucatu, São Paulo, Brazil [2]Tecsapp Solutions, Campo Grande, Mato Grosso do Sul, Brazil

7.1 Introduction

The tradition of fermenting cassava roots with natural fermentation, for the production of "puba" flour, has its origin in an indigenous process to soften the roots, to facilitate their disintegration when they did not have efficient grinders. This floor is a typical food from the north of Brazil (Paixão & Silva et al., 2021). For more information on the different cassava flours produced in Brazil, see the upcoming Volume 4 in this series.

Sour starch is a derivative of cassava starch found in every country in South America. It is mainly produced in Brazil, Colombia, and Paraguay (Alarcón & Dufour, 2002; Cereda & Vilpoux, 2004, 2006; Cereda et al., 2001; Díaz et al., 2018; Westby & Cereda, 1994). The information from Putri et al. (2012) and Díaz et al. (2018a) that sour starch could be obtained from other types of starch and that it would be produced in Indonesia is erroneous and may confuse the readers.

As a traditional rural industry product, sour starch is commercially known as *polvilho azedo*, *almidón agrio*, and *amido azedo*, depending on the region of origin (Díaz et al., 2018a).

Lacerda et al. (2005) pointed out that sour cassava starch is a traditional fermented food used in South America to obtain expanded products without the use of raising agents, such as expanded snacks (*biscoito de polvilho*) and rolls. That is why many of the documents that deal with this product and its derivatives are in Portuguese or Spanish.

In Brazil, sour starch is used for the preparation of fried foods and baked goods such as traditional cheese breads (*pão-de-queijo*). The bread substitute made with sour starch is a soft and crunchy product, recognized as a safe food for consumption with several fillings and side dishes. In Brazil, it is now catering to fast-food consumption (for more information on sour starch derivatives, see Chapter 8). This traditional product, made only from starch, has expanded its consumption throughout Latin

DOI: https://doi.org/10.1016/B978-0-323-90842-9.00014-5

America. Cassava sour starch is widely used as a basic ingredient in Colombia to manufacture *pan de yucca* and *pandebono* (Figueroa-Flórez et al., 2018).

Cheese bread is usually eaten as a complement to coffee. Because it is considered a healthy food, sour starch can also be consumed in the form of cookies, being sold along highways, in bakeries, and in supermarkets. Due to consumption in specialized stores, bars, and fast-food outlets, cheese bread production demanded the development of frozen products, ready to be baked and complemented with fillings and flavors, sweet and salty, which encouraged sales in Europe and the United States, largely to cater to Latin consumers (Cereda & Vilpoux, 2004).

In addition, sour starch derivatives are gluten-free products, with potential application in the baking market for people with celiac syndrome (see Chapter 8). These derivatives are prepared as biscuits, snacks, and as a coating material for peanuts that is very popular in some countries of South America (Cereda & Vilpoux, 2004).

In Brazil, frozen dough for the fast preparation of cheese bread in the oven was popularized for use at home or commercially. The first change resulting from this increase in demand was to alter the proportion of sour starch used in cheese bread preparation, replacing half or all the sour starch with native cassava starch. This development facilitated the increase in consumption, as it reduced the cost and made it easier to obtain sufficient raw material, since the traditional process of sour starch production is artisanal and could not meet the demand.

The international dissemination of information on the expansion characteristics of this cassava starch led to an increase in global interest, with scientific articles in which corn starch and other exotic starches, other acids, and other sources of heat or radiation were tested and with results that were not conclusive. Two countries that produce and consume this product stand out in research and publications, Brazil and Colombia. For this reason, they will be cited more often in this chapter.

7.2 Importance of sour starch in South America

Sour starch is very popular in South America. It is produced on a small scale, exclusively with cassava starch, by natural fermentation where lactic bacteria predominate (Cereda & Brito, 2017; Cereda & Vilpoux, 2006). It differs from native cassava starch and other native starches for its unique expansion capacity in bakery products, without the addition of baking soda or yeasts and from the wheat flour because of the absence of gluten. Garcia and Leonel (2005) showed that this property is only possible with native cassava starch. It does not occur with cereal starches and is not very evident for starches extracted from other underground crops (sweet-potato, canna, and cocoyam).

Sour starch is obtained from a spontaneous fermentation conducted in fermentation tanks. However, the end of fermentation is evaluated empirically, and the process occurs without standardization, which results in products of heterogeneous physicochemical and sensory quality (Cereda & Giaj-Levra, 1987; Cereda & Lima, 1985; Cereda & Vilpoux, 2004, 2006; Cereda et al., 1982, 2001; Cereda, 1975, 1987; Matta et al., 2019; Penido et al., 2018; Westby & Cereda, 1994).

In Colombia, *almidón agrio*, as it is known, is obtained by wet extraction of the starch from the roots, followed by natural fermentation and sun drying, mainly in Valle del Cauca (Mora Pareja et al., 2017). The manufacturing process can be considered small-scale and ranges from extraction to sun-drying, which is done in terraces close to the ground, unlike in Brazil, where it is done in tables (jiraus—Fig. 7.2). Figueroa-Flórez et al. (2018) added that in Colombia, the production of *almidón agrio* is carried out with a traditional and low-tech method in small companies called *rallanderías*. Sour starch is a high added-value product that is obtained through spontaneous fermentation with native starch, without control of process variables and which gives rise to a product of heterogeneous quality. Mora Pareja et al. (2017) highlighted its widespread domestic food consumption throughout the country, where it is used as an ingredient in the preparation of several typical bakery products (*pan de yuca, pan de bono, almojabanas, buñuelos*, etc.) and more recently in the industrial manufacture of snacks, as in Brazil.

Another producing country in South America where sour starch is very important for the local diet is Paraguay. Its most typical product is the *chipa*, a U-shaped bread roll, with a formulation very close to the Brazilian cheese bread, but with the addition of corn flour and other complements. The same product, with variations, is called *cuñapé* in Bolivia. Paraguay also exports sour starch to supply the Paraguayan community living in Argentina (Cereda et al., 2001).

In Brazil, consumer demand for sour starch is high (Cereda, 1983a). The magazine Food Ingredients (Pão de queijo, 2001) emphasizes that cheese bread has been historically linked to traditional Brazilian cuisine for more than 200 years, and since the mid-1980s, its consumption has expanded to the entire country, thanks to its manufacture on an industrial scale. Bakeries, snack bars, restaurants, hotels, fast-food chains, and especially cheese bread chains offer a variety of products that are gaining more and more consumers, even abroad, to where it is being exported in growing volumes. Many researchers are striving to present ways to improve or stabilize the quality of products, with solutions ranging from the use of coadjuvants to substitutes for raw materials, such as cheese powder, flavorings, and preservatives.

However, the biggest problem is with the availability of sour starch. To get around this difficulty, the cheese bread makers have eliminated the scalding process, which, in all countries, is traditionally and mandatorily used in the elaboration of food products made with sour starch. The scalding consists of pouring a mixture of boiling water, milk, and fat over the sour starch, allowing part of the starch to gel. The popularity of cheese bread has greatly boosted the production of sour starch, one of the main raw materials for this product. As the process is still artisanal even with production in medium and large starch processing plants, the difficulty of supplying a homogeneous quality has led to the development of substitute products. The concern with cheese bread, evidenced by technical publications in specialized magazines, demonstrates the increase of commercial interest in sour starch.

Next, the process is presented with its problems.

7.3 Artisanal and industrial production

The traditional sour starch manufacturing described by Westby and Cereda (1994), Cereda et al. (2001), and Cereda and Vilpoux (2004, 2006) comprises the extraction of starch from cassava roots, natural fermentation, and sun drying. The authors also describe the fermentation of commercial cassava starch leaving the product in water, but this does not always allow for sufficient lactic fermentation. According to Brito et al. (2021), the artisanal process can be adapted to be continuous, without losing its properties.

Acid modification by natural means of fermentation has been demonstrated to be the most effective technique of altering the functionality of cassava starch (Onitilo et al., 2007). In natural fermentation, lactic acid-producing bacteria predominate (Cereda & Lima, 1985; Cereda & Vilpoux, 2004). In all cases, the cassava starch impregnated with lactic acid is then exposed to the sun in a natural drying process, an operation that actively participates in the starch modification process. However, sun drying brings a series of inconveniences that range from product loss and high dust contamination to the lack of product standardization (Cereda, 1983a).

In Colombia, the production of sour starch is made in higher regions, with cultivation of varieties adapted to the local climate, which depends on altitude. The Department of Cauca is the main producer in this country and represented almost 80% of the national production in 2002 (Fernández et al., 2002).

Traditional sour starch production is depicted in Fig. 7.1. The natural fermentation takes place in tanks (Cardenas & Buckle, 1980; Cereda & Vilpoux, 2004, 2006; Cereda, 1987; Martinez & Zapata, 1981). At this stage, a sequence of groups of microorganisms appears, first the aerobes, followed by the anaerobes, and finally the microaerobic lactic bacteria. This sequence was found both in Colombia and Brazil. During fermentation, the tanks should always remain with supernatant water to avoid browning of the fermented starch by oxidation and to protect the mass from impurities (Cereda & Vilpoux, 2006; Cereda et al., 2001; Westby & Cereda, 1994).

Once natural fermentation is complete, the fermented mass deposited in the tanks is removed at about 45%–50% moisture, crumbled on bamboo frames lined with thin white cloth, and allowed to dry in the sun (Fig. 7.2). The picture also shows the product losses,

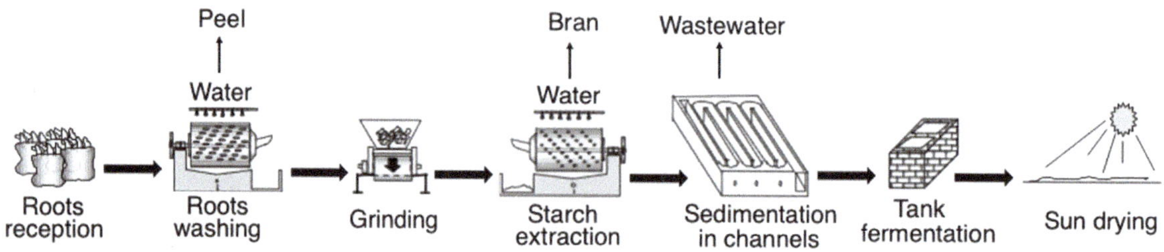

FIGURE 7.1 Stages of cassava starch extraction and sour starch production in Valle del Cauca, Colombia. *Based on Alarcón, M. F., & Dufour, D. L. (2002). Almidón agrio de yuca en Colombia. In P. B. Ospina, H. Ceballos, E. Alvarez, A. C. Bellotti, L. A. Calvert, V. B. Arias, L. L. F. Cadavid, L. B. Pineda, R. G. A. Llano, M. I. Cuervo, (Eds.),* La yuca en el Tercer Milenio: Sistemas modernos de producción, procesamiento, utilización y comercialización. *Centro Internacional de Agricultura Tropical (CIAT); Consorcio Latinoamericano para la Investigación y el Desarrollo de la Yuca; Proyecto IP-3 de Mejoramiento de Yuca, Cali, CO. p. 470–502. https://cgspace.cgiar.org/bitstream/handle/10568/55257/capitulo26.pdf?sequence = 1&isAllowed = y. Consultado em: 4/abril/2023.*

FIGURE 7.2 Sun drying of sour starch in Minas Gerais state, southeastern region of Brazil.

with the ground covered in white. The losses occur when placing and removing the product, with the action of the wind, and in case of sudden rainfall, the quick removal of the material is necessary (Cereda & Vilpoux, 2004; Cereda et al., 2001; Cereda, & Vilpoux, 2006; Westby & Cereda, 1994).

Periodically, the starch needs to be stirred with the hands by the employees of the production units, to even out the humidity and break the clumps. Some companies have mechanized this operation. The drying places should be isolated by vegetation to avoid the fall of dirt and dust, without hiding the sun. The soil should be covered with sand or crushed stone (Cereda, 1987).

Another factor that can contribute to the presence of sandy or earthy material in the final product is the fermentation tanks. These tanks are usually made of masonry, covered with flagstones or tiles, or painted with a waterproofing material that is not harmful to health. With fermentation, the acidity increases, and it is common for the tank walls and lining joints to be attacked, with material dragged into the fermented product (Fig. 7.3).

According to Demiate et al. (1997), the presence of sandy or earthy material was detected in the analysis of sour starch samples obtained by the artisanal process in Brazilian industries. Another source of contamination of sour starch can come from exposure to environmental dust in unpaved places or close to roads. Wind can be an important vector and can be circumvented with the use of windbreaks or live fences around the processing plant.

Over time, the sour starch industries have evolved toward greater mechanization and automation. More information about equipment used in this process is presented in Chapter 3.

Freitas (2020) described the sour starch produced in the Makmil Ltda industry, located in Conceição dos Ouros, a mountainous region of

FIGURE 7.3 Sour starch fermentation tank wall eroded by acidity.

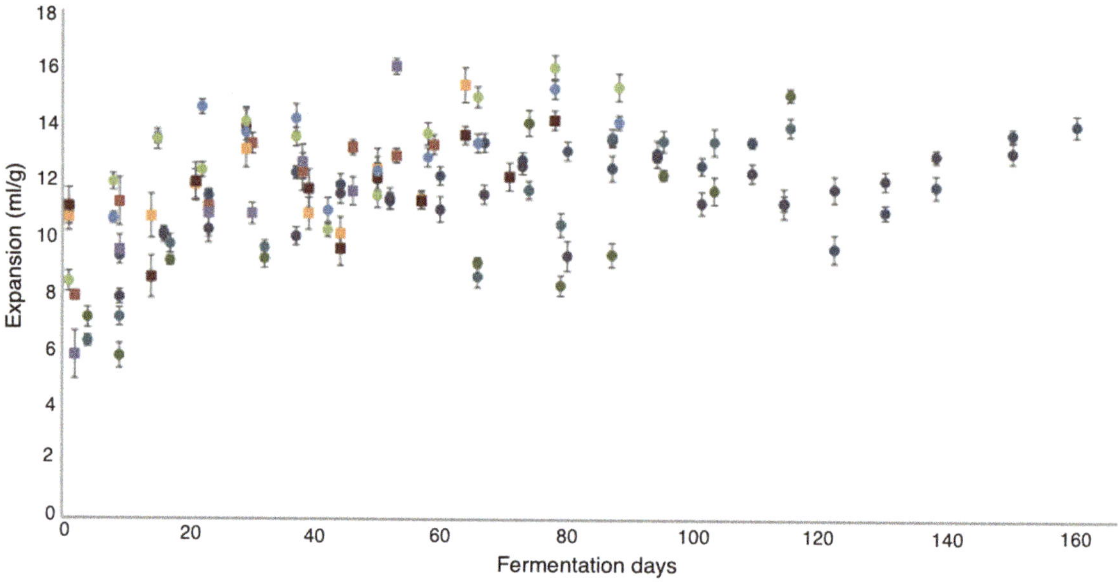

FIGURE 7.4 Variation of measured expansion in the product from different fermentation tanks, expressed in mL/g, as a function of the number of fermentation days and acidity (color and shape of the dots represent different fermentation tanks). *From Freitas, M. C. G. D. (2020). Characterization of the fermentation profile of the sour cassava starch (sour starch) produced by the commercial industry Makmil Ltda de Conceição dos Ouros, MG. 2020. 71 p. Thesis de final de Curso (College os Chemical Engeneer, USP, Lorena). (In Potuguese).*

the state of Minas Gerais in southeastern Brazil. The author found that the expanding power of the produced sour starch stabilizes after 40 days of fermentation (Fig. 7.4), which is frequent in natural fermentation, and can even reach 60 days (Cereda & Vilpoux, 2004; Cereda et al., 2001; Cereda, & Vilpoux, 2006; Westby & Cereda, 1994).

It is important to differentiate the expansion obtained by sour starch, directly in the oven and without the addition of raising agents, from extruder-processed starch that also gains expansion (Jiamjariyatam et al., 2015; Mariotti et al., 2006). Sour starch can mimic extruded products without the use of equipment.

Freitas (2020) explained that the industry controls the quality of its product by analyzing the expansion in the oven, to adjust the best fermentation time. It proceeds in batches, to standardize the quality of the final product for its customers. The expansion is measured by the method developed by Maeda and Cereda (2001), described and commented in item 7.5. Although the increase in acidity is easier to determine than the expansion volume, the company justifies its choice because of a better evaluation of the final quality of the product. The criterion used is an average of 14 cm^3/g, with batches varying from 12 to 16 cm^3/g. As can be seen in Fig. 7.5, the correlation between acidity and expansion is not high enough for acidity to be considered a reliable quality criterion, which had already been reported by Cereda and Lima (1985), who only achieved a good correlation between expansion and lactic acid content.

In the case of natural fermentation, Freitas (2020) recommended as inoculum the yeast itself that is deposited in the fermentation tanks, a solution that had already been reported by Cereda and Lima (1985). Freitas (2020) justified that the inoculated tanks showed greater

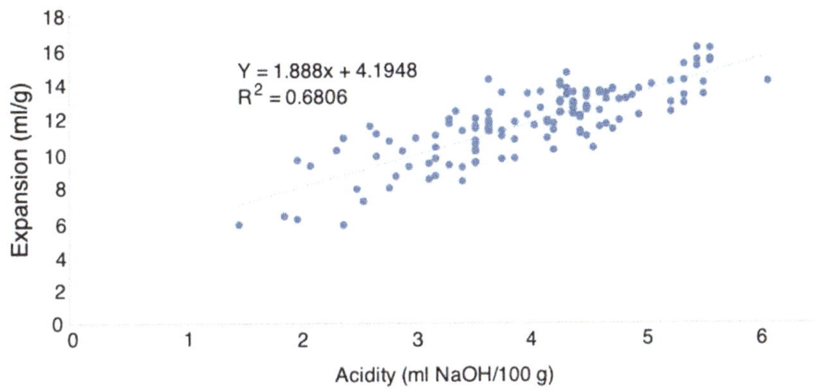

FIGURE 7.5 Variation of the expansion, in cm^3/g, as a function of acidity, expressed in mL NaOH N.100/g. *From Freitas, M. C. G. D. (2020). Characterization of the fermentation profile of the sour cassava starch (sour starch) produced by the commercial industry Makmil Ltda de Conceição dos Ouros, MG. 2020. 71 p. Thesis de final de Curso (College os Chemical Engeneer, USP, Lorena). (In Potuguese).*

expansion than the others, even after less time of fermentation.

Although Freitas (2020) described a company that dominates the complete production cycle, in artisanal production, it is not uncommon for companies to buy cassava starch from large industries, which provide a cheaper product (for more details on starch extraction processes in large companies, we suggest reading Chapter 2).

It is not uncommon for sour starch producers to complain about delayed fermentation when using commercial starch. Cereda et al. (2001) explained that the use of centrifugation to purify and concentrate the cassava starch can cause wash-out of the natural inoculum existing in the starch, requiring reinforcement with the cassava starch itself or fermented material left in the tanks.

Some large cassava starch processing units in Brazil and Paraguay have adapted to produce sour starch in an area reserved for fermentation tanks, to which part of the starch suspension is diverted. In this process, the natural fermentation lasts from 5 to 10 days, depending on the ambient temperature, without control of the acids generated. These companies also use a sun-drying process that promotes, at the same time, the UV-C treatment and the drying, which is done in tables that can be easily collected and protected from rain and dirt.

From the farmers' information, it was possible to select two parameters as important in the production process of a more standardized sour starch: natural fermentation and sun drying (Cereda, 1987). Some publications on the subject have been concerned with these parameters, trying to introduce more hygienic, faster, and standardized processes. Fermentation is very important because it is the source of lactic acid production.

7.3.1 Natural sour starch fermentation

Several papers and research reports have studied the routes and agents of natural fermentation of sour starch (Cereda & Vilpoux, 2006, Cereda, 1975), which have several challenging aspects since the substrate consists of insoluble granular starch in the form of glucose chains linked by α-(1,4) and α-(1,6) bonds.

The intense natural fermentation in the presence of varied microorganisms, as reported by

Cereda (1975), who established phases in this fermentation, occurs without nutritional supplementation. The description of the occurrence of phases in the natural fermentation of sour starch with lactic fermentation occurring at the end, when dissolved oxygen is depleted, favoring microaerobic lactic bacteria, is provided. Natural fermentation of cassava starch can occur even in the decanted starch mass with approximately 50% moisture, after removing the surface water. Native starch, as the sole substrate of the process, has about 98% starch in the granular form and about 0.06% total nitrogen, expressed as dry matter. However, a suspension in tanks of freshly extracted and decanted cassava starch, or dry starch suspended in water, presents within 24 h a visible turbidity of the supernatant water in the presence of bubbles in the decanted starchy mass, which subsequently rise to the surface forming foam. The first microorganisms identified were aerobic sporulated bacteria producing α-amylase, followed by lactic bacteria and finally oxidative yeasts.

The first microorganisms to develop in the cassava starch suspension have enzymes capable of releasing glucose. The source of nitrogen is the biological fixation of atmospheric N_2, confirmed by the identification of groups of nonsymbiotic atmospheric N_2-fixing microorganisms. At the same time, an increase in total nitrogen content from 1.4% to 5.6% (w/v) in the liquid supernatant was observed by Cereda and Giaj-Levra (1987).

These results were confirmed by Lacerda et al. (2005), who analyzed 30 samples of sour cassava starch collected from two industries in the Brazilian state of Minas Gerais. The authors detected lactic acid bacteria (LAB), yeasts, mesophilic microorganisms, *Bacillus cereus*, and fecal coliforms. *Lactobacillus plantarum* and *Lactobacillus fermentum* were the prevalent LAB in product from both industries, at numbers between 6.0 and 9.0 log cfu/g. *Lactobacillus perolans* and *Lactobacillus brevis* were present in a minor fraction of the population. *Galactomyces geothricum* and *Issatchenkia* sp. were the prevalent yeasts at numbers of 5.0 log cfu/g. The results of Carvalho et al. (1996) and Cardenas and Buckle (1980) confirm the predominance of lactic fermentation, responsible for 60% of the total acidity and 66%–82% of lactic acid over the total acids produced, the remainder being a mixture of acetic and butyric acids.

Penido et al. (2018) also reported the presence of LAB and yeasts isolated, enumerated, and grouped by Restriction Fragment Length Polymorphism and polymerase chain reaction fingerprinting, respectively. One isolate of each molecular profile was identified by sequencing of the rRNA gene. LAB were prevalent throughout the entire process. *Lactobacillus brevis* (21.5%), which produced the highest values of acidity, and *Lactobacillus plantarum* (13.9%) were among the most frequent species. *Pichia scutulata* (52.2%) was the prevalent oxidative yeast and showed amylolytic activity. These species were tested as single and mixed starter cultures in a pilot-scale fermentation process for 28 days. *L. plantarum* exhibited better performance as a starter culture, which suggests its potential to produce sour cassava starch.

For Figueroa, Davila, and Pourquié (1995), the acidification process is initiated by heterofermentative bacteria, mostly *Leuconostoc* spp, and is succeeded by the growth of homofermentative bacteria, predominantly *Lactobacillus plantarum*. Fermentation leading to the production of sour starch is accomplished by two microbial communities belonging to LAB, which are heterofermentative and homofermentative. Throughout starch fermentation, the lactic flora predominates and remains stable until the end of the fermentation process. Sour starch undergoes a fermentation by different microorganisms and has different organoleptic properties than if it had only undergone a lactic fermentation.

To explain how such a poor medium can support a fermentation as complex as that of microaerobic lactic bacteria, Cereda (1975) described corrosion marks on the cassava starch granules, which would be the effect of enzymes in the release of simple sugars, which in turn are substrates for the other microorganisms. This corrosion has been confirmed by several researchers. This damage is apparent as holes or pores in the granules, and with increasing acidity, erosion traces become more obvious (Garcia et al., 2016). Other authors have reported different traces of erosion of cassava starch granules that occur during fermentation, where amylose was found to be more degraded than amylopectin (Franco et al., 2010). These fermented sugars make the medium acidic and favor other groups of microorganisms, creating conditions for the development of lactic bacteria (Cereda & Giaj-Levra, 1987).

The formation of pores decreases the particle density of sour cassava starch compared to that of unfermented cassava starch. Furthermore, a negative correlation was found between particle density and baking expansion performance (Marcon et al., 2009).

7.4 Legislation and sour cassava starch characteristics

Sour starch is opaque white in color and is not much different from native cassava starch. When produced by the traditional process, besides having a cheesy smell, it exhibits a very characteristic granulation that can be felt on the fingers. Another characteristic of sour starch is that it has less paste viscosity, in the same concentration, of cassava starch (Cereda, 1983c).

According to Brazilian legislation (Brasil, 2005), sour starch is the acidified starch product from cassava, traditionally produced by natural fermentation and dried in the sun. In this legislation, the sour starch is differentiated from cassava starch only by the acidity content, without considering its main feature, the property of expansion without the need for chemical or biological yeast.

The Brazilian Ministry of Agriculture, Livestock and Supply does not register or inspect sour starch, under the allegation that, despite efforts to analyze the expansion, so far it has not been possible to present a method that has minimal repeatability and reproducibility, as well as conditions for products differentiation (native starch and sour starch, as well as their various degrees of mixture).

Therefore, according to the Brazilian 2005 regulations (Brasil, 2005), the limits for sour starch in Brazil are as follows:

- **Physical chemistry:**
  - Humidity % w/w: maximum 14.00.
  - Starch % w/w: minimum 80.00.
  - Ash % w/w: maximum 0.5.
  - Acidity expressed as mL NaOH N.100/g maximum of 5.00.
- **Microbiology:**
  - Total plate count: less than or equal to 5.10^4/g.
  - Fecal coliform bacteria: 0 in 1 g.
  - Sulfite-reducing Clostridium (44°C): maximum 20/g.
  - *Staphylococcus aureus*: absent in 0.1 g.
  - Molds and yeasts: maximum 10^3/g.
  - Salmonella: absent in 25 g.

The 2005 regulation, which was regulated in 2022, does not consider that when dealing with a natural fermentation, therefore a biotechnological process; it is natural that there is variation in acidity and, as a result, expansion, since these analyses depend on two factors hardly controllable, the content of lactic acid produced by microorganisms and UV radiation from the sun. In this case, it would be better to use ranges of values, as mentioned by Freitas (2020).

Most of the acid generated in natural fermentation is lactic acid, but because it is a natural fermentation, there is logically a variation in these levels, from 30% to 80%, which explains that the expansion method is also subject to variations. This problem disappears when lactic acid is directly incorporated in native cassava starch in well-established quantities, without going through fermentation (Cereda & Catâneo, 1986; Cereda & Lima, 1985; Cereda et al., 2001; Nunes & Cereda, 1994) and when the product is then subjected to UV-C under standardized conditions (Brito et al., 2021).

Besides the Brazilian legislation not foreseeing the expansion analysis, it presents another problem, which is the limitation of acidity, foreseen in a maximum of 5.00 mL NaOH N.100/g. One of the main characteristics of sour starch is its acidity, considered by many as a quality criterion. The setting of limits for this parameter is not based on any technical or health protection criteria. Lactic acid, which is mainly responsible for the high acidity of sour starch, appears in fermented dairy products.

The second most frequent acid is acetic acid, also found in vinegar, and butyric acid, responsible for the aroma of cheese and butter (Cereda & Lima, 1985; Cereda & Vilpoux, 2004, 2006).

The proof of this statement is in the results of HPLC analysis on 29 samples of commercial sour starch carried out by Demiate et al. (1999) in the south and southeast regions of Brazil. The authors found concentrations between 120 and 830 mg/kg of lactic acid, 680 mg/kg of acetic acid, 130 mg/kg of propionic acid, and 570 mg/kg of butyric acid. Organic acids are present in sour cassava starch and contribute with organoleptic and physical characteristics like aroma, flavor, and the exclusive baking property, that differentiate this product from the native cassava starch. The average value found by the authors for acidity was 6.0, with pH minimum of 3.9 and maximum of 4.7. The variations in pH and acidity are shown in Fig. 7.6.

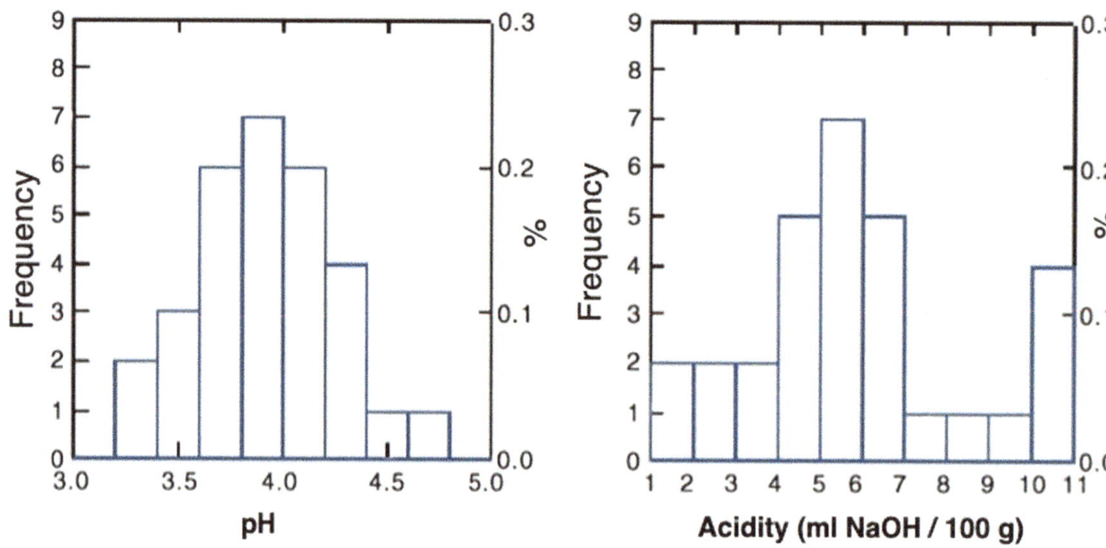

FIGURE 7.6 Variation of pH and titratable acidity in 30 commercial Brazilians sour starch samples. *From Cereda, M. P. & Vilpoux, O. E. (2004). Sour starch, quality criteria for use in food products. In: M. P. Cereda & O. E. Vilpoux (Eds.).* Technology, use and potentialities of Latin American starchy tuber crops *(Vol. 3, pp. 333–376). Cargill Foundation. Chapter 12. (In Portuguese).*

TABLE 7.1 Mean count results in two commercial sour starch samples from southeastern Brazil, in FCU.100/g.

Microorganisms	Sample 1	Sample 2
Total plate count	<10	3100
Fecal coliform bacteria	<10	<10
Thermoresistant coliforms (44°C)	<10	<10
Sulfite-reducing clostridium (44°C)	-*-	-*-
Staphylococcus aureus	<100	<100
Molds and yeasts	360	120
Salmonella	-*-	-*-

FCU, Forming Colonies Unities; -*- Not detected.
From Cereda, M. P., & Vilpoux, O. E. (2004). Sour starch, quality criteria for use in food products. In: M. P. Cereda & O. E. Vilpoux (Eds.), Technology, use and potentialities of Latin American starchy tuber crops *(Vol. 3, pp. 333–376). Cargill Foundation. Chapter 12. (In Portuguese).*

For Demiate et al. (1997), the pH values define sour starch as weakly acid, with an average of 4.0. The pH and acidity values are consistent with the natural fermentations that occur.

In other aspects, the composition of sour starch is very close to native cassava starch where the starch content is high and the protein content low so that even with fermentation, sour starch is still considered a calorie product.

The water activity of the sour starch ranges from 0.6 to 0.7 at 25°C, very close to the 0.6 obtained for native cassava starch. The values are sufficient to prevent the growth of microorganisms, besides the fact that acidity and the action of solar radiation control the growth of microorganisms, as can be seen in Table 7.1.

It is observed that despite the problems of the lack of uniformity, the microbial quality of the samples falls within Brazilian legal limits, probably due to the factors already pointed out.

7.5 Specific analyses for sour starch

Expansion power is the main characteristic of sour starch. The first expansion evaluations were made in home kitchens, using various ingredients such as milk, butter, eggs, and cheese.

The variability in composition of these ingredients made it difficult to compare performance between different batches of sour starch. Cereda (1983b) established a basic formulation for expanded snacks (*biscoitos*), similar to that used in baking. The standard formulation consisted of sour starch 100%, hydrogenated fat 25%, salt 4%, and water around 80%, with variation according to the sour starch used. The amount of water should be sufficient to allow for the formation of a dough similar with toothpaste, equivalent to 60 Farinographic Units (FU), adjusted by Nunes and Cereda (1994) based on Cereda's method (1983b) and inspired by the method of the American Association of Cereal Chemists (1978) with the instrumental use of a Brabender Farinograph. The use of the equipment allowed for the establishment of a standard consistency dough.

Fernández et al. (2002) agreed that it is impossible to achieve uniformity when the quality analysis, in the case of expansion, does not use standardized ingredients. In these procedures, the characteristics of the baked products were dependent to a large extent on the quality, quite variable, of the other ingredients used, especially cheese and fat. Because of this, the results of the evaluation carried out by the producers of sour starch and that of the consumers often do not coincide, giving the product a poor image in the market.

Besides being time-consuming, this method requires specialized training of the analyst and investment in expensive equipment. The biscuit-shaped test specimens are made with the aid of a confectioner's apparatus directly on an ungreased aluminum baking sheet and baked at 200°C for 10–20 minutes. The evaluation of the sour starch quality is measured from the specific volume, which is determined by seed displacement. With that, the method for measuring oven expansion developed by Cereda (1983c) allows for greater repeatability.

Using this method, Plata-Oviedo and Camargo (1995) evaluated commercial sour starch samples and obtained a range of 11–15 cm^3/g. Demiate et al. (1997) analyzed 52 commercial sour starch samples collected on the market in the southeastern states of Brazil. Table 7.2. presents the average results obtained with dough consistency adjusted by visual evaluation and with the use of the Brabender Farinograph.

The results in Table 7.2 highlight that there was a difference between the two ways of adjusting the consistency of the dough used for the test, with results always higher when the adjustment was made in Brabender equipment. Despite the differences, because it does not require specialized equipment, the visual adjustment method is more suitable for dissemination to small producers, intermediaries, and buyers.

To evaluate the starch expansion property of sour starch, Demiate et al. (2000a,b) used 12 g of starch mixed with near 10 mL of boiling distilled water, to obtain a partially gelatinized, soft, and moldable dough. The sample was then manually mixed using a glass stirring rod, until it became homogeneous and detached from the walls of the container. The obtained dough was divided into three equal portions, rounded to spheres with near 3 cm o diameter and baked using a preheated electric oven, for 30 minutes at 200°C.

The optimum oven temperature was 200°C. Doughs baked in silicone molds needed 25 minutes to provide an average expansion of 8.9 cm^3/g, with 1.46% variation. The Teflon molds were ready after 20 minutes, with an average expansion of 8.2 cm^3/g and a greater variation of 7.59%. The samples proved to be sufficiently resistant to handling to allow measurements, without the need to be coated with wax, as suggested by Matta et al. (2019).

This method circumvented most of the expansion evaluation problems. Maeda and Cereda (2001) analyzed a large number of samples with this method and adjusted the expression of expansion results to only one place after the comma.

This simple method was adopted by other researchers from Brazil and abroad (Fernández et al., 2002; Qi et al., 2021). Qi et al. (2021) adapted this method, mixing 50 g of the sour starch samples (d.b) with 40 mL of boiling water using a stand mixer at intermediate speed for 2 minutes. The dough was divided into 10 g samples with a rectangular shape and 1 cm thickness and baked at 200°C for 20 minutes in an electric oven.

The results obtained by Qi et al. (2021) confirm that alone, the fermentation, sun, and UV lamp drying treatments do not provide expansion. The expansion of native cassava starch was 3.1 cm^3/g, similar to that of 3.41 cm^3/g, obtained by oven-dried fermented cassava starch, but after fermentation and submitted to sunlight, it showed an expansion of 6.7 cm^3/g, and the same starch when treated with a UV

TABLE 7.2 Classification of the specific volume of the expansion of 52 sour starch samples, with dough consistency adjusted by visual evaluation or with Brabender Farinograph.

	General average in cm^3/g	
	Visual consistency	**Consistency of 60 UF**
Average (Standard Deviation)	14, 0 (1.5)	18, 9 (3.6)
Type A (high expansion)	15, 8	23, 4
Type B (medium expansion)	12, 2–15, 7	14, 4–23, 3
Type C (low expansion)	12, 1	14, 3

FU, Brabender Farinographic Units.

lamp showed an expansion of 5.7 cm^3/g. The results obtained by these and other authors (Bashir & Aggarwal, 2019; Díaz et al., 2019; Qi et al., 2021) corroborate those of Cereda et al. (1982) on the need for joint fermentation treatment with UV light.

Matta et al. (2019) also adapted the method reported by Demiate et al. (2000a,b). After cooling, the expanded samples were impermeabilized by immersion in liquid paraffin. Then, their volume was measured by water displacement using a graduated cylinder.

Despite the improvements, the need to add boiling water hinders the application of the method described by Demiate et al. (2000a,b). The cooling of this water during manipulation may cause variation of the gelatinization content, with a change in the expansion, besides the risk of accident. Thus, Brito et al. (2021) changed this method by adding cold water, which facilitates standardization. They also evaluated two types of baking molds, silicone molds and metal coated with polytetrafluoroethylene (Teflon). The samples were placed directly into the molds and baked under temperature-controlled conditions at 200°C and 250°C for 20 or 25 minutes. The selection of the best method took into account the highest values for specific volume, measured in cm^3/g, and the variability of the obtained values.

7.5.1 Other methods of analysis for sour starch

Although expansion is the main tool for evaluating sour starch, many other methods are available, many of them international and well standardized, some adjusted for use in sour starch, such as those cited by Qi et al. (2021).

7.5.1.1 *Pasting properties*

Vatanasuchart et al. (2005) found a linear relationship between peak viscosity and the baking expansion of sour cassava starch. However, Marcon et al. (2006) reported that the peak viscosity of cassava starch is not strongly related to its baking expansion ability, but rather to the structural changes caused by the fermentation and drying processes. The mechanisms underlying these changes have not been elucidated. It has also been proved that the viscosity of starch paste is related to the content of lactic acid or butyric acid that accumulates during fermentation. However, no correlation was found between the viscosity of starch paste and baking expansion (Díaz et al., 2018).

7.5.1.2 *Infrared spectroscopy*

The physicochemical methods used for starch analysis are time-consuming and use large samples, but mid-infrared (MIR) spectroscopy allows a small sample, identifying most modifications quickly and accurately, as proven in many starchy products, including starches with expansion properties (Demiate et al., 2000a,b; Dolmatova et al., 1998; Dupuy & Demiate, 2022).

If infrared spectroscopy presents many advantages over traditional physicochemical analyses, it requires good standardization with well-established samples. This analysis allows also to explore the molecular structure as a function of the functional groups present but is hampered by the chemical composition of starch consisting substantially of glucose molecules, which dilutes the few radicals that give modified starch its special properties (Kizil et al., 2002).

To circumvent this problem, Shokey and Hanna (1993) proposed the use of enzymatic hydrolysis of starch as an auxiliary process to infrared analysis.

The possibility of sour starch being a modified starch has been little explored by research (Cereda & Vilpoux, 2004, 2006; Cereda et al., 2003; Macêdo et al., 2023). The presence of free radicals produced by oxidation was emphasized by Demiate et al. (2000a,b) and

Cereda et al. (2003), who investigated the mechanism of modification that occurs and identified the oxidation reaction. So, in this respect, sour starch can already be considered a modified starch, although this type of modification does not produce a lasting expansion.

In Fourier transform infrared (FTIR) spectroscopy used by Macêdo et al. (2023), the authors also had problems to relate sour starch to modified starches. They evaluated sour starch using MIR analysis as a tool. The authors used prior treatment by enzymatic hydrolysis to remove excess glucose, as recommended by Kizil et al. (2002) because, according to Chough et al. (2006), glucose molecules hinder the analysis of infrared spectrometry. The technique described by the authors is based on the specificity of the enzymes' reaction with the substrate. As modifications prevent the action of the enzymes, the hydrolysis allows for the elimination of unmodified glucose chains, making it easier to identify the changes that have occurred. The precautions mentioned by Uthumporn et al. (2010) were followed, with pH and temperature control in the ranges indicated by the enzymes manufacturers to remove only the areas of the structure of the starch granules that did not offer obstacles to the enzymes.

To compare modified cassava starches commercialized in Brazil, Macêdo et al. (2023) used samples of 10 g (DM) of starch dispersed in 500 mL of deionized water under agitation and adjusted pH to the range of 6.0–8.0, as recommended by Novozymes as suitable for the action of α-amylase. The adjustment was made with acetic acid (0.01 M) or sodium hydroxide (0.01 M). The suspension was then kept under stirring in a water bath, and 30 μl of thermal resistant Thermamyl Novozymes was added as recommended, when temperature ranged from 90°C to 100°C. After 60 minutes, the pH of the slurry was again adjusted to the range of 4.0–4.5 and the temperature between 57°C and 70°C. After that, 30 μL of AMG Novozyme (gluco-amylase synthesized by *Aspergillus Niger*) was added. The suspension remained under stirring at 600 rpm for 60 minutes according to Novozyme recommendation. After the enzymatic treatment, the samples were put into dialysis bags (Inlab 33 mm × 21 mm × 30 cm) with a porosity of 25 Å (Angstroms), with demineralized water at 4°C for 15 days. The solubilized sugars were passed through the pores of the dialysis tubes, and a solid residue, resistant to enzymes, remained inside the tube. Periodic exchanges of deionized water were carried out to eliminate the sugars. Tests to detect reducing sugars were performed daily to confirm that washing had been sufficient. The residue was then recovered by drying in an oven at 50°C with air circulation. The dried material was weighed and prepared for MIR spectroscopy analysis.

Using this method, Macêdo et al. (2023) evaluated samples of native cassava starch, commercial sour starch, and five of the most common modified starches in Brazilian market (oxidized, acetylated, acid-modified, etherified, and esterified). The MIR analysis was performed after hydrolysis of the starches, and the results were evaluated by principal component analysis. The analysis identified a clustering between the sour starch sample, acid-modified, which was expected, and oxidized starches, showing a proximity between these starches. The presence of lactic acid has already been described in this chapter and may explain the proximity with acid-modified starch. Modification by oxidation was mentioned by Demiate et al. (2000a) and Cereda et al. (2003). The results confirmed the possibility of MIR analysis to identify modified starches and the need to treat the samples by enzymes to remove glucose molecules. This analysis can be conducted quickly and with greater certainty and is an important tool in distinguishing the native starches from the modified ones (Smith, 1967; Trela et al., 2020).

Another important result found by Macêdo et al. (2023) using MIR analysis was that the residual mass of the hydrolyzed sour starch sample did not differ from the hydrolyzed oxidized starch and the hydrolyzed etherified samples. One explanation for this result is the hypothesis that the change that occurs in cassava sour starch is a graft. Contributing to this hypothesis is the fact that the lactic acid-modified starch sample showed a lower percentage of carboxyl. In this case, the literature explains that some of the carbonyl groups presented are converted into carboxyls, which are responsible for the polymerization of lactic acid by the inclusion of double oxygens and the formation of the lactide molecule, composed of two carbonyls (Takizawa et al., 2004; Wang et al., 2012). This hypothesis will be developed further in Section 7.7.

The residual mass corresponds to a new approach in the analysis of modified starches and is strongly recommended. The greater the weight of the residue, the greater the area of the starch granule refractory to the action of the enzymes, which means that it resists enzymatic digestion.

7.5.1.3 *Carboxyls or reducing power analysis*

Demiate and Cereda (2000) and Cereda et al. (2003) highlighted the % of carboxyls as important in the characterization of sour starch. Macêdo et al. (2023) also highlighted this analysis, identifying the % of carboxyls and the hydrolysis residues of starch samples as tools to analyze sour starch. Carboxyls measure the free radicals created at the nonreducing extremity of amylose and amylopectin, the main polymers that starch is composed of. The highest values were observed for the oxidized starch, but also in sour starch. Later, these molecules, partially reduced from their molecular mass, can reaggregate, but the carboxyl percentage can indicate both oxide and acid modifications.

7.5.1.4 *Use of amylolytic enzymes*

The treatment of samples with amylolytic enzymes is a powerful tool to reduce the amount of glucose, concentrate, and highlight the points of modification. In modified starches, a cross-link is present for every 100–3000 glucose units (Wurzburg, 1986), which makes identification in analysis difficult.

Macêdo et al. (2023) showed that native cassava starch has the highest percentage of mass removal, precisely because its $\alpha-1,4$ bonds are all potentially subject to enzymatic action. Thus, the native starch sample obtained a residue weight of 0.01%, while it was 0.40% for the etherified samples and 0.30% for oxidized starches, similar to the one of sour starch.

Therefore, the use of enzyme treatment coupled with IR analysis was judged appropriate for the evaluation of the sour starch production experiments.

7.5.1.5 *Analyses performed by Qi et al. (2021)*

Qi et al. (2021) reported results of fermentation conducted in the laboratory to simulate the traditional sour starch production process and follow the changes that should result in the main characteristic of the product, which is expansion without the use of raising agents. The authors inoculated a suspension of native starch in water of 50 g.100/g (d.b) with pure cultures of lactic bacteria. After fermentation, the decant was subjected to drying procedures, while being exposed to sunlight (2 mm thick layer) or ultraviolet radiation (100 μW/cm^2) for 8 h. A control sample was heat-dried for 8 h in an oven at 45°C. The results obtained, with the exception of the expansion results, which have already been presented, are presented below.

7.5.1.5.1 Scanning Electron Microscopy

Qi et al. (2021) hypothesized that treatments such as sunlight and UV form a network of gelatinized starch after scalding and that

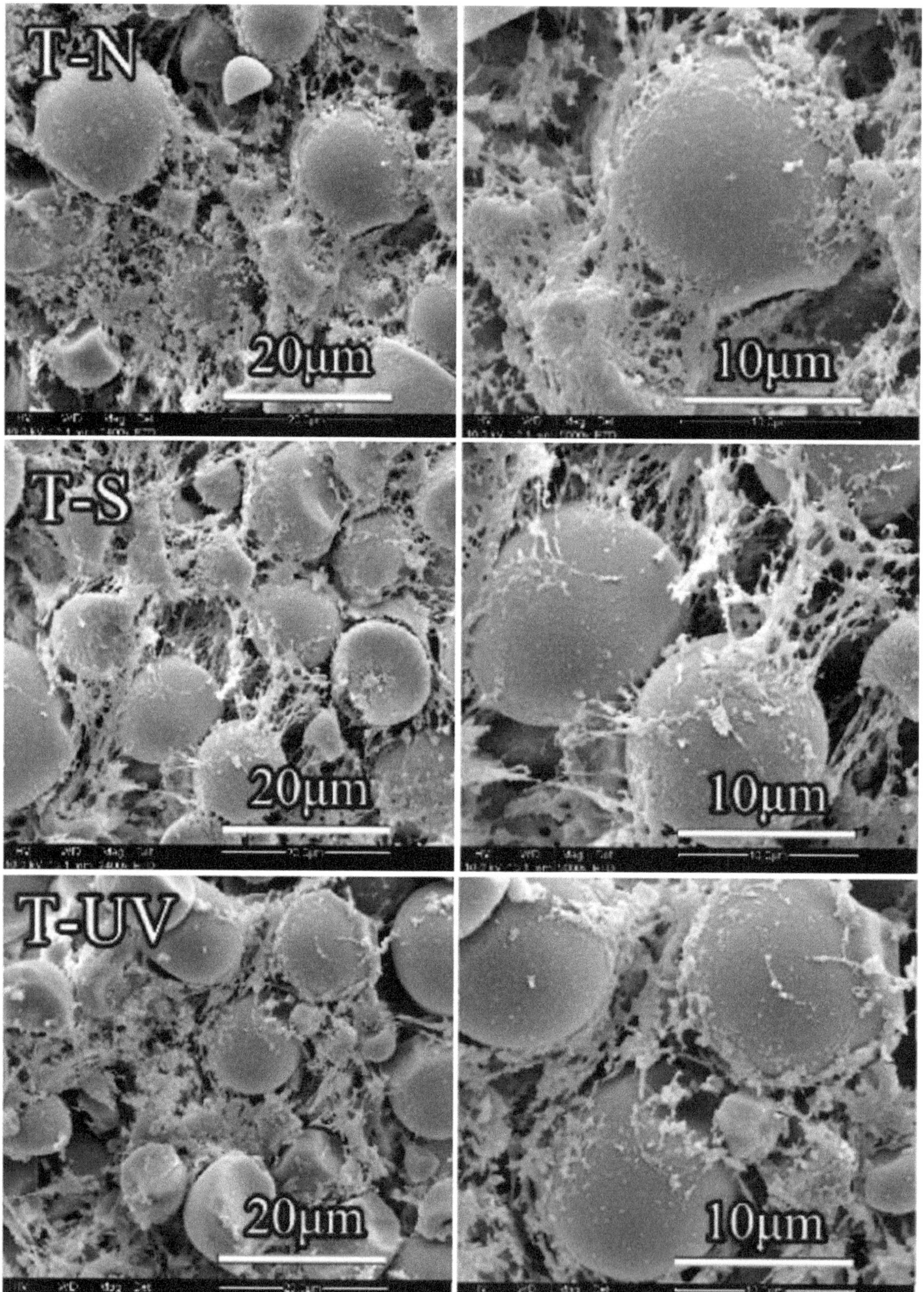
T-N
20μm
10μm
T-S
20μm
10μm
T-UV
20μm
10μm

this network would be responsible for the gas holding capacity, which is characterized by expansion in the oven. The scanning electron microscopy method was used to try to locate and explain this network (Fig. 7.7).

The analysis, even under light microscopy, can be relevant to get an overview about the samples, as well presented by Qi et al. (2021).

7.5.1.5.2 Starch granules morphologies by polarized light to identify the polarized cross

Polarized light showed that the crystal structure of cassava starch granules was little altered by the treatments that conferred expansion to the samples (Qi et al., 2021). The modified samples differed little from the untreated native starch. Thus, in doughs with fermentation + sunlight and fermentation + UV treatments, the starch with a granular shape still had a clear polarized cross. Furthermore, the X-ray diffraction results revealed that none of the treatments affected the crystal form of cassava starch, which still exhibited an A-type structure with peaks at $2\theta = 15.1°$, $17.1°$, $18.1°$, and $23.6°$. This result had already been reported by Mestres and Rouau (1997).

However, Matta et al. (2019), in research very similar to that of Qi et al. (2021), obtained discrepant results in crystallinity and increased disordered crystalline regions with lower molecular weight in the two types of starches, fermented with sunlight and with UV light treatment. Both appeared to have a continuous network structure and polarized cross in scalded dough, which retains gas, thus promoting high baking expansibility.

7.5.1.5.3 Pasting properties

The analysis of scalded doughs under different treatments was the major contribution of the results obtained by Qi et al. (2021) to explain the properties of sour starch. For Machado et al. (2010), there is a lack of research on the structure of hot scalded dough prepared with starch and hot water and about the need for the scalding process to produce starch with high expandability.

The results showed that scalding of native cassava starch resulted in cassava dough with formation of a continuous network structure. This result is similar to the one obtained with wheat dough by Zhu et al. (2019). The continuous gel structure and lower paste-forming temperature of starch doughs are both conducive to gas holding and expansion (Föste et al., 2017).

After treatment with sunlight and UV light, the cassava starch dough contained both starch granules with a relatively intact crystalline structure and large fragmented gel, indicating that this structure was created in the process of hot scalding. However, while part of the starch was completely gelatinized, the remaining starch retained the complete structure. However, under the same conditions, the persistence of gelatinized, nongelatinized, and intermediate starch granules occurs in all types of starches and cannot be considered a characteristic of sour starch.

7.5.1.5.4 Gel exclusion chromatography

The results of the determination of molecular weight distribution obtained with gel exclusion chromatography revealed two peaks corresponding with the cassava starch molecule. Peak 1 corresponds to amylopectin and peak 2 to amylose. The samples that showed expansion (fermented + sunlight and fermented + UV radiation) showed a significant reduction of relative molecular weight of branched amylopectin,

FIGURE 7.7 Image of scalded dough from native cassava starch (T-N), fermented and subjected to treatment by sun exposure (T-S) and to UV radiation (T-UV). *Photos from Q. Qi, Y. Hong, Y. Zhang, Z. Gu, L. Cheng, Z. Li, & C. Li. (2021). Effect of cassava starch structure on scalding of dough and baking expansion ability.* Food Chemistry, *352, 129350. https://doi.org/10.1016/j.foodchem.2021.129350.*

while the polydispersity index was significantly increased ($P < .05$). Alvarado et al. (2013) already mentioned that sunlight was responsible to change the molecular content of amylose and amylopectin.

The result of Qi et al. (2021) that the amylopectin mass was reduced after treatment with lactic acid and solar or artificial radiation was also confirmed by X-ray diffraction and does nothing to explain the expansion property that characterizes sour starch.

In general, it can be stated that the method that best evaluated the results to obtain starch with expansion property, with focus on reproducing the sour starch, was the expansion in the oven. The appearance of the starch granules under microscopy cannot be related to the treatments, the crystal structure was slightly damaged, and the polarized cross was detected in undestroyed granules. The analysis of moldable doughs obtained after scalding showed that water and heat used on the starch during the scalding process produced images of the dough morphology with intact starch granules rather than absolute gelatinization. In doughs obtained from cassava starch subjected to lactic fermentation and treatment under sun or UV radiation, the starch with a granular shape still retained a clear cross-polarized crystalline structure.

The results do not explain why, according to Brito and Cereda (2015), the sour starch presents expansion even without previous scalding, only with the addition of cold water and without forming an elastic and consistent dough. In this case, the reports about the importance of scalding and the formation of an elastic dough should be reviewed.

It would have been desirable to dose carboxyls in all treatments and especially in those that presented expansion in the oven, because research relates the content of carboxyls and carbonyls with the application of UV radiation, natural or artificial. The gelatinization phenomenon of the granules was evaluated only by images, without the use of lugol.

If the theory of grafting lactic acid fragments into the structure of cassava starch is correct, with the formation of an ether group, it is likely that the cross-link and grafting have tied up the starch structure and do not allow major changes in the crystal structure, which has been observed in several researches.

The results of analyses to explain gas retention are not conclusive and have not advanced much in the explanations already available for understanding the phenomenon that gives sour starch expansion. Another problem found in the literature on sour starch is the insistence that lactic fermentation is mandatory in the process. However, once the importance of lactic acid in this biological process is known, it is possible to bypass this step. Thus, results published by Brito et al. (2021) demonstrate that it is possible to obtain cassava starch with equal or greater expansion property than traditional sour starch by replacing lactic fermentation (natural or inoculation-induced) simply by adding commercial, food grade lactic acid in a starch-in-water suspension, with subsequent submission to UV-C provided by commercial lamps. More details of this method are described in Section 7.7, which deals with innovations in the sour starch production process.

It should also be noted that the term drying should not be confused with treatment. In the case of sour starch, the important aspect of keeping the product in the sun is not drying, which could be done in a flash-dryer, as for the other starch derivatives, but the UV-C treatment.

According to Cereda et al. (2001), the mechanism that regulates the expansion of sour starch has not yet been fully explained, but it is known that the product has lower gelatinization temperature than native cassava starch (Cereda, 1983c). Sour starch is a modified starch, so it is important to know the types of modifications that could explain the expansion property. For this, it is recommended to read Chapter 6.

7.6 Innovations to improve the traditional production process

Innovations are essential for product development, to better serve markets and consumers. They can come from academic research or from companies. Academic research disseminates innovations mainly in the format of scientific articles, which should be in the public domain. However, when there is financial interest or the possibility of implementation as an industrial process, innovations are often disclosed in patents.

A patent is a technical document, specifically developed with the purpose of both describing the innovation and preventing it from being copied, for a period established according to the intended innovation. Even under this period of protection, the patent allows another user to learn about the described process and enhance its knowledge for further improvements. Filing the patent simultaneously reports the stage already achieved and encourages further development into new innovations. In general, the texts are short and to the point and describe in detail the innovation developed, in the certainty of having protection for the process or equipment, or for both.

The Google patent search in Portuguese using the word sour starch provided 58 references, while the English search was necessarily broader because there is no single term for this product, sometimes called fermented cassava starch, fermented tapioca starch, sour starch, and so forth.

The words to characterize cassava starch offer at least two possibilities in English, cassava & starch or tapioca & starch. The search using the words cassava & starch resulted in 40 pages with approximately 11,000 abstracts of submitted patents available in Google patents. When replacing the word cassava with tapioca & starch, 10 pages were found.

To better identify patents linked to the production of sour starch, the two forms were consulted, plus the word expansion. Therefore, the searches were made with the terms cassava or tapioca starch expansion and sour starch, resulting in patents for processes, products, and uses. These patents were complemented by scientific papers to identify innovations in sour starch production.

From the research developed by Marney Pascoli Cereda since 1973 (PhD Thesis), the first premise was that for cassava starch to exhibit the property of expansion in the oven, the following steps were necessary: (1) contact of the starch with lactic acid in a concentration and time to be established and (2) exposure to UV-C radiation, which predominates in solar radiation in periods when the commercial product is exposed to the sun, for a period to be established.

These assumptions started to be adopted from the moment it was published that sour starch is the physically modified starch traditionally prepared from cassava starch by fermentation followed by sun-drying (Alvarado et al., 2013; Nunes & Cereda, 1994; Zhu, 2015). This process is mostly artisanal, difficult to control, and highly heterogeneous. Consequently, the safety and quality of the final product is subject to wide variations (Camargo et al., 1988; Freire et al., 2015).

From these findings, several lines of sour starch development can be found, one that aimed to speed up the fermentation process, another that sought to improve UV-C treatment, and one that developed physicochemical modifications, mainly by replacing UV with oxidation treatment.

7.6.1 Improving natural fermentation

The fermentation of the cassava root is a process that was already used by the Amerindians to preserve food, especially during their wanderings through the forests. To feed the travelers, it was common to put mass of grated and naturally fermented cassava root, which thus preserves better. In these spontaneous fermentations, the lactic acid is always detected (Cereda & Vilpoux, 2006).

Sour starch production primarily involves three main steps: isolation of starch, fermentation by natural microflora, and drying under the sun. Native starch after extraction undergoes spontaneous anaerobic fermentation for about 20–30 days at 15°C–25°C, predominantly by LAB, followed by subsequent drying under sunlight for about 1–2 days depending upon the weather conditions (Cereda & Giaj-Levra, 1987; Cereda & Lima, 1985; Cereda et al., 1982; Cereda, 1975; Dias et al., 2011; Díaz et al., 2018b; Gomes et al., 2005).

Some research considered the possibility of the existence of cassava varieties with more susceptible starch for sour starch production. Onitilo et al. (2007) evaluated the influence of cassava varieties on the physicochemical and functional properties of sour starches. The characteristics of the extracted starches highlighted by the authors were the ash content, pH, amylose, amylopectin, starch damage, total titratable acidity, sugar, and starch content. The results showed that the selected criteria showed no correlation with fermentation or expansion, so the concept that there could be cassava varieties more suitable than others for sour starch production remains to be proven.

Schindler (2001) filed a patent for accelerating the fermentation process. The cassava starch suspension with 40% solids is pumped directly into a double walled metal reactor for heating with hot water or steam, with continuous agitation with a paddle mixer of the starch suspension in the reactor. At this stage, the lactic bacteria would develop faster to accelerate the starch fermentation process, with a reduction from 45–60 days to 03–07 days, depending on the type of product desired. However, Schindler forgot that the lactic bacteria are microaerophilic and grow at room temperature, as is the case with traditional sour starch.

With the same objective of making the fermentation process faster, Marcon et al. (2006) evaluated the effect of glucose syrup addition on sour cassava starch fermentation. Four treatments were tested: traditional, 0.10%, 0.25%, and 0.50% of glucose syrup added to the cassava suspension volume. The results showed that a glucose syrup concentration of 0.50% contributed to doubling the fermentation yield. The objective was to observe the effect of the cassava starch fermentation process on starch granule microstructure with glucose addition, as compared with industrial processing. However, the authors did not explain how they evaluated the effect or why to add glucose if the lactic bacteria have amylolytic enzymes, since it makes more sense to add lactic acid directly on the starch.

Patent BR102115030137A2 filed by Almeida-Starling (2017) used native cassava starch, lactic acid, and inoculum (fermented starch dough) to accelerate the fermentation and sun-drying process. Hong et al. (2021) submitted a patent for an industrial process to use fermented tapioca in baking and make it faster by adding selected LAB, in this case *L. plantarum*.

During starch fermentation, a diverse microflora develops onto the submerged starch. As a result, amylolytic enzymes and different organic acids are produced that alter the starch molecules' molecular structure and surface morphology by providing small corrosion onto the surface of the starch granules. Under direct sunlight, starch further confers the sour starches with improved functional properties such as a greater capacity of gas retention and dough expansion during baking (Díaz et al., 2018a; Garcia et al., 2016; Sena Aquino et al., 2015).

The succession of microorganisms originating from the raw material and the fermentation tanks is mainly responsible for the natural fermentation of starches, with LAB predominating. The microorganisms from fermentation include *Leuconostoc mesenteroides, Lactobacillus plantarum*, Lactobacillus *brevis*, *Lactobacillus fermentum*, and yeasts (*Galactomyces geothricum* and *Issatchenkia* sp.) (Kostinek et al., 2007; Lacerda et al., 2005; Penido et al., 2018).

Therefore, the fermentation of sour starch is not just a case of preserving or developing aroma and flavor. Lactic acid is an integral part of the process and, in addition to the presence of lactic acid, its amount is also important. The only exception comes from Sena Aquino et al. (2015), who found no predominance of lactic acid in fermentations for sour starch production. However, the authors do not explain at which moment of the natural fermentation the samples were collected.

In Penido et al. (2018), applied for a patent for a process in which the succession of microorganisms from raw materials and fermentation tanks occurs naturally during cassava fermentation and results in a microbiota with a prevalence of LAB, such as *Lactobacillus plantarum*, *Lactobacillus fermentum*, *Lactobacillus brevis*, and *Leuconostoc mesenteroides*. In the production of sour cassava starch in Brazil, Lactobacillus occurs in association with yeasts, such as *Galactomyces geothricum* and *Issatchenkia* (now *Pichia*).

It is important to remember that these microorganisms appear naturally in the fermentation of cassava starch, which hardly explains the purpose of this patent.

Figueroa-Flórez et al. (2018) evaluated the effect of the solid/liquid ratio (1:1, 1:2, 1:3 w/w) and the enrichment of the fermentation medium by adding glucose syrup and yeast extract (0.75% w/w) on the physicochemical properties of the medium, such as pH and titratable acidity, during the fermentation step. In addition, the quality of the fermented starch was evaluated with FTIR spectroscopy, along with analysis of the paste viscosity behavior during the heating and cooling cycles with a viscoamylograph and the dough expansion capacity. The results showed a significant increase in the titratable acidity of the fermentation medium ($P < .05$) that was linked to a pH decrease during the first 15 days of the fermentation.

According to Bangar et al. (2022), sour starches are physically modified by natural fermentation or LAB, followed by drying. In addition, compared to native starch, sour starches show improved swelling and solubility, are thermally stable, and are less susceptible to hydrolysis. Moreover, the fermentation process is responsible for increased gelatinization potential and decreased aging ability of starches, which could benefit starch-based products, including noodles and bread. It is important to note that the authors do not relate the results to the main characteristic of sour cassava starch, which is expansion in the oven.

However, despite its importance, the improvement of cassava starch fermentation lost its meaning when it became clear the importance of lactic acid in the expansion property of sour starch and not the fermentation itself. Thus, Plata-Oviedo (2010) described the process for obtaining cassava starch with expansion property that consists of treating a suspension of native cassava starch with organic acid. However, he did not treat the acidified starch with UV.

7.6.2 Improving UV treatment of starch

The incidence of ultraviolet radiation from the sun has the advantage of reducing the counts of contaminating microorganisms. An example of this can be seen in Table 7.1, where the total plate counts of fecal coliforms, thermoresistant coliforms, sulfite-reducing clostridium, *Staphylococcus aureus*, and salmonella are below or within the legal limits. Only yeasts and molds have higher counts, but are still within the legal limits (Demiate et al., 1997). However, despite this advantage, the UV influence of the sun has the disadvantage of providing an uneven expansion, in case of rainy or even cloudy weather, compromising the production.

The hypothesis is that the modification in the starch structure caused by UV-C radiation is instantaneous, which characterizes a physically and chemically modified starch. This hypothesis allows thinking about the design of an equipment that would allow for the treatment of cassava starch for the confection of sour starch, in starch granules impregnated with lactic acid and exposed to UV-C radiation. It is worth noting that with this process, a modified starch can be produced by the same concept as traditional sour starch, but in a faster and more controlled process.

Nunes (1994) evaluated the three UV radiation bands available in special lamps, UV-A, U-V-B, and UV-C, and established a process replacing sun exposure by special lamps that emit UV-C radiation with results that led to the proposal of a continuous production process of sour starch. Nunes and Cereda (1994) reported that to meet the requirement of exposure to UV-C radiation, mechanical revolving systems of the cassava starch paste already impregnated with 6% lactic acid were included, which allowed circumventing the problem of exposing the starch to solar radiation.

The first Brazilian patent application on sour starch was also the first to address the mechanization of the process of treatment of macerated starch with lactic acid, with subsequent submission to UV-C radiation. It was deposited in Portuguese under the number BR 94022303 A (Nunes & Cereda, 1995). The text already cited the replacement of the fermentation process by the addition of lactic acid. The nearly dry mass was to be stirred periodically by means of paddles, while lamps provided UV radiation. The band cited in the patent was *UV-C* (200–290 nm) because it is the one that predominates in sunlight in the periods preferred by producers. However, the expansion tests with the developed process resulted in great variability, which was explained by the difficulty of the UV radiation to reach a sufficient and constant number of cassava starch granules impregnated with lactic acid. Although the reaction is instantaneous, the penetration of UV-C into the starch is very superficial and reaches a small amount of starch granules. Also, it was evidenced that UV-C lamp treatment of low moisture starch mass under mechanical stirring raised dust, which reduced the effect of the radiation.

In this case, mechanization was done by stirrers adapted to the roof of a tunnel where the UV lamps were also fixed. This patent was rejected by the Brazilian INPI (National Institute of Patents and Innovation), for considering that it was an artificial drying project and therefore did not represent innovation. Even so, this process was initially disclosed in Portuguese and, with the approval by the European Union of the project "Valorisation des produits, sous-produits et déchets, de la petite et moyenne industrie de transformation du manioc en Amérique Latine," it began to be disclosed in English.

As a result, articles began to be published on the use of ultraviolet radiation (Durán-Álvarez et al., 2016; Rincón & Pulgarin, 2004). This radiation is of low penetration, hence the need to regularly stir the fermented dough exposed to the sun in the artisanal process of producing sour starch. Each revolving exposes more starch granules to solar radiation, which is very high in energy, but low in penetration.

Several articles have also begun to relate the production of sour starch to the oxidation of native cassava starch. Lactic acid, which develops due to fermentation, is oxidized by UV rays during sun drying and modifies the functionality of the starch, showing aptitude for expansion during baking in an oven or frying (Dias et al., 2007; Putri et al., 2012).

Atrous et al. (2016) applied for a process patent in which they substituted UV radiation with gamma radiation (3, 5, 10, 20, 35, and 50 kGy). Although all the analysis methods cited in Section 7.5. were evaluated, the authors do not report whether they evaluated the expansion in the oven.

Matta et al. (2019) analyzed the effect of fermentation combined with exposure to sunlight or UV light treatment. The authors observed a decrease in crystallinity and increased disordered crystalline regions with lower molecular weight in starches, fermented with sunlight or fermented with UV light treatment. Both treatments appeared to have a continuous network structure and a polarized cross in the scalded dough, which favors gas retention and water loss, thus promoting high baking expansibility.

Hong et al. (2021) subjected a fermented cassava starch to 150,000 lux illumination for 8 hours, but reported no expansion tests on the final product.

7.6.3 Modified starches as sour starch substitutes

With the increasing consumption of sour starch and a traditional and still rudimentary production process, it was predictable that companies would look for substitutes, especially among modified starches. The biggest problem lies in the variation in expansion, which can be explained by variations in lactic acid content and solar radiation, dependent on climatic conditions.

As already reported, the first solution found was to use less sour starch and the second one, to replace the scalding process, traditional in cheese bread and expanded snacks (*biscoitos*), with commercial pregelatinized starch. Often the product, food grade, was already included in the ready mixes for cheese bread.

Then, companies producing or selling modified starches created product lines as potential substitutes for sour starch. The most publicized and still available in the market is Expandex, developed by Corn Products and that promises to replace sour starch as an enhancer of the expansion property. Avebe, when it was present in Brazil, claimed to have a line of modified starches to meet the various formulations and industrial processes of cheese bread manufacturing and, more specifically, the frozen cheese bread dough. According to the cooperative, these starches improved the stability of products when subjected to partial thawing cycles, common in supermarket shelves.

Ingredion Brazil also offers modified starches specially developed for the cheese bread market, including pregelatinized cassava starch, such as National 75NB and other modified cassava starches, such as National 75 H and Precisa Bake 11, specific for cheese bread, and Snow Flake XP, of more general use for bakery products. There are also balanced starch compositions to replace the traditional mixture of sour starch with cassava starch or even formulations that use only cassava starch.

These products are primarily used in cheese bread formulations, but not in expanded snakes (*biscoitos de polvilho*), where the expansion characteristic of sour starch is more important. Although they represent an important percentage of the market, they serve as substitutes for sour starch, but do not replicate this starch. In recent years, much research has been done with oxidation processes, to try to reproduce the characteristics of sour starch.

7.6.3.1 Oxidation of starch

In his Master's and Doctoral research, Ivo Motin Demiate highlighted the importance of oxidation modifications in sour starch, which result in the increase of free radicals. Demiate's research was a watershed in the sour starch production process, which highlighted the importance of carboxyls and carbonyls present in sour starch samples and evaluated their importance in the expansion process.

Demiate and Cereda (2000) highlighted the expansion of cassava starch subjected to the oxidation process. Even though the authors reported that the property was temporary, and that the expansion was reduced with storage time, the results were disseminated, generating other papers and process and use patents.

According to the authors, the starch fermentation and subsequent solar drying (through ultraviolet radiation) is a complex process that promotes the starch oxidation and its consequent molecular depolymerization. This complex pattern also provides a good oven-expansion property to the obtained product, which is not achieved by a normal cassava starch or other starch sources. Thus, the authors proved that corn starch, even undergoing the same fermentation and sun-drying process as cassava starch, does not develop the same properties.

With the assumption that only oxidation would solve the problem, Demiate et al. (1997) tested potassium permanganate and lactic acid, with good expansion obtained. However, potassium permanganate is not food grade. Thus, they started using hydrogen peroxide and, under similar conditions, obtained a specific volume around 10.8 cm^3/g. Despite dispensing with the treatment by ultraviolet radiation, the expansion property in the oven by the oxidation reaction with lactic acid is not permanent and the expansion regresses with storage time, which does not occur with the traditional sour starch.

Always with the conviction that oxidation was the solution, patents began to promote this technology. In his patent, after accelerated fermentation, as seen in Item 7.6.1, Schindler (2001) advocated a chemical oxidation reaction of the cassava starch that would eliminate the sun exposure phase. The fermented and chemically oxidized sour starch suspension is then dehydrated in vacuum filter, before being dried in industrial dryer, following the traditional starch processing process (see Chapter 2). The process thus provides a commercial product that is hygienic and free of bacterial contamination.

This process is clearly an adaptation of the results reported by Demiate and Cereda (2000). Despite knowing that the process was only temporary, many researchers came to believe that the expansion was only an interaction between organic acids, especially lactic acid, and oxidation from exposure to UV radiation, either natural from the sun or artificial with lamps. Thus, the valorization of the oxidation process can be found in several patents, such as the U.S. Patent Application 17/108,382 developed in Jiangnan University (Hong et al., 2021), which describes the production process and application of fermented tapioca starch for baking, the Brazilian patent BR00002805A, which describes the use of sour starch as an expansion agent for baking food products.

Dias et al. (2007) tested the expansion properties of corn starch. They fermented cassava and corn starches in the laboratory at 20°C, with one fraction of each starch sun-dried and another oxidized with hydrogen peroxide and artificially dried. Fermentation of the two starches at 0, 10, 30, and 50 days was also evaluated. The expansion property at baking was evaluated as well as the paste property. It was found that the fermentation promotes modification that aids in the oxidation of cassava and corn starches raising the titratable acidity of the product. The fermented and oxidized cassava starch with solar exposure developed the expansion property, but not the corn starch. The best results for the expansion property were in the cassava starch oxidized with hydrogen peroxide after 50 days of fermentation. The authors do not follow the evolution of the expansion over time.

Also influenced by oxidation modification, Granza (2016) submitted an industrial process patent with the purpose of developing starch-based products with expansion and low retrogradation properties for application in the manufacture of cheese breads and expanded snacks. The proposal aimed to improve the shelf-life quality of gluten-free baked goods due to rapid aging, sensorially perceived by hardening in the first hours after baking the doughs, and in the case of frozen products, due to water release (syneresis). Instead of using sour starch, the formulation cited was based on oxidized cassava starch, justified as having a high expansion property.

Matta et al. (2019) proposed the ozone as an oxidation agent as an alternative process to obtain cassava starch with good oven-expansion property. The effects of ozonation on starch granular and molecular structure and on the main starch properties were evaluated. The ozone processing reduced the starch molecular size, also forming carbonyl and carboxyl groups (oxidation process). The structural changes led to pastes with higher clarity, better oven expansion, and softer texture. The improvement in starch properties and technological aspects were related with the molecular changes. The authors state that ozone oxidation proved to be a viable and easier alternative for the conventional process, but do not cite results of shelf life or sensory evaluation of the final product.

7.6.4 Synthesis of the acidification, oxidation, and UV methods

The analysis shows that many patents were based on the hypothesis that oxidation of cassava starch simplifies the process. Thus, it was initially considered that sour starch was a starch modified by oxidation, but research results showed that the oven expansion, even with lactic acid treatment, does not hold during storage. Starch fermentation and its subsequent solar drying (through ultraviolet radiation) is a complex process that promotes the starch oxidation and its consequent molecular depolymerization. This complex pattern also provides a good oven-expansion property to the product, which is not achieved by a native cassava starch or other starch sources.

In the search for a more controlled sour starch production process, the first and most important innovation was the realization that the natural fermentation process, with the formation of various acids, provided only lactic acid for the process that generates the oven expansion property that characterizes and differentiates sour starch.

Analysis of fermented cassava starches exposed to the sun at different times of the day confirmed the importance of UV-C radiation, which in southeastern Brazil predominates between 9 am and 3 pm. This discovery allowed us to define the second innovation, that the reaction that develops the expansion property in the oven is the reaction of the lactic acid found in cassava starch with UV-C radiation from the sun. In a patent of the Nestlè company, Chicaroni (2015) highlighted the need for the presence of lactic acid and UV-C radiation in the necessary amount to obtain a process for preparing a cassava starch-based food product having a substantial controlled expansion.

Santos et al. (2021), in a similar work to that of Nunes and Cereda (1994), evaluated the joint effect of UV-C radiation and lactic acid concentration on the native cassava starch sample. As a reference, the authors used a commercial modified starch, Expandex, and three commercial sour starch samples. The UV-C device had 5 UV light lamps (254 nm, 30 W), which emitted radiation to 0.45 m^2 ($\pm$6144 cd/m^2) under constant mechanic revolving at room temperature. The distance between the lamps and the starch was 8 cm. For the treatment with lactic acid, the native starch was dispersed in lactic acid solution at ambient temperature (23°C $\pm$ 1°C). Then, it was kept under stirring in a thermostatic water bath (30°C) for 15 minutes. Later, the dispersion was filtered with vacuum filtration and the starch was spread over, in a layer approximately 0.5 cm thick, and exposed to the UV-C device for irradiation exposition. After the treatment, the samples were placed in drying ovens set at 45°C for 7 h. The modification was carried out with three repetitions.

The characterization of the samples included, besides the expansion by the method of Maeda and Cereda (2001), the determination of pH, carbonyl, and carboxyl, amylose content, X-ray diffraction, thermal properties, and pasting properties (these and other starch-specific analyses can be found in Chapters 12 and 13 in Volume 1 of this series).

The results showed that the lactic acid concentration did not show significant effects on the specific volume. The highest specific volume was detected in cassava starch treated with 20 g/kg of lactic acid. Therefore, the combination of lactic acid concentration of 20 g/kg with 75 minutes exposure to UV-irradiation was selected as the optimal condition for photochemical modification. The physicochemical properties of cassava native starch have been changed, showing lower viscosity, a tendency to retrograde, and greater internal stability. The Expandex sample obtained the highest expansion among all samples, with higher carboxyl content, while the three commercial sour starch samples showed great variation in all evaluations.

Although they followed the principle of reacting lactic acid with UV-C, the authors used mechanical revolving, reproducing only the results of Nunes and Cereda (1994), with the same problems.

These problems were circumvented in a patent proposed by Brito and Cereda (BR 10201402768), with results published by Brito et al. (2021). They introduced the UV-C treatment of a starch suspension in water and treated with lactic acid for 1 hour. The starch concentration in water was 6% (dry basis), and the lactic acid content was 4% (dry basis). The results showed that only the combination of acidification and ultraviolet radiation allows for an expansion (11.0 cm^3/g) higher than the commercial sour starch (8.2 cm/g) adopted as standard. It was concluded that it is possible to achieve a substitute for fermented cassava starch by acidification of native cassava starch with lactic acid in a continuous flow system using a medium-pressure mercury vapor lamp of 1600 W, which emits radiation between 235 and 280 nm, starting with a dose around 5000 J/L of UV-C radiation. These results proved that fermentation can be substituted by addition of food grade lactic acid and solar radiation by UV-C (BR 10201402768) in a continuous flow. The system allows for a much larger number of granules to be reached by UV radiation and reacts with lactic acid compared to static systems with agitation.

Moreover, the proposed system proves that in suspension at 6% starch in water, the UV-C action is instantaneous. Keeping the starch in suspension allows for a much larger number of granules already impregnated with lactic acid to be reached by irradiation, when compared to static systems or tables with mechanical stirrers.

However, it is yet to be established what kind of modification this process originated in the cassava starch, in order to transform it into sour starch.

7.7 Theory to explain the results obtained

As mentioned by Dufour et al. (1995) and Alvarado et al. (2013), the nature of the changes in sour cassava starch remains to be clarified. Two hypotheses have been suggested by the authors to explain the link between starch degradation and breadmaking ability:

- At the molecular level, free-radical formation results in starch oxidation and depolymerization (Bertolini et al., 2000, 2001; Demiate et al., 2000b; Dias et al., 2011; Vatanasuchart et al., 2005).
- At the supramolecular level, the structure of the starch granules is altered and results in modified gelatinization behavior (Camargo et al., 1988, Gomes et al., 2005, Onitilo et al., 2007).

Before coming to any conclusions, it is important to summarize the knowledge collected about sour starch production.

- Results from the use of commercial UV-C on cassava starch without addition of lactic acid proved that from radiation of 2000 J/L occurs oxidation of the starch with formation of carboxylases (0.124%–0.158% of the starch weight). So, depending on the power applied, UV-C reduces the microbial load or sterilizes the starch, while causing oxidation and generating a whiter starch. However, it was clear that even at doses of 4000 J/L, expansion in the oven does not occur or is very low without the addition of lactic acid.
- Experiments have shown a relationship between UV-C radiation and free radical formation as an oxidation effect (Demiate & Cereda, 2000).
- Unpublished results obtained by Nunes and Cereda show, with data from lactic acid dosage by HPLC, that when lactic acid is added to cassava starch, part of this acid is not recovered in the sample after irradiation, even after washing. This allows for the conclusion that part of the lactic acid remains in the starch and that the oven expansion property is related to the attachment of part of the lactic acid molecule on the starch, inclusion caused by UV-C radiation.

To explain the expansion property of the sour starch, it is necessary to link all these factors, which allows establishing the theory of formation of a graft in the starch structure.

7.7.1 Grafting models in starch

For grafting to occur, free radicals and molecular excitation are required. Ultraviolet radiation, gamma radiation, and oxidizing agents (oxide-reducer pairs) are almost always involved in grafted starch generating reactions, making it more evident that there is a commonality between sour starch and grafted starches modification processes.

7.7.1.1 *Grafting acids with starch*

Mostafa (1995) addressed the grafting of rice starch with acrylic acid in the production of a grafted starch for application in textile industry. He used the potassium permanganate/citric acid oxide-reducer pair to promote the reaction between acrylic acid and rice starch. The author established the reaction conditions such as reagent concentration, reaction time, and temperature. The modified starch was not evaluated for the expansion property.

The acrylic acid (H2 C=CH-COOH), in the reaction described by Mostafa (1995), is a compound that has a structure very close to the lactic acid (H3 C-CHOH-COOH). The reaction proposed by the author allows for grafting the acrylic acid onto the rice starch molecule and obtaining a product for application in the textile industry. Demiate et al. (1997) performed the same tests with native cassava starch, replacing the acrylic acid by lactic acid. As a pair of oxide-reducer behavior, the authors used potassium permanganate/citric acid. The concentrations were those used by Mostafa (1995), but later lactic acid was added in concentrations up to 1% w/w, as calculated by Nunes (1994) and Nunes and Cereda (1994). This concentration was obtained using 25 mL of 85% lactic acid for each kg of native cassava starch. The samples oxidized with potassium permanganate and treated with lactic acid showed specific volumes around 17 cm^3/g (Demiate et al., 1997). It was possible to establish correlation between expansion and carboxyl content to predict the degree of expansion in the modified starches. However, as shown in Section 7.6.3.1, the expansion does not maintain with time.

Other authors to explore starch grafting were Wang et al. (2012). They used corn starch, and the grafting was intended to produce a composite with waterproofing characteristics (for more information regarding starch derived composites, Chapter 14 is suggested). The grafted composite in this case is a lactic acid derivative, a lactyl.

7.7.1.2 Grafting of lactic acid with cassava starch

One hypothesis is that the treatment with lactic acid alone is not relevant for the modification but is relevant if the lactic acid is present during the treatment with *UV-C*, with a known oxidizing effect. The grafting of the lactide formatted on the starch structure as an ether (Takizawa et al., 2004) removes acid radicals from the starch, which could explain the increase and stability of pH values in samples subjected to the joint action of the addition of lactic acid and *UV-C* treatment (Dupuy et al., 1997).

According to Wang et al. (2012), lactitol formation as an inclusion in the starch structure is reported in samples of native corn starch grafted with lactic acid. According to the authors, this inclusion cannot be identified without subjecting the sample to enzymatic treatment.

In the experiment by Wang et al. (2012), using starch and lactic acid (weight) 4:1, a degree of replacement of 58.9% per graft was obtained. The reaction takes place in two steps. The first ring is opened with the addition of an extra oxygen and the lactic acid is polymerized to a lactide. Because of grafting, another oxygen is added to form the connection between the amylose or amylopectin chain and the grafted radical. The same authors point out that first the reaction produced a starch ester, which then progresses to a starch ether group. The level of the grafted starches is expressed in grafting degree and uses the same tools mentioned for the identification of starch esters and ethers, such as percentage of carboxyl groups and IR peak profile. The polylactate polymer can retain gases.

7.7.2 Grafting in sour starch production

Data not yet published allowed establishing the graph of Fig. 7.8, with the peaks that characterize the region of fingerprinting of cassava starch subjected to treatment with lactic acid followed by UV-C as in the process described by Brito et al. (2021). All samples were subjected to hydrolysis with amylolytic enzymes, as suggested by Shokey and Hanna (1993), Kizil et al. (2002), and Wang et al. (2012).

Fig. 7.8 highlights two regions of response for the two starches treated with lactic acid, the region between 1585–1595 and 945/cm. Fairbrother et al. (1991) and Rhiel et al. (2002) identified the lactate band region around 1575/cm, close to the region identified in Fig. 7.8 and which appears to show the grafting reaction between cassava starch and lactic acid.

In the same analysis with corn starch, the peak in the lactate band region (1585–1595/cm) does not appear, while the peak at 945/cm appears in all samples. During the extraction of starch, corn undergoes a lactic acid treatment, which may come from natural fermentation, which explains this result. Thus, it can be concluded that the band at 945/cm in the MIR analysis represents the presence of lactic acid in the starch, while the band 1585–1595/cm identifies the grafting reaction between this acid and the cassava starch. This grafting of lactic acid explains the gas retention, since lactyl is impermeable to water.

The fact that this grafting appears with or without the use of UV does not explain the importance of this treatment. However, it is possible to infer that an oxidation reaction increases the free radicals of the starch, which allows for a greater reaction with lactic acid.

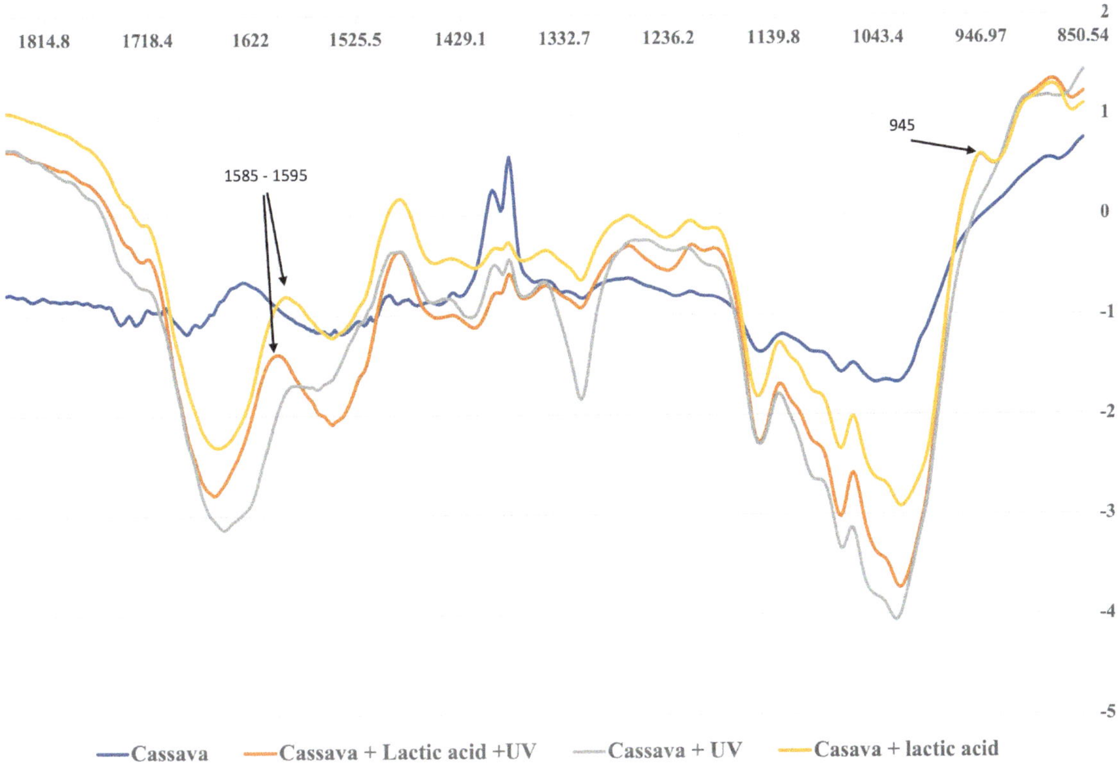

FIGURE 7.8 MIR analysis of hydrolyzed samples of native cassava starch (cassava), native cassava starch with lactic acid (cassava + lactic acid), native cassava starch with UV-C treatment (cassava + UV), and complete treatment (cassava + lactic acid + UV).

In the case of oxidation, this reaction is temporary, while the energy released by UV-C treatment allows for a definitive fixation of these bonds.

Besides the expansion property, it is also necessary to explain the other available analysis results. A modification by grafting explains the characterization results of sour starch, such as the permanence of the polymerization cross and the little disorder caused in the crystalline region of the starch granules. The cross-links strengthen the bonds in the starch granule reducing its swelling. The graft not only consolidates the area in which it was implanted, but also allows for its expansion without granule collapse (Macêdo et al., 2023). However, these results are far from explaining why only cassava starch develops the expansion property.

7.8 Final considerations

The consumption of sour starch derivatives as fast food, such as cheese bread, led to a high increase in consumption, but production did not keep pace. Cheese bread has become a more sophisticated food, consumed with salty and sweet fillings. The production of frozen dough (Fig. 7.9) spread consumption and facilitated distribution. However, the formulation of cheese bread began to be altered due to the

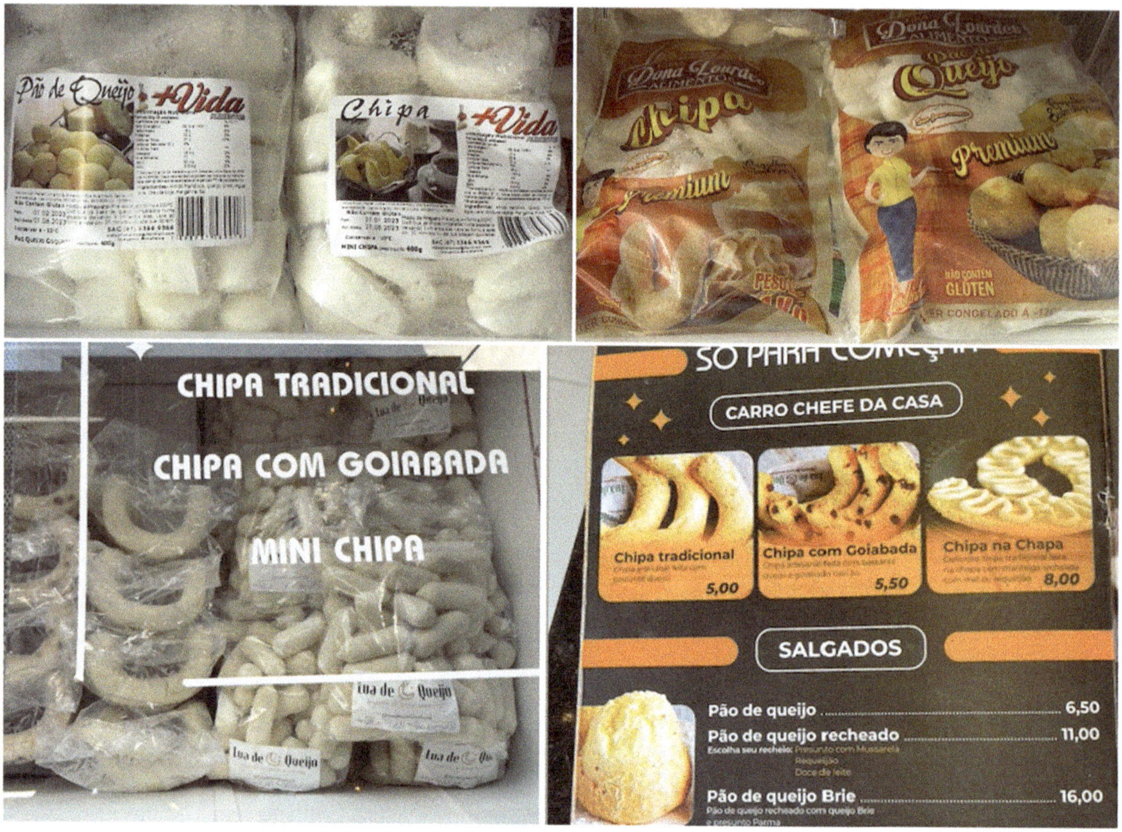

FIGURE 7.9 Frozen and valorized products made with sour starch found in the Midwest region of Brazil: Chipa (Paraguayan influence) and cheese bread.

difficulties of sour starch to meet the demand, the great variations in quality of this product, and the higher price in relation to cassava starch. There was a shift from formulations consisting entirely of sour starch to partial or total substitution by cassava starch, which made the final product heavier and more compact.

Despite these problems, sour starch is still a highly sought-after product, with consumption potential throughout Latin America and in countries in the northern hemisphere, with the possibility of supplying the market for celiacs and other food allergies (see Chapter 8).

Although the technologies for starch extraction and modification have evolved in South America, the production of sour starch still coexists with artisanal processes. In the large processing units in the southeast of Brazil and Paraguay, or in some small units in central Brazil, there is a more modern extraction and better fermentation follow-up, but the sun-drying part is still traditional.

Much of the research conducted in recent years has focused on replacing traditional sour starch with oxidized starch, with storage problems. So, to meet the market, the product Expandex from Ingredion, commercialized for many years and whose process is kept secret, is still the most commercially used. Other modified starches of more dubious confection can also be found in the Brazilian market. It is interesting to note that almost all the patents on sour starch were developed by educational and research institutions, without the involvement of starch processing companies.

As the modification phase of sour starch is still artisanal, there is a great heterogeneity of quality determined by the expansion in the oven that is specific for sour starch and that differentiates it from native starch. However, some modified starches, such as oxidized starches, can show expansion, although with gradual loss over time. After the expansion, the best method to point out the differences in structure in starch and especially in sour starch is MIR analysis. Results for the fermented cassava starch samples approximated oxidized and acid-modified starch with similar mass loss of the residue. Among the commercial modifications of classic starch, the closest to commercial fermented cassava starch was a modification by oxidation and the formation of ether groups (Macêdo et al., 2023). This proximity is justified by the fact that ultraviolet radiation generates free ends in the starch where lactic acid molecules can attach themselves.

Although there are published articles that report the possibility of obtaining sour starch continuously and with standardized expansion power, the process is not known to have been installed. In this process, it is possible to use food-grade lactic acid directly as an ingredient without its production by fermentation. The cassava starch suspension can be treated under a UV-C lamp on a continuous basis. The process can be implemented directly in a starch processing company, like any other type of modified starch.

This continuous process, with the addition of food-grade lactic acid, seems to be the best solution to obtain a product of good eating quality and standardized expansion.

The fact that only cassava starch acquires the property of oven expansion after a natural fermentation or maceration in lactic acid for at least 1 hour (Brito et al., 2021) at room temperature and that this property remained unchanged for years is surprising and needs to be better explained. Moreover, the same process for obtaining sour starch can generate two other types of modified starches, oxidized and low microbial content.

Thus, the use of commercial UV-C radiation would avoid the use of commercial reagents with hypochlorite or peroxide (hydrogen peroxide) in commercial production, which, besides being chemicals, can modify the properties of starch in excessive amounts. The use of ultraviolet radiation is a physical process, more easily accepted by consumers.

Still, it was not possible to establish with certainty what type of modification generates the oven expansion property. The most recent results point to an oxidized-acetylated starch, which indicates a grafting of part of the lactic acid molecule in the structure of cassava starch, caused by the reaction of lactic acid with UV-C radiation. This hypothesis would explain the fact that traditional sour starch or sour starch elaborated in the continuous process do not lose their expansion property in the oven, which happens with the starch only acidified or oxidized. The use of MIR proved to be essential to understand this complex modified starch, which arose from a fermentative process.

With the internationalization of the topic, sour starch has become an important subject for research, with unanswered questions and discoveries that may open other paths for research.

References

Alarcón, M. F., & Dufour, D. L. (2002). Almidón agrio de yuca en Colombia. In P. B. Ospina, H. Ceballos, E. Alvarez, A. C. Bellotti, L. A. Calvert, V. B. Arias, L. L. F. Cadavid, L. B. Pineda, R. G. A. Llano, & M. I. Cuervo (Eds.), *La yuca en el Tercer Milenio: Sistemas modernos de producción, procesamiento, utilización y comercialización*. Centro Internacional de Agricultura Tropical (CIAT);

Consorcio Latinoamericano para la Investigación y el Desarrollo de la Yuca; Proyecto IP-3 de Mejoramiento de Yuca, Cali, CO. p. 470–502. https://cgspace.cgiar.org/bitstream/handle/10568/55257/capitulo26.pdf?sequence=1&isAllowed=y. Consultado em: 4/abril/2023.

Almeida-Starling, C. A. (2017). Mistura amido-goma com propriedade de expansão e baixa retrogradação para aplicação em produtos panificados livres de glúten (pão de queijo e biscoitos de polvilho). BR BR102014031045A2.

Alvarado, P. M., Grosmaire, L., Dufour, D., Toro, A. G., Sánchez, T., Calle, F., Santander, M. A. M., Ceballos, H., Delarbre, J. L., & Tran, T. (2013). Combined effect of fermentation, sun-drying and genotype on breadmaking ability of sour cassava starch. *Carbohydrate Polymers, 98* (1), 1137–1146. Available from https://doi.org/10.1016/j.carbpol.2013.07.012.

American Association of Cereal Chemists. (1978). *Approved methods of the american association of cereal chemists*. St. Paul.

Atrous, H., Benbettaïeb, N., Hosni, F., Danthine, S., Blecker, C., Attia, H., & Ghorbel, D. (2016). Gamma radiation induced modifications of tapioca starch. *Asian Academic Research Journal of Multidisciplinary, 3*(3). Available from https://all3dp.com/2/fused-deposition-modeling-fdm-3d-printing-simply-explained/.

Bangar, S. P., Whiteside, W. S., Singh, A., Ozogul, F., Gupta, A., & Gahlawat, S. K. (2022). Properties, preparation methods, and application of sour starches in the food. *Trends in Food Science & Technology, 121*, 44–58.

Bashir, K., & Aggarwal, M. (2019). Physicochemical, structural and functional properties of native and irradiated starch: A review. *Journal of Food Science and Technology, 56*(2), 513–523.

Bertolini, A. C., Mestres, C., & Colonna, P. (2000). Rheological properties of acidified and UV- irradiated starches. *Starch/Starke, 52*, 340–344.

Bertolini, A. C., Mestres, C., Raffi, J., Buléon, A., Lerner, D., & Colonna, P. (2001). Photodegradation of cassava and corn starches. *Journal of Agricultural and Food Chemistry, 49*, 675–682.

Brasil. (2005). Regulamento técnico para produtos de cereais, amidos, farinhas e farelos. Agência Nacional de Vigilância Sanitária (ANVISA) Ministério da Saúde, Resolução-RDC N° 263, de 22 de Setembro de 2005.

Brito, V., Santos, R., Narcisa-Oliveira, J., Demiate, I., & Cereda, M. P. (2021). Continuous flow process to make fermented cassava starch substitute: I. Physicochemical and functional properties. *Brazilian Journal of Food Technology, 24*, e2020017. Available from https://doi.org/10.1590/1981-6723.01720.

Brito, V. H. D. S., & Cereda, M. P. (2015). Method for determining specific volume as a quality standard for sour starch and substitutes. *Brazilian Journal of Food Technology, 18*(1), 14–22. (In Portuguese).

Camargo, C., Colonna, P., Buleon, A., & Richard-Molard, D. (1988). Functional properties of sour cassava (Manihot utilissima) starch: Polvilho azedo. *Journal of the Science of Food and Agriculture, 45*(3), 273–289.

Cardenas, O. S., & Buckle, T. S. D. (1980). Sour cassava starch production: A preliminary study. *Journal of Food Science, 45*, 1509–1528.

Carvalho, E. P. d., Canhos, V. P., Ribeiro, V. E., & Carvalho, H. P. d. (1996). Polvilho azedo: Aspectos físicos, químicos e microbiológicos. *Pesquisa Agropecuária Brasileira, 31*(2), 129–137.

Cereda, M. P. (1975). Microorganisms and organic acids occurring in the fermentation of cassava starch. *Anais da Academia Brasileira de Ciências, Rio de Janeiro, 47*, 361–362. (In Portuguese).

Cereda, M. P. (1983a). Standardization for quality tests of fermented cassava starch. I. Biscuit formulation and preparation. *Boletim SBCTA, 17*(3), 287–295. (In Portuguese).

Cereda, M. P. (1983b). Quality evaluation of two samples of fermented cassava starch (polvilho azedo). *Boletim Sociedade Brasileira Ciência Tecnologia de Alimentos, 17*(3), 305–320. (In Portuguese).

Cereda, M. P. (1983c). Deterinação de viscosidade de fécula fermentada de mandioca (polvilho azedo). *Boletim da Sociedade Brasileira de Ciências e Tecnologia de Alimentos, 17*(1), 15–24.

Cereda, M. P. (1987). Technology and quality of sour starch. *Informe Agropecuário, Belo Horizonte, 13*(145), 63–68. (In Portuguese).

Cereda, M. P., & Brito, V. H. D. S. (2017). *Fermented foods and beverages from cassava* (Manihot esculenta *Crantz) in South America*. In *Fermented foods of Latin America* (pp. 202–223). CRC Press.

Cereda, M. P., & Catâneo, A. (1986). Avaliação de parâmetros de qualidade do amido fermentada de mandioca. *Revista Brasileira de Mandioca, Cruz das Almas, 5*(2), 55–62.

Cereda, M. P., & Giaj-Levra, L. A. (1987). Finding non-symbiotic nitrogen-fixing bacteria in natural fermentation of cassava starch. Revista Brasileira da Mandioca. *Cruz das Almas, 61*, 29–33. (In Portuguese).

Cereda, M. P., & Lima, U. D. A. (1985). Aspects of cassava starch fermentation III. Determination of organic acids. *Turrialba, 35*(1), 19–24. (In Portuguese).

Cereda, M. P., & Vilpoux, O. (2006). Fermented cassava starch. In R. C. Ray (Ed.), *Advances in microbiology*. New Delhi: Hoticultural Narosa Publishing House.

Cereda, M. P., & Vilpoux, O. E. (2004). Sour starch, quality criteria for use in food products. In M. P. Cereda, & O. E. Vilpoux (Eds.), *Technology, use and potentialities of*

Latin American starchy tuber crops (Vol. 3, pp. 333–376). Cargill Foundation, Chapter 12. (In Portuguese).

Cereda, M. P., Bonassi, I. A., Lima, U. D. A., & Wosiacki, G. (1982). Cassava starch fermentation tests using substrate sterilized with methyl bromide. In *NationaL Fermentation Symposium. Viçosa*. Anais. Viçosa, Brasil. (In Portuguese).

Cereda, M. P., Chuzel, G. C., Vilpoux, O., & Nunes, O. L. D. S. (2001). Modificação da fécula por fermentação. In *Lima, Aquaroni, Borzani & Schmidell, Edtora Edgard Blücher Ltada, São Paulo, Biotecnologia Industrial, Processos Fermentativos e Enzimáticos* (Vol. 3, pp. 413–464). (In Portuguese).

Cereda, M. P., Vilpoux, O. F., & Demiate, I. M. (2003). *Modified starches, Technology, uses and potentialities of South American starchy crops* (Vol. 3, pp. 246–332). São Paulo: Cargill Foundation, Chapter 12. (In Portuguese).

Chicaroni, E. D. P. (2015). *Cassava starch baked food product. WO- WO2015177277A1, Nestec (Nestlé) S.A.*

Chough, S. H., Kim, K. S., Hyung, K. W., Cho, H. I., Park, H. R., Choi, O. K. J., Song, S. H., & Rogers, K. R. (2006). Identification of botanical origins of starches using a glucose biosensor and amyloglucosidase. *Sensors & Actuators B, 114*, 573–577.

Demiate, I. M., & Cereda, M. P. (2000). Some physicochemical characteristics of modified cassava starches presenting baking property. *Energia na Agricultura (Brazil), 15*(3), 36–46.

Demiate, I. M., Dupuy, N., Huvenne, J. P., Cereda, M. P., & Wosiacki, G. (2000a). FTIR spectroscopy of modified cassava starches presenting expansion properties. *Hzywnosc, Krakow, 7*(2), 49–58. (Suplement, 2).

Demiate, I. M., Dupuy, N., Huvenne, J. P., Cereda, M. P., & Wosiacki, G. (2000b). Relationship between baking behavior of modified cassava starches and starch chemical structure determined by FTIR spectroscopy. *Carbohydrate Polymers, 42*(2), 149–158.

Demiate, I. M., Barana, A. C., Cereda, M. P., & Wosiacki, G. (1999). Organic acid profile of commercial sour cassava starch. *Food Science and Technology, 19*, 131–135.

Demiate, I. M., Sanger, S. A., Vogler, Z., Cereda, M. P., & Wosiacki, G. (1997). Characteristics of quality of sour starch samples produced or commercialized in Paraná State. *Arquivos de Biologia e Tecnologia, 40*(2), 321–330.

Dias, A. R. G., Elias, M. C., Oliveira, M., & Helbig, E. (2007). Oxidação dos amidos de mandioca e de milho comum fermentados: Desenvolvimento da propriedade de expansão. *Ciência e Tecnologia de Alimentos, 27*(4), 794–799. Available from https://doi.org/10.1590/S0101-20612007000400019.

Dias, A. R. G., Zavareze, E. D. R., Elias, M. C., Helbig, E., da Silva, D. O., & Ciacco, C. F. (2011). Pasting, expansion and textural properties of fermented cassava starch oxidised with sodium hypochlorite. *Carbohydrate Polymers, 84*(1), 268–275.

Díaz, A., Dini, C., Viña, S. Z., & García, M. A. (2018). Technological properties of sour cassava starches: Effect of fermentation and drying processes. *Lwt, 93*, 116–123.

Díaz, A., Dini, C., Viana, S. Z., & García, M. A. (2018a). Fermentation and drying effects on bread-making potential of sour cassava and ahipa starches. *Food Research International, 116*, 620–627.

Díaz, A., Dini, C., Viña, S. Z., & García, M. A. (2018b). Technological properties of sour starches: Effect of fermentation and drying processes. *LWT- Food Science and Technology, 93*, 116–123. Available from https://doi.org/10.1016/j.lwt.2018.03.029.

Díaz, A., Dini, C., Viña, S. Z., & García, M. A. (2019). Fermentation and drying effects on bread-making potential of sour cassava and ahipa starches. *Food Research International, 116*, 620–627.

Dolmatova, L., Ruckebusch, C., Dupuy, N., Huvenne, J., & Legrand, P. (1998). Identificationof modified starches using infrared spectroscopy and artificial neural network processing. *Applied Spectroscopy, 52*, 329–338.

Dufour, D., Brabet, C., Zakhia, N., & Chuzel, G. (1995). *Influence of fermentation and sun-drying on the breadmaking capacity of cassava sour starch. Transformation alimentaire du manioc* (pp. 399–416). Paris: ORSTOM.

Dupuy, N., & Demiate, I. M. (2022). Exploring the scientific interess for search oriegin in South America from 2000 to 2020: Focus on non-destructiva analytical methods. In M. P. Cereda & O. F. Vilpoux (Eds.), *Starchy crops morphology, extraction, properties and applications* (pp. 289–304). New York.

Dupuy, N., Wojciechowskic, T. A., Huvenn, J. P., & Legrand, P. (1997). Mid-red spectroscopy nachemormetrics in corn starch classification. *Journal of Molecular Structure, 410–411*, 551–554.

Durán-Álvarez, J. C., Avella, E., Ramírez-Zamora, R. M., & Zanella, R. (2016). Photocatalytic degradation of ciprofloxacin using mono-(Au, Ag and Cu) and bi-(Au–Ag and Au–Cu) metallic nanoparticles supported on TiO2 under UV-C and simulated sunlight. *Catalysis Today, 266*, 175–187.

Fairbrother, P., George, W. O., & Williams, J. M. (1991). Whey fermentation: On-line analysis of lactose and lactic acid by FTIR spectroscopy. *Applied Microbiology and Biotechnology, 35*, 301–305.

Fernández, A., Zakhia, N., Ruiz, R., & Trujillo, J. M. (2002). Desarrollo de un método sencillo para medir la calidad del almidón agrio de yuca: Impacto del método sobre la agroindustria rural en el departamento del Cauca (Colombia).

Figueroa, C., Davila, A. M., & Pourquié, J. (1995). Lactic acid bacteria of the sour cassava starch fermentation. *Letters in Applied Microbiology, 21*(2), 126–130.

Figueroa-Flórez, J., Cadena, E. M., & Rincón, C. C. (2018). Modification of the fermentation process for sour cassava starch with expansion properties. *Advance Journal of Food Science and Technology, 16*(SPL), 55–61.

Föste, M., Jekle, M., & Becker, T. (2017). Structure stabilization in starch-quinoa bran doughs: The role of water availability and gelatinization. *Carbohydrate Polymers, 174*, 1018–1025.

Franco, C. M. L., Ogawa, C., Rabachini, T., Rocha, T. D. S., Cereda, M. P., & Jane, J.-L. (2010). Effect of lactic acid and UV irradiation on the cassava and corn starches. *Brazilian Archives of Biology & Technology, 53*(2), 443–454.

Freire, A. L., Ramos, C. L., & Schwan, R. F. (2015). Microbiological and chemical parameters during cassava based-substrate fermentation using potential starter cultures of lactic acid bacteria and yeast. *Food Research International, 76*(3), 787–795.

Freitas, M. C. G. D. (2020). Characterization of the fermentation profile of the sour cassava starch (sour starch) produced by the commercial industry Makmil Ltda de Conceição dos Ouros, *MG*. 2020. 71 p. Thesis de final de Curso (College os Chemical Engeneer, USP, Lorena). (In Potuguese).

Garcia, A. C. D. B., & Leonel, M. (2005). Efeito da concentração de ácido lático sobre a propriedade de expansão em amidos modificados fotoquimicamente. *Ciência e Agrotecnologia, 29*, 629–634. Available from https://doi.org/10.1590/s1413-70542005000300018.

Garcia, M. C., Franco, C. M. L., Júnior, M. S. S., & Caliari, M. (2016). Structural characteristics and gelatinization properties of sour cassava starch. *Journal of Thermal Analysis and Calorimetry, 123*(2), 919–926.

Gomes, A. M., da Silva, C. E. M., & Ricardo, N. M. (2005). Effects of annealing on the physicochemical properties of fermented cassava starch (polvilho azedo). *Carbohydrate Polymers, 60*(1), 1–6.

Granza, A. G. (2016). Mistura amido-goma com propriedade de expansão e baixa retrogradação para aplicação em produtos panificados livres de glúten (pão de queijo e biscoitos de polvilho). Titular(es): Universidade Estadual de Ponta Grossa. Br 102014031045-2 a2.

Hong, Y., Qiaoting, Q. I., Gu, Z., Cheng, L., Li, Z., & Li, C. (2021). *Production Process and Application of Fermented Tapioca Starch for Baking*. Jiangnan University. U.S. Patent Application 17/108,382.

Jiamjariyatam, R., Kongpensook, V., & Pradipasena, P. (2015). Effects of amylose content, cooling rate and aging time on properties and characteristics of rice starch gels and puffed products. *Journal of Cereal Science, 61*, 16–25.

Kizil, R., Irudayaraj, J., & Seetharaman, K. (2002). Characterization of irradiated starches by using FT-Raman and FTIR spectroscopy. *Journal of Agricultural and Food Chemistry, 50*(14), 3912–3918. Available from https://doi.org/10.1021/jf011652p.

Kostinek, M., Specht, I., Edward, V. A., Pinto, C., Egounlety, M., Sossa, C., Mbugua, S., Dortu, C., Thonart, P., Taljaard, L., & Mengu, M. (2007). Characterisation and biochemical properties of predominant lactic acid bacteria from fermenting cassava for selection as starter cultures. *International Journal of Food Microbiology, 114*(3), 342–351.

Lacerda, I. C., Miranda, R. L., Borelli, B. M., Nunes, Á. C., Nardi, R. M., Lachance, M. A., & Rosa, C. A. (2005). Lactic acid bacteria and yeasts associated with spontaneous fermentations during the production of sour cassava starch in Brazil. *International Journal of Food Microbiology, 105*(2), 213–219. Available from https://doi.org/10.1016/j.ijfoodmicro.2005.04.010.

Macêdo, I. Y. L., Dupuy, N., Brito, V. H. D. S., Gil, E. D. S., Silva, G. N. D. M. E., Moreno, E. K., Demiate, I. M., & Cereda, M. P. (2023). Modified cassava starches' identification through mid-infrared spectroscopy and exploratory analysis. *Analytica, 4*(2), 126–140.

Machado, A. V., Araújo, F. M. M. C. D., & Pereira, J. (2010). Caracterização física, química e tecnológica do sour starch. *Revista Verde de Agroecologia e Desenvolvimento Sustentável, 5*(3), 4.

Maeda, K. C., & Cereda, M. P. (2001). Avaliação de duas métodos de expansão ao forno do sour starch. *Boletim da Sociedade Brasileira de Ciência e Tecnologia de Alimentos, Campinas, 21*(2), 139–143.

Marcon, M. J. A., Kurtz, D. J., Raguzzoni, J. C., Delgadillo, I., Maraschin, M., Soldi, V., Reginatto, V., & Amante, E. R. (2009). Expansion properties of sour cassava starch (polvilho azedo): Variables related to its practical application in bakery. *Starch-Stärke, 61*(12), 716–726. Available from https://doi.org/10.1002/star.200900132.

Marcon, M. J. A., Vieira, M. A., Santos, K., De Simas, K. N., Amboni, R. R. D. D. M. C., & Amante, E. R. (2006). The effect of fermentation on cassava starch microstructure. *Journal of Food Process Engineering, 29*(4), 362–372.

Mariotti, M., Alamprese, C., Pagani, M. A., & Lucisano, M. (2006). Effect of puffing on ultrastructure and physical characteristics of cereal grains and flours. *Journal of Cereal Science, 43*(1), 47–56.

Martinez, A., & Zapata, L. E. (1981). Aislamiento y caracterizacion dela microflora presente durante la fermentacion del almidon de yuca. *Technologia, 174*, 23–30.

Matta, M. D., Jr., Castanha, N., Dos Anjos, C. B. P., Augusto, P. E. D., & Sarmento, S. B. (2019). Ozone technology as an alternative to fermentative processes to improve the oven-expansion properties of cassava starch. *Food Research International, 123*, 56–63. Available from https://doi.org/10.1016/j.foodres.2019.04.050.

Mestres, C., & Rouau, X. (1997). Influence of natural fermentation and drying conditions on the physicochemical characteristics of cassava starch. *Journal of the Science of Food and Agriculture, 74*, 147–155. RSC. Avaiable in: https://agritrop.cirad.fr/389099/. Consulted on: 30/April/2023.

Mora Pareja, J. A., Ramírez, C. A., & Barrera, B. F. (2017). Delicias del Valle: Pandebono valluno.

Mostafa, K. M. (1995). Graft polymerization of acrylic acid onto starch using potassium permanganate acid (redox system). *Journal of Applied Polymer Science, 56*(2), 263–269.

Nunes, O. L. G. S. (1994). Effects of drying process on the development of expantion in cassava starch hydrolyzed by lactic acid. 1994 87f. Dissertação (Mestrado em Agricultura/Energia na Agricultura) Faculdade de Ciências Agronômicas, Universidade Estadual Paulista, Botucatu, 1994. (In Portuguese).

Nunes, O. L. G. D. S., & Cereda, M. P. (1994). Effects of drying process on the development of expantion in cassava starch hydrolyzed by lactic acid. In *International Meeting on Cassava Flour and Starch, 1., 1994, Cali. Anais. p.* (p. 110). Cali: Centre de Cooperation Internationale en Recherche Agronomique pour le Devellopment/Centro Internacional de Agricultura Tropical.

Nunes, O. L. G. S. D., & Cereda, M. P. (1995). Aperfeiçoamento de processo de obtenção de sour starch com emprego de amido de mandioca através de efeito fotoquímico, BR 94022303A, Depositada em 1994-07-25 e published em 1995-03-07.

Onitilo, M. O., Sanni, L. O., Daniel, I., Maziya-Dixon, B., & Dixon, A. (2007). Physicochemical and functional properties of native starches from cassava varieties in Southwest Nigeria. *Journal of Food Agriculture and Environment, 5*(3/4), 108. Available from https://doi.org/10.1080/10942910601048994.

Paixão Silva, G. D. L., Bento, J. A. C., Oliveira, A. R., Garcia, M. C., Soares Júnior, M. S., & Caliari, M. (2021). Pasting and thermal properties of fermented cassava (*Manihot esculenta* Crantz). *Journal of Food Science and Technology, 58*, 1441–1448. Available from https://doi.org/10.1007/s13197-020-04656-3.

Pão de queijo. (2001). The flavor from Minas Gerais that, in the industrial version, conquered the consumer. *Food Ingredientes*, 36–40, São Paulo, n.12.

Penido, F. C. L., Piló, F. B., Sandes, S. H. D. C., Nunes, Á. C., Colen, G., Oliveira, E. D. S., Rosa, C. A., & Lacerda, I. C. A. (2018). Selection of starter cultures for the production of sour cassava starch in a pilot-scale fermentation process. *Brazilian Journal of Microbiology, 49*, 823–831.

Plata-Oviedo, M. S. V. (2010). Process for obtaining cassava starch with expansion property. PI0002805-3B116/11/2010. Invention Patent (In Portuguese).

Plata-Oviedo, M. S. V., & Camargo, C. R. O. (1995). Determinação de propriedades físico-químicas e funcionais de duas féculas fermentadas de mandioca (sour starch). *Boletim da Sociedade Brasileira de Ciência e Tecnologia de Alimentos, Campinas, 4*(1), 59–65.

Putri, W. D. R., Haryadi, M. D. W., & Cahyanto, M. N. (2012). Role of lactic acid bacteria on structural and physicochemical properties of sour cassava starch. *APCBEE Procedia, 2*, 104–109. Available from https://doi.org/10.1016/j.apcbee.2012.06.019.

Qi, Q., Hong, Y., Zhang, Y., Gu, Z., Cheng, L., Li, Z., & Li, C. (2021). Effect of cassava starch structure on scalding of dough and baking expansion ability. *Food Chemistry, 352*, 129350. Available from https://doi.org/10.1016/j.foodchem.2021.129350.

Rhiel, M., Ducommun, P., Bolzonella, I., Marison, I., & Von Stockar, U. (2002). Real-time in situ monitoring of freely suspended and immobilized cell cultures based on mid-infrared spectroscopic measurements. *Biotechnology and Bioengineering, 77*(2), 174–185.

Rincón, A. G., & Pulgarin, C. (2004). Field solar *E. coli* inactivation in the absence and presence of TiO2: Is UV solar dose an appropriate parameter for standardization of water solar disinfection? *Solar Energy, 77*(5), 635–648.

Santos, T. P. R., Leonel, M., Mischan, M. M., & Cabello, C. (2021). Study and application of photo-modified cassava starch with lactic acid and UV-C irradiation. *LWT, 139*, 110504. Available from https://doi.org/10.1016/j.lwt.2020.110504.

Schindler, F. S. (2001). Processo para o fabrico de sour starch, BR BR9905066A, submeeted: 1999-07-30. Published 2001-03-13.

Sena Aquino, A. C. d. M., Azevedo, M. S., Ribeiro, D. H. B., Costa, A. C. O., & Amante, E. R. (2015). Validation of HPLC and CE methods for determination of organic acids in sour cassava starch wastewater. *Food Chemistry, 172*, 725–730.

Shokey, A. S., & Hanna, M. A. (1993). Properties of irradiated starches. *Food Structure, 12*, 397–410.

Smith, R. J. (1967). Characterization and analysis of starches. In R. L. Whistler, & E. F. Paschall (Eds.), *Starch: Chemistry and technology*. New York: Academic Press:.

Takizawa, F. F., Silva, G. D. O. D., Konkel, F. E., & Demiate, I. M. (2004). Characterization of tropical starches modified with potassium permanganate and lactic acid. *Brazilian Archives of Biology and Technology, 47*, 921–931.

Trela, V. D., Ramallo, A. L., & Albani, O. A. (2020). Synthesis and characterization of acetylated cassava starch with different degrees of substitution. *Brazilian Archives of Biology and Technology, 63*.

Uthumporn, U., Zaidul, I. S. M., & Karim, A. A. (2010). Hydrolysis of granular starch at sub-gelatinization temperature using a mixture of amylolytic enzymes. *Food and Bioproducts Processing*, *88*, 47–54.

Vatanasuchart, N., Naivikul, O., Charoenrein, S., & Sriroth, K. (2005). Molecular properties of cassava starch modified with different UV irradiations to enhance baking expansion. *Carbohydrate Polymers*, *61*(1), 80–87.

Wang, Q., Hu, Y., Hu, J., Liu, Y., Yang, X., & Bian, J. (2012). Convenient synthetic method of starch/lactic acid graft copolymer catalyzed with sodium hydroxide. *Bulletin of Materials Science*, *35*, 415–418. Available from https://doi.org/10.1007/s12034-012-0295-5.

Westby, A., & Cereda, M. P. (1994). Production of fermented cassava starch (sour starch) in Brazil. *Tropical Science, Amsterdam*, *34*(2), 203–210.

Wurzburg, O. B. (1986). *Modified starches: properties and uses*. CRC Press.

Zhu, F. (2015). Composition, structure, physicochemical properties, and modifications of cassava starch. *Carbohydrate Polymers*, *122*, 456–480.

Zhu, T.-W., Zhang, X., Li, B., & Wu, H. (2019). Effect of interesterified blend-based fast-frozen special fat on the physical properties and microstructure of frozen dough. *Food Chemistry*, *272*, 76–83.

CHAPTER

8

Starch-based pasta and bakery products for traditional markets and celiac syndrome consumers—examples of the traditional process of noodles in China

Olivier François Vilpoux[1,2] *and Marney Pascoli Cereda*[2,3]

[1]Federal University of Rondonopolis, Post-graduate Program in Environmental Management and Technology - PPGTA, Rondonopolis, Mato Grosso, Brazil [2]Tecsapp Solutions, Campo Grande, Mato Grosso do Sul, Brazil [3]Retired from São Paulo State University, Botucatu, São Paulo, Brazil

8.1 Introduction

Breads and pasta are common products made of wheat flour and which are consumed daily in a large part of the world. As mentioned by Smith et al. (2004), bakery products are an important part of a balanced diet and, today, a wide variety of such products can be found on supermarket shelves. This includes unsweetened goods such as bread, rolls, buns, crumpets, muffins and bagels, sweet goods, such as pancakes, doughnuts, waffles and cookies, and filled goods, such as fruit and meat pies, sausage rolls, pastries, sandwiches, cream cakes, pizza, and quiche.

The disadvantage of most bakery products is the low biological and physiological value and high calorie content (Konovalova et al., 2020). Modern production of bread and bakery products is aimed at their enrichment with vitamins and minerals to increase biological value and reduce calorie content, as they are products with daily consumption.

In many tropical countries, the search for ingredients and processes that allow for obtaining bakery products without the use of wheat flour has often been seen to reduce, or even eliminate, the dependence on imports from North American and European countries. It is a fact that the introduction of wheat derivatives in South America, Central America, Caribbean, and Africa, sometimes forced, changed consumption habits, reducing or eliminating traditional products and leading some countries that were not self-sufficient in wheat production to depend on imports.

Yano (2019) stated that wheat bread is consumed globally and has played a critical role in the story of civilization since the development of agriculture. While the aroma and flavor of

Starch Industries
DOI: https://doi.org/10.1016/B978-0-323-90842-9.00005-4

this staple food continue to delight and satisfy most people, some individuals have a specific allergy to wheat or a genetic disposition to celiac syndrome. Celiac syndrome is an autoimmune disorder caused by genetic as well as environmental factors (Singh et al., 2018).

The problem with celiac patients is that the intake of gluten leads to small intestinal damage, typically leading to malabsorption because the gluten protein has protease-resistant regions in its structure (Shan et al., 2005). A food allergy to wheat may cause a life-threatening reaction, such as anaphylaxis (Matsuo et al., 2005). In contrast, repetitive exposure to wheat flour may cause baker's asthma or rhinitis, mostly characterized as occupational allergic diseases (Salcedo et al., 2011). This type of food allergy usually occurs hours or even days after ingestion of wheat products and are characterized by chronic eosinophilic inflammation of the gastrointestinal tract (Nowak-Węgrzyn et al., 2015).

Its average occurrence in the United States and Europe is estimated at about 1% (Biagini et al., 2004; Di Sabatino et al., 2012; Young et al., 1994). Other data indicate a frequency of about 6% in the United States, 0.9% in the United Kingdom (Biagini et al., 2004; Young et al., 1994), and 0.2% in Japan (Morita et al., 2012). A new group of non-celiac gluten sensitivity has been recently proposed and became an increasingly recognized clinical condition in patients in whom celiac disease and wheat allergy have been ruled out (Barbaro et al., 2018). So far, a gluten-free diet is the only safe and effective treatment for the abovementioned conditions of gluten intolerance (Cianferoni, 2016).

To improve the quality of life of these patients from a dietary standpoint, food-processing researchers have been seeking to develop high-quality gluten-free products. As the quality of wheat breads depends largely on the viscoelastic properties of gluten, various ingredients have been employed to simulate its effects, such as hydrocolloids, transglutaminase, and proteases. Recent attempts have included the use of redox regulation as well as particle-stabilized foam as advancements in the development of gluten-free bread.

Bangar et al. (2022) reported that starch noodles are gaining interest between occidental people due to the massive popularity of gluten-free foods. Raw materials, key processing steps, additives, cooking, and textural properties determine the quality of starch noodles. The introduction of traditional or novel chemical additives used in starch noodles also impacts their quality. Due to the increasing demand of consumers for safe food items with a long shelf life, the development of starch noodles and other convenience food products has increased. Also, the incorporation of modified starches overcomes the shortcomings of native starches, such as lack of viscosity and thickening power, retrogradation characteristics, or hydrophobicity. Starch can improve the stability of the dough structure but reduce the strength and resistance to deformation of the dough. Some technological, sensory, and nutritional challenges also impact the production process.

Besides being classified as calorific foods, pasta and bakery products can affect other consumers with food intolerance syndrome (FIS), which has only grown due to new dietary problems. With this, products capable of meeting the increasing number of people who avoid cereal flours are gaining prominence worldwide. In this case, it is very important to develop cereal-free products, which can fill a market that is currently highly underserved and undersupplied.

Because of the importance of replacing wheat flour with gluten-free products, due to production difficulties in many tropical countries and the large number of coeliac's, this chapter deals with the production of pasta and different types of bread made from starch extracted from the underground organs of crops native from Latin America.

8.2 Preparation processes of gluten-free starch pastas

It is necessary to remember that for the elimination of gluten, the foundations of traditional baking must be rethought. Cereal-based doughs are characterized by a well-established protein network (gluten), which depends on the content and quality of gliadin and glutenin, in a well-established model. Starch-based doughs, on the other hand, are made up of glucose molecules. Although well used for making oriental pasta, these doughs have a structure and functionality that is little addressed in research and depends on the viscosity and consistency of glucose networks. The advantage of using this type of network is that there are still no reports of known food allergies and intolerances related to this type of structure; on the other hand, there is a lack of knowledge regarding its theoretical basis and mechanism of action.

Asia produces many kinds of starch noodles, manufactured according to different processes and raw materials. Each starch source creates noodles with totally different taste, color, and texture. For example, noodles made from canna starch (*Canna edulis*) are much more flexible and whiter than noodles made with sweet-potato starch (*Hipomoeu batatas*) (Fig. 8.1). Kim and Wiesenborn (1996) studied pastas produced with potato starch of different genotypes, examining aspect and texture characteristics such as sheen, transparence, firmness, stickiness, and elasticity. Galvez and Ressurrección (1992) studied factors related to the quality of mung-bean (*Vigna radiata*) pastas. As a result, the most important features and quality indicators were color, sheen, and transparence of the dry product, besides the texture of the cooked product that must present firmness and gum properties. Galvez et al. (1995) examined the acceptability of mung-bean pastes in terms of raw and cooked-paste aspect, texture, and palatability.

FIGURE 8.1 (A) Canna and (B) sweet-potato starches noodles.

In Asian countries, the production of starch pasta can be both an artisanal and a high-tech process. Starch pastas are essentially cultural products, as Asia has enough wheat to make traditional pastas. In China, a wide variation of starch pastas can be found, made from rice, sweet potato, canna, and arrowroot (Fig. 8.2). Wang et al. (2020) described a production unit in China producing annually 70,000 tons of noodles and 10,000 tons of vermicelli made with sweet potato starch. It is also suggested to consult Chapter 3 on the extraction of starch to make these noodles.

The literature cites several technologies for making unconventional pasta, obtained by replacing wheat flour with starch. Milatovic and Ballini (1986) described techniques based on heating the starch suspension at high temperature, before being vigorously mixed with the other ingredients, while Pagani et al. (1981) reported a process that promotes the pregelling of 5%–7% of the raw material, with the subsequent mixing of this portion with the total ingredients of the dough at room temperature. In this process, the amylose in the gelled portion favors polymerization with the untreated

FIGURE 8.2 Traditional starch-based noodles sold in local Chinese market.

starch fraction, creating a structure that performs functions similar to those of gluten in wheat flour-based doughs. Protein is used to strengthen the structure of starch doughs.

The following items describe different types of noodles processed with canna starch commercialized in China and Vietnam. Besides raw material, the process may greatly influence the noodle obtained. Many manufacturers of canna starch noodles were visited in the Guangzhou province of China, which borders Vietnam. Four different noodles formation technologies were observed in relation to starch gelatinization and pasta extrusion. In addition to Canna noodles, we also visited a company manufacturing noodles from arrowroot (*Maranta arundinacea*) starch, using different technology.

8.2.1 Gelatinization of starch after noodles formation

The production of starch noodles consists of the formation, by extrusion, of a gelatinized starch-paste. The gelatinization and the formation are the two main stages and may be performed in a different order and separate stages. In simpler processes, the wet starch is placed in a container, which has a perforated base for the formation of noodles (Fig. 8.3). The paste is pressed against the holes by a screw, with an extruder effect. When exiting the extruder, noodles fall into a boiling-water tank, where starch is quickly gelatinized.

Right after that, noodles are pulled out from hot water and submerged in a cooling tank, under cold water (Fig. 8.4). In addition to being cooled, the pasta are loose at this stage, because when the starch gel cools, its adhesive properties are reduced.

Once cool, the noodles are removed from the water and hung on bamboo poles to remove excess moisture (Fig. 8.5). To facilitate separation, a wide-toothed comb is passed between the noodles that are drying.

Drying can be completed in greenhouses or outdoor (Fig. 8.6) and can last from 1 to several days, depending on the relative humidity of the air and the ambient temperature.

In this process, the starch gelatinization is done in water, which allows for the absorption of a greater amount of water. This absorption continues when cooling the noodles, which is also done in water. The large amount of moisture increases the retrogradation, leaving the product with a milky white color. The high retrogradation of starch allows obtaining more

FIGURE 8.3 Boiling-water tank and noodles extruder for ungelatinized paste.

FIGURE 8.4 Noodles in a cooling tank with cold water.

granular noodles, with a consistency that approaches that of wheat noodles.

8.2.2 Gelatinization in a cooking tank before the extrusion process

In this process, starch is gelatinized before the formation of noodles. It is placed in a closed cooking tank (Fig. 8.7A), heated from below with firewood and isolated with refractory bricks. After gelatinization, starch is extruded to form noodles (Fig. 8.7B). In this system, noodles are directly placed to dry, without going through a cooling bath.

The noodles are put out to dry in the open air. Fig. 8.8 shows noodles drying on a cloudy day without sun. The noodles made by this process have not passed through the water tank to cool down In this case, the final product contains less moisture and undergoes a lower retrogradation than the noodles from the process described in Section 8.2.1. The low retrogradation allows for more transparent noodles with a less granular texture after cooking.

In the region visited, the production of noodles is a family activity, done by dozens of families in the same municipality.

8.2.3 Gelatinization and extrusion in two separate extruders

To simplify the gelatinization process, some families use a heated extruder just to gelatinize

FIGURE 8.5 Elimination of excess water in noodles.

FIGURE 8.6 Outdoor drying of canna noodles in South of China.

the starch. Afterward, the gelatinized starch is placed in an unheated extruder to form the noodles (Fig. 8.9).

When they come out of the formation extruder, noodles are cooled in water (Fig. 8.10). This stage enables noodles to re-absorb water, increasing the retrogradation level of the end product. Since the starch has not gelatinized in water, as in the process described in item 8.2.1, the retrogradation is lower, allowing a product to be obtained with color and texture intermediate between the noodles processed in items Sections 8.2.1 and 8.2.2.

Following the cooling stage, noodles are placed to dry like in the other processes.

8.2.4 Process of noodles using just an extruder

To economize time and work labor, the larger company of the region uses a conventional extruder that enables the gelatinization of starch and the formation of noodles in just one stage (Fig. 8.11B).

This equipment is a classic extruder that allows for the use of drier raw material, obtaining a noodle with low moisture content, which does not need to be dried as in previous processes. The lower moisture content in the material reduces the degree of retrogradation, obtaining a more transparent noodle.

FIGURE 8.7 Closed cooking tank for starch gelatinization (A) and noodles extruder (B).

FIGURE 8.8 Open air drying of noodles produced from extrusion of gelatinized canna starch.

FIGURE 8.9 Formation of noodles using two extruders, one for gelatinization (right) and one for formation (left).

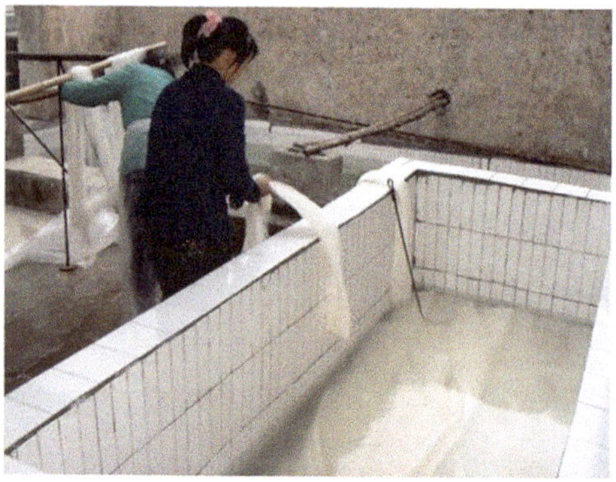

FIGURE 8.10 Cooling of noodles processed with two extruders in cold water tank.

Because it is made from dried canna starch, this process allows for better control of moisture at the extruder inlet. For a good distribution of the added water, the use of a mixer is necessary (Fig. 8.11A).

The dry starch is put into the mixer, with the addition of just enough water to form the pasta and for the starch to be partially gelatinized during extrusion. The classic extruder allows for the use of drier raw material, obtaining a low moisture content noodle that, like wheat noodles, can be air- or oven-dried.

The same pasta formation process can be adjusted by using pregelatinized starch in a mixture with dry starch, or even by adding boiling water to the dry starch during mixing (for more information on starch properties and gel formation, please see Chapter 10 of Volume 1 in this Series).

8.2.5 Noodle production with manual cutting

There is a noodle production process that uses only the gelatinization of the starch, without going through an extruder. This process has been described by Hermann et al. (1997) for canna starch and observed in China from arrowroot starch, but other raw materials are also used, such as mung bean starch.

In the first stage of the process, arrowroot dry starch is mixed with a liquid suspension of low-concentration gelatinized starch (1%–2%), which increases the viscosity of the mixture and prevents the granular starch from settling out. Next, a layer about 1 cm thick of the starch suspension, with between 15% and 20% dry matter, is placed in a covered cooker for gelatinization (Fig. 8.12A). The cooker is heated by water bath to prevent burning of the suspension. The fact that the lid is closed allows for a homogeneous gelatinization with a high concentration of solids.

After gelatinization, the film is removed from the cooker. To be removed, the film is rolled onto a piece of wood. After removal, the film is stretched over a bamboo platform (Fig. 8.12B). According to Hermann et al. (1997), when stretched, the film increases in size by 2.5 times. At the same time as it stretches, the starch film gains transparency

After losing moisture and reaching between 42% and 45% dry matter (Hermann et al., 1997), the films are wound for cutting. Traditionally, a guillotine system is used (Hermann et al., 1997), but knives or scissors can also be used too (Fig. 8.13). To be rolled and cut, the noodles

FIGURE 8.11 Conventional mixer (A) and extruder (B) for the formation of low moisture starch noodles.

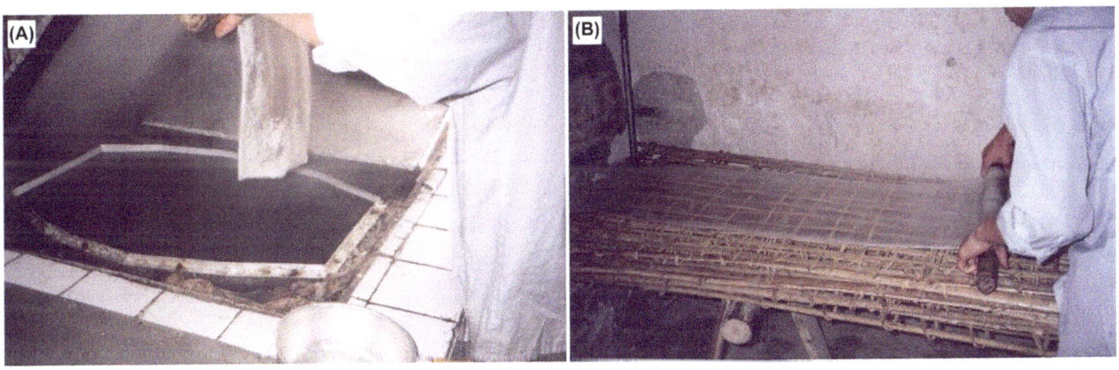

FIGURE 8.12 Arrowroot film removed from the cooker (A) and tightened over a bamboo platform (B).

must remain moist. After cutting, the noodles are stored until the end of drying.

8.3 Process for other types of gluten-free baked dough

Besides noodles, starch also enables the production of special gluten-free starchy breads. The importance of cassava as an energetic source in human food has already been established. However, little has been said about the use of cassava derivatives in bakery products, such as Brazilian cassava flour, cassava chips flour, native starch, and sour starch. According to El-Dash (1991), cassava chips flour was practically the only cassava derivative, as flour had already been used in bread-making, even so, in mixtures with wheat flour.

According to Yano (2019), the aroma emanating from a bread bakery is unmistakably enticing. The flavor and crunchy texture of wheat breads sharpen our appetite and satisfy our basic human cravings for comfort as well

FIGURE 8.13 Manual cutting of noodles after they have been rolled out.

as nutrition. Simply stated, bread-making is composed of three steps: mixing/sheeting, fermenting, and baking processes (Dobraszczyk & Morgenstern, 2003). Wheat bread is now one of the most representative food in the world.

Property of wheat gluten is responsible for the production of breads with high quality. However, some genetically predisposed people cannot eat wheat bread because gluten causes harmful reactions in them (Yano, 2019). In this short review, we summarize the gluten-dependent swelling mechanism of wheat bread and the recent scientific effort to make bread without gluten.

In a wheat flour dough, the gluten matrix is composed mainly of the protein network of gluten, starch granules, and water. This matrix encloses the fermentation gas and making small "balloons." Thus, the dough rises as the fermentation proceeds. On the other hand, hydration of flour from gluten-free cereals, such as rice, results in a kneaded mass rather than viscoelastic "dough," as the proteins of these mass do not possess the network-forming properties typically found in gluten (Renzetti & Rosell, 2016). Therefore, the CO_2 from fermentation rises to the surface, while starch granules and yeast settle (Schober, 2009). Generally, a gluten-free batter without a thickening agent, such as hydrocolloids, becomes foamy (Hamada et al., 2013; Yano, 2010).

Various starchy flours have been evaluated to replace wheat flour, first to reduce its price when imported, but increasingly often to eliminate gluten. The need to produce gluten-free bread has broadened the understanding of bread quality and the role of starch in the process. Other ingredients can be added to improve the volume, aroma, flavor, texture, crust, crumb, and nutritional value of breads, such as locust gum and sodium alginate, mono-stearate glyceryl, polyoxi-ethylene-stearate, hydroxi-propyl-methyl-cellulose, calcium stearoyl-lactylate, carboxy-methyl-cellulose, surfactant agents like mono- and di-glyceride, pentosan, apple pectin, albumen, $NaHCO_3$, egg white, and margarine

(Dickinson, 2018; Machado, 1996; Schober, 2009). Typical gluten-free breads contain hydrocolloids (e.g., xanthan gum and guar gum), which increase the viscosity of the liquid phase, keeping the starch granules, yeast, and gas bubbles suspended in the fermentation process (Dickinson, 2018; Schober, 2009).

The development of new gluten-free products or the substitution for other food (eggs and milk) is one of the tendencies (Holub, 1990; Kaczmarski & Kurzatkowska, 1988; Kokke et al., 1994; Metcalfe, 1984; Wintz & Kuntze, 2000). Another possibility is the modification of the protein structure of gluten with enzymes such as proteinase (Tanabe et al., 1996).

Confocal laser scanning images of the breadcrumbs suggested that the gelatinized starch phase was the main structure component in the protease-treated bread. Thus, protease may partially degrade the large macromolecular protein complex embedding starch granules (Derycke et al., 2005; Hamaker & Griffin, 1993), resulting in improved continuity of the starch phase as well as better rheological properties of the batter (Yano, 2019).

In this sense, some authors established that gluten-free bread could be made using many elements, isolated or jointly, such as wheat, corn, cassava, potato and rice starches, soybeans, dwarf beans, and sour cassava starch. These formulations were intended to improve the volume, flavor, texture, crust, crumb, and nutritional value of bread (Yano, 2019). Better bread quality (flavor, texture, and volume), reduced production cost, and wider availability are all important issues (Pickering, 1907).

Several starchy flours (Brazilian cassava flour, cassava chips flour, yam flour, whole-corn fat-free flour, and sorghum flour) and protein flours (soybeans flour, lupin flour, and amaranth flour) were examined to partially substitute wheat flour. Protein flours would have a significant value in areas where animal protein is lacking, increasing the nutritional value of bread. Feasibility research was done on starchy foods, mainly to increase added value of roots, tubers, and cereals that are produced and slightly used in Brazil (Machado, 1996). One of the most studied flours as a partial or total substitute for wheat flour is the rice flour (El Khoury et al., 2018). Legally, these foods should not be marketed as bread, but are better designated as bread substitutes.

The demand for the development of gluten-free foods is growing (Gobbetti et al., 2018). This demand is not limited to the gluten-intolerant population. Although it is not clear whether a gluten-free diet is beneficial for health, some gluten-tolerant consumers believe that gluten-free food products are healthier (Christoph et al., 2018; Rej et al., 2018).

Other studies claim that despite recent improvements in the formulation and availability of gluten-free foods, they still are less available and more expensive than gluten-containing versions. They generally have adequate levels of fiber and sugar, but lower levels of protein and higher levels of fat compared with their gluten-containing counterparts (Allen & Orfila, 2018).

8.3.1 Structure of a starch network

The gluten network is essentially protein- and water-soluble. Gluten is made from two major wheat proteins, which together represent 85% of wheat endosperm protein: gliadin and glutenin. Kneading of wheat dough promotes the hydrogen bonding and disulfide cross-linking interactions of these proteins, eventually producing a viscoelastic and highly conformational protein network termed "gluten" (Wieser, 2007). In wheat dough, the gas is confined in the continuous "gluten matrix" (Gan et al., 1990). After the reaction of gliadin and glutenin in salt solution (NaCl), the gluten remains insoluble.

Starch, on the other hand, is characterized by many glucose units joined by intermolecular hydrogen bonds (for more details on starch structure, see Chapters 9 and 10 in Volume 1 of this series). In the beginning of 90's, El-Dash (1991) mentioned that a starch structure, under the appropriate conditions, may present enough viscoelastic properties to support gases produced during paste fermentation, or even substitute the gluten net. Starch molecules, formed by amylose and amylopectin and having many units of glucose, may be bound between themselves by intermolecular hydrogen links. As many molecules of amylose and amylopectin are linked, they form a three-dimensional net, capable of retaining gases and water in a similar way to glutens, as seen in Fig. 8.14. As this structure is maintained just by hydrogen links, it is weaker than gluten, which has ionic interactions and disulfide links between layers, achieving a strongly linked structure. To obtain a bread substitute, it is necessary to do more than generate a dough. It must form alveolus, retain the gases and air incorporated in the dough, and expand in the oven, generating the flavor, aroma, and appearance of cereal bakery products.

Replacing a protein network composed of glutenin and gliadin with a carbohydrate network, in this case glucose molecules, brings serious problems for gas retention and aroma development during baking. To circumvent these problems, one alternative is to reinforce the structure by adding protein to the formulation.

Pagani (1986) confirmed that the addition of protein substances to the formulation allows for obtaining an unconventional dough of good quality, with formation of a network during cooking. This protein additive will be most effective when there is perfect initial solubilization and rapid coagulation during thermal treatment (drying or cooking). High temperature drying favors the formation of the protein network, which will wrap the gelatinized starch during cooking, preventing it from dispersing in the aqueous phase. This reduces the stickiness that characterizes starch gels.

Another problem is that in the starch network, the oxidizing and reducing agents normally employed in traditional baking to strengthen the gluten structure do not work. However, it is possible to strengthen the starch network by increasing hydrogen bonds, even if the water and gas absorption capacity is also reduced, which requires specific conditions to allow for the development of a strong viscoelastic structure with adequate functional properties.

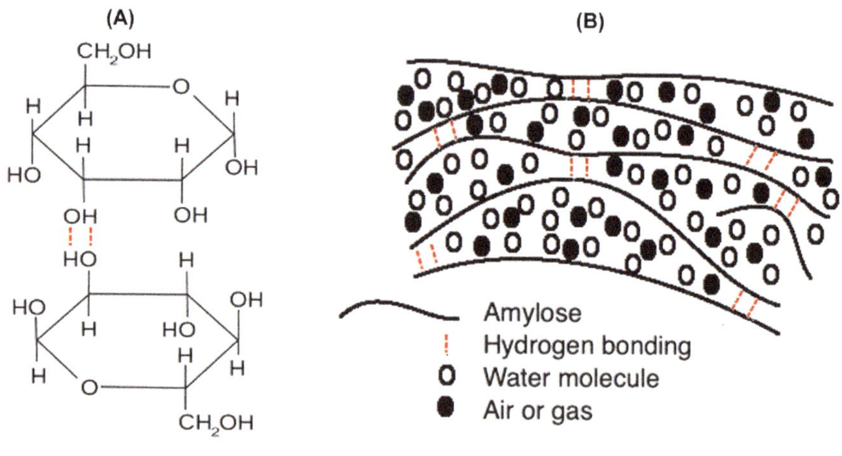

FIGURE 8.14 Structure of the starch (polysaccharides) network. (A) Intermolecular hydrogen links between glucose molecules; (B) film formation capable of retaining gasses and water.

If the importance of cassava as an energy source in human food is well established, the use of derivatives such as cassava flour, flour from chips, starch and "fermented cassava starch," or sour cassava starch, is little publicized outside Latin America.

In research on the development of alternative breads with cassava flour fortified with soy flour, Eggleston et al. (1992) evaluated the addition of egg white, margarine, and xanthan gum and concluded that all increased the amount of air retained by the dough at the mixing stage, but also the amount of gas retained after 60 minutes of dough fermentation. As the final volume of the bread also depends on the stability of the dough, the authors supplemented the addition of egg white with the use of margarine, which reduced gas loss. The authors considered that both protein and lipids act as stabilizers, reducing the gelatinization rate and starch solubility in the bread. Besides being well accepted by consumers in Nigeria, these bread substitutes maintained good characteristics during storage.

8.3.2 The South American "sour cassava starch" as a gluten-free bread solution

Cereda et al. (1995) described the process of obtaining sour starch, a traditionally fermented and sun-dried South American product used to prepare typical bread-like expansion products. They emphasized the importance of sun exposure to develop the expansion capacity of sour starch in baking. Mestres et al. (1996) also evaluated sour starch in comparison with cornstarch under the same treatment and found that cornstarch did not exhibit expansion, unlike cassava. The authors emphasized the expansion during baking for sun-dried sour starch, with observation of honeycomb structure in the kernel. The authors also confirmed that sun drying is essential to confer the expansion ability of sour starch upon cooking and rheological properties. Later, Brito et al. (2021) highlighted the UV-C of solar radiation as responsible for the modification that ensures and modulates the oven expansion. For more detailed information about the sour cassava starch, it is suggested to read Chapter 7.

Sour cassava starch is a modified starch which can be found in many South American countries. It presents popular names such as "almidón agrio" and "polvilho azedo." Sour cassava starch is used to make small "breads" mixed with cheese (Fig. 8.15), which justifies the name "cheese bread," products present in all countries that produce sour starch ("pão de queijo"; "pan de bono"; "chipa"), which guarantees a good market in most South American countries. Besides dispensing with

FIGURE 8.15 (A) Cheese bread (left) and "chipa" (right) in a snack bar in Campo Grande, Midwestern Brazil; (B) internal structure of a cheese bread.

gluten, which circumvents the problem of allergic consumers, the use of sour cassava starch, due to its rare expansion property, also dispenses with the use of raising agents such as baking soda and yeast. The protein in the cheese may explain the strengthening of the starch network but mixing just 5 grams of sour cassava starch in 5 mL of distilled water can represent an expansion of 8–12 cm^3/g (Brito & Cereda, 2015). Nine commercially available ready mixes of cheese bread were evaluated by Aplevicz and Demiate (2007) who found great variability, with expansion ranging from 4.1 to 10.8 cm^3/g.

Few consumers, including those with food intolerance syndromes, realize that "cheese bread" has the characteristics of baked bread dough, with gas-retaining cells, elasticity, crispy crust, and expansion in the oven. Provided the bread substitutes are properly baked and cooled, the gas is trapped in the dough structure. Researching the performance of sun-exposed fermented starch (sour starch) in baking, Shen et al. (1998) produced breads with good viscoelasticity from 80% sour starch and 20% soy flour, which were found to be more nutritious and economical than those with regular wheat flour.

Fig. 8.16 shows the difference in the expansion test of sour cassava starch in an oven because of the formation of a network expanding during baking that is able to trap gas, even without the addition of baking soda or yeasts. When making dough with sour cassava starch, other ingredients such as grated cheese and eggs can be used to make "cheese bread".

The use of sour cassava starch to make cheese bread opens the South American market, where it is presented as a traditional product, but with more global potential, in the case of serving consumers with FIS.

8.3.3 Development of gluten-free bread premix

However, the development of a gluten-free formulation does not solve all the problems of people with celiac syndrome, or food allergies. Although gluten-free bread has been well accepted by consumers who have no memory of consuming bread made from wheat flour,

FIGURE 8.16 Starch–water mixture expansion test to differentiate mixtures with cassava starch (A) from mixtures with sour cassava starch (B).

acceptance is lower among other consumers. The problem is that celiac syndrome usually affects only one member of a family, but all the family members eventually adopt a gluten-free diet. That is why it is important to have a product that is well accepted even by people accustomed to consuming bread made from wheat flour.

Another problem is the difficulty of finding gluten-free bakery products. For this reason, it is important to develop a loaf bread formulation that, if properly packaged, has a shelf life of up to 15 days, which facilitates distribution in less accessible places.

The use of premix in industrial bakery is a commercial reality that provides savings in time and resources, especially for professionals outside the major centers. They can be commercialized in large volumes or for home use, in one-kilo packages. To provide a fresh bread to consumers with celiac syndrome or allergic to cereals, Escouto and Cereda (1999, 2000) developed a dry premix that, when used, requires only the addition of liquid, be it water, milk, or similar. In this process, the dough was obtained with cassava flour and sour cassava starch, salt, and sugar and, if necessary, lipids and biological yeast. The process uses the same infrastructure as a bakery to make a loaf of bread.

The best breads had the following formulation, 42% sour starch, 20% cassava flour, 15% dehydrated egg white, 3% hydrogenated vegetable fat, 12% sugar, 10% fresh yeast, and 1% salt. This formulation makes it possible to create a dough with good expansion power and enough honeycomb to make the product look like a bread. In this case, yeast was added just to enhance flavor and disguise the scent of sour cassava starch. The manufacturing process adopted with the better response in dough texture was that of direct dough, with production of breads with direct baking, with support of a mixer follow by cylinder. The oven was heated up to 220°C for 30 minutes. The final product had the internal aspect of French bread with a gummy but alveolate crumb (Fig. 8.17). These variables were selected from previous tests, as they have the most influence on the final product.

When the dehydrated egg white content was increased to 27.85%, the bread characteristics were even better (Fig. 8.17C) and the weight loss between the raw dough and the baked bread was reduced by 14%. The weight

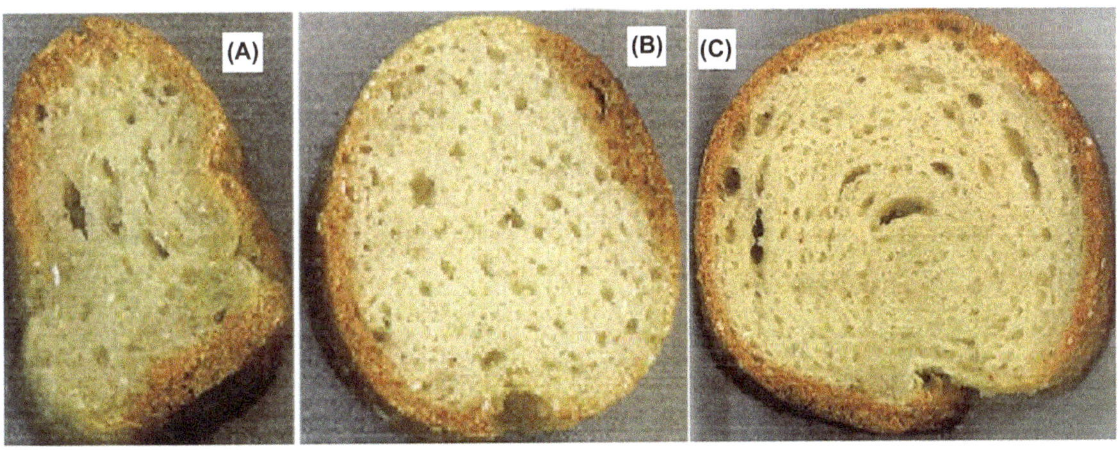

FIGURE 8.17 Aspect of gluten-free bread slices and effect of increasing the addition of dehydrated egg white. (A) 8.92%; (B) 18.57; (C) 27.85% of dehydrated egg white.

loss of the loaves during baking was 18.5% due to evaporation of water, which was not retained by the gluten network as occurs with wheat flour. The aroma and flavor were characteristic of bread, with a reduction of the typical aroma of cassava products. The bread aroma was adequate, and the taste became sweet. This formulation can also be used in conventional bakeries.

After the release of the articles by Escouto and Cereda (1999, 2000), other articles were published on gluten-free bread premix. Coronel et al. (2021) evaluated gluten-free bread premix with the objective to develop a novel alternative gluten-free premix for production of a bread product. Eight formulations of premixes were tested, and a commercial gluten-free premix was used as control. The best bread exhibited a higher protein, ash, crude fiber, and polyunsaturated fatty acid content and a higher antioxidant activity.

8.4 Final considerations

As highlighted in this chapter, starch can produce noodles that compete with wheat flour and are traditionally used for food in Asian countries, particularly those from sweet potato, arrowroot, and canna. Other non-wheat products are also possible to be made with starch from various sources.

In addition to their cultural value, these starch-based products characterize a group of gluten-free foods that are valued worldwide for serving consumers with sensitization syndromes that require lifelong exclusion of gluten from their foods. This chapter describes these products and draws attention to the fact that other food syndromes have recently been identified, such as the still poorly understood FIS, which further enhances the value of the gluten-free market.

One of the most sought-after products in this market is bread without gluten because of the widespread consumption of this staple food. This chapter highlights sour starch, a typical South American product, that after fermented and subjected to UV-C radiation from the sun acquires the property of expansion without the use of baking soda or yeasts.

Other gluten-free bakery dough products can be prepared with cassava derivatives, such as cassava flour, which, besides being gluten free, is a good source of dietary fiber. A good bread dough results from the presence of pregelatinized starch, formed due to the low gelatinization temperature of cassava starch (see Chapter 10 in Volume 1 in this series), which results from oven or hot air drying. This pregelatinized starch is responsible

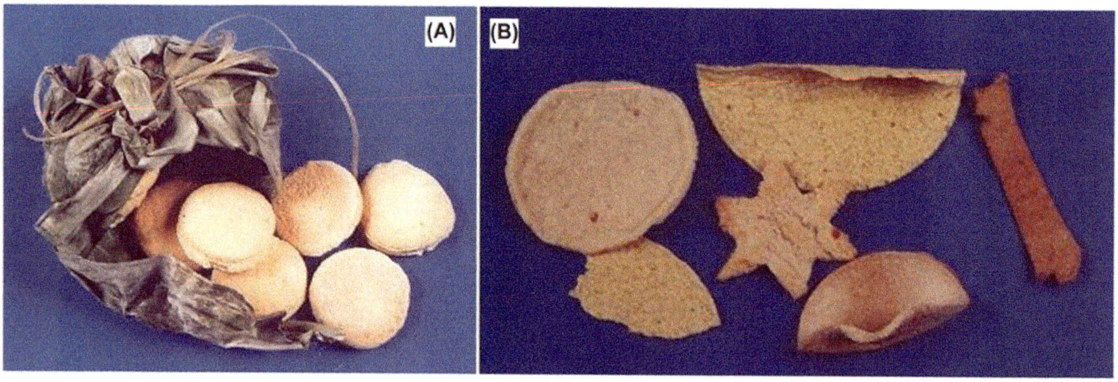

FIGURE 8.18 Sweet (A) and savory cookies of the appetizer and taco type (B), obtained by traditional baking from cassava flour.

for the starch network capable of retaining the gas formed during fermentation, in the making of bread substitute products. It also allows for the elaboration of cookies and crackers, both salty (snacks) and sweet. The preparation is simple, and the products can be flavored (Fig. 8.18).

In addition to the traditional culturally related market, the market for gluten-free products is growing worldwide, in a mechanism with no turning back. Celiac alone account for about 1% of the world's population and are forced to exclude wheat, barley, and rye from their food for life. Although starch is used in this market, its importance has not been properly appreciated. For this to happen, more information needs to be brought to the industry.

The processes used in Asia for noodles highlight different types of formulation used in making starch-based pasta. The preparation is simple, and the products can be flavored and other ingredients included.

References

Allen, B., & Orfila, C. (2018). The availability and nutritional adequacy of gluten-free bread and pasta. *Nutrients*, *10*(10), 1370. Available from https://doi.org/10.3390/nu10101370.

Aplevicz, K. S., & Demiate, I. M. (2007). Characterization of native and modified cassava starches and their use in baked products. *Sociedade Brasileira de Ciência e Tecnologia de Alimentos*, *27*(3), 478–484. Available from https://doi.org/10.1590/S0101-20612007000300009. (In Portuguese).

Bangar, S. P., Ali, N. A., Olagunju, A. I., Pastor, K., Ashogbon, A. O., Dash, K. K., Lorenzo, J. M., & Ozogul, F. (2022). Starch-based noodles: Current technologies, properties, and challenges. *Journal of Texture Studies*, *54* (1), 21–53. Available from https://doi.org/10.1111/jtxs.12730.

Barbaro, M. R., Cremon, C., Stanghellini, V., & Barbara, G. (2018). Recent advances in understanding non-celiac gluten sensitivity. *F1000Research*, *7*. Available from https://doi.org/10.12688/f1000research.15849.1.

Biagini, R. E., MacKenzie, B. A., Sammons, D. L., Smith, J. P., Striley, C. A., Robertson, S. K., & Snawder, J. E. (2004). Evaluation of the prevalence of antiwheat-, anti-flour dust, and anti-α-amylase specific IgE antibodies in US blood donors. *Annals of Allergy, Asthma & Immunology*, *92*(6), 649–653. Available from https://doi.org/10.1016/S1081-1206(10)61431-7.

Brito, V., Santos, R., Narcisa-Oliveira, J., Demiate, I., & Cereda, M. (2021). Continuous flow process to make fermented cassava starch substitute: I. Physicochemical and functional properties. *Brazilian Journal of Food Technology*, *24*. Available from https://doi.org/10.1590/1981-6723.01720.

Brito, V. H. D. S., & Cereda, M. P. (2015). Method for the determination of specific volume as a quality standard for fermented cassava starch and substitutes. *Brazilian Journal of Food Technology*, *18*(1), 14–22. Available from https://doi.org/10.1590/1981-6723.0214.

Cereda, M. P., Nunes, O. L. G. S., & Westby, A. (1995). Brazilian fermented cassava starch: A low cost acidic starch with modified functional properties. *Science, Technology & Development, London*, *13*(1), 43–49.

Christoph, M. J., Larson, N., Hootman, K. C., Miller, J. M., & Neumark-Sztainer, D. (2018). Who values gluten-free? Dietary intake, behaviors, and sociodemographic characteristics of young adults who value gluten-free food. *Journal of the Academy of Nutrition and Dietetics*, *118*(8), 1389–1398. Available from https://doi.org/10.1016/S0733-5210(09)80153-7.

Cianferoni, A. (2016). Wheat allergy: Diagnosis and management. *Journal of Asthma and Allergy*, *9*, 13. Available from https://doi.org/10.2147/JAA.S81550.

Coronel, E. B., Guiotto, E. N., Aspiroz, M. C., Tomás, M. C., Nolasco, S. M., & Capitani, M. I. (2021). Development of gluten-free premixes with buckwheat and chia flours: Application in a bread product. *Lwt*, *141*, 110916.

Derycke, V., Veraverbeke, W. S., Vandeputte, G. E., De Man, W., Hoseney, R. C., & Delcour, J. A. (2005). Impact of proteins on pasting and cooking properties of nonparboiled and parboiled rice. *Cereal Chemistry*, *82*(4), 468–474. Available from https://doi.org/10.1094/CC-82-0468.

Di Sabatino, A., Vanoli, A., Giuffrida, P., Luinetti, O., Solcia, E., & Corazza, G. R. (2012). The function of tissue transglutaminase in celiac disease. *Autoimmunity Reviews*, *11*(10), 746–753. Available from https://doi.org/10.1016/j.autrev.2012.01.007.

Dickinson, E. (2018). Hydrocolloids acting as emulsifying agents-How do they do it? *Food Hydrocolloids*, *78*, 2–14. Available from https://doi.org/10.1016/j.foodhyd.2017.01.025.

Dobraszczyk, B. J., & Morgenstern, M. P. (2003). Review: Rheology and the breadmaking process. *Journal of Cereal Science*, *38*, 229–245.

Eggleston, G., Omoaka, P. E., & Ihedioha, D. O. (1992). Development and evaluation of products from cassava flour as new alternatives to wheaten breads. *Journal of Science and Food Agriculture, 59*(3), 377–385.

El Khoury, D., Balfour-Ducharme, S., & Joye, I. J. (2018). A review on the gluten-free diet: Technological and nutritional challenges. *Nutrients, 10*(10), 1410. Available from https://doi.org/10.3390/nu10101410.

El-Dash, A.A. (1991). Molecular structure of gluten and viscoelastic properties of dough: A new concept. In *Brazilian Congress of protein* (Vol. 1, p. 511).

Escouto, L., & Cereda, M. (1999). Ajustes de formulação para pães sem glúten. *I Simpósio em Energia na Agricultura, 1*, 251–256. (In Portuguese).

Escouto, L.F. S., & Cereda, M.P. (2000). Use of fermented cassava starch in the development of a formulation for bread without gluten. In *Proceedings of the Twelfth Symposium of International Society of Tropical Root Crops (ISTRC)* (pp. 159–161).

Galvez, F. C. F., & Ressurrección, A. V. A. (1992). Rehabilitee of the focus group technique in determining the quality characteristics of mungbean (*Vigna radiata (L.) wilczec*) noodles. *Journal of Sensory Studies, 7*, 315–326.

Galvez, F. C. F., Resurreccion, A. V. A., & Ware, G. O. (1995). Formulation and process optimization of mungbean noodles. *Journal of Food Processing and Preservation, 19*(3), 191–205.

Gan, Z., Angold, R. E., Williams, M. R., Ellis, P. R., Vaughan, J. G., & Galliard, T. (1990). The microstructure and gas retention of bread dough. *Journal of Cereal Science, 12*(1), 15–24.

Gobbetti, M., Pontonio, E., Filannino, P., Rizzello, C. G., De Angelis, M., & Di Cagno, R. (2018). How to improve the gluten-free diet: The state of the art from a food science perspective. *Food Research International, 110*, 22–32. Available from https://doi.org/10.1016/j.foodres.2017.04.010.

Hamada, S., Suzuki, K., Aoki, N., & Suzuki, Y. (2013). Improvements in the qualities of gluten-free bread after using a protease obtained from *Aspergillus oryzae*. *Journal of Cereal Science, 57*, 91–97. Available from https://doi.org/10.1016/j.jcs.2012.10.008.

Hamaker, B. R., & Griffin, V. K. (1993). Effect of disulfide bond-containing protein on rice starch gelatinization and pasting. *Cereal Chemistry, 70*, 377–380. Available from http://pascal-francis.inist.fr/vibad/index.php?action = getRecordDetail&idt = 3812492.

Hermann, M., Quynh, N. K., & Peters, D. (1997). Reappraisal of edible canna as a high-value starch crop in Vietnam. *CIP Program Report, 98*, 415–424.

Holub, S. (1990). Development of gluten free extruded bakery products. *Ernaehrung, 14*, 208–209.

Kaczmarski, M., & Kurzatkowska, B. (1988). The contribution of some environmental factors to the development of cow's milk and gluten intolerance in children. *Roczniki Akademu medycynei em Juliana mrchlewskigo w Biolymatocu, 33*, 151–165.

Kim, Y. S., & Wiesenborn, D. P. (1996). Starch noodle quality as related to potato genotypes. *Journal of Food Science, 61*(1), 248–252.

Kokke, F. T., Van-Elburg, R. M., Van-Overbeek, F. M., & Heymans, H. S. (1994). A new biscuit free of cow's milk, chicken egg protein, lactose and gluten for children with food hipersentivity. *Nederlands Tijdachrift voor Geneeskunde, 17*, 2549–2552.

Konovalova, I.D., Orobinskaya, V.N., Kholodova, E.N., Pisarenko, O.N., & Galdin, E.V. (2020). The use of combined composites of non-traditional raw materials of plant origin. In the production of functional bakery products. In *IOP Conference Series: Materials Science and Engineering* (Vol. 941, No. 1, p. 012058). IOP Publishing.

Machado, L.M. P. (1996). *Pão sem glúten*: Otimização de algumas variáveis de processamento. 186f. Dissertação (Mestrado em Tecnologia de Alimentos) – Faculdade de Engenharia de Alimentos – Universidade de Campinas, Campinas, 1996. (In Portuguese).

Matsuo, H., Morimoto, K., Akaki, T., Kaneko, S., Kusatake, K., Kuroda, T., Niihara, H., Hide, M., & Morita, E. (2005). Exercise and aspirin increase levels of circulating gliadin peptides in patients with wheat-dependent exercise-induced anaphylaxis. *Clinical & Experimental Allergy, 35*(4), 461–466. Available from https://doi.org/10.1111/j.1365-2222.2005.02213.x.

Mestres, C., Boungou, O., Zakhia, N., Rouau, X., & Faure, J. (1996). Comparison of the ability of fermented maize flourand cassava starch for making bread-like products.

Metcalfe, D. D. (1984). Food hypersensitivity. *Journal of Allergy and Clinical Immunology, 73*(6), 749–762.

Milatovic, L., & Ballini, N. (1986). Un nuovo approccio tecnologico e nutrizionale alla produzione della pasta. *Tecnica Molitoria, 37*(10), 801–815.

Morita, E., Chinuki, Y., Takahashi, H., Nabika, T., Yamasaki, M., & Shiwaku, K. (2012). Prevalence of wheat allergy in Japanese adults. *Allergology International, 61*(1), 101–105. Available from https://doi.org/10.2332/allergolint.11-OA-0345.

Nowak-Węgrzyn, A., Katz, Y., Mehr, S. S., & Koletzko, S. (2015). Non–IgE-mediated gastrointestinal food allergy. *Journal of Allergy and Clinical Immunology, 135*(5), 1114–1124. Available from doi.org/10.1016/j.jaci.2015.03.025.

Pagani, A., Resmini, P., & Dal Bon, G. (1981). Formulazione e produzione di paste alimentari a partire da materie prime non convenzionali. *Tecnica Molitoria, 32*(1), 1–24.

Pagani, M.A. (1986). Pasta products from non-conventional raw materials. *Pasta and Extrusion Cooked Foods. C. Mercier & C. Cantarelli, ed. Elsevier Applied Science, London*, 52-68.

Pickering, S. U. (1907). Emulsions. *Journal of the Chemical Society*, *91*, 2001–2021.

Rej, A., Kurien, M., Tosi, P., Trott, N., & Sanders, D. (2018). The role of a gluten free diet in 'lifestylers'? the first double blind randomised study. Gut 67, abstr. A220, PWE-145. doi.org/10.1136/gutjnl-2018-BSGAbstracts.439.

Renzetti, S., & Rosell, C. M. (2016). Role of enzymes in improving the functionality of proteins in non-wheat dough systems. *Journal of Cereal Science*, *67*, 35–45. Available from https://doi.org/10.1016/j.jcs.2015.09.008.

Salcedo, G., Quirce, S., & Diaz-Perales, A. (2011). Wheat allergens associated with Baker's asthma. *Journal of Investigational Allergology and Clinical Immunology*, *21*(2), 81, PMID: 21462797.

Schober, T. J. (2009). In E. Gallagher (Ed.), *Gluten-free food science and technology* (1st Ed.). Chichester: Wiley-Blackwell, Ch. 9.

Shan, L., Qiao, S. W., Arentz-Hansen, H., Molberg, Ø., Gray, G. M., Sollid, L. M., & Khosla, C. (2005). Identification and analysis of multivalent proteolytically resistant peptides from gluten: Implications for celiac sprue. *Journal of Proteome Research*, *4*(5), 1732–1741. Available from https://doi.org/10.1021/pr050173t.

Shen, H., Wang, M., Ntagenyera, C., & Manirakiza, C. (1998). Applying cassava sour food starch in bread making. *Zhongguo Liangyou Xuebao*, *13*(6), 19–20.

Singh, P., Arora, A., Strand, T. A., Leffler, D. A., Catassi, C., Green, P. H., Kelly, C. P., Ahuja, V., & Makharia, G. K. (2018). Global prevalence of celiac disease: Systematic review and meta-analysis. *Clinical gastroenterology and hepatology*, *16*(6), 823–836. Available from https://doi.org/10.1016/j.cgh.2017.06.037.

Smith, J. P., Daifas, D. P., El-Khoury, W., Koukoutsis, J., & El-Khoury, A. (2004). Shelf life and safety concerns of bakery products—a review. *Critical Reviews in Food Science and Nutrition*, *44*(1), 19–55. Available from https://doi.org/10.1080/10408690490263774.

Tanabe, S., AraI, S., & Watanabe, M. (1996). Modification of wheat flour brome lain and baking hypoallergenic bread with added ingredients. *Bioscience, Biotechnology, and Biochemistry*, *60*, 1269–1272.

Wang, X., Wang, J., Liu, H., Zhao, L., Wang, Y., Wu, X., & Liao, X. (2020). Improving the production efficiency of sweet potato starch using a newly designed sedimentation tank during starch sedimentation process. *Journal of Food Processing and Preservation*, *44*(10), e14811.

Wieser, H. (2007). Chemistry of gluten proteins. *Food Microbiology*, *24*, 115–119.

Wintz, E., & Kuntze, R. (2000). Specifics in a production dietetic bakery products for certain consumers groups. *Žito-hleb*, *27*(4-5), 143–150.

Yano, H. (2010). Improvements in the bread-making quality of gluten-free rice batter by glutathione. *Journal of Agricultural and Food Chemistry*, *58*, 7949–7954. Available from https://doi.org/10.1021/jf1003946.

Yano, H. (2019). Recent practical research in the development of gluten-free breads. *NPJ Science of Food*, *3*(1), 1–8. Available from doi.org/10.1038/s41538-019-0040-1.

Young, E., Stoneham, M. D., Petruckevitch, A., Barton, J., & Rona, R. (1994). A population study of food intolerance. *The Lancet*, *343*(8906), 1127–1130. Available from https://doi.org/10.1016/S0140-6736(94)90234-8.

CHAPTER

9

Traditional Brazilian foods based on partial gelatinization of cassava starch: tapioca pearls ("sagu"), Brazilian tapioca ("tapioquinha"), broken tapioca pearls ("tapioca"), and popped tapioca pearls ("farinha de tapioca")

Olivier François Vilpoux[1,2] *and Marney Pascoli Cereda*[2,3]

[1]Federal University of Rondonopolis, Post-graduate Program in Environmental Management and Technology - PPGTA, Rondonopolis, Mato Grosso, Brazil [2]Tecsapp Solutions, Campo Grande, Mato Grosso do Sul, Brazil [3]Retired from São Paulo State University, Botucatu, São Paulo, Brazil

9.1 Introduction

Before starting this chapter, it is necessary to describe the products to be discussed and their names. In many countries, the terms cassava starch, tapioca flour, and tapioca starch are confounded. However, in this chapter, the term "tapioca" refers to products derived from native partially gelatinized cassava starch. The products to be covered, with their typic names in Portuguese, are:

- Gum: *goma*, the humid starch used to produce the other four products.
- Brazilian tapioca: *tapioquinha* or *tapioca*.
- Tapioca pearl: *sagu*.
- Broken tapioca pearl: *tapioca*.
- Popped tapioca pearl: *farinha de tapioca* (tapioca flour).

First, it is important to point out that all these products are manufactured from partially gelatinized native cassava starch. Brazilian tapioca is called *tapioquinha*, or *tapioca*, in Brazil. It is a product that assimilates to a cassava pancake. Until the early 2000s, it was a typical regional product of the North and Northeast regions of Brazil, produced essentially by small units in local markets.

Starch Industries
DOI: https://doi.org/10.1016/B978-0-323-90842-9.00003-0

In these regions, it is sold in the streets or served in hotels as part of the breakfast.

In the last few years, this product has seen a great market development, and today the gum, the humid starch used to produce Brazilian tapioca, is present in most of the supermarkets of the country, sold in 1 kg packages for home use. With the expansion of consumption, large industries in southern Brazil began to rehydrate dry starch to be sold in supermarkets in the form of wet starch ready for home production of Brazilian tapioca. With this, they conquered most of the market.

In Brazil, another product is also named tapioca. It is produced and consumed in the southern region of the country and is known for its appearance of a broken tapioca pearl. To differentiate this product from Brazilian tapioca, the *tapioca* produced in southern Brazil was named broken tapioca pearl.

Tapioca pearl is a derivative of cassava starch of great importance in many countries. In many places, despite its commercial importance, it is produced essentially by hand. It comes in different shapes, granulometry, and colors. In Brazil, it is known as *sagu* and is most diffused in the southeastern region, as a dessert and looks like small white pearls.

Collado and Corke (1998) explained that the tapioca pearls were originally made from starch extracted from the sago palm tree (*Metroxylon sagu* and *Metroxylon rumphii*) (Fig. 9.1), a plant found in Asia and the Pacific. In the 19th century, sago played an important role in the crews' diet on the ships that transported tea on their trips back to Europe and ended up being incorporated into the diet in England's industrial north. In the second half of the 20th century, it began to be used in good-looking desserts such as puddings, blanc manger, and others.

According to Collado and Corke (1998), sago was given to children and convalescents due to its easy digestibility. The consumers developed a fascination for its elastic and gummy texture, a characteristic of the product. Due to the increase in demand and the difficulty of

FIGURE 9.1 Grinding of sago palm kernel (A) and extraction of starch from the grinded material (B). *Photos from Wikimedia Commons: A: https://commons.wikimedia.org/wiki/File:Sagu-01_081119-3840_sim.JPG; B: https://commons.wikimedia.org/wiki/File:Sagu-05_081119-3856_sim.JPG.*

production with sago palm tree, sago is now produced mainly from cassava starch and is called tapioca pearl. In Brazil, the product has kept its original name and continues to be called *sagu*.

Finally, similar to tapioca pearls, however expanded like popcorn, popped tapioca pearls (*farinha de tapioca*[1]) are only consumed in Pará, in the Brazilian Amazon region. For Godoy (1940), popped tapioca pearls and tapioca pearls are nothing more than the product derived from the heating of humid starch. Until today, Europe has an attractive market of tapioca pearls to thicken soups, broths, and milk preparations, with a creamy consistency, known as "ambrosia."

This chapter gathers the information available on these different products, based on available literature and reports from site visits.

Information in literature concerning tapioca pearls, Brazilian tapioca, and popped tapioca pearls production is rare. Although the consumption of tapioca pearls and Brazilian tapioca is relatively expressive, available literature is obsolete, and the processing equipment is made to order and not in line. On the other hand, popped tapioca pearls are a hand-processed product from the State of Pará, which consist of expanded tapioca pearls and are very popular in desserts such as ice creams, cakes, creams, and puddings.

9.2 Starch extraction process in small units—gum production

Gum is the raw material used to produce all the other products discussed in this chapter. The gum used to produce tapioca pearls (*sagu*), broken tapioca pearls (*tapioca*), and the gum sold in supermarkets, for the production of Brazilian tapioca (*tapioquinha*), are made in large companies located in the South region, which use the starch extraction technologies presented in Chapter 2. The gum used to produce Brazilian tapioca (*tapioquinha*) in the North and Northeast regions of Brazil and to process popped tapioca pearls (*farinha de tapioca*) is made in small companies, from the process and equipment described in this chapter.

The process of cassava starch extraction for companies with daily capacity higher than 50 tons of cassava was described in Chapter 2, and the processes for companies with capacity between 5 and 50 tons were described in Chapter 3. Thus, in this chapter, only the extraction process found in small sized companies, with capacity under 5 tons of roots per day and with little or no mechanization, is described.

9.2.1 Peeling and washing

The roots are piled under the processing plant shed and manually peeled with the aid of knives (Fig. 9.2A). The roots are usually peeled within 2 days. In addition to washing by hand, the peeled roots can be collected in baskets and discharged into tanks with water. The washing is done by shaking the roots with sticks or by a person stepping on them while still in the water tank. In some larger units, cassava can be peeled and washed in a rotary peeling machine, like the one presented in Chapter 3 (Fig. 9.2B).

9.2.2 Grating

In small units, grating is a dry operation (Fig. 9.3A), different from the water grinding, which is generally used in the modern starch extraction process. Most graters are cylindrical, made of wood, with metal saws or choppers (Fig. 9.3B). Then, the grated mass is received in a tank, made of wood or concrete, for starch

[1] In Brazil this product is called tapioca flour, but to avoid confusion with cassava flour or cassava starch, it was decided to call it popped tapioca pearl, which better defines the product's characteristics.

FIGURE 9.2 Manual peeling of cassava (A) and use of a rotary batch peeler (B).

FIGURE 9.3 Typical grater of a small unit (A); wooden cylinder with metal saws (B); and wooden cylinder covered with a metal foil pierced with nails (C).

extraction. The most rudimentary graters are made of wooden cylinders covered with a metal foil pierced with nails, like a cheese grater (Fig. 9.3C).

9.2.3 Extraction

In many small units, the extraction is done on frames covered with cotton fabric, supported on the sides of the tanks, which can be made of masonry (Fig. 9.4B), wooden troughs, or any other easily accessible material (Fig. 9.4A). The processing is manual, unlike the larger companies (Chapters 2 and 3), consisting of successive washings of the pulp. In the smaller units, the extraction is done in hanging cloth bags (Fig. 9.4A).

The mass retained on the sieve (bran) is removed, squeezed manually, and piled outside the processing unit. In samples collected in Pará, the bran contained about 70% of water and from 72% to 75% of starch, in dry basis. This starch percentage is similar to that found in some more modern starch plants (see Chapter 2).

FIGURE 9.4 Gum extraction in hanging cloth bags (A) and in frames covered with cotton fabric (B).

9.2.4 Purification and dewatering

The starch in suspension is left decanting for about 3 hours. According to information obtained in some communities in the Northern region of Brazil, it is common to add salt to water to make decantation easier.

After the decantation is finished, the residual water is drained. In some units, the water is removed through holes located on three levels in the tanks, which are opened one by one. If they are all opened at the same time, the current formed by the water drags part of the starch already decanted. The surface of the deposited starch has a firm consistency, facilitating its washing and the removal of surface impurities. The water containing lighter impurities and a small amount of starch is decanted into another tank to obtain a lower quality product.

After the water containing impurities has been removed, the deposited starch surface is broken up and clean water is added. The starch is put back into suspension for further cleaning and is allowed to decant. The next day, the tanks are drained of the surface water to remove the clean starch. This starch is white and shiny, without streaks and surface dirt.

The starch is then removed from the tank in the form of pieces that look like fresh cheese, due to the thixotropic property that causes, at a 50%–55% humidity, the mass to change from solid to liquid state, when stirred, and return in solid form, when at rest. The removed pieces of starch are placed in cotton bags or braided plastic fiber (raffia) where they are left to drain (Fig. 9.5).

After this step, the starch is ready to be processed into tapioca pearls, Brazilian tapioca, popped tapioca pearls, or simply being commercialized in the form of gum.

9.3 Products presentation and processes

From the wet starch obtained according to the process described in Section 9.2, for small units, or by more modern processes (Chapters 2 and 3), it is possible to produce four types of products: Brazilian tapioca (*tapioquinha*), broken tapioca pearls (*tapioca*), popped tapioca pearls (*farinha de tapioca*), and tapioca pearls (*sagu*).

FIGURE 9.5 Bags used to drain starch.

9.3.1 Brazilian tapioca

Brazilian tapioca, named "tapioca" or "tapioquinha" in Brazil, is a traditional product typical of the North and Northeast regions of the country, where it was originally produced from starch originating from small units (gum). In recent years, it has entered the consumer's taste and the gum, or wet cassava starch, the raw material used for its preparation, can now be found in most markets throughout the country (Fig. 9.6A).

Gum is sold with a moisture content of 45%–50%, with or without preservatives, according to the brand. The high humidity of the starch gives it a granular aspect, as shown in Fig. 9.6B.

To produce Brazilian tapioca, the humid starch is sieved or crushed over a heated plate (Fig. 9.7), where it forms a kind of thin pancake or crepe. Due to its moisture, the starch gelatinized on the surface, keeping the opaque white color in its interior. Once the tapioca is gelatinized, it becomes more consistent and can be turned to be well gelatinized on both sides. Most of the market is in humid starch form (gum), with preparation of the Brazilian tapioca done by the final consumer (home, hotel, and restaurant). However, instead of the humid starch, some companies commercialize the tapioca already formed and rolled, in the form of cigars (Fig. 9.7A and C).

After it is ready, the tapioca can be filled with various kinds of stuffing, like coconut, cheese, butter, the most traditional of the Northeast region, or less traditional stuffing, like Calabrese type sausage, chicken, ham, milk candy, banana, or chocolate.

9.3.2 Broken tapioca pearls

In Brazil, tapioca can also define a product with irregularly shaped starch aggregates,

FIGURE 9.6 Type of packaging found in supermarkets in Brazil for preparation of Brazilian tapioca (A) and granular starch structure (B).

FIGURE 9.7 Tapioca production in small industrial units (A and C) and in a hotel in the northeastern region of Brazil (B).

FIGURE 9.8 Ovens for broken tapioca pearls used in the South of Brazil. *Photo from Liviston teles Sena, Paranavai Maquinas.*

similar to a broken tapioca pearl. There are industries that produce only broken tapioca pearls, mainly cassava starch units from the South of the country, in equipment different from the ones used for tapioca pearls.

In this case, cassava starch with a humidity close to 50%, as for tapioca pearls, is crumbled over a heated oven (Fig. 9.8) like the one used to dry cassava flour (see the upcoming Volume 4 in this series). The oven temperature is kept around 70°C–80°C, to allow the starch to gelatinize. The starch is stirred while gelatinized and dried, allowing aggregation into irregular particles, characteristic of broken tapioca pearls.

The oven drying, at higher temperatures than for tapioca pearls, leaves the product with a transparent appearance, while the tapioca pearls have a milky appearance.

TABLE 9.1 Physicochemical composition of various broken tapioca pearls samples produced in three Brazilian states.

	Paraná (South)	São Paulo (South-east)	Maranhão (North-east)
Moisture (%)	9.04	13.55	15.00
% Dry matter			
Starch	96.02	96.80	83.32
Protein	0.28	-*-	-*-
Lipids	0.17	0.08	-*-
Ashes	0.16	0.13	-*-
Fibers	0.14	0.40	2.48
Others			
pH	4.64	4.73	4.54
Acidity (*)	1.19	1.27	2.00

(*) mL of NaOH N/100 g; -*- data not available.

9.3.2.1 Characteristics of broken tapioca pearls' commercial product

The characterization of the commercial products from different Brazilian states is displayed in Table 9.1.

The results from Table 9.1 reveal a great similarity between the products from São Paulo and Paraná states, produced in modern starch plants. On the other hand, the broken tapioca pearls produced in Maranhão (North-east of Brazil), made with starch from small units, shows a lower starch content and greater fiber content, denoting a less efficient extraction and purification system.

In microbiological analyses of broken tapioca pearls samples from two modern units from South of Brazil, 1920 and 100 CFUs (Colony-

TABLE 9.2 Granulometry for commercial broken tapioca pearls, expressed in % retained, from a sample produced in a modern cassava starch unit in South of Brazil.

Sieve number	Opening (mm)	Retention (%)
4	4.76	16.73
6	3.36	68.66
8	2.38	13.86
10	2.00	0.05
12	1.69	0.26
16	1.19	0.03
18	1.00	0.01
Bottom	<1.00	0.51

Forming Units) were counted, respectively, for the total count in plates. No other microorganism was detected. The values found are low and satisfy the limits established by Brazilian legislation.

Most of the broken tapioca pearls analyzed in Table 9.2, which were produced in a modern unit in southern Brazil, are between the values of 3.36 and 4.76 mm, with almost all of them larger than 2.38 mm. The material retained on sieves smaller than 2 mm is probably due to imperfections or broken granules.

9.3.3 Popped tapioca pearls processing

In Brazil, popped tapioca pearls is produced by small units in the State of Pará, in the Amazon region. This product is regional, being found mainly in this state, but has a great potential for expansion in other regions. In most part of the process, popped tapioca pearls are prepared like tapioca pearls. This product could also be called expanded tapioca pearls because it is a tapioca pearl that has undergone an explosion process similar to that of popcorn. As for Brazilian tapioca, starch with 45%–50% moisture is used.

The popped tapioca pearls production described in this chapter explains the process as it is performed in Vila dos Americanos, in the municipality of Santa Isabel and in the area surrounding the city of Castanhal, about 70 km from Belém, in the State of Pará, North of Brazil.

According to information obtained from the production sites, popped tapioca pearls production technology arose around 1940, in the *Vila dos Americanos*, being introduced by João Miguel, a local producer. In 1988, in the areas surrounding this site, there were about 23 production units, of which 4 produced only the humid starch (gum) and 19 produced only the popped tapioca pearls, acquiring the gum from the other units. In 2004, all the starch was bought from cassava starch industries in South of Brazil, where the product is cheaper. Before being used, this starch had to be moistened. In 2020, the companies changed again and were producing their own starch, in a similar process to the one described in Section 9.2.

There is no official statistic on the amount produced or commercialized. Due to the scale in which popped tapioca pearls are produced, it is difficult to establish the yields and the processing capacity. Data from the producers themselves inform a weekly production from 800 to 1000 kg.

Guimarães et al. (1988) reported that popped tapioca pearls are a hand-processed product from the State of Pará and that information from the literature concerning this product is rare. As with other cassava derivatives, it is characterized by the high starch level and low protein, lipid, and mineral levels, which make the popped tapioca pearls a highly caloric food.

Popped tapioca pearls are consumed in the Amazon region in the form of porridges, cakes, puddings, ice creams, as well as an accompaniment for the palm tree fruit extract called *açaí*. It is commercialized in the open markets in the city of Belém, capital of the state of Para, where the unit of measure is the liter.

FIGURE 9.9 Appearance of popped tapioca pearls commercialized in the state of Pará (A and B) and cake made with popped tapioca pearls (C).

After the production process, the popped tapioca pearls present characteristics that are similar to those of expanded rice, such as that used in breakfast cereals (Fig. 9.9). The color varies from white to cream, as a function of starch or gum purity, with the same mushroom shape observed in popcorn.

The mass balance and the flowchart are shown in Fig. 9.10. As the companies are small, with manual extraction, productivity is very low, obtaining only 12 kg of popped tapioca pearls for every 100 kg of cassava, productivity less than half of that obtained in large cassava starch companies (Chapter 2).

The phases of popped tapioca pearls production are disaggregation of the humid starch, lumping, scalding, expansion, classification, and packaging.

9.3.3.1 *Disaggregation*

The wet starch, or gum, is passed through a bamboo sieve with a mesh size between 2.0 and 2.5 mm, which has the sole purpose of disaggregating the partially dehydrated starch (Fig. 9.11A). The starch falls into a wooden trough 3.0 m long, 0.70 m wide, and 0.60 m high. This process can also be mechanized, with the use of a disaggregation machine (Fig. 9.11B). This machine is similar to the one shown in Fig. 9.3A and B, but with the addition of spikes to facilitate the disaggregation of the starch.

9.3.3.2 *Lumping (pearl formation)*

The manual operation is done on cotton cloth stretched over a wooden frame (Fig. 9.12A). The person works standing with the palms of the hands open on the cloth, on which he/she places about 1 kg of the crumbled wet starch. The light and rotating movement of the hands on the wet starch creates lumps or granules. Their size depends on the humidity of the starch. If it is too wet, it may stick to the cloth, and if too dry, the granules fall apart. To increase the diameter of the granules, a new amount of wet starch is added and pressed lightly against the cloth, forming a second layer on top of the first.

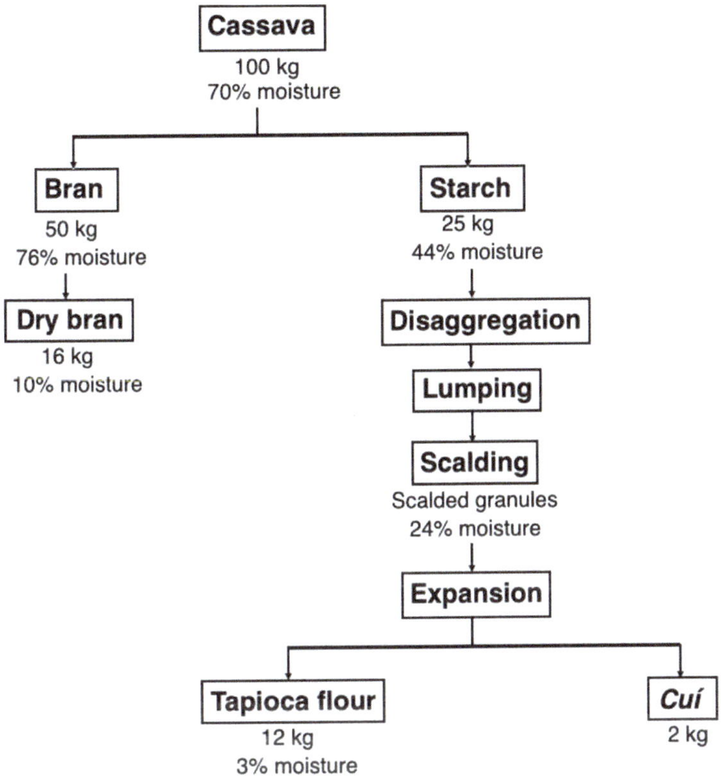

FIGURE 9.10 Flowchart and mass balance for tapioca flour production, in a flour house in the municipality of Vila dos Americanos, Pará, Brazil.

FIGURE 9.11 Manual disaggregation of moist starch (A) and approximate view of the cylinder of a disaggregation machine (B).

FIGURE 9.12 Wooden frame used for the pearl formation in the manual process (A) and concrete mixer type machine for the mechanized process (B).

The pearl formation can also be made with an equipment similar to a concrete mixer (Fig. 9.12B). The granules formed are identical to the tapioca pearls. Whistler, BeMiller, and Paschall (2012) described starch extraction from the core of the palm tree and the making of the sago pearls. The humidified starch is pressed against an iron sieve, or woven wire screen, and shaken, to produce the granules. In India, manual processing of sago pearls is still very present. Once the pearls are obtained, they are selected by passing them through a sieve and toasted in shallow pans, to obtain partial gelatinization and drying.

When the granules reach the desired diameter, they are selected on a sieve with a mesh size of about 2.80 mm. This process can be manual or mechanized (Fig. 9.13). In this step, the finer and coarser particles are separated before being returned to the process.

Part of the smaller granules are put on the cotton cloth used in the lumping process to aid in forming the next lot. The selected pearls are placed in a wooden trough, where they await the scalding phase, which is performed the same day.

9.3.3.3 Scalding

The tapioca pearls are scalded in an oven similar to the flour ovens used in the Amazon region. According to local information, the pan in the oven must be made of iron, as copper is not

FIGURE 9.13 Sieve system (one coarse and one fine sieve) for the selection of tapioca pearls (intermediate part) in the production of popped tapioca pearls.

considered suitable. The granules must be stirred constantly, at the beginning of the operation with a fiber broom (Fig. 9.14), and after 5–8 minutes with a wooden squeegee. The stirring with fiber brooms, often from palm leaf, prevents the granules from breaking. The entire operation lasts 15–20 minutes, enough time for the pearls to become partially gelatinized, more resistant, and with a tendency to stick together.

After being scalded, the pearls are removed and placed in wooden troughs where they remain resting until the next day. Without this last operation, the number of defective or non-expanding granules can be as high as 20%. Before going on to the next operation, the scalded granules are sieved through a 2.80-mm sieve to loosen the sticky granules and increase yield. At this point in the processing, the scalded granules have around 24% humidity.

9.3.3.4 *Popping or expansion of the tapioca pearls*

This operation is done in the same oven as for scalding, but with a stronger fire. To

FIGURE 9.14 Manual scalding of the tapioca pearls.

prevent the material from sticking or burning on the surface of the pan, this one should not be coated with oil or other fatty products, the burning of which can alter the color and flavor of the product. The oven should also not be used for roasting or toasting another product. The popping operation is performed by placing the blanched granules on the hot surface of the oven. The granules are stirred with the help of wooden squeegees by two people. After a few minutes, the temperature is increased with more wood and the granules expand like popcorn, becoming white and opaque, with an expanded polystyrene appearance. The volume doubles, and the white granules become crispy but remain tasteless, which is a characteristic feature of the product.

After popping, the popped tapioca pearls have around 3% moisture, but during storage, the moisture increases until it reaches equilibrium with air (around 12% moisture). About 1/2 m^3 of firewood is consumed to scald the equivalent of 100 kg of starch, in the scalding and expansion operations.

9.3.3.5 Classification

After popping, the popped tapioca pearls are classified in sieves, with the finest fraction being called *cuí*, which is mostly made up of nonexpanded or partially expanded tapioca pearls.

9.3.3.6 Packaging and commercialization

Commercialization is done in bulk at markets and fairs. The local market demands products with uniform expansion. The tapioca pearls partially expanded or without expansion are marketed with lower prices in less demanding markets. Popped tapioca pearls can also be marketed added with a grated coconut. Guimarães et al. (1988) clarified that in this case, the grated coconut is added after the popped tapioca pearls are passed through the sieve and while they are still warm.

The authors established the equivalence between the volume and weight of popped tapioca pearls with and without a coconut. They concluded that 1 liter of natural popped tapioca pearls is equivalent to 184.96 g and the same product with coconut is equivalent to 246.26 g. The popped tapioca pearls have a density of 0.18 kg/L, while the popped tapioca pearls with a coconut have a higher density, of 0.25 kg/L.

9.3.3.7 Parameters for popped tapioca pearls expansion

The manufacturing technology of popped tapioca pearls is based on the same expansion processes that are applied to other food products, the best-known being popcorn. Griswold (1972) described the factors involved in the process of making popcorn, explaining that popping results from cooking the starch under special conditions. According to the author, as the kernels of popcorn are heated, the water vapor contained within expands, increasing the internal pressure, until it reaches sufficient energy to explode the kernel. At the same time, the starch granules are instantly gelatinized as in flash vaporization. Among the desirable characteristics, the author cites the power of expansion, which is the ratio between the final volume of the popcorn popped and the initial volume of the kernels. This property depends on the moisture content of the kernel and the cooking method. Popcorn is charac-terized by a hard, horny endosperm. When popped or expanded, the rounded kernels take on the shape of a mushroom.

As the popping of popcorn depends on the water vapor expansion of the kernel during heating, the correct moisture content is one of the most important factors in determining popcorn expansion. The recommended moisture content of popcorn is 13%–15%. The expansion of expanded kernels decreases if the moisture content is higher or lower than optimal. When the humidity is excessive, the popcorn pops with rough fragmentation. At the right moisture, the surface is slightly rough and when the moisture is too low the surface is smooth. The temperature should be adjusted to ensure a 60–90 second popping time (Griswold, 1972).

The evaluation of the moisture content of the scalded granules of tapioca pearls used in the production of popped tapioca pearls was 24%, higher than that cited by Griswold (1972) for popcorn. In parallel, the time of popping is longer than that of popcorn. As there are no data in the literature about the ideal temperature and humidity for the popping of popped tapioca pearls, an experiment was established for this evaluation.

The scalded pearls were brought from the producing region to be evaluated in the Southeastern region of Brazil. With the transportation, the granules dehydrated slightly, presenting 22.50% of humidity. A stainless-steel container was immersed in a thermostatic oil bath, with an error of plus or minus 5°C.

In a previous test, it was determined that the minimum temperature for the start of popping of scalded tapioca pearls is 200°C. From this temperature, samples of pearls were placed in the container and the average time for the start of setting was measured, varying the temperature from 200°C to 240°C. The volume of tapioca flour (final volume) was determined in a measuring cylinder, the initial volume being fixed at 25 mL. The data were obtained in duplicate.

The expansion was calculated as the percentage increase over the volume of the scalded granules. According to the information from the producers, this increase would be 100%, a value that was never reached in the test (Fig. 9.15). Under the conditions under which the test was performed, heating above 240°C was not possible, due to the burning of the oil.

The results of the experiments presented in Fig. 9.15 indicate a sensitive temperature effect up to 230°C, with a decrease in the spraying time from 300 seconds to 200 and an increase from 30% to 56% in the percentage of expansion. Between 230°C and 240°C, the effect of the temperature increase is less clear, while the popping time continues to decrease, the percentage of popping, averaged over the two repetitions, remains stable.

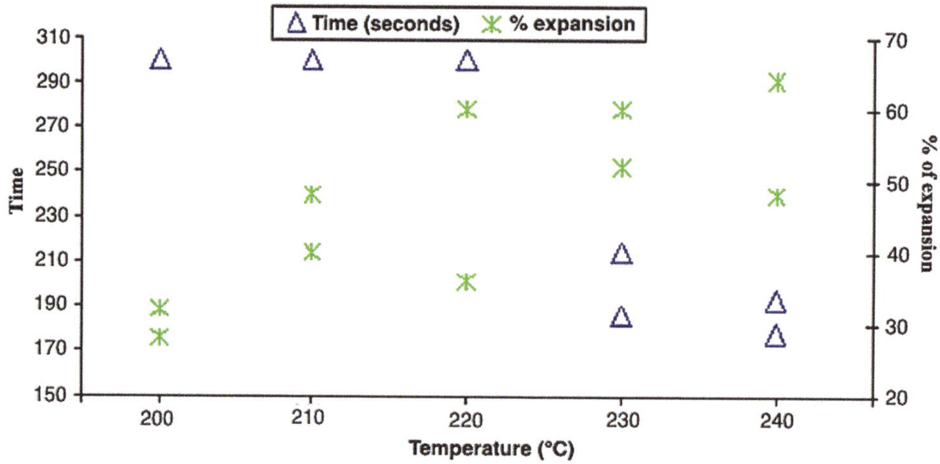

FIGURE 9.15 Volume of popped tapioca pearls after popping, in percentage of expansion, as a function of oven temperature and time.

Another important result of the experiment is the great irregularity in the percentage of expansion, with results that vary greatly at the same temperature.

Other experiments should be performed with scaled pearls of different humidity and temperature. These tests will be very important to better understand the production parameters of popped tapioca pearls, allowing for the production of a more homogeneous product.

9.3.3.8 *Characterization of popped tapioca pearls*

The characterization of the product is important to know the nutritional value, its quality as a food and for health, as well as the points that need to be improved in processing, such as the presence of dirt and contaminants. The characterization was done using samples collected from local units and commerce in the producing regions (Table 9.3). More detailed studies concerning popped tapioca pearls composition are described in Guimarães et al. (1988) and Carmo and Pena (2021).

The moisture levels of the samples analyzed in Table 9.3 present values with greater variability than those found by Guimarães et al. (1988) and Carmo and Pena (2021). Since the product is hygroscopic, equilibrium with atmospheric humidity is expected, especially in older samples. The starch levels found are coherent with those found by Carmo and Pena (2021), whereas the other values present great

TABLE 9.3 Physicochemical composition of commercial popped tapioca pearls in Belém, North Brazil.

	Samples			
	1	2	3	4
Moisture	**8.35**	**4.92**	**14.07**	**10**
% in dry matter				
Starch	95.7	96.04	92.62	93.3
Protein	0.08	0.01	0.21	0.02
Lipids	0.09	0.03	0.09	0.4
Ashes	0.31	0.11	0.11	0.33
Fibers	0.1	0.13	0.16	0.58
Others				
Amylose (% starch)	17.21	17.12	19.22	-*-
pH	4.39	4.62	5.03	4.38
Acidity (*)	2.02	0.98	1.34	0.14

Subtitles: (*) mL of NaOH N/100 g; -*-: no information.

variability, probably due to hand processing for extraction, which also took place in the samples analyzed by Guimarães et al. (1988). The low pH and most especially the high acidity are evidence of the development of fermentation processes due to the long decantation period for the starch. The pH and acidity values are similar to those found by Carmo and Pena (2021).

The acidity is compatible with the microbiological content of the samples, which show low bacteria counts and a higher presence of molds and yeasts, more adapted to a slightly acid pH. Despite being a handmade process, no toxigenic microorganisms were detected, probably due to the acidity, which creates conditions not suitable for the development of these agents.

Guimarães et al. (1988) analyzed commercial products collected in fairs in Belém, Pará state. A total of 28 samples were collected and analyzed, 20 common and eight with the addition of grated coconut. Evaluations of flavor, consistency, and uniformity of granulation were performed, classified in small, medium, and large granules. Of the commercial popped tapioca pearls samples analyzed, 30% were classified as composed of small granules, 20% as medium, and 45% as large. In the samples with the added coconut, none of the eight samples presented large granules, and samples with medium and small granules were found in equal amounts (50%).

The tasting evaluation characterized the product as suitable or not suitable, separated in starch flavor and fermented odor. The consistency was classified as toasted, slightly toasted, slightly hard, hard, and softened (Table 9.4).

Table 9.4 shows that most common popped tapioca pearls samples analyzed presented a toasted consistency and their own proper flavor, which the authors did not define. A significant part, or 25%, still had a starch flavor.

TABLE 9.4 Consistency and flavor characterization of commercial samples of tapioca flour.

	Percentage
Consistency	
Toasted	40.0
Slightly toasted	5.0
Softened	10.0
Slightly hard	5.0
Hard	40.0
Flavor	
Proper	70.0
Fermented	5.0
Starch flavor	25.0

Based on data from Guimarães, M. D. F., Barbosa, W., Oliveira, M., & de Lima, C. L. S. (1988). Caracterização tecnológica e química do produto "farinha de tapioca".

Guimarães et al. (1988) obtained characteristic values of an amylaceous product with a high caloric content. Their result indicate that the quality of the popped tapioca pearls suffer interference from the purification process, with highly variable protein, ash, and fiber levels. The caloric content of the product with a coconut is slightly higher because of the higher fat and protein levels. The caloric content is proportionate by the starch, which represents most of the popped tapioca pearls components and the increase in protein and fat has little influence.

9.3.4 Tapioca pearls processing

Collado and Corke (1998) and Cereda and Vilpoux (2003) indicated that tapioca pearls can be prepared from wet or dry starch. In this case, the first step is to moisten the starch to reach 40%–45% moisture content[2]. The starch

[2] Some large companies work with the starch coming out of the piller, with less than 40% moisture.

is then granulated into small balls, which can be done manually or by mechanized granulators in small units (see Fig. 9.12B). The spheres, or pearls, are stirred for 8–10 minutes on a heated plate with a greased surface to prevent sticking. Mechanical ovens have been developed to speed up the process and improve the quality of the tapioca pearls. The tapioca pearls are then dehydrated by dry heat between 40°C and 60°C, blown with hot air to about 10% humidity, cooled, and packaged.

The oldest Brazilian citation is that of Godoy (1940), and more recent citations add little to this reference, which describes and illustrates some equipment. According to Lima (1982), the starch is heated on a metal surface of a special cooker and cooked to agglomerate the granules, evaporating water and gelatinizing the starch, leading to the formation of translucent grains that characterize the product. The cooker is a concave-cylindrical, double-jacketed pan, through which steam circulates with 3–5 atmospheres. The starch can be distributed through a sieve or a device consisting of a tube with one end closed by a perforated plate. For Godoy (1940), this plate is provided with holes of diameter varying from 3 to 5 mm. A piston forces the starch to come out in the form of rods, which, falling on the cooker, take an approximately spherical shape.

Lima (1982) added that after a few moments, the small granules are detached from the plate where they are stuck and are vigorously stirred, so that the cooking is done in a perfect and uniform way. When the material has a glassy appearance, the cooking is stopped, and the material is sent to drying ovens where it is dried completely on double-jacketed steam tables. Drying is completed in drying rooms with air circulation at a temperature not exceeding 60°C.

A type of oven described by Godoy (1940) consists of one or more large metal tables, 5 m long, with double walls between which steam circulates. When the production is larger, it is better to use closed ovens where the tables are replaced by overlapping shelves, for better use of space, regulating the exit of hot air saturated with moisture. The tapioca pearls are distributed in thin layers on the heated surface that should be periodically turned to prevent sticking. The drying takes about 4 hours, with the temperature set at 60°C. Another way to dry the tapioca pearls is to distribute the product on trays arranged in wagons that travel through a tunnel heated by hot air.

Ciacco and Cruz (1982) described tapioca pearls as a partially gelatinized, spherical starch product. It is usually consumed as a dessert or in the form of porridge. The literature describes the manufacture by using a concentrated dispersion of cassava starch (45% dry matter weight/weight) placed in a rotating cylinder, whose surface is perforated. As the cylinder rotates, the starch particles acquire a spherical shape, pass through the hole, and fall onto a plate heated to a temperature of 60°C–80°C. In contact with the hot plate, the starch on the surface of the particles is partially gelatinized and stabilizes in the spherical form. After gelatinization, the spherical particles are dried and acquire an opaque color.

Today, in the modern starch production units in the South of Brazil, a continuous process is used, with several rotating drums in sequence (Fig. 9.16). In the first section of the first drum, the tapioca pearls are formed. The starch with 40%–45% moisture is fed into the upper part of a rotating drum where it is granulated (Fig. 9.16A). In the lower section of the same drum, or in another drum, as in Fig. 9.16B, the walls are made of sieves of two different mesh sizes to select the product with the desired granulometry. At the end of the drum, the larger particles are eliminated, which, after grinding, return to the beginning of the process. The size of the tapioca pearls can be controlled according to the opening size of the sieve in the classifying drum. Several sieves mounted in series allow to obtain tapioca pearls of various sizes.

FIGURE 9.16 Sequence of drums for tapioca pearls production: (A) Drum for formation of tapioca pearls; (B) drum for selection of tapioca pearls; (C) drum for gelatinization of tapioca pearls. *Photo from Liviston teles Sena, Paranavai Maquinas.*

The selected tapioca pearls go into the second drum or the third when the selection is made on a separate drum, as shown in Fig. 9.16C. It is a double-jacketed drum heated with steam at a sufficient temperature (120°C–200°C, depending on the size of the pearls) to gelatinize the starch (Fig. 9.16C). The sago goes through this rotating drum and comes out gelatinized on the surface, while the inside keeps its white color.

After this step, the product is forwarded to another drum, or a circular oven, like the one shown in Fig. 9.8, where drying is completed. At this stage, the product stays around 1 hour to come out completely dry. The temperature and time must be adjusted so that complete gelatinization of the starch does not occur, a situation that would leave the pearls transparent. The importance of not completely gelatinizing the pearls makes companies in Brazil prefer the circular oven to the drum, because the temperature control is easier.

9.3.4.1 *Characteristics of tapioca pearls' commercial product*

Tapioca pearls are mainly consumed in Asia, where they are part of numerous dishes. In Brazil, they are used more in the south, where they are prepared with wine, orange juice, fruits, and in other ways (Fig. 9.17A and B). To be consumed, tapioca pearls are put in water to hydrate and grow. They are then cooked in fruit juice or red wine.

Collado and Corke (1998) reported that in normal Brazilian culinary applications, the tapioca pearls are boiled in water for 15–20 minutes, covered and left to cool, washed in cold water to avoid having them stick together, and then used

FIGURE 9.17 Dessert of tapioca pearls prepared in the south of Brazil with wine (A); ready-to-use tapioca pearls, sold in supermarkets in the South and Southeast regions of Brazil, with orange flavor (B); and tapioca pearls products used as desserts in India (C). *Photo A from Wikimedia Commons: https://commons.wikimedia.org/wiki/File:SaguDeVinho.jpg.*

as garnish for sweet hot drinks, or cooked in pans or in oven, with coconut milk and/or milk to make porridge or pudding (Fig. 9.17A and B). In India, sago pearls can be agglutinated into disks that are used in dessert preparations (Fig. 9.17C). Tapioca pearls in these disks are slightly expanded, indicating production over a hot plate while they are still wet, similar to the popped tapioca pearls production process.

Collado and Corke (1998) studied the quality and the culinary characteristics of the pearls commercialized in Hong Kong, the Philippines, and Singapore, made with sago (palm tree), cassava (tapioca), and sweet potato, among other starches. To simulate preparation conditions, they studied the viscosity profiles of native (raw) starch and the gelatinized granules. Furthermore, since the pearls are consumed as a dessert, the effects of sugar on their functional properties were established. The form of consumption was also simulated.

The authors discovered notable variations in the paste properties as a function of the raw material used, variations that were attributed to the different pH values. The ground granules that presented an alkaline pH had high levels in viscosity of the cold paste and in retrogradation. However, the addition of 65% sugar, calculated over the pearls sample weight, drastically reduced the retrogradation values, whereas the retrogradation values for the starch pastes with an acid pH were little affected.

In laboratory evaluations that simulated processing, Collado and Corke (1998) observed that the production process of the granules apparently introduced a moderate modification, with an increase in stability of the ground granule paste for pearls made of tuberous starch, compared to nonprocessed starch.

Collado and Corke (1998) characterized Hong Kong's commercial product (Table 9.5). The granule samples analyzed for raw and cooked diameter weighed 5 g.

Table 9.5 indicates variations between the pearls made with different raw materials. In all parameters evaluated, the tapioca pearls had the highest values. In the statistical analysis between sago and cassava, the difference was significant only for pH (P of 0.02). Collado and Corke (1998) did not explain the high pH levels in tapioca pearls. In general, cassava starch pH is between 6 and 7 and can fall even more in the event of hand-processed extraction. One possible explanation comes from the size of the companies producing the tapioca pearls. Larger companies, with modern starch extraction, achieve a higher pH product than smaller, less mechanized units.

TABLE 9.5 Pearls characteristics from different raw materials, commercialized in Hong Kong.

Origin	Moisture	pH	Raw diameter (mm)	Cooked diameter (mm)
Cassava (Tapioca)	15.0	9.4	3.01	5.0
	13.7	9.4	3.01	5.0
	13.8	9.4	4.39	7.0
	13.3	9.3	2.6	4.5
	11.4	4.4	6.68	9.0
	12.0	5.6	2.13	4.5
	13.5	4.4	6.8	9.0
Cassava average	**13.23**	**7.78**	**4.16**	**6.57**
Sago	14.3	4.7	2.16	5.0
	9.4	5.6	1.95	3.5
	9.2	5.2	4.28	6.5
	8.0	5.5	2.01	4.0
Sago average	**10.23**	**5.25**	**2.60**	**4.75**
Potato and bean	**10.7**	**4.8**	**2.13**	**4.5**
Sweet potato	**9.3**	**5.4**	**2.21**	**5.0**

From Collado, L. S., & Corke, H. (1998). Pasting properties of commercial and experimental starch pearls. Carbohydrate Polymers, *35(1–2), 89–96.*

There is a big variation among the different tapioca pearls, denoting a lack of quality standardization. Regarding sago, the results are more homogenous, but there is also great heterogeneity.

9.4 Final considerations

All the products discussed in this chapter have great market potential, even if, for most of them, this potential is still underexploited. Brazilian tapioca is only consumed in Brazil. It is a product that in the early 2000s was known only in the northeast of the country, but which has since become one of the main cassava starch derivatives in the country.

Tapioca pearls are consumed essentially in Asia and a little bit in Europe. In Brazil, they are found essentially in the south of the country, in populations of German colonization, and are practically unknown in other regions. As for Brazilian tapioca, this product could be better developed to reach a larger market in Brazil or even in other Latin American countries. Same thing for broken tapioca pearls, which have a similar use.

Perhaps the most interesting product, popped tapioca pearls are only consumed in one state in Brazil, in a very limited market. This product assimilates to an expanded rice, with many possible uses, such as addition to breakfast cereals, energy bars, accompaniment to ice cream and other desserts, and so forth. It could be an interesting product for the large tapioca pearl companies, which already have the necessary equipment for its production, needing only a different temperature regulation in the final drying.

Acknowledgments

The authors thanks EMBRATER, in the person of its National Cassava Coordinator, Agronomist Jairo Ribeiro da Silva, for the effort in creating new opportunities for the diffusion of knowledge; extension technicians from Pará, for the collaboration and trust demonstrated; the EMBRAPA researchers in Pará for the cooperation in the analyses and samplings, as well as for technical information rendered; and Mr. Livinston, from the company Paranavai Maquinas, for the photos about the tapioca pearls production process.

References

Carmo, J. D., & Pena, R. (2021). Caracterização físico-química de farinhas de tapioca produzidas e comercializadas em diferentes localidades no estado do Pará. *Brazilian Journal of Development, Curitiba, 7*(8), 83874–83883. Available from https://doi.org/10.34117/bjdv7n8-546.

Cereda, M. P., & Vilpoux, O. F. (2003). Cap. 11. Sago, tapioca, and tapioca flour production processes. In (1st ed.). M. P. Cereda, & O. F. Vilpoux (Eds.), *Tecnologia, usos e potencialidades de tuberosas amiláceas Latino Americanas* (Vol. 3). São Paulo: Fundação Cargill.

Ciacco, C. F., & Cruz, R. (1982). Fabricação de amido e sua utilização. sd: Secretaria da Indústria e Comércio.

Collado, L. S., & Corke, H. (1998). Pasting properties of commercial and experimental starch pearls. *Carbohydrate Polymers, 35*(1–2), 89–96.

Godoy, J. D. (1940). *Fecularia e amidonaria.* Rio de Janeiro: Tomiti & Lanzara.

Griswold, R. M. (1972). *Estudo experimental dos alimentos/The experimental study of foods.* São Paulo: Edgard Blucher/EDUSP, 469 p.

Guimarães, M. D. F., Barbosa, W., Oliveira, M., & de Lima, C. L. S. (1988). Caracterização tecnológica e química do produto "farinha de tapioca".

Lima, U. D. A. (1982). Manual técnico de beneficiamento e industrialização da mandioca. *São Paulo: Secretaria de Ciência e Tecnologia, 2.*

Whistler, R. L., BeMiller, J. N., & Paschall, E. F. (Eds.), (2012). *Starch: Chemistry and technology.* Academic Press.

CHAPTER

10

Starch-phytochemical complex: the case of tropical starchy sources

Luis A. Bello-Perez[1], *Monserrat Hernandez-Bautista*[1], *Apolonio Vargas-Torres*[2] *and Hector A. Romero-Hernandez*[1]

[1]National Polytechnic Institute, CEPROBI, Yautepec, Morelos, Mexico [2]Autonomous University of the State of Hidalgo, Institute of Agricultural Sciences, Tulancingo de Bravo, Hidalgo, Mexico

10.1 Introduction

Starch is the main macronutrient of foods, and its complete hydrolysis in the small intestine supplies glucose to blood. Starch is present in cereals, pseudocereals, tubers, roots, rhizome, legumes, and unripe fruits. Usually, these starchy sources are cooked and consumed or dried to obtain a flour that is used as raw material in foods as bakery and breakfast products, snacks, and pasta, among others (Fig. 10.1).

In recent years, starch digestion has garnered considerable interest because the excess of consumption of starchy foods with high digestion rate has been associated to health problems, such as obesity, diabetes, cardiovascular diseases, and metabolic syndrome. Intrinsic factors as granule size, amylose/amylopectin ratio, molar mass of starch components, and chain-length (CL) distribution are associated to starch hydrolysis. Extrinsic factors as gelatinization and retrogradation degree, structure modification (chemical, physical, enzymatic, and combinations of these), interaction with molecules (lipid, protein, nonstarch polysaccharides, and polyphenols), and the food matrix type (Miao & Hamaker, 2021) are also related.

Dietary phytochemicals (phenolic acids, polyphenols, flavonoids, and anthocyanidins) are bioactive compounds present in plants. Their bioactivity modulates oxidative and inflammatory stress, stimulates the immune system, alters macronutrient digestion, and has a prebiotic-like effect on gut microbiota. Some evidence, such as epidemiological and clinical studies, suggests that the consumption of foods rich in polyphenols can reduce the risk of age-related chronic diseases (Shahidi & Yeo, 2018). Recent interest has focused on food processing (Debelo et al., 2020) since it may alter stability, bioaccessibility, bioavailability, and fermentation in the colon.

Interactions between phytochemicals and starch binding via noncovalent bonds and hydrogen bonds alter starch digestion (Bordenave et al., 2014; Miao & Hamaker, 2021; Shahidi & Pan, 2022).

Starch Industries
DOI: https://doi.org/10.1016/B978-0-323-90842-9.00002-9

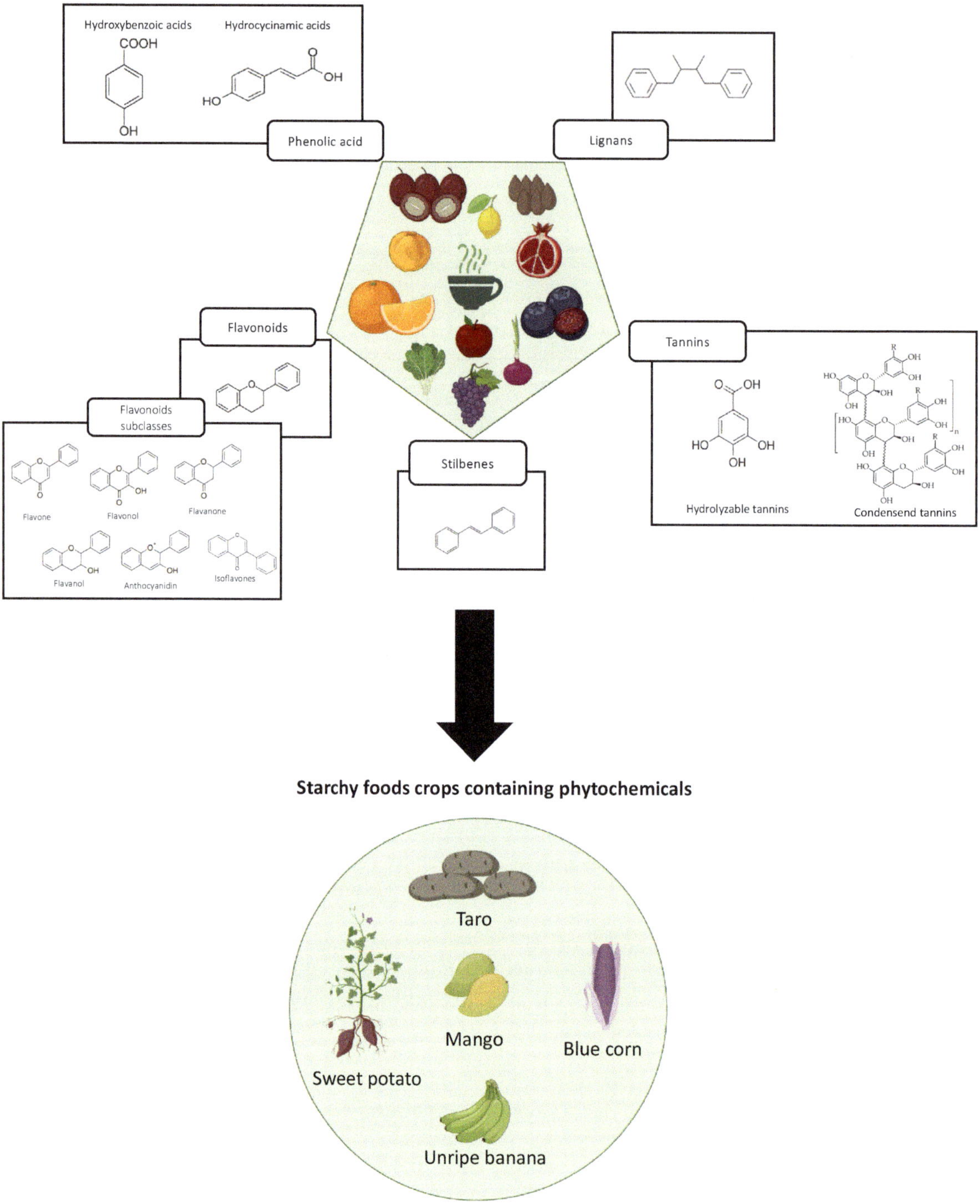

FIGURE 10.1 Traditional foods with different types of phytochemical compounds and alternative foods crops rich in starch and phytochemicals (created with Biorender).

TABLE 10.1 Starches sources traditionally involved in the formation of starch-phytochemical complex and its effect on starch digestion.

Starch source	Phytochemical type	Formation complex and the effect on starch digestibility	References
Debranched cassava starch	Ferulic acid	The phytochemical was entrapped in the hydrophobic core of amylose. Digestibility did not show significant difference	Hung et al. (2013)
Rice starch	Dodecyl gallate	The phytochemical hydrophobic segment resided in the helix of the amylose cavities, with the hydrophilic part left outside; this forms inclusion complexes. Decreased digestibility	Chi et al. (2018)
	Gallic acid	Phytochemicals help starch reassembly by noncovalent interaction, decreased digestibility. The formation of a V-type single helix structure, decreased digestibility	Chi et al. (2019), Liu et al. (2019)
(11.2%, apparent amylose)	Apple phytochemicals	The starch formed a network with phytochemicals due to the treatment	Chou et al. (2020)
(17%, apparent amylose)	Ferulic, gallic acid, and quercetin	FA and GA had more significant effects on physicochemical properties of starch; this would form starch-phytochemicals interactions	Han et al. (2020)
(19.39%, amylose content)	Anthocyanins	Formation of starch-anthocyanins complexes through noncovalent bonds, which causes a decrease in starch digestibility	Miao, and Xu, Jia, et al. (2021)
(21.2%, apparent amylose)	Apple phytochemicals	The phytochemicals interacted with starch chains through intermolecular interactions	Chou et al. (2020)
Cynanchum auriculatum starch	Quercetin	The process was efficient for the synthesis of phytochemical-starch conjugates, decreased digestibility	Liu et al. (2018)
Lotus seed starch	Green tea phytochemicals	Hydrogen bonding, the main force of interaction, decreased digestibility	Guo et al. (2019)
Lotus seed starch	Green tea phytochemicals	Noncovalent interaction between starch and phytochemicals; this formed single helix inclusion complexes with hydrophobic effects and noninclusive, decreased digestibility	Zhao et al. (2019)
Purple sweet potato starch	Ferulic acid, isoferulic acid, 4-hydroxybenzoic acid, and caffeic acid	Formation of noncovalent hydrogen bonds between phenolic acids and starch. Decreased digestibility; hydroxybenzoic acid was the most significant	Su et al. (2020)
Tartary buckwheat starch	Quercetin	Noninclusive complex between starch and quercetin through noncovalent interactions, decreased digestibility	Gao et al. (2021)
(41.5%, apparent amylose)	Rutin and quercetin	Phytochemicals compounds showed to be hard bound to starch although steric hindrance might exist	Wang et al. (2021)

Tropical starchy sources have gained interest because they provide starch with different or new functional, physicochemical, and digestibility characteristics, unlike that from conventional sources (maize, potato, rice, and wheat, among others) (Table 10.1).

Furthermore, underground starchy raw materials are particularly rich in phenolic compounds (for more details on the composition of Latin American starchy crops, we suggest reading Chapter 8 of Volume 1 in this collection).

Starch-phytochemical interactions have been studied in model systems, and starches from different sources are used to form complexes with phytochemicals (Table 10.1). Rice starch interacts with apple phytochemicals and specific ones, such as ferulic and gallic acids, quercetin, and dodecyl gallate through noncovalent interactions (Chi et al., 2018; Chi et al., 2019; Chou et al., 2020, 2020b; Han et al., 2020). Phytochemicals can produce inclusion and noninclusion complexes with amylose and reduce starch digestion (Hung et al., 2013; Liu et al., 2019; Zhao et al., 2019).

The application of underused starchy sources is an alternative for farmers in tropical countries because weather conditions (hurricanes, cyclones, and high temperatures), pests, deficient postharvest management, and low sale prices produce loss of crops and income. These issues reduce farmers' interest in growing tropical crops.

The use of tropical starchy sources is an interesting alternative to produce functional foods where the presence of phytochemicals or the addition of a plant extract with phytochemicals generate interactions that depend on the starch structure and the phytochemical type (Table 10.2).

Some tropical crops, such as unripe banana, mango, and sweet potato, are starchy sources rich in polyphenols. During cooking, they can produce starch-polyphenol complexes with a concomitant reduction in starch digestion. Some flours from those tropical starchy crops can be used to produce foods as pasta, bread, porridges, cookies, and other snacks with a reduced starch digestibility, increased resistant starch content, and low glycemic index (Ovando-Martinez et al., 2009; Pico et al., 2019) evaluated the resistant starch (RS) content in raw and cooked sweet potato flour at high hydrostatic pressures and reported a higher RS in cooked flour. This was mainly attributed to the greater retrogradation and, therefore, a higher fraction of RS (information on so-called resistant starch can be found in Chapter 15 of Volume 1 in this series).

10.2 Starch structure and phytochemical complex

Starch is the main polysaccharide in cereals, pseudocereals, legumes, tubers, roots, and unripe fruits. In the tropical regions of the world, these food crops are harvested and used boiled and fried or to prepare bakery products, snacks, pasta, and porridges, among others. The starch structure plays a key role in the physicochemical, functional, digestibility, and sensory characteristics of those foods. The study of starch granules starts with the evaluation of the morphology (shape and size) and granule size distribution (see Chapter 9 of Volume 1 in this series). These features are associated with the characteristics of the starchy foods mentioned above.

Amylose and amylopectin, the two major components of starch granules, have been widely reported in terms of their structure (essentially linear and branched, respectively) and their arrangement in granules. The effects of both structures on the functionality and physicochemical, rheological, and digestibility features of starch and starchy foods have been analyzed. Advances in analytical methods, including instruments with high resolution, have yielded information on the structural features of starch components, such as molar mass, gyration and hydrodynamic radius, dispersity, and CL distribution of amylopectin. The structural features of starch components are related to the rheological, thermal,

TABLE 10.2 Starchy flours of tropical sources in the food matrix to evaluate resistant starch content.

Starch/flour	Food	Phytochemical content	Resistant starch (%)	References
Unripe banana flour	Pasta	EPC in cooked spaghetti (control): 1.33 ± 0.05 mg GAE/g EPC in enriched cooking spaghetti with unripe banana flour: 1.68 ± 0.09mg GAE/g	12.42	Ovando-Martinez et al. (2009)
	Noodle	TPC after the cooking process: 28.6 ± 0.84 mg GAE/kg TPC after the cooking process: 90.4 ± 2.20 mg GAE/kg	12.3	Choo and Aziz (2010)
	Processed flour	TPC in banana flours: Oven-dried: 337.79 ± 9.79 mg GAE/100 g dm Freeze-dried: 335.66 ± 9.79 mg GAE/100 g dm Oven-dried: 1195.97 ± 14.28 mg GAE/100 g dm Extruded: 515.82 ± 2.65 mg GAE/100 g dm	>5	Pico, and Xu, Guo, et al. (2019)
Banana peel flour	Noodle	TPC in control noodle: 121.5 ± 15.2 mg GAE/100 g TPC in noodle enriched with banana peel powder: 121.5 ± 15.2 mg GAE/100 g	–	Pasha et al. (2022)
Sorghum flour	Processed flour	Higher content of tannins in treated flours (1.8–50.6 mg CE/g)	15.4	Mkandawire et al. (2013)
Sweet potato flours (purple-flesh)	Processed flour	TPC of control flour: 206.23 ± 1.3 mg GAE TPC in extruded flour: 294.35 ± 0.9 mg GAE	–	Soison et al. (2014)
(Purple-flesh)	Bread	TPC in the mixture: 858 ± 28 mg GAE/kg TPC after baking: 1016 ± 12 mg GAE/kg	–	Cui and Zhu (2022)
(Purple-flesh)	Noodle	TPC in purple sweet potato noodles: 195.58 ± 2.06 mg GAE TPC in boiled purple sweet potato: 357.33 ± 1.28 mg GAE	14.29	Nurdjanah et al. (2022)
(Orange sunset, purple dawn and red)	Processed flour	TPC in unprocessed flour: Orange sunset: 1603 ± 3 mg GAE/g db Purple dawn: 906 ± 3 mg GAE/g db Purple red: 288 ± 5 mg GAE/g db TPC in processed flour: Orange sunset: 1956 ± 7 mg GAE/g db Purple dawn: 1117 ± 3 mg GAE/g db Purple red: 408 ± 2 mg GAE/g db	3.7	Cui and Zhu (2019)
Mango flour (mango processing by-product)	Muffin	TSP control muffins: 1.86 ± 0.28 g GAE/100 g dw TSP-enriched muffin with mango flours: 5.36 ± 0.46 g GAE/100 g dw	–	Ramírez-Maganda et al. (2015)

(Continued)

TABLE 10.2 (Continued)

Starch/flour	Food	Phytochemical content	Resistant starch (%)	References
Blue corn flour	Porridges	EP in blue maize flour: 165 ± 14 mg GAE/g dm EP in white maize flour: 127 ± 7 mg GAE/g dm	3.2	Camelo-Méndez et al. (2017)
	Pasta	TPC in pastas with blue maize flour: 82.5 mg GAE/g dm TPC in pastas with chickpea flour: 176.0 mg GAE/g dm TPC in pastas with unripe plantain flour: 210.4 mg GAE/g dm	3.5	Camelo-Méndez et al. (2018)
	Pasta	TPC in pasta without blue corn flour: 75.42 ± 3.64 mg GAE/g TPC in pasta with blue corn flour: 210.41 ± 3.04 mg GAE/g	5.03	Camelo-Méndez et al. (2018)
Pigeon pea flour	Pasta	TPC in control pasta: 132.69 ± 2.65 mg GAE/100 g TPC in pasta enriched with Pigeon pea flour: 182.76 ± 2.52 mg GAE/100 g	–	Sharma et al. (2021)

EPC: Extractable polyphenols content; TPC: total phenolic content; GAE: gallic acid equivalents; CE: catechin equivalents; TSP: total soluble polyphenol; EP: extractable polyphenols, dw: dry weight; dm: dry matter; db: dry base.

functional, and digestibility characteristics of starchy foods, and more recently, the interactions with other macromolecules (proteins, lipids, and nonstarch polysaccharides) and phytochemicals. The structural features of starch components can provide data on better end-uses of starchy sources to produce modifications (chemical, physical, enzymatic, and dual) that improve the characteristics of the polysaccharide and create new foods and applications.

The study of the starch structure in tropical crops as plantain, mango (Millan-Testa et al., 2005), and banana (Agama-Acevedo et al., 2015) includes the evaluation of their molecular structure: mango starch shows a molar mass and gyration radius of 13.4×10^6 g/mol and 145 nm, respectively, and those of plantain are 6.7×10^6 g/mol and 106 nm, respectively. These data were obtained using high-performance size-exclusion chromatography (HPSEC) coupled with the MALS detector (Dawn 8+ Heleos, Wyatt Technology Corporation, CA) and refractive index detector (Agilent 1100, model G1362A, Agilent Technology, UK), with a Shodex OHpak SB-807 HQ column (Showa Denko K.K., Tokyo, Japan). Four species were analyzed with molar mass between 2.9 and 3.1×10^8 g/mol and gyration radius between 286 and 302 nm. A similar HPSEC and detector system, ultrahydrogel guard column, and two OHpak SB-807 HQ and OHpak SB-804 HQ connected in series were used to determine the molar mass of six Musa genotypes. The molar mass (g/mol) ranged between 0.78 and 3.77, important data for the physicochemical, functional, and rheological characteristics of each banana starch and choosing a Musa genotype for specific applications. Aroid cultivars were analyzed with asymmetrical flow field fractionation (AF4) coupled with the MALS and refractive index; the molar mass of the three cultivars ranged between 5.7 and 13 g/mol, with gyration radius between 311 and 410 nm (Hoyos-Leyva et al., 2017). The starch molecular structure of each tropical food crop depends on the method used for starch solubilization and separation (HPSEC and AF4) as well as the column type in HPSEC.

Other structural studies on starch have used debranching with enzymes as pullulanase and isoamylase to produce linear chains and determine the CL distribution. High-performance anion-exchange chromatography (HPAEC) equipped with pulsed amperometric detection (PAD, Dionex, Sunnyvale, CA) and CarboPac PA 200 anion-exchange column was used. The four banana varieties showed high content of B1 chains with a degree of polymerization (DP) of 13–24, followed by A chains with DP $\leq$ 12; the maximum DP determined was 82–90 (Agama-Acevedo et al., 2015). Aroid cultivars were analyzed using the HPAEC-PAD system, and the highest content of chains was found in those with DP 13–24, followed by chains with DP 6–12. The molecular characteristics largely determine the physicochemical, functional, rheological, and digestibility characteristics of aroid starches (Hoyos-Leyva et al., 2017).

The structural results of the mentioned tropical food crops agree with the cluster model. Recently, the structure of amylopectin was analyzed in agreement with the backbone model based on internal chain composition. Banana cultivars were used for starch extraction, and their structure was identified following Bertoft (2004): the first step was identifying the ϕ, β-limit dextrins produced after amylopectin separation from the starch granule. The amylopectin was dispersed in 90% dimethyl sulfoxide in a water bath (70°C) for 10 min and the stirring continue overnight at room temperature. To produce ϕ-limit dextrins, the sample was treated with the enzyme phosphorylase (2 days), and to produce the ϕ,β-limit dextrins, the enzyme β-amylase was added and incubated overnight at room temperature under a magnetic stirrer. The hydrolyzed starch was freeze-dried and prepared for HPAEC-PAD analysis by debranching with isoamylase and pullulanase. The information of the amylopectin structure included CL, ϕ, β-limit CL, external CL, and internal CL (Yee et al., 2021). The amylopectin of Musa genotypes showed a significant difference in the fine structure, and they were classified as type 4 (high number of B3-chains, few BS-chains, and low S:L and BS:BL ratios) (Bertoft, 2018; Vamadevan & Bertoft, 2018).

10.3 Phytochemical compounds

Usually, phytochemical compounds are considered bioactive compounds, secondary metabolites of plants widely distributed in plant foods. Epidemiologic studies have shown that polyphenols in fruits and vegetables provide some health benefits during chronic-degenerative diseases related to oxidative stress (obesity, cancer, diabetes, and cardiovascular diseases, among others) (Cladis et al., 2020; Shahidi & Pan, 2022). Polyphenols are benzene derivatives with one or more hydroxyl groups attached to a phenolic ring. In the vegetable kingdom around 8,000 molecules with these chemical characteristics are found. Phenolic compounds are classified based on their botanical origin and biological functions, and mostly the structure of their carbon skeleton and number and position of substituents (Fig. 10.1). The main types of polyphenols are flavonoids, lignans, phenolic acids, and stilbenes (Cianciosi et al., 2022; Tsao, 2010).

10.3.1 Polyphenols in fruits

Tropical food crops are a good source of phytochemicals, which interact with starch when they are used for food preparations. Ripe and unripe banana provides large amounts of resistant starch dietary fiber and phenolic compounds. This fruit can contain up to 90.4 mg/100 g fresh weight of total phenols (Sun et al., 2002). Banana flour from the varieties *Musa acuminata* L. and *Musa acuminata* Colla AAA, obtained from pulp and peel, has a significantly higher total phenolic content (373.88–685.57 and 91.25–160.77 mg GAE/100 g dry weight, dw) when made with unripe banana than when produced from its ripe counterpart, 230.21–586.29 and 75.01–91.90 mg GAE/100 g dw, respectively (Fatemeh et al., 2012). In recent years, researchers have developed food products from unripe plantain flour. Ovando-Martinez et al. (2009) prepared spaghetti with plantain flour (*Musa paradisiaca* L.) and determined the polyphenol content and antioxidant capacity (extractable polyphenols [EP], condensed tannins, hydrolyzable polyphenols [HPP], ABTS). Semolina was substituted for plantain flour (15%, 30%, and 45%), and the pasta showed a higher content of polyphenols compared to the controls. These bioactive compounds were related to anthocyanins and proanthocyanidins (condensed tannins). Additionally, unripe plantain flour (*Musa cavendish*) has been subjected to extrusion, presenting a higher content of EP and HPP. Therefore, extrusion is reported to promote the bioaccessibility of flavonols and phenolic acids due to the breakdown of cell compartments (Pico, & Xu, Guo, et al., 2019).

Jackfruit (*Artocarpues heterophyllus* L.) has a high concentration of polyphenols (carotenoids, flavonoids, sterols, and tannins) that has gained interest as a fast-growing fruit tree with high availability for planting and harvesting (Zhang et al., 2021). The pulp has a total phenolic content of 0.18–0.46 mg GAE/g dw and total flavonoids of 0.19–1.20 m RE/g dw (Jagtap et al., 2010). Jackfruit seed is a new source of starch that should be studied in more detail since it is related to β-carotene, α-carotene, β-zeacarotene, α-zeacarotene, and a carotenoid, dicarboxylic acid, crocetin, and β-carotene-5,6α-epoxide (Chandrika et al., 2004; Baliga et al., 2011).

Mango (*Mangifera indica* L.) is the second tropical fruit produced worldwide after banana. It mainly contains ascorbic acid, β-carotene, and other phytochemical compounds. Several carotenoids, such as β-carotene, luteoxanthin, neochrome, 9-cis-violaxanthin, and violaxanthin dibutyrate ester, have been isolated (Pott et al., 2003). In addition, by-products such as peel and seeds are a source of polyphenols often used in bakery products (Ramírez-Maganda et al., 2015). Patiño-Rodríguez et al. (2020) obtained 99.71 mg GAE/g dw of phytochemical compounds from mango pulp flour and 16.51 mg GAE/g dw from peel flour.

10.3.2 Polyphenols in underground stachy organs

Taro or Malanga (*Colocasia esculenta* L.) is a tuber from tropical regions consumed mainly as a carbohydrate source. It has a total phenolic content of around 187 mg GAE/100 g fresh weight and a total flavonoid content of 61 mg CAE/100 g fresh weight (Simsek & El, 2015). It is used in flours, pasta, canned goods, and beverages. Taro flour has a total phenolic content of 1.20 and 4.17 g GAE/kg dw (Arıcı et al., 2016). Cassava (*Manihot esculenta*) is a tropical tuber with a high content of carbohydrates and bioactive compounds (Chandrasekara & Josheph Kumar, 2016). Its total phenolic content (4.58–29.82 g GAE/g extract) varies depending on the extraction solvent (Yi et al., 2011). Although cassava flour has a high potential as a raw material due to its antioxidant capacity, its total phenolic content is 2.69–4.44 mmol AG/100 g dw (Nilusha et al., 2021), and the content of phytochemical compounds is affected by the variety and the farming methods.

Sweet potato (*Ipomoea batatas* L.) is a fundamental crop for the diet of many developing countries and contains a wide variety of phytochemicals, such as flavonoids, carotenoids, and anthocyanins (Escobar-Puentes et al., 2022). The tuber is rich in anthocyanins and is of particular interest for extraction at an industrial level. Bridgers et al. (2010) determined the effect of solvent, solid charge, and incubation temperature of anthocyanin extraction from purple-fleshed sweet potato and obtained a range of total polyphenols from 350 to 550 mg CAE/100 g dw and anthocyanins from 50 to 190 mg cyanidin-3-glucoside/100 g dw. Sweet potato flour has been studied for the development of new food products; its total phenolic content is 115.98–279.78 mg GAE/100 g dw, affected by the flour production process, while its anthocyanin concentration is 1.71–27.91 mg cyanidin-3-glucoside/100 g dw (Soison et al., 2014). The concentration of phytochemical compounds changes when sweet potato flour is treated with high hydrostatic pressure, and the total phenolic and β-carotene concentrations increase significantly along with pressure, attributed to the rupture of cell membranes (Cui & Zhu, 2019). Therefore, these flours can be used as alternative ingredients.

In the next paragraphs, a general description of the chemical characteristics of bioactive compounds is presented. It was reported that phenolics compounds can interact with starch during processing of raw materials to produce

food matrix (Shahidi and Pan, 2022), during processing of raw materials as drying and freeze drying (Pico, & Xu, Guo, et al., 2019) or during coingestion of dishes that contain starch and phenolics (Kan et al., 2020). However, interactions in situ are not well reported. The interaction starch and phenolics depend on the structure of phenolics compounds and impact on starch digestibility (Chen et al., 2020).

10.3.2.1 Phenolic acids

Phenolic acids are nonflavonoid polyphenol compounds found in two groups: derivatives of benzoic acid and cinnamic acid. Gallic, vanillic, and syringic acids are classified as hydroxybenzoic acids, while hydroxycinnamic acids include ferulic, sinapic, caffeic, and coumaric acids. In fruits and vegetables, phenolic acids are free; in cereals, they are attached to other macromolecules as non-starch polysaccharides (Durazzo et al., 2019; Tsao, 2010).

10.3.2.2 Flavonoids

They are the main group of phenolic compounds and impart the yellow, orange, red, blue, and purple colors to fruits and plant leaves. Their general structure is C_6-C_3-C_6, and they present two aromatic rings (A and B) and carbons identified as C5, C6, C7, and C8 and C2′, C3′, C4′, C5′, and C6′, respectively, joined by a heterocyclic ring with oxygen (ring C, with three main carbons C2, C3, and C4). Generally, flavonoids are present as free monomers, but they are found as β-glucoside in foods from plants. They include six subclasses: anthocyanins, flavones, flavonols, isoflavones, flavanones, and flavanols (Bordenave et al., 2014), which differ by the central heterocyclic ring and due to the number and position of OH-groups (Cladis et al., 2020; Cosme et al., 2020; Pan et al., 2022). These metabolites are responsible for biological functions that include signaling during nodulation, protection against ultraviolet radiation, and color of flowers (important during pollination) (Proença et al., 2022).

- Flavanols

  Also named flavan-3-ols, flavanols are commonly known as catechins. In their structure, they do not have double bonds between C2 and C3; therefore, the hydroxylation in C3 leads to two chiral centers in the molecule (in C2 and C3) and four possible diastereoisomers (Tsao, 2010). Flavanols are present in foods as monomers and polymers called catechins and proanthocyanidins, respectively (Bordenave et al., 2014; Martín & Ramos, 2021). Catechins and epicatechins can form some conjugates, especially with gallic acid, to form epicatechin gallate, epigallocatechin, and epigallocatechin gallate (Durazzo et al., 2019).
- Flavonols

  The most widespread type of flavonoids in plant foods, flavonols are characterized by having a hydroxyl group in C3. Among the most common flavonols are quercetin, kaempferol, and myricetin (Cosme et al., 2020). They are found in several fruits and vegetables as well as some commonly consumed beverages like wine and tea (Cianciosi et al., 2022).
- Flavones

  Flavones are compounds responsible for the white-yellow color of plants and contain a double bond between C2 and C3 in their structure. They are very similar to flavonols and only differ in the hydroxylation in C3 (Zuiter, 2014). Among the most common compounds are lutein, chrysin, and apigenin (Cianciosi et al., 2022).
- Isoflavones

  Isoflavones are characterized by having a phenolic group in C3 instead of C2; they are phytoestrogen compounds found mainly in legumes, and the most common are genistein, daidzein, and glycitein. These types of compounds have an impact on human health since they are reported to have antioxidant activity (Martín & Ramos, 2021; Truzzi et al., 2021).
- Flavanones

  Flavanones are predominant structures in citrus fruits, such as oranges, lemons, and limes, so they can also be named citroflavonoids or bioflavonoids. In the human diet, they are found in the form of β-glucosides, and among the most common are naringenin, hesperetin, and eriodictyol (Najmanová et al., 2020).
- Anthocyanins

  Anthocyanins are water-soluble pigments responsible for the blue-violet and red-orange color of fruits and flowers; they have a basic structure constituted by the flavylium ion (Cianciosi et al., 2022; Durazzo et al., 2019). Most anthocyanins are found acylated by organic acids through ester bonds. There are around 500 anthocyanins, and among the most common are cyanidin, pelargonidin, delphinidin, petunidin, malvidin, and peonidin (Cianciosi et al., 2022).

10.3.2.3 Lignans

Lignans are compounds characterized by the presence of two phenolic rings and are linked by a four-carbon bridge. These compounds are found in greater amounts in seeds, such as flax seeds, sesame seeds, and soybeans, while smaller proportions are observed in fruits, particularly berries. The intake of these compounds is related to the prevention of cardiovascular diseases and cancer (Cladis et al., 2020; Durazzo et al., 2019).

10.3.2.4 Stilbenes

Stilbenes are present in plants and are produced after a physical injury. They have a basic structure consisting of two phenolic rings connected by a two-carbon bridge. These compounds are present in grapes, almonds, blueberries, blackberries, vines, plums, and wine (Cianciosi et al., 2022; Cladis et al., 2020). Resveratrol, the most common stilbene, has been shown to exhibit cardioprotective effects (Reinisalo et al., 2015).

10.3.3 Bioaccessibility and bioavailability of polyphenols

Phenolic compounds have shown various effects on health. They have low bioavailability since they are poorly absorbed at the tissue level, so their biological effects strictly depend on bioaccessibility and bioavailability (Rein et al., 2013; Comunian et al.,2022; Shahidi & Pan, 2022). Bioavailability is defined as the fraction of compound that manages to reach the target tissue where the biological action will be exerted. This will depend on bioaccessibility, the digestion and absorption processes, and the components present in the food matrix. In general, the processes involved in bioavailability are the release of phenolic compounds in ingested food, the chemical-structural changes that occur after digestion and in the small intestine, the absorption of compounds in epithelial cells, the modifications produced by the bacterial flora, phase I/II reactions of human metabolism, and the

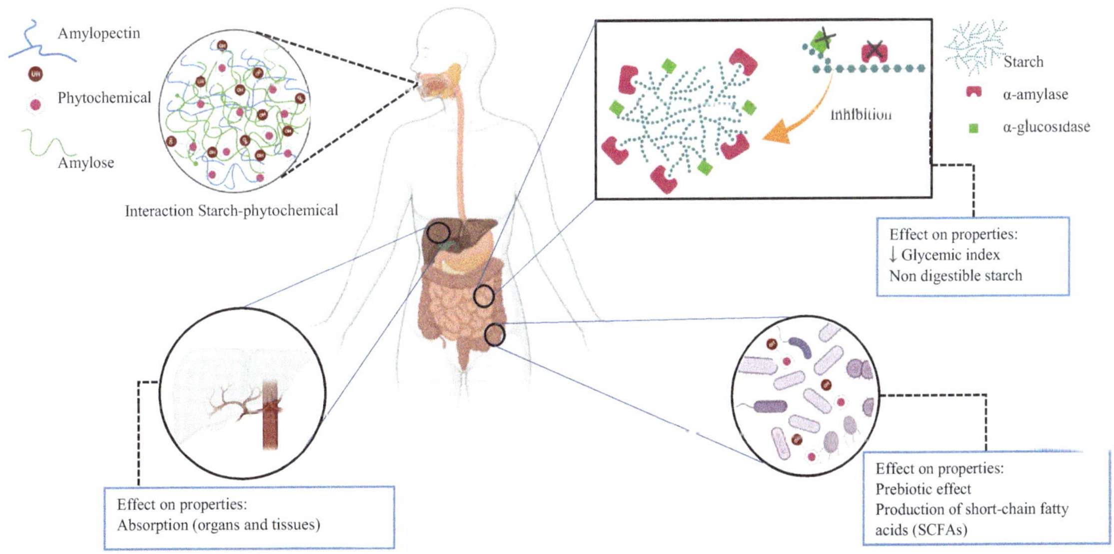

FIGURE 10.2 Effect of starch-phytochemical interactions on starch digestibility.

transport of compounds through the bloodstream to various tissues and organs (Cianciosi et al., 2022). For its part, bioaccessibility is defined as the fraction of compounds released from the food matrix after the digestion process that can cross the intestinal barrier and are available for absorption (Cianciosi et al., 2022; Saura-Calixto et al., 2007). The factors that affect bioaccessibility are given by variables, such as solubility, pH, viscosity, some food matrix treatments, and individual consumers' factors (age, sex, microbiota, BMI, state of health) (Fig. 10.2).

The interactions between polyphenols and food components remain unclear. These compounds can be mixed with the different components that constitute a food matrix, such as carbohydrates, fiber, lipids, proteins, and water, among others (Bordenave et al., 2014).

10.4 Starch-phytochemical complex

Phytochemicals and macromolecules are raw materials used in food formulations and processed to produce food matrices where they interact with each other (Bordenave et al., 2014; Ferruzzi et al., 2020; Kan et al., 2020; Miao & Hamaker, 2021; Shahidi & Pan, 2022). The interactions starch-phytochemical are of special interest because this polysaccharide is the main structure-building macromolecule in many foods (bakery products, pasta, snacks, breakfast cereals, and others). In recent years, there has been a growing interest in regulating the starch digestion rate because a high digestion and rapid hydrolysis of starch are associated with the development of diverse health issues, such as overweight and obesity, and chronic-degenerative diseases as diabetes, metabolic syndrome, cardiovascular problems, and cancer.

10.4.1 Phytochemical-starch complex

When deepening the studies on the starch digestion rate, several self-assembled starch V-type complexes between this polysaccharide and other molecules and macromolecules have emerged (Fig. 10.3). They are also useful for managing health problems related to the hydrolysis and glucose absorption of the polysaccharide, suggested as a novel type of resistant starch, RS type 5 (Gutiérrez & Tovar, 2021).

The V-type complex includes the starch-lipid complex, but recently, the starch-polyphenol complex has also been suggested as a V type. The V_I type complex is produced by the formation of a hydrophobic cavity where the polyphenol is reserved. Anther complex, the V_{IIa} type, shows a lamellae-like semicrystalline structure (Gutiérrez & Tovar, 2021), and recently an additional V-type complex named V_{IIb} type has been proposed. Its amylose helix or linear chains can be attached to both ends of the polyphenol. The change in the starch structure due to the complex formation is not recognized by digestive enzymes and is not hydrolyzed. Furthermore, the interaction starch-polyphenol delays the starch–starch interactions, hindering retrogradation (for more details on the interaction between lipids and starches, it is suggested to read Chapter 9 in Volume 1 of this series).

10.4.2 Effect of processing

Raw materials or food formulations to create foods with high starch contents undergo milling. To make flour, roots and tubers, which have high moisture contents, also go through a drying phase in addition to grinding. Starchy food crops such as maize, potato, wheat, rice, and cassava are used for starch extraction and purification, where some phenolics compounds remain in the extracted and purified starch (see chapter 8 in Volume 1 of this series).

The mechanical disruption of raw materials affects the original structure where the different components are present in compartments. Unit operations such as milling release the components that are more susceptible to interact

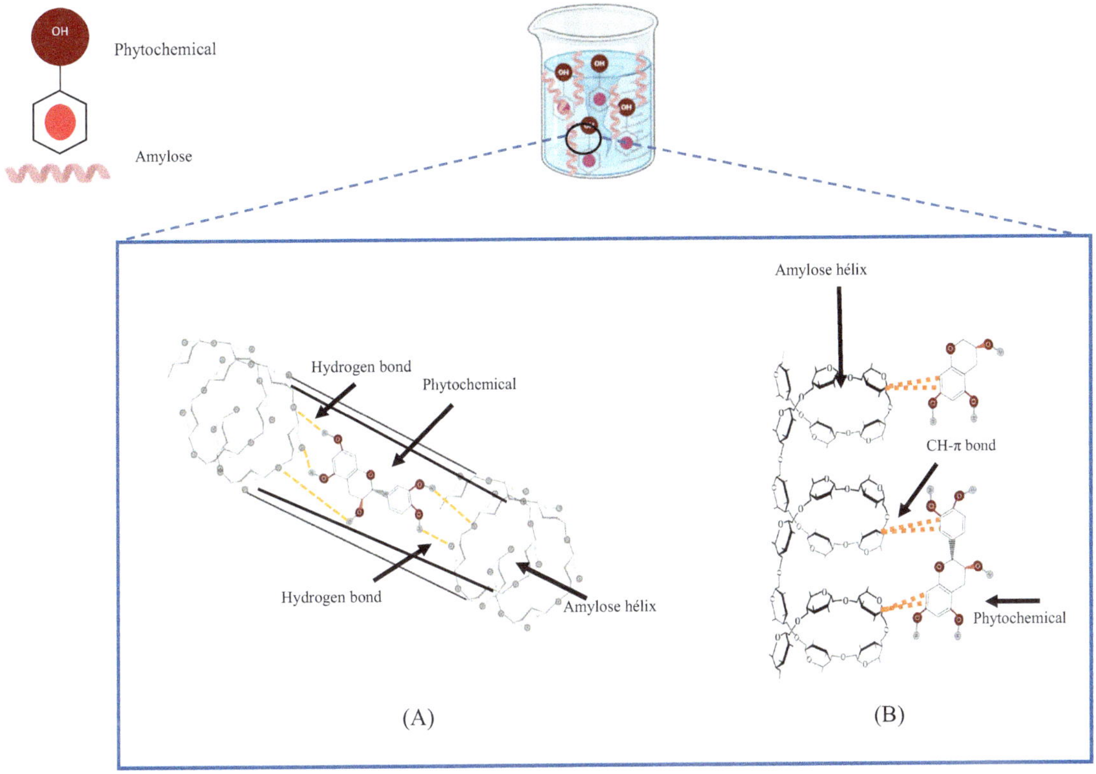

FIGURE 10.3 Models of interactions between starch and phytochemicals: (A) Starch-phytochemical V_I-type complex and (B) starch-phytochemical V_{II}-type complex. *Adapted from Romero-Hernández, H. A., Gutiérrez, T. J., & Bello-Pérez, L. A. (2022). Can starch-polyphenol V-type complexes be considered as resistant starch?* Food Hydrocolloids, *124, 107226. https://doi.org/10.1016/j.foodhyd.2021.107226*

between them during processing. In roots and tubers, because of the high moisture content before drying, this interaction may be favored. In addition, it is well known that the starch in coarse particles of whole grains is more slowly hydrolyzed that its fine counterpart. However, interactions between phytochemical and starch tend to decrease. The effect could be associated with the gelatinization degree of starch that depends on conventional processes, such as baking, extrusion, autoclaving, pressure-cooking, and parboiling, along with nonconventional processing as ultrasonication, high-pressure, and microwaves. For amylaceous raw materials originating from roots and tubers, this process can occur even more easily due to the lower gelatinization temperatures, in the range of 60°C–65°C (for more in-depth information on starch/water/heat relationships, we suggest reading Chapter 10 in Volume 1 of this series).

In general, phytochemicals are sensible to temperature during conventional processing, a characteristic that must be considered (Fig. 10.4).

10.4.3 Metabolism of phytochemicals

The free phytochemicals that are not complexed with macromolecules as starch are released from the food matrix in the first steps of digestion (mouth and stomach), and some are absorbed into the circulatory system (D'Archivio et al., 2010). These phytochemicals

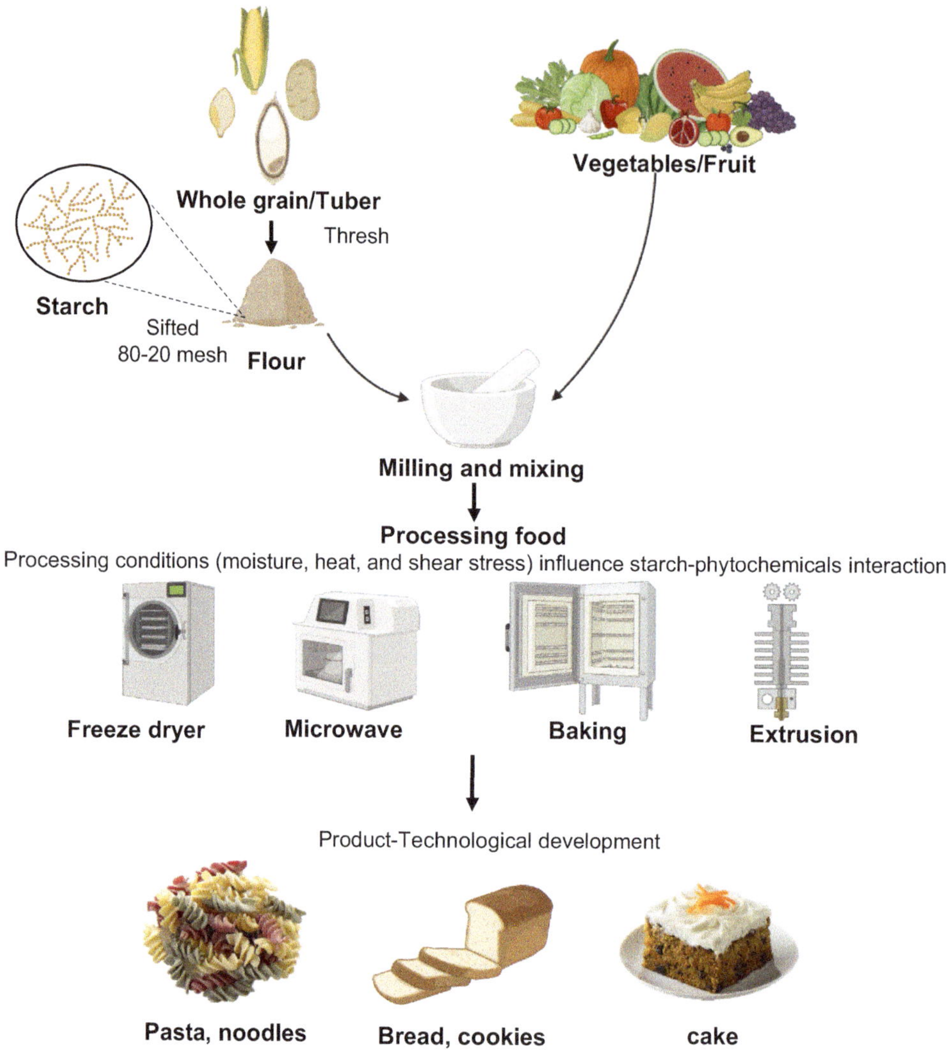

FIGURE 10.4 Effect of processing of starchy raw materials on the interactions starch-phytochemicals (creating with Biorender).

are metabolized in the liver, and metabolites are produced and released back into the bloodstream. The phytochemicals that are not absorbed in the small intestine continue toward the colon and become substrate to the microbiota. Similarly, the complex phytochemical-starch cannot be digested in the small intestine and continues to the colon. Both free phytochemicals and the complex phytochemical-starch reach the colon, where they are fermented by the microbiota, producing smaller derivatives and short-chain fatty acids (SCFAs) from starch. The smaller phytochemicals and SCFAs are released back into the circulatory system and distributed to extravascular space (body tissues) or excreted from the body through the kidneys or feces.

10.5 Starch digestion rate and health

Starch is one of the most important dietary carbohydrates and the major energy source of most people's diets. Nowadays, there is a growing interest in evaluating the enzymatic hydrolysis of starches to understand how to control their digestion, which ends with the glucose absorption, as an encouraging strategy for the prevention and treatment of metabolic disease (Bello-Perez et al., 2020). Some of the main strategies to reduce the digestion rate of starch are chemical, thermal, enzymatic, and mechanical modifications. Recently, the chemical modification of starch has been performed by adding extracts from different plant or fruit sources with high phytochemical concentrations (Papoutsis et al., 2021). Several reports show that phytochemicals (e.g., polyphenols) can inhibit amylase activity and thus delay starch digestion (Giuberti et al., 2020; Gutierrez et al., 2020; Sun & Miao, 2020). Bioactive compounds play a key role in starch digestibility since they can be used as an effective strategy to design functional foods and contribute to the control of metabolic disease. However, further studies are necessary to fully understand the complex between phenolic compounds and starch, which could contribute to improving their use in the food industry.

10.5.1 Effect of polyphenol compounds on the starch digestibility rate

The main sources of starches in many tropical countries are grains, roots, and tubers, such as maize, cassava, potato, wheat, and sweet potatoes. Yam (*dioscorea* sp.), canna (*Canna indica*), arrowroot (*Maranta arundinacea*), and unripe fruits, such as banana (Neela & Fanta, 2019; Wani et al., 2017), and the starch extracted from these phenolic sources may also be used. Several agricultural products with high starch contents are used in the development of functional foods, focused on impacting gastrointestinal health. However, in modern life, excessive nutrient and energy intake is prevalent in many countries and communities. It is thought to be the reason behind the high incidence of obesity, diabetes, cardiovascular diseases, and many other chronic diseases (Church & Martin, 2018).

Consequently, different modifications must be made before applying natural starch for industrial purposes with the aim to decrease the digestibility rate and increase slowly digestible and resistant to hydrolysis starches content. Amylose and amylopectin chains are the main polymers present in starch, and the chains can form hydrogen bonds (-OH) with a variety of small molecules, as phytochemical compounds (polyphenols) (Camelo-Méndez et al., 2017; Chen et al., 2020; Han et al., 2020; Ibironke et al., 2021; Miao, & Xu, Jia, et al., 2021; Xu et al., 2021).

Recently, plant foods rich in polyphenolic compounds have become popular worldwide due to their various biological activities: antioxidant capacity, glycemic index control, and α-amylase-inhibitor activities (Giuberti et al., 2020). The inhibition of digestive enzymes (α-amylase and α-glucosidase) by a polyphenol is highly related with its molecular structure, as it results from binding interactions between polyphenols and the enzyme (Jakobek, 2015; Sun et al., 2016). Other studies have reported a direct interaction between starch and phenolic compounds through hydrogen bonds and hydrophobic and electrostatic interactions (Bordenave et al., 2014), which also contribute to decrease the digestibility rate of starch.

The chemical interaction starch-polyphenol is classified into self-assembled and assembled starch-polyphenol V-type complexes (Deng et al., 2021) that may affect the metabolism of starch digestibility rate through three different mechanisms: (1) starch-polyphenol V-type complexes certainly escape the enzymatic action of α-amylase under in vitro and in vivo conditions (Romero-Hernández et al., 2022); (2) enzymatic inhibition is the result of hydroxyl and galloyl substituents and conjugated polyphenol systems on α-amylase tryptophan molecules in the hydrolysis of type V starch-polyphenol complexes and apparently increases resistant to hydrolysis starches content and reduces starch digestibility (Nyambe-Silavwe et al., 2015; Pico & Martínez, 2019; Romero-Hernández et al., 2022; Sun & Miao, 2020; Sun, & Wang, & Miao, 2020; Zheng et al., 2020); and (3) the polyphenols loaded in starch-polyphenol V-type complexes inhibit glucose transporters 2 under in vitro and in vivo conditions due to the competition glucose versus polyphenols (Hanhineva et al., 2010). They interfere with glucose transport across the intestinal brush border, and they can reduce the glycemic index (Romero-Hernández et al., 2022).

10.6 Final considerations

The use of nonconventional starchy food crops has increased due to sustainability, where raw materials should be used, including those considered as by-products (e.g., peel). Tropical food crops constitute an important but underused food source in many regions of the globe (see Chapter 14 in Volume 1 of this series).

The presence of different molecules and phytochemicals is key to developing functional foods and fighting health problems associated with the consumption of foods with high lipid and digestible carbohydrate content. The study of the interactions between phytochemical compounds and starch will in turn support the development of new or improved healthy foods. The starch structure and type of phytochemicals present in tropical food crops are important to understand the mechanisms producing the interactions and design starchy foods with slowly digestible starch and resistant to hydrolysis starches digestion.

In general, starchy tropical food crops have been little divulged, little researched, and underused. They are traditionally consumed, but they can also be used as raw materials for the development of new functional foods. Tropical starchy food crops contain phytochemicals that can produce complexes with different macromolecules present in a food matrix and regulate the metabolism of lipids, minerals, proteins, and carbohydrates. This issue has gained interest in recent years since tropical starchy food crops are an alternative to produce foods with health benefits, in countries where other starchy crops, such as wheat and potatoes, are more difficult to produce. There is an opportunity for countries in this region of the globe to take advantage of the climate conditions and increase the production of those food crops with economic benefits for the farmers and the population.

References

Agama-Acevedo, E., Nuñez-Santiago, M. C., Alvarez-Ramirez, J., & Bello-Pérez, L. A. (2015). Physicochemical, digestibility and structural characteristics of starch isolated from banana cultivars. *Carbohydrate Polymers, 124*, 17–24. Available from https://doi.org/10.1016/j.carbpol.2015.02.003.

Arıcı, M., Yıldırım, R. M., Özülkü, G., Yaşar, B., & Toker, O. S. (2016). Physicochemical and nutritional properties of taro (*Colocasia esculenta* L. Schott) flour as affected by drying temperature and air velocity. *LWT, 74*, 434–440.

Baliga, M. S., Shivashankara, A. R., Haniadka, R., Dsouza, J., & Bhat, H. P. (2011). Phytochemistry, nutritional and pharmacological properties of *Artocarpus heterophyllus* Lam (jackfruit): A review. *Food Research International, 44* (7), 1800–1811.

Bello-Perez, L. A., Flores-Silva, P. C., Agama-Acevedo, E., & Tovar, J. (2020). Starch digestibility: Past, present, and future. *Journal of the Science of Food and Agriculture, 100* (14), 5009–5016. Available from https://doi.org/10.1002/jsfa.8955.

Bertoft, E. (2004). Lintnerization of two amylose-free starches of A- and B-crystalline types, respectively. *Starch-Stärke, 56*(5), 167–180. Available from https://doi.org/10.1002/star.200300255.

Bertoft, E. (2018). Analyzing starch molecular structure. *Starch in food* (pp. 97–149). Woodhead Publishing.

Bordenave, N., Hamaker, B. R., & Ferruzzi, M. G. (2014). Nature and consequences of non-covalent interactions between flavonoids and macronutrients in foods. *Food & Function, 5*(1), 18–34. Available from https://doi.org/10.1039/C3FO60263J.

Bridgers, E. N., Chinn, M. S., & Truong, V. D. (2010). Extraction of anthocyanins from industrial purple-fleshed sweetpotatoes and enzymatic hydrolysis of residues for fermentable sugars. *Industrial Crops and Products, 32*(3), 613–620.

Camelo-Méndez, G. A., Agama-Acevedo, E., Tovar, J., & Bello-Pérez, L. A. (2017). Functional study of raw and cooked blue maize flour: Starch digestibility, total phenolic content and antioxidant activity. *Journal of Cereal Science, 76*, 179–185. Available from https://doi.org/10.1016/j.jcs.2017.06.009.

Camelo-Méndez, G. A., Agama-Acevedo, E., Rosell, C. M., de, J., Perea-Flores, M., & Bello-Pérez, L. A. (2018). Starch and antioxidant compound release during in vitro gastrointestinal digestion of gluten-free pasta. *Food Chemistry, 263*, 201–207. Available from https://doi.org/10.1016/j.foodchem.2018.04.075.

Camelo-Méndez, G. A., Flores-Silva, P. C., Agama-Acevedo, E., Tovar, J., & Bello-Pérez, L. A. (2018). Incorporation of whole blue maize flour increases antioxidant capacity and reduces in vitro starch digestibility of gluten-free pasta. *Starch-Stärke, 70* (1–2)1700126. Available from https://doi.org/10.1002/star.201700126.

Chandrasekara, A., & Josheph Kumar, T. (2016). Roots and tuber crops as functional foods: A review on phytochemical constituents and their potential health benefits. *International Journal of Food Science, 2016*.

Chandrika, U. G., Jansz, E. R., & Warnasuriya, N. D. (2004). Analysis of carotenoids in ripe jackfruit (*Artocarpus heterophyllus*) kernel and study of their bioconversion in rats. *Journal of the Science of Food and Agriculture, 85*, 186–190.

Chen, N., Gao, H.-X., He, Q., Yu, Z.-L., & Zeng, W.-C. (2020). Interaction and action mechanism of starch with different phenolic compounds. *International Journal of Food Sciences and Nutrition, 71*(6), 726–737. Available from https://doi.org/10.1080/09637486.2020.1722074.

Chi, C., Li, X., Feng, T., Zeng, X., Chen, L., & Li, L. (2018). Improvement in nutritional attributes of rice starch with dodecyl gallate complexation: A molecular dynamic simulation and in vitro study. *Journal of Agricultural and Food Chemistry, 66*(35), 9282–9290.

Chi, C., Li, X., Zhang, Y., Chen, L., Xie, F., Li, L., et al. (2019). Modulating the in vitro digestibility and predicted glycemic index of rice starch gels by complexation with gallic acid. *Food Hydrocolloids, 89*, 821–828.

Choo, C. L., & Aziz, N. A. A. (2010). Effects of banana flour and β-glucan on the nutritional and sensory evaluation of noodles. *Food Chemistry, 119*(1), 34–40. Available from https://doi.org/10.1016/j.foodchem.2009.05.004.

Chou, S., Li, B., Tan, H., Cui, H., Zhang, S., Wang, H., & Meng, X. (2020). Effect of ultrahigh pressure on structural and physicochemical properties of rice and corn starch incomplexes with apple polyphenols. *Journal of the Science of Food and Agriculture, 100*(15), 5395–5402. Available from https://doi.org/10.1002/jsfa.10589.

Chou, S., Li, B., Tan, H., Zhang, W., Zang, Z., Cui, H., Wang, H., Zhang, S., & Meng, X. (2020). The effect of pH on the chemical and structural interactions between apple polyphenol and starch derived from rice and maize. *Food Science & Nutrition, 8*, 5026–5035. Available from https://doi.org/10.1002/fsn3.1800.

Church, T., & Martin, C. K. (2018). The obesity epidemic: A consequence of reduced energy expenditure and the uncoupling of energy intake? *Obesity, 26*(1), 14–16. Available from https://doi.org/10.1002/oby.22072.

Cianciosi, D., Forbes-Hernández, T. Y., Regolo, L., Alvarez-Suarez, J. M., Navarro-Hortal, M. D., Xiao, J., Quiles, J. M., Battino, M., & Giampieri, F. (2022). The reciprocal interaction between polyphenols and other dietary compounds: Impact on bioavailability, antioxidant capacity and other physico-chemical and nutritional parameters. *Food Chemistry, 375*131904. Available from https://doi.org/10.1016/j.foodchem.2021.131904.

Cladis, D. P., Weaver, C. M., & Ferruzzi, M. G. (2020). Poly) Phenol metabolism: A primer for practitioners. *Nutrition Today, 55*(5).

Comunian, T. A., Drusch, S., & Brodkorb, A. (2022). Advances of plant-based structured food delivery systems on the in vitro digestibility of bioactive compounds. *Critical Reviews in Food Science Nutrition, 62*(23), 6485–6504. Available from https://doi.org/10.1080/10408398.2021.1902262.

Cosme, P., Rodríguez, A. B., Espino, J., & Garrido, M. (2020). Plant phenolics: Bioavailability as a key determinant of their potential health-promoting applications. *Antioxidants, 9*(12), 1263.

Cui, R., & Zhu, F. (2019). Physicochemical properties and bioactive compounds of different varieties of sweetpotato flour treated with high hydrostatic pressure. *Food Chemistry, 299*125129.

Cui, R., & Zhu, F. (2022). Changes in structure and phenolic profiles during processing of steamed bread enriched with purple sweetpotato flour. *Food Chemistry, 369*130578. Available from https://doi.org/10.1016/j.foodchem.2021.130578.

D'Archivio, M., Filesi, C., Varì, R., Scazzocchio, B., & Masella, R. (2010). Bioavailability of the polyphenols: Status and controversies. *International Journal of Molecular Sciences, 11*(4), 1321–1342. Available from https://doi.org/10.3390/ijms11041321.

Debelo, H., Li, M., & Ferruzzi, M. G. (2020). Processing influences on food polyphenol profiles and biological activity. *Current Opinion in Food Science, 32*, 90–102. Available from https://doi.org/10.1016/j.cofs.2020.03.001.

Deng, N., Deng, Z., Tang, C., Liu, C., Luo, S., Chen, T., & Hu, X. (2021). Formation, structure and properties of the starch-polyphenol inclusion complex: A review. *Trends in Food Science & Technology, 112*, 667–675. Available from https://doi.org/10.1016/j.tifs.2021.04.032.

Durazzo, A., Lucarini, M., Souto, E. B., Cicala, C., Caiazzo, E., Izzo, A. A., Novellino, E., & Santini, A. (2019). Polyphenols: A concise overview on the chemistry, occurrence, and human health. *Phytotherapy Research, 33*(9), 2221–2243. Available from https://doi.org/10.1002/ptr.6419.

Escobar-Puentes, A. A., Palomo, I., Rodríguez, L., Fuentes, E., Villegas-Ochoa, M. A., González-Aguilar, G. A., & Wall-Medrano, A. (2022). Sweet potato (*Ipomoea batatas* L.) phenotypes: From agroindustry to health effects. *Foods, 11*(7), 1058.

Fatemeh, S. R., Saifullah, R., Abbas, F. M. A., & Azhar, M. E. (2012). Total phenolics, flavonoids and antioxidant activity of banana pulp and peel flours: Influence of variety and stage of ripeness. *International Food Research Journal, 19*(3).

Ferruzzi, M. G., Hamaker, B. R., & Bordenave, N. (2020). Phenolic compounds are less degraded in presence of starch than in presence of proteins through processing in model porridges. *Food Chemistry, 309*125769. Available from https://doi.org/10.1016/j.foodchem.2019.125769.

Gao, S., Liu, H., Sun, L., Cao, J., Yang, J., Lu, M., et al. (2021). Rheological, thermal and in vitro digestibility properties on complex of plasma modified tartary buckwheat starches with quercetin. *Food Hydrocolloids, 110*106209. Available from https://doi.org/10.1016/j.foodhyd.2020.106209, Article.

Giuberti, G., Rocchetti, G., & Lucini, L. (2020). Interactions between phenolic compounds, amylolytic enzymes and starch: An updated overview. *Current Opinion in Food Science, 31*, 102–113. Available from https://doi.org/10.1016/j.cofs.2020.04.003.

Guo, Z., Zhao, B., Chen, J., Chen, L., & Zheng, B. (2019). Insight into the characterization and digestion of lotus seed starch-tea polyphenol complexes prepared under high hydrostatic pressure. *Food Chemistry, 297*124992. Available from https://doi.org/10.1016/j.foodchem.2019.124992, Article.

Gutierrez, A. S. A., Guo, J., Feng, J., Tan, L., & Kong, L. (2020). Inhibition of starch digestion by gallic acid and alkyl gallates. *Food Hydrocolloids, 102*105603. Available from https://doi.org/10.1016/j.foodhyd.2019.105603.

Gutiérrez, T. J., & Tovar, J. (2021). Update of the concept of type 5 resistant starch (RS5): Self-assembled starch V-type complexes. *Trends in Food Science & Technology, 109*, 711–724. Available from https://doi.org/10.1016/j.tifs.2021.01.078.

Han, X., Zhang, M., Zhang, R., Huang, L., Jia, X., Huang, F., & Liu, L. (2020). Physicochemical interactions between rice starch and different polyphenols and structural characterization of their complexes. *LWT, 125*109227. Available from https://doi.org/10.1016/j.lwt.2020.109227.

Hanhineva, K., Törrönen, R., Bondia-Pons, I., Pekkinen, J., Kolehmainen, M., Mykkänen, H., & Poutanen, K. (2010). Impact of dietary polyphenols on carbohydrate metabolism. *International Journal of Molecular Sciences, 11*(4), 1365–1402. Available from https://doi.org/10.3390/ijms11041365.

Hoyos-Leyva, J. D., Bello-Pérez, L. A., Yee-Madeira, H., Rodriguez-Garcia, M. E., & Aguirre-Cruz, A. (2017). Characterization of the flour and starch of aroid cultivars grown in Mexico. *Starch-Stärke, 69*(9–10)1600370. Available from https://doi.org/10.1002/star.201600370.

Hung, P. V., Phat, N. H., & Phi, N. T. L. (2013). Physicochemical properties and antioxidant capacity of debranched starch–ferulic acid complexes. *Starch-Stärke, 65*(5-6), 382–389. Available from https://doi.org/10.1002/star.201200168.

Ibironke, O., Ajayi, I. O., Otemuyima, A. A., & Olumuyiwa, S. F. (2021). Vegetable polyphenols inhibit starch digestibility and phenolic availability from composite carbohydrate foods *in-vitro*. *Journal of Agriculture and Food Research*, 3100116, In this issue. Available from https://doi.org/10.1016/j.jafr.2021.100116.

Jagtap, U. B., Panaskar, S. N., & Bapat, V. A. (2010). Evaluation of antioxidant capacity and phenol content in jackfruit (*Artocarpus heterophyllus* Lam.) fruit pulp. *Plant Foods for Human Nutrition, 65*(2), 99–104.

Jakobek, L. (2015). Interactions of polyphenols with carbohydrates, lipids and proteins. *Food Chemistry, 175*, 556–567. Available from https://doi.org/10.1016/j.foodchem.2014.12.013.

Kan, L., Oliviero, T., Verkerk, R., Fogliano, V., & Capuano, E. (2020). Interaction of bread and berry polyphenols affects starch digestibility and polyphenols bio-accessibility. *Journal of Functional Foods, 68*103924. Available from https://doi.org/10.1016/j.jff.2020.103924.

Liu, J., Wang, X., Yong, H., Kan, J., Zhang, N., & Jin, C. (2018). Preparation, characterization, digestibility and antioxidant activity of quercetin grafted *Cynanchum auriculatum* starch. *International Journal of Biological Macromolecules, 114*, 130–136. Available from https://doi.org/10.1016/j.ijbiomac.2018.03.101.

Liu, Y., Chen, L., Xu, H., Liang, Y., & Zheng, B. (2019). Understanding the digestibility of rice starch gallic acid complexes formed by high pressure homogenization. *International Journal of Biological Macromolecules, 134*, 856–863. Available from https://doi.org/10.1016/j.ijbiomac.2019.05.083.

Martín, M. Á., & Ramos, S. (2021). Impact of dietary flavanols on microbiota, immunity and inflammation in metabolic diseases. *Nutrients*, *13*(3), 850.

Miao, L., Xu, Y., Jia, C., Zhang, B., Niu, M., & Zhao, S. (2021). Structural changes of rice starch and activity inhibition of starch digestive enzymes by anthocyanins retarded starch digestibility. *Carbohydrate Polymers*, *261*117841. Available from https://doi.org/10.1016/j.carbpol.2021.117841.

Miao, M., & Hamaker, B. R. (2021). Food matrix effects for modulating starch bioavailability. *Annual Review of Food Science and Technology*, *12*, 169–191. Available from https://doi.org/10.1146/annurev-food-070620-013937.

Millan-Testa, C. E., Mendez-Montealvo, M. G., Ottenhof, M. A., Farhat, I. A., & Bello-Perez, L. A. (2005). Determination of the molecular and structural characteristics of okenia, mango, and banana starches. *Journal of Agricultural and Food Chemistry*, *53*(3), 495–501. Available from https://doi.org/10.1021/jf048862x.

Mkandawire, N. L., Kaufman, R. C., Bean, S. R., Weller, C. L., Jackson, D. S., & Rose, D. J. (2013). Effects of sorghum (*Sorghum bicolor* (L.) Moench) tannins on α-amylase activity and in vitro digestibility of starch in raw and processed flours. *Journal of Agricultural and Food Chemistry*, *61*(18), 4448–4454. Available from https://doi.org/10.1021/jf400464j.

Najmanová, I., Vopršalová, M., Saso, L., & Mladěnka, P. (2020). The pharmacokinetics of flavanones. *Critical Reviews in Food Science and Nutrition*, *60*(18), 3155–3171. Available from https://doi.org/10.1080/10408398.2019.1679085.

Neela, S., & Fanta, S. W. (2019). Review on nutritional composition of orange-fleshed sweet potato and its role in management of vitamin A deficiency. *Food Science & Nutrition*, *7*(6), 1920–1945. Available from https://doi.org/10.1002/fsn3.1063.

Nilusha, R. A. T., Jayasinghe, J. M. J. K., Perera, O. D. A. N., Perera, P. I. P., & Jayasinghe, C. V. L. (2021). Proximate composition, physicochemical, functional, and antioxidant properties of flours from selected cassava (*Manihot esculenta* Crantz) varieties. *International Journal of Food Science*, *2021*.

Nurdjanah, S., Nurdin, S. U., Astuti, S., & Manik, V. E. (2022). Chemical components, antioxidant activity, and glycemic response values of purple sweet potato products. *International Journal of Food Science*, *2022*7708172. Available from https://doi.org/10.1155/2022/7708172.

Nyambe-Silavwe, H., Villa-Rodriguez, J. A., Ifie, I., Holmes, M., Aydin, E., Jensen, J. M., & Williamson, G. (2015). Inhibition of human α-amylase by dietary polyphenols. *Journal of Functional Foods*, *19*, 723–732. Available from https://doi.org/10.1016/j.jff.2015.10.003.

Ovando-Martinez, M., Sáyago-Ayerdi, S., Agama-Acevedo, E., Goñi, I., & Bello-Pérez, L. A. (2009). Unripe banana flour as an ingredient to increase the undigestible carbohydrates of pasta. *Food Chemistry*, *113*(1), 121–126.

Pan, Y., Li, H., Shahidi, F., Luo, T., & Deng, Z. (2022). Interactions among dietary phytochemicals and nutrients: Role of cell membranes. *Trends in Food Science & Technology*, *124*, 38–50. Available from https://doi.org/10.1016/j.tifs.2022.03.024.

Papoutsis, K., Zhang, J., Bowyer, M. C., Brunton, N., Gibney, E. R., & Lyng, J. (2021). Fruit, vegetables, and mushrooms for the preparation of extracts with α-amylase and α-glucosidase inhibition properties: A review. *Food Chemistry*, *338*128119. Available from https://doi.org/10.1016/j.foodchem.2020.128119.

Pasha, I., Basit, A., Ahsin, M., & Ahmad, F. (2022). Probing nutritional and functional properties of salted noodles supplemented with ripen banana peel powder. *Food Production, Processing and Nutrition*, *4*(1), 22. Available from https://doi.org/10.1186/s43014-022-00100-5.

Patiño-Rodríguez, O., Bello-Pérez, L. A., Agama-Acevedo, E., & Pacheco-Vargas, G. (2020). Pulp and peel of unripe stenospermocarpic mango (*Mangifera indica* L. cv Ataulfo) as an alternative source of starch, polyphenols and dietary fibre. *Food Research International*, *138*109719.

Pico, J., & Martínez, M. M. (2019). Unraveling the inhibition of intestinal glucose transport by dietary phenolics: A review. *Current Pharmaceutical Design*, *25*(32), 3418–3433. Available from https://doi.org/10.2174/1381612825666191015154326.

Pico, J., Xu, K., Guo, M., Mohamedshah, Z., Ferruzzi, M. G., & Martinez, M. M. (2019). Manufacturing the ultimate Unripe banana flour: Impact of drying and extrusion on phenolic profile and starch bioaccessibility. *Food Chemistry*, *297*124990.

Pott, I., Breithaupt, D. E., & Carle, R. (2003). Detection of unusual carotenoid esters in fresh mango (*Mangifera indica* L. cv. 'Kent'). *Phytochemistry*, *64*, 825–829.

Proença, C., Ribeiro, D., Freitas, M., & Fernandes, E. (2022). Flavonoids as potential agents in the management of type 2 diabetes through the modulation of α-amylase and α-glucosidase activity: A review. *Critical Reviews in Food Science and Nutrition*, *62*(12), 3137–3207. Available from https://doi.org/10.1080/10408398.2020.1862755.

Ramírez-Maganda, J., Blancas-Benítez, F. J., Zamora-Gasga, V. M., García-Magaña, M. D. L., Bello-Perez, L. A., Tovar, J., & Sáyago-Ayerdi, S. G. (2015). Nutritional properties and phenolic content of a bakery product substituted with a mango (*Mangifera indica*)'Ataulfo'processing by-product. *Food Research International*, *73*, 117–123.

Rein, M. J., Renouf, M., Cruz-Hernandez, C., Actis-Goretta, L., Thakkar, S. K., & da Silva Pinto, M. (2013). Bioavailability of bioactive food compounds: A challenging journey to bioefficacy. *British Journal of Clinical*

Pharmacology, 75(3), 588–602. Available from https://doi.org/10.1111/j.1365-2125.2012.04425.x.

Reinisalo, M., Kårlund, A., Koskela, A., Kaarniranta, K., & Karjalainen, R. O. (2015). Polyphenol stilbenes: Molecular mechanisms of defence against oxidative stress and aging-related diseases. *Oxidative Medicine and Cellular Longevity, 2015*340520. Available from https://doi.org/10.1155/2015/340520.

Romero-Hernández, H. A., Gutiérrez, T. J., & Bello-Pérez, L. A. (2022). Can starch-polyphenol V-type complexes be considered as resistant starch? *Food Hydrocolloids, 124*107226. Available from https://doi.org/10.1016/j.foodhyd.2021.107226.

Saura-Calixto, F., Serrano, J., & Goñi, I. (2007). Intake and bioaccessibility of total polyphenols in a whole diet. *Food Chemistry, 101*(2), 492–501. Available from https://doi.org/10.1016/j.foodchem.2006.02.006.

Shahidi, F., & Pan, Y. (2022). Influence of food matrix and food processing on the chemical interaction and bioaccessibility of dietary phytochemicals: A review. *Critical Reviews in Food Science and Nutrition, 62*(23), 6421–6445. Available from https://doi.org/10.1080/10408398.2021.1901650.

Shahidi, F., & Yeo, J. (2018). Bioactivities of phenolics by focusing on suppression of chronic diseases: A Review. *International Journal of Molecular Sciences, 19*(6), 1573.

Sharma, R., Dar, B. N., Sharma, S., & Singh, B. (2021). In vitro digestibility, cooking quality, bio-functional composition, and sensory properties of pasta incorporated with potato and pigeonpea flour. *International Journal of Gastronomy and Food Science, 23*100300. Available from https://doi.org/10.1016/j.ijgfs.2020.100300.

Simsek, S., & El, S. N. (2015). In vitro starch digestibility, estimated glycemic index and antioxidant potential of taro (*Colocasia esculenta* L. Schott) corm. *Food Chemistry, 168*, 257–261.

Soison, B., Jangchud, K., Jangchud, A., Harnsilawat, T., Piyachomkwan, K., Charunuch, C., & Prinyawiwatkul, W. (2014). Physico-functional and antioxidant properties of purple-flesh sweet potato flours as affected by extrusion and drum-drying treatments. *International Journal of Food Science & Technology, 49*(9), 2067–2075.

Su, J., Tan, C., Gao, Y., & Feng, Y. (2020). Four phenolic acids from purple sweet potato and their effects on physicochemical, digestive and structural characteristics of starch. *International Journal of Food Science & Technology, 56*(4), 1896–1904.

Sun, J., Chu, Y. F., Wu, X., & Liu, R. H. (2002). Antioxidant and antiproliferative activities of common fruits. *Journal of Agricultural and Food Chemistry, 50*(25), 7449–7454.

Sun, L., & Miao, M. (2020). Dietary polyphenols modulate starch digestion and glycaemic level: A review. *Critical Reviews in Food Science and Nutrition, 60*(4), 541–555. Available from https://doi.org/10.1080/10408398.2018.1544883.

Sun, L., Wang, Y., & Miao, M. (2020). Inhibition of α-amylase by polyphenolic compounds: Substrate digestion, binding interactions and nutritional intervention. *Trends in Food Science & Technology, 104*, 190–207. Available from https://doi.org/10.1016/j.tifs.2020.08.003.

Sun, L., Warren, F. J., Netzel, G., & Gidley, M. J. (2016). 3 or 3′-Galloyl substitution plays an important role in association of catechins and theaflavins with porcine pancreatic α-amylase: The kinetics of inhibition of α-amylase by tea polyphenols. *Journal of Functional Foods, 26*, 144–156. Available from https://doi.org/10.1016/j.jff.2016.07.012.

Truzzi, F., Tibaldi, C., Zhang, Y., Dinelli, G., & D'Amen, E. (2021). An overview on dietary polyphenols and their biopharmaceutical classification system (BCS). *International Journal of Molecular Sciences, 22*(11), 5514.

Tsao, R. (2010). Chemistry and biochemistry of dietary polyphenols. *Nutrients, 2*(12), 1231–1246.

Vamadevan, V., & Bertoft, E. (2018). Impact of different structural types of amylopectin on retrogradation. *Food Hydrocolloids, 80*. Available from https://doi.org/10.1016/j.foodhyd.2018.01.029.

Wang, L., Wang, L., Li, Z., Gao, Y., Cui, S. W., Wang, T., & Qiu, J. (2021). Diverse effects of rutin and quercetin on the pasting, rheological and structural properties of Tartary buckwheat starch. *Food Chemistry, 335*127556. Available from https://doi.org/10.1016/j.foodchem.2020.127556.

Wani, S. M., Amin, S., Javaid, I., Masoodi, F. A., Mir, S. A., Ganai, S. A., & Yildiz, F. (2017). Minimal processing of tropical and subtropical fruits, vegetables, nuts, and seeds. In F. Yildiz, & R. C. Wiley (Eds.), *Minimally processed refrigerated fruits and vegetables* (pp. 469–512). Boston, MA: : Springer US.

Xu, T., Li, X., Ji, S., Zhong, Y., Simal-Gandara, J., Capanoglu, E., Xia, J., & Lu, B. (2021). Starch modification with phenolics: Methods, physicochemical property alteration, and mechanisms of glycaemic control. *Trends in Food Science & Technology, 111*, 12–26. Available from https://doi.org/10.1016/j.tifs.2021.02.023.

Yee, J., Roman, L., Pico, J., Aguirre-Cruz, A., Bello-Perez, L. A., Bertoft, E., & Martinez, M. M. (2021). The molecular structure of starch from different Musa genotypes: Higher branching density of amylose chains seems to promote enzyme-resistant structures. *Food Hydrocolloids, 112*106351. Available from https://doi.org/10.1016/j.foodhyd.2020.106351.

Yi, B., Hu, L., Mei, W., Zhou, K., Wang, H., Luo, Y., & Dai, H. (2011). Antioxidant phenolic compounds of cassava (*Manihot esculenta*) from Hainan. *Molecules (Basel, Switzerland), 16*(12), 10157–10167.

Zhang, Y., Li, B., Xu, F., He, S., Zhang, Y., Sun, L., & Tan, L. (2021). Jackfruit starch: Composition, structure, functional properties, modifications and applications. *Trends in Food Science & Technology, 107*, 268–283.

Zhao, B., Wang, B., Zheng, B., Chen, L., & Guo, Z. (2019). Effects and mechanism of high pressure homogenization on the characterization and digestion behavior of lotus seed starch–green tea polyphenol complexes. *Journal of Functional Foods*, *57*, 173–181. Available from https://doi.org/10.1016/j.jff.2019.04.016.

Zheng, Y., Tian, J., Yang, W., Chen, S., Liu, D., Fang, H., Zhang, H., & Ye, X. (2020). Inhibition mechanism of ferulic acid against α-amylase and α-glucosidase. *Food Chemistry*, *317*126346. Available from https://doi.org/10.1016/j.foodchem.2020.126346.

Zuiter, A. S. (2014). *Proanthocyanidin: Chemistry and biology: From phenolic compounds to proanthocyanidins. Reference module in chemistry, molecular sciences and chemical engineering*. Elsevier.

CHAPTER

11

Starch: a natural, safe, and multifunctional ingredient for cosmetic formulations

Juliana Bonametti Olivato

Pharmaceutical Sciences Department, State University of Ponta Grossa, Ponta Grossa, Parana, Brazil

11.1 Cosmetics—definitions and overview

Cosmetic is defined by Food and Drug Administration (FDA) as "a product (excluding pure soap) intended to be applied to the human body for cleansing, beautifying, promoting attractiveness, or altering the appearance" (section 201 (i) of the Federal Food, Drug, and Cosmetic Act). According to the European Union Cosmetics Directive (European Comission, 2009), a cosmetic is "any substance or preparation intended to be placed in contact with the various external parts of the human body (epidermis, hair system, nails, lips and external genital organs) or with the teeth and the mucous membranes of the oral cavity with a view exclusively or mainly to cleaning them, perfuming them, changing their appearance and/or correcting body odors and/or protecting them or keeping them in good condition." Based on the FDA and EU definitions, a cosmetic is developed exclusively for external uses and is fundamental to be safe for the consumer (U.S. Food & Drug Administration (FDA), 2022).

The global cosmetics market size exceeded US $350 billion in 2022, growing annually by almost 5%. The market for cosmetics, toiletries, and fragrances is dominated by the United States, which holds 19.4% of the global market. In this sector, the consumption market of Brazil is the fourth, with 4.3% (Brazilian Association of the Cosmetic, Toiletry, & Fragrance Industry ABIHIPEC, 2022). It is estimated at over 8000 cosmetic ingredients, voluntary registered for the FDA. It presents a challenge in investigating and regulating the safety of cosmetics. Prospective studies were published in 1980 and 1983 to collect and assess reactions and adverse effects of the cosmetics (Adams et al., 1983). Recently, the Cosmetic Ingredient Review (CIR) Expert Panel is responsible for a systematic review of the scientific data available for cosmetic ingredients, categorizing each ingredient for its safety.

EC Cosmetics Regulation 1223/2009 (UNION TEPATCOTE, 2009) states that the safety of finished cosmetic products can be assured based on knowing the safety of the ingredients that they contain. Component safety information is based

Starch Industries
DOI: https://doi.org/10.1016/B978-0-323-90842-9.00007-8

on bibliographic data and in vitro and human tests, since in the European Union (EU), finished cosmetics are no longer tested on animals. The Council Directive (Annex 1 of Regulation EC 1223/2009) also established common rules for producing a report on the safety of cosmetic products. This current environment focuses attention on starch, with approved safe and multifunctional properties to be used in cosmetic products.

Underground starch crops remain unexplored concerning their use in cosmetic applications. This chapter will explore new potential applications of these starches, enhancing their commercial application, which creates a demand for its inclusion in international compendiums that guarantee the minimum quality requirements for pharmaceutical and cosmetic ingredients.

11.2 Starch: a potential ingredient for biocosmetics

Increased customer demand for green cosmetics and concern about the toxicity of fossil ingredients, sustainability, costs, and environmental concerns have diverted the commercial focus from fossil-based ingredients to biobased ingredients. In this way, the so-called biocosmetics has emerged, and the requirements for their production include they be made from 100% natural ingredients, understood as derived from plants, animals, microorganisms, enzymes, and organic crops, and used for topical skin, hair, face, and oral care (Bom et al., 2019; Goyal & Jerold, 2023). In the same way, vegan cosmetics, made with animal-free and cruelty-free ingredients, have earned attention over the years.

Some difficulties are associated with the choice of molecules to be used in biocosmetics. The desired characteristics of these components are high availability, nontoxicity, biodegradability, and biocompatibility. Furthermore, by choosing a biocosmetic, the consumer expects the quality and efficiency of an "unsustainable" alternative, and this has been one of the biggest challenges for the cosmetics industry (Bom et al., 2019). In this context, starch represents a potential substitute for synthetic polymers and can be used as a native, or for better performance, some modifications may be considered.

Cosmetics containing starch are developed for application to the skin and/or hair, which require frequent use for varying periods depending on the application that may extend over many years. In some instances, these cosmetics may have contact with the eyes and mucous membranes. The CIR has completed an evaluation of the safety of polysaccharides used in cosmetic products, with an examination of relevant animal and human data on these materials and concluded that starch is safe for cosmetic uses. The native and modified starches with proven safety for cosmetic use, according to CIR, and listed by the FDA and the Voluntary Cosmetic Registration Program (VCRP) as presently used starches for cosmetics, are presented in Table 11.1. Voluntary submissions conducted by VCRP provide FDA with the best estimate of information available about cosmetic ingredients.

As shown in Table 11.1, most starches used in cosmetics come from cereals, mainly corn and wheat, with corn starch being the most consumed in the world. Among the roots and tubers, potato and cassava stand out, both being of South American origin. Potato is an important source of starch in Europe (as is wheat), and cassava starch is the second most produced starch in the world and the most exported (for more details on starch production, see Chapter 3 of Volume 1 in this series).

Except for cassava, which is a cheap and widely available starch, all the raw materials listed in Table 11.1 are products from developed countries, rice and Kudzu being produced mainly in Japan. Thus, many raw materials originating from tropical countries, such as roots and tubers of South American origin, are

TABLE 11.1 Starches presently used in cosmetics (Cosmetic Ingredient Review CIR, 2023)

Botanical name	Common name	Starch type	Type of modification
Cereals			
Avena sativa	Oat	Native	None
Oryza sativa	Rice	Native	None
Triticum vulgare	Wheat	Native	None
		Hydrolyzed starch	Hydrolysate of wheat starch obtained by acid enzyme or other method of hydrolysis
		Hydroxypropyltrimonium hydrolyzed wheat starch	Quaternary ammonium salt based on the hydrolyzed wheat starch
		Laurdimonium hydroxypropyl hydrolyzed wheat starch	Quaternary ammonium chloride based on the hydrolyzed wheat starch
Zea mays	Corn	Native	None
		Modified	3-Dodecenyl dihydro-2,5-furandione DS < 0.1
		Partially cross-linked	Glyceryl
		Hydrolyzed starch octenylsuccinate	Reaction product of octenylsuccinic anhydride with hydrolyzed corn starch
		Hydroxypropyltrimonium hydrolyzed corn starch	Quaternary ammonium salt based on the hydrolyzed corn starch
		Sodium starch octenylsuccinate	Sodium salt of the reaction product of octenylsuccinic anhydride with corn starch.
Roots and tubers			
Pueraria lobata	Kudzu	Native	None
Solanum tuberosum	Potato	Native	None
		Modified starch	Ether from the reaction of haloethylaminodipropionic acid and potato starch (DS < 0.1).
Manihot esculenta	Cassava	Native	None
Not specified			
Not specified[a]		Starch diethylaminoethyl ether	Product obtained by conversion of some hydroxyl groups in starch to diethylaminoethyl ether groups
		Starch hydroxypropyltrimonium chloride	Quaternary ammonium formed by the reaction of starch with 2,3- epoxypropyltrimethylammonium chloride
		Hydrogenated starch hydrolysate	End-product of the controlled hydrogenation of hydrolyzed starch
		Hydroxypropyl starch	Propylene glycol ether of starch
		Sodium carboxymethyl starch	Sodium salt of a carboxymethyl derivative of starch

(Continued)

TABLE 11.1 (Continued)

Botanical name	Common name	Starch type	Type of modification
		Sodium oxidized starch acetate/succinate	Sodium salt as product of the esterification of oxidized starch with acetic acid and succinic acid anhydrides
		Starch acetate	Product obtained by the reaction of starch with acetic anhydride

[a]*Starches from commercial botanical sources (corn, potato, wheat, and rice) modified and evaluated by Cosmetic Ingredient Review Expert Panel.*
Data adapted from Cosmetic Ingredient Review (CIR) (2023). Available from https://www.cir-safety.org/.

not used due to a lack of knowledge and research. Among the roots and tubers of South American origin with the greatest potential for starch use in cosmetics are sweet potatoes (*Ipomea batatas*), with large production in many countries, as well as canna (*Canna edulis*) and arrowroot (*Maranta arundinacea*), with small starch units in several countries in Asia and Latin America (see Chapter 3).

11.3 The role of starch in cosmetic formulations

The growing interest in cosmetic products, particularly botanical polysaccharides, largely reflects consumer preferences for natural cosmetic products. Starch is an abundant, low-cost, renewable, and nontoxic natural polymer. It is also a multipurpose component widely used in food and non-food materials, such as biodegradable packaging, biomedical and pharmaceutical applications, and cosmetics (See Chapter 3 in Volume 1 of this series for more information about starch markets).

The use of starch in cosmetics is remarkable for its natural and renewable properties, also favoring better sensorial properties. Starch can be used in cosmetics in the granular (native and modified) or soluble forms. Different applications of starch can be considered based on the distinct properties resulting from the different shapes and sizes of the granules obtained from several botanical sources. This versatile, safe, and low-cost raw material has made starch one of the main excipients used in the cosmetic industry (to learn more about starch modifications, Chapter 6 is suggested).

The cosmetic applications of starch include gel forming (hydrophilic gels), thickener, dusting powders, stabilizer of emulsions or promoter of viscosity of the product, and modifier of the sensory properties of the formulations, as summarized in Fig. 11.1.

Two forms of starch are used to prepare cosmetic formulations: soluble starches (pregelatinized) or in granular form. Soluble starch is mainly used as a thickener, contributing to viscosity due to hydrodynamic volume exclusion, rheology modification, and emulsion stabilization, and provides a sensation of softness and conditioning for hair and skin. The granular structure of the starch is also responsible for interesting sensory properties like a soft and silky feel and can be used as oil absorber, reducing greasiness and/or caking on the skin (Lochhead, 2007).

The particle size is crucial for the applications of starch in different cosmetics. Small particles typically provide the accumulation of the product on the skin, and larger granules provide good fluidity and softness. Starch granules extracted from South America underground organs vary greatly in size, and some of them could have interesting properties in this market (for more information and details, we suggest reading Chapter 10 in Volume 1 of this series).

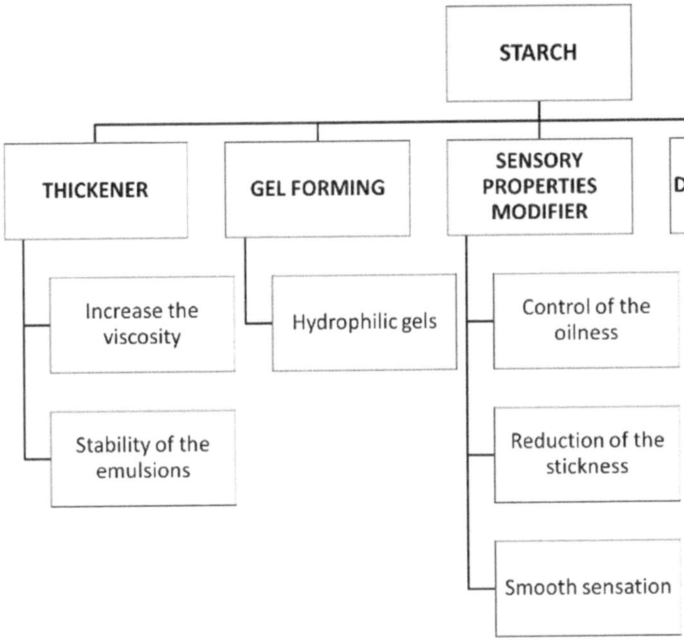

FIGURE 11.1 Main applications of starch in cosmetics.

Novel forms of starch application, such as microcapsules, nanoparticles, and composites, have recently emerged in the cosmetic industry. Due to their small size, fine biological compatibility, and environmental-friendliness, starch-based nanoparticles are a promising biomaterial for new applications in cosmetics, acting as nanocarriers for bioactive ingredients or to improve their absorption and bioavailability (Kanlayavattanakul & Lourith, 2015; Qiu et al., 2019). Starch nanocrystals have proved to be effective in stabilizing emulsions via the Pickering mechanism.

11.3.1 Thickeners

Thickeners consist of natural and synthetic polymers that are important auxiliary materials in the cosmetic industries to achieve some viscosity in systems containing surfactants, for example, emulsions or cleansing products as hair cosmetics and shampoo (Lehmann et al., 2008). Starch is one of the most widely used nonionic natural polysaccharides as a thickener. There are already different starch-based thickeners known in the cosmetics industry, mainly with hydrolyzed corn starch and hydrolyzed wheat starch. Lehmann et al. (2008) evaluated the 2-hydroxyalkyl carboxymethyl corn starch, which has been shown to increase the viscosity of the system by up to 50% when used in selective aqueous surfactant systems.

11.3.1.1 Starch as emulsion stabilizer

An emulsion is a system containing two immiscible liquids, where one liquid is dispersed as spherical droplets in another liquid phase. The instability of the emulsions is caused by the unfavorable contact between the two immiscible phases. Emulsions can be classified as oil-in-water (O/W) and water-in-oil (W/O), reflecting the dispersed and continuous phases, respectively. In both cases, it is necessary to stabilize the emulsions to prevent coalescence of the dispersed drops. Stabilization of emulsion droplets is often achieved by the addition of surfactants or emulsifiers, which act by different mechanisms: (1) decreasing the interfacial tension and (2) increasing the steric hindrances and/or electrostatic repulsion between the droplets (Rayner et al., 2012).

The hydrophilic portion of an emulsion consists of water and other substances, which are soluble in this phase. The stability of an emulsion is influenced by the viscosity of the dispersed phase. When considering O/W emulsion, the presence of polymeric substances that may increase the viscosity of the dispersed phase (aqueous) is necessary. In this context, starch is the most used natural polymer for this purpose. Native and modified starches are used to improve stability of emulsions. It is important to consider the relationship between concentration x viscosity to ensure the reproducibility of the process and the stability of the cosmetic, preventing phase separation.

The process of choosing the best starch for application in cosmetic emulsions depends on the physicochemical performance of each starch, their stability in the presence of electrolytes, pH ranges, high temperatures and the pasta clarity, texture, and gelatinization temperature. When a clear pasta is required, this favors the use of underground starch crops.

11.3.2 Gel forming

Hydrophilic gels consist essentially of an aqueous phase containing a gel forming agent, water, and adjuvants to reduce stickiness. Gel is an important vehicle for incorporating active ingredients for applications on face and body, including pigments and fragrances. Cosmetic preparations based on gels include dermatologic base for hydrophilic actives; hair gels; bath gels; aftershave gels; and toothpaste.

For this purpose, pregelatinized starches are frequently used to facilitate the cold formation of the gel. The main objectives of the use of starch for gel production are the good spreadability, sensorial characteristics, and good stability over a wide pH range. Moreover, when compared to other gel-forming agents (gums, cellulose derivatives, carboxyvinyl polymers, etc.), starch represents a low-cost alternative.

11.3.2.1 Hydrogels

Hydrogels are defined as superabsorbent materials that are able to retain large amounts of water in their structure. This is achieved through appropriate gelling agents, such as colloidal substances and polymers. Polymeric hydrogels are formed by a three-dimensional and hydrophilic network, composed of crosslinked polymers thermodynamically compatible with water, allowing them to swell in aqueous media (Peppas et al., 2000).

Starch is a natural polymer constructed from simple sugar building blocks that are hydrated in an aqueous environment, thereby creating a hydrogel into which water can enter and consequently forming strong and stable interconnected structures (Kanlayavattanakul & Lourith, 2015; Larrea-Wachtendorff et al., 2020). In the case of starch, water absorption capacity, volume expansion, and granular strength are directly linked to the amylose/amylopectin proportion, which usually occurs in cereals at a ratio of 1:3, respectively.

Hydrogels can hold between 20 and 100% of water relative to their total weight, and the quantity of water adsorbed by the hydrogels is related to the hydrophilicity of the chains. In this system, water is immobilized by the polymers and produces a moisturizing effect. The properties of starch-based hydrogels are affected by various parameters: starch type, vegetable source, the gelatinization process, and the presence of plasticizer and/or crosslinking additives. In cosmetic applications, hydrogels are mainly used for skin care purposes due to their important benefits, such as long residence times on the application site and reduced frequency of product administration (Mitura et al., 2020). An example of modified starch that can absorb large quantities of water is starch graft copolymer, but it still needs research considering its application in cosmetics (Taylor et al., 2007).

Starch-based hydrogels are used for the development of novel cosmetic products, mainly containing active ingredients, and applied as facial masks (sheet masks). According to Almeida et al. (2022), cassava starch-based facial masks can be used to moisturize the skin and promote antiaging performance in adding bioactive ingredients obtained from natural sources (malt bagasse), which confer antioxidant properties to the materials.

Blends of poly vinyl alcohol, starch, and bentonite were prepared using the solution casting method, with the aim of obtaining a basis for facial mask formulations. The results indicated good physical stability over a 12-week period, without skin irritation using facial masks and indicating an effective use of these compositions in facial formulations (Chakraborty et al., 2019).

11.3.3 Sensory properties modifier

The cosmetic sensory profiler is based on the texture, as one key sensory aspect, the sense of spreadability, and how the skin stays after applying the product. Consumer demand today is based on easier and more enjoyable application, without a whiting effect and sticky or heavy layer.

Native and modified starches are frequently used to enhance the sensory properties and achieve a more pleasant application of cosmetics. The principal contributions of starch in sensory effects include:

- Control of the oiliness.
- Reduction of tackiness/stickiness.
- Smooth skin after feel (give the skin a young, smooth and soft appearance).

These characteristics are beneficial for different cosmetic formulations, like compact makeup, lotions, and creams.

The main modification of starch for cosmetic applications involves the incorporation of hydrophobic groups along the backbone or cross-linking. In this way, a hydrophobic modification reduces friction, and the velvety sensory property is achieved. In addition to this effect, a superior performance of starch in water repellency and oil absorption is attained, which is essential in formulations containing high proportions of oil.

The main sensory properties of starch as a raw material in cosmetic formulations have been assessed by Vieira et al. (2020). The authors evaluated cassava starch in a gel formulation with 20% (wt.) of starch, which was related to the attributes of opacity and freshness.

11.3.4 Dusting powders

Dusting powders are thinly divided powders intended to be applied to the skin for cosmetic, therapeutic, prophylactic, or lubricating purposes. Dust elements must absorb moisture and lipids, adhere to the surface, be nontoxic, and have specific rheological properties. Starches are good absorbers and largely used as dusting/absorbing powders. Considering this, the smaller size of quinoa granules' starches (*Chenopodium quinoa*), as low that 1 μm, promotes unique applications, mainly as dusting agents. Furthermore, certain varieties of cassava starch may be used in the cosmetic industry as a dusting agent, due to their small granules (4–25 μm), such as some of the varieties evaluated by Gladys et al. (2014).

Recently, concern about the presence of asbestos in talc powder, particularly for childcare purposes, has stimulated the search for natural and safe alternatives. In this way, starch has been considered as a substitute for talc in personal care formulations, due to their safety and improved oil and moisture absorption.

11.3.5 Stabilizer

In general, starch is used as a stabilizing agent primarily in emulsions and it requires modification by inserting hydrophobic lateral chains, which enable it to act as an amphiphilic molecule. However, native gelatinized corn starch and high-amylose starch can be used as stabilizers in O/W/O double emulsions, with the aim of developing a green approach using natural starch, which is a renewable resource with high biocompatibility (Yang et al., 2021). Additionally, native granular starch can be used to stabilize Pickering emulsions, when the goal is to produce a surfactant-free system.

11.3.5.1 Particle-stabilized emulsions or Pickering emulsions

Pickering emulsion was described for S.U. Pickering (1907) and corresponds to an emulsion stabilized for solid particles, which are adsorbed in the interface of the two phases (oil and water) and avoid the coalescence of the oiled particles, stabilizing the emulsion. Particle-stabilized emulsions (Pickering emulsions) are a way to obtain a surfactant-free system; using them, it is possible to avoid skin irritation, which can sometimes be linked to surfactants, and promote a high stability against coalescence (Marku et al., 2012).

Starch granules are edible, abundant, and derived from natural sources, contrary to most other particles used to produce Pickering emulsions. Starch, therefore, represents attractive properties in terms of safety and consumer acceptance. However, native starch granules are hydrophilic and not suitable to be adsorbed at the oil–water interface of an emulsion. Thus, modification of starch molecules is needed to improve their hydrophobicity. Due the existence of three active hydroxyl groups on glucose molecules (C_2, C_3, and C_6), it is possible to perform the chemical modifications of starch to increase its hydrophobicity (Wang et al., 2020).

Besides the hydrophobic character, the size and shape of the particles influenced the stability of the emulsion. This fact has attracted interest in the cosmetic industries aimed at producing nanoparticles that are capable of stabilizing emulsions and act as carriers of active ingredients at the epidermal level. The application of nanoparticles in cosmetics is a controversial topic, despite the efforts of the FDA and other regulatory agencies to assess and identify the risks of a systemic absorption of the components.

11.3.5.2 Modified starch as emulsions stabilizers

Aluminum starch octenylsuccinate (ASO) is prepared by treating granular starch with octenylsuccinic anhydride in the presence of alkali, and then the mixture is treated with aluminum sulfate. This method enables the addition of hydrophobic side chains to the starch molecule, which reduces the tendency of the starch to associate in solution, lose clarity, and form gels. According to CIR Expert Panel (Nair & Yamarik, 2002), ASO is safe in cosmetic formulations if established limits on concentrations of heavy metals are not exceeded.

ASO is used in cosmetic formulations as an anticaking and thickener. In emulsions, the amphiphilic property of ASO-starch could stabilize the structure of the emulsion, acting as an emulsifying agent. By combining the role of ASO-starch as thickener, increasing the viscosity of the continuous phase, and emulsifying (surfactant), the cost of the final product is reduced (Choi et al., 2020; Nair & Yamarik, 2002; Tesch et al., 2002).

Tesch et al. (2002) tested ASO as a stabilizer for O/W emulsions and proved that this modified starch is a surface-active substance, acting as an effective emulsifying agent. The authors conclude that steric hindrance, resulting from repellent forces between overlapping electron clouds, is the main stabilizing effect of ASO starches in emulsions.

Starch-based vehicles containing aluminum starch octenylsuccinate or starch nanocapsules were evaluated by Marto et al. (2018). The results proved that formulations induced skin hydration and microcirculation. Based on studies involving human volunteers, the formulations did not induce any sensitization. As a result, it was concluded that these ingredients are safe for use in cosmetics at the proposed levels.

11.4 Underground starch crops for cosmetic applications

The role of starch in cosmetic formulations depends on their source and how it is modified. A global consumer movement for healthier ingredients, replacing modified starches with native ones, with the same or similar properties, highlights the potential of new starches as cosmetic ingredients with enhanced properties and obtained from a variety of botanical sources. In this way, the underground starchy crops represent an unexploited way of obtaining starch with similar or even improved properties when compared to corn starch.

In this way, Table 11.2 presents information to guide the application of new starches in cosmetics. In this sense, efforts for the commercial application of starches from natural sources must guarantee their availability in the market, at accessible prices.

Underground starch crops represent an economically important and sustainable supply of starch, with potential for a variety of applications in the cosmetic industry. Moreover, this would be a new market for starches that are already traditionally used as food. However, functional properties must be improved by chemical, physical, enzymatic modifications, and their combinations, to achieve the performance of commercially used starches.

Between the underground starchy crops from Latin America, potato and cassava starches are the most widely used and commercialized for cosmetic applications, but some potential substitutes could be found (for more information and details, we suggest reading Chapter 10 in Volume 1 of this series). Yam (*Dioscorea esculenta*) and Taro (*Colocasia esculenta*) present granules smaller than 10 μm (Cereda, 2023). Small granules are responsible for the mattifying appearance and fixation of dusting powders. Native starch extracted from mango had a great capacity of oil/sebum absorption and can be an alternative for use in dry shampoos (Table 11.2).

In general, the high viscosity and transparency of the starches from underground crops favor their application as a gel former, especially in hair products (Table 11.2). Botanical source does not interfere with these characteristics when elevated starch concentrations ($>20\%$ by weight) are used (for more details, see Chapter 10 in Volume 1 of this series).

11.4.1 Cassava starch

Cassava starch has lower gelatinization temperatures and elevated water-binding, which make this starch interesting for cosmetic applications (Li et al., 2017). Cassava starch is recommended for use in the powder form in body powders, in pressed powders, in fragranced balms, in aftershave, and after bath products and is frequently used as an oil absorbent, with a superior capacity to absorb oil compared to talc and very similar to corn starch. This character is interesting in decorative cosmetics to promote a matte appearance on the skin.

In recent advances, cassava starch was complexed with ferulic acid, producing a modified starch with good antioxidant ability, which suggests an interesting potential application in cosmetic, as antiaging formulations (Fang et al., 2020).

TABLE 11.2 Functions, required characteristics, and applications of starch for cosmetic uses.

Starch functions	Cosmetics	Role of starch in the formulation	Required characteristics	Commercial starch currently used
Thickener	Shampoo, liquid soaps, and shower gels	– Viscosity agent – Contributes to improve foam structure – Moisturizing	– High swelling capacity	Cross-linked starches
	Emulsions (creams and lotions)	– Emulsions stabilizer when added before emulsification, contributing to increase the viscosity of the formulations	– High swelling capacity	
Sensory properties modifier	Cosmetics in general	– Enhance sensorial properties if added after in granular form in emulsions – Pleasant and silky-soft touch – Moisturizing effect – Improved spreadability	– Resistance to gelatinization	Cross-linked starches
Dusting powders	Makeup	– Mattifying appearance and fixation – High oil/sebum absorption – Enhanced skin feel	– Small granules – High surface area	Native rice and corn starches
	Dry shampoos	– High oil/sebum absorption – Add volume to hair	– No perceived after-hair feel	– Native rice starch – Distarch phosphate (grainy)
Gel former	Hair styling products	– Long lasting hold – Pleasant hair feel	– High transparency – Film forming – Compatibility with ethanol	Hydrolyzed corn starch
	Shampoo, liquid soaps, shower gels	– Gel former	– High clarity	Hydrolyzed corn starch

11.5 Trends in cosmetic industry: noncommercial starches

Commercial starches are essentially obtained from corn, wheat, potatoes, and cassava, and they dominate the global starch markets for the food and pharmaceutical industries (Odeku, 2013). There is also relevant production of rice and sweet potato starches, mainly in Asia (for more information and details on starch markets, we suggest reading Chapter 3 in Volume 1 of this series).

The versatility of new botanical sources of starch represents the current innovation in the cosmetics sector, in response to an increase in demand for natural ingredients as a raw material, with properties that correspond to the requirements of new cosmetic applications. The properties of the new starches must be in accordance with the reference standards listed in the pharmacopeias (provided in section 11.6 of this chapter).

In this way, noncommercial starches have been the subject of several investigations aimed at exploring the possibility of application in cosmetics. It should be noted that the main research on alternative botanical starch sources focuses on food applications. The main applications/

studies of noncommercial starches in cosmetics are listed in Table 11.3.

Starch granules isolated from quinoa have been used as emulsion stabilizing particles, because the granules are relatively small (0.5–3 μm in diameter), reducing the amount required (mg starch per mL of oil) to stabilize an emulsion droplet interface (Matos et al., 2013; Rayner et al., 2012). Emulsion stabilized with native quinoa starch granules may have applications in the paper, cosmetic, and pharmaceutical drug formulations, in which starch is an approved excipient.

"Pinhão" is the seed of the *Araucaria angustifolia*, a pine from the south of Brazil, with about 70% (d.b.) of starch, whose granules have a granulometry of about 15 μm, with rounded and oval shapes and gelatinization temperatures between 50°C and 60°C. This starch represents a potential raw material in the field of cosmetics, to be used as a viscosity agent (thickener) and sensory enhancer (Daudt et al., 2014, 2015; Wosiacki & Cereda, 1985).

Daudt and coworkers (2015) evaluated topical gel and emulsion formulations with 3% (w/w) "pinhão" starch and observed improved sensory properties (spreadability) and increased viscosity of basic formulations. Furthermore, all formulations maintained the physical aspects and rheological behavior, with proven stability. In this study, the potential of skin-specific formulations was assessed by 10 untrained judges, women aged between 20 and 30 years. Volunteers showed that the formulations were well tolerated, without causing any visible skin irritation. Additionally, the formulations examined were appropriate for cosmetic use and fall into the range of healthy skin (pH 4.6–5.8).

Starch granules of Myrosma (*Myrosma cannifolia*) were extracted and characterized by Rincón et al. (2005) and presented a different shape mainly oval and polyhedron, with a smooth surface and size ranging between 9 and 15 μm. Compared to talc granulometry, no significant difference in the diameter of *Myrosma* starch granules was observed, which indicates a potential substitute for talc in cosmetic formulations. In this study, the safety of *Myrosma* starch for cosmetic applications was demonstrated by dermotoxicological studies.

Thanyapanich et al. (2021) extracted starch from unripe banana (*Musa* spp.) and observed oval and round-shaped granules of 10–39 μm in diameter, amylose content around 25%, and B-type crystalline structure. Banana starches have been used in cosmetic powder formulations to provide a smooth sensation and good spreading properties, and the results showed potential for talc substitution.

Despite this potential, it is not possible to assert that these starches can be commercialized. Price and availability are other factors that must be considered, as indicated by Vilpoux and Siveira Junior (2023). Table 11.3 confirms the scarcity of work with starches from roots and tubers of South American origin.

TABLE 11.3 Noncommercial starches evaluated for cosmetic uses.

Botanic name	Common name	Origin	References
Myrosma cannifolia	Guapo or cokui	Venezuela	Rincón et al. (2005)
Chenopodium quinoa	Quinoa	Peru, Bolívia, Equador (Andes)	Ahamed et al. (1996), Matos et al. (2013), Marku et al. (2012), Rayner et al. (2012)
Araucaria angustifolia	"Pinhão"	Brazil	Wosiacki and Cereda (1985), Daudt et al. (2014, 2015)
Musa spp.	Banana	Thailand	Thanyapanich et al. (2021)

11.6 Quality parameters of starch for cosmetic products

Official international standards provide a scientific basis for quality control of a product, which includes raw materials and final products. They are legally binding and served as reference to other regulation agencies, as the Brazilian Health Regulatory Agency—ANVISA. This quality parameters are applied for starches from different sources, and the attributes to be analyzed in the starches destined for pharmaceutical and/or cosmetic applications include the following:

- Identification by microscopic analysis (considering the botanical characteristics);
- pH (between 5.0 and 8.0);
- Microbial limits (described in Table 11.4);
- Loss on drying (it loses not more than 20.0% of its weight, at 130°C for 90 minutes);
- Residue on ignition (maximum of 0.6%);
- Limit of iron (limit of 10 μg of iron per g);
- Limit of oxidizing substances (not more than 1.4 mL of 0.002 N sodium thiosulfate is required);
- Limit of sulfur dioxide (not more than 50 μg per g).

Brazilian regulation (ANVISA) provides the parameters for microbiological control of personal hygiene products, cosmetics, and perfumes. These parameters are described by RDC/ANVISA N° 630/2022 standard (Agência Nacional de Vigilância Sanitária, 2022), which includes the Mercosul standard 51/1998 (Mercado Comum do Sul, 1998), and are summarized in Table 11.4. They are common to international pharmacopeias.

The starches destined for cosmetic products are sterilized (gamma irradiation; heat treatment) in most cases, which guarantee the microbiological safety of this raw material.

11.7 Final considerations

Underground starch crops represent new potential applications in cosmetics, which creates a demand for its inclusion in international compendiums that guarantee the minimum quality requirements for pharmaceutical and cosmetic ingredients.

Cosmetics Regulation states that the safety of finished cosmetic products can be assured based

TABLE 11.4 Microbiological parameters for cosmetics and raw materials.

Cosmetic products	Microbiological parameters
T-1	**1.** Total aerobic mesophilic microorganisms: not more than 10^2 UFC/g or mL, with the maximum limit of 5×10^2 UFC/g or mL; **2.** Absence of *Pseudomonas aeruginosa* in 1 g or 1 mL; **3.** Absence of *Staphylococcus aureus* in 1 g or 1 mL; **4.** IV—absence of total and fecal coliforms in 1 g or 1 mL; **5.** Absence of sulfite-reducing clostridia in 1 g (exclusively for talc).
T-2	**1.** Total aerobic mesophilic microorganisms: not more than 10^8 UFC/g or mL, with the maximum limit of 5×10^8 UFC/g or mL; **2.** Absence of *Pseudomonas aeruginosa* in 1 g or 1 mL; **3.** Absence of *Staphylococcus aureus* in 1 g or 1 mL; **4.** Absence of total and fecal coliforms in 1 g or 1 mL; **5.** Absence of sulfite-reducing clostridia in 1 g (exclusively for talc).

T-1: products for children's use; products for the eye area; products that are in contact with mucous membranes.
T-2: other cosmetic products susceptible to microbiological contamination.
Adapted from RDC/ANVISA No 630/2022 standard.

on knowing of the safety of the ingredients that they contain. Starch is an abundant, low-cost, renewable, and nontoxic natural polymer, a versatile ingredient to be used in cosmetics. The increasing demand for green and vegan cosmetics has diverted commercial attention from fossil/animal to biobased ingredients. In this way, the so-called biocosmetics has emerged, and starch represents a potential substitute for synthetic polymers.

Starch is a versatile ingredient and performs several functions in the cosmetic formulation as a gel former (hydrophilic gels), thickener, dusting powder, promoting the stabilization of emulsions or contributing to the viscosity of the product and modifying the sensory properties of the formulations. To guarantee the safety of starch to be used in cosmetics, international pharmacopeias collect and standardize the microbiological and physicochemical analysis of starch as raw material.

Underground starch crops represent a possibility of using new starches as cosmetic ingredients, with similar or even improved properties when compared to commercial starch. However, they remain poorly known and new research should be conducted to further their development.

References

Adams, W. R. M., Maibach, H., Jordan, W. J., et al. (1983). Special article A five-year study of cosmetic reactions*. *Journal of the American Academy of Dermatology, 13*(6), 1062–1069. Available from https://doi.org/10.1016/S0190-9622(85)70258-7.

Ahamed, N. T., Singhal, R. S., Kulkami, P. R., et al. (1996). Physicochemical and functional properties of *Chenopodium quinoa* starch. *Carbohydrate Polymers, 31*, 99–103.

Agência Nacional de Vigilância Sanitária (ANVISA). (2022). RDC no 630/2022 – Produtos de higiene pessoal, cosméticos e perfumes.

Almeida, L. C. M., Carvalho, F. K., & Olivato, J. B. (2022). Production and characterization of starch/PVA blends with bioactive compounds from malt bagasse. *Macromolecular Symposia, 406*(2200013). Available from https://doi.org/10.1002/masy.202200013.

Bom, S., Jorge, J., Ribeiro, H. M., et al. (2019). A step forward on sustainability in the cosmetics industry: A review. *Journal of Cleaner Production, 225*, 270–290. Available from https://doi.org/10.1016/j.jclepro.2019.03.255.

Brazilian Association of the Cosmetic, Toiletry and Fragrance Industry (ABIHIPEC). (2022). Available from https://abihpec.org.br/site2019/wpcontent/uploads/2023/01/Panorama_do_Setor_Atualizado_Ingles_VF_17_01_23-1.pdf.

Cereda, M. P. (2023). Chapter 10: Application properties of starches extracted from underground starchy crops of South America. *Starchy crops morphology, extraction, properties and applications* (pp. 245–287). Academic Press, ISBN: 978-0-323-90058-4.

Chakraborty, S., Anoop, V., George, N., et al. (2019). Physicochemical stability evaluation of cosmetic formulations of PVA, starch and MMT clay nanocomposites. *SN Applied Sciences, 1*(6). Available from https://doi.org/10.1007/s42452-019-0607-0, Springer Nature.

Choi, H. D., Hong, J. S., Pyo, S. M., et al. (2020). Starch nanoparticles produced via acidic dry heat treatment as a stabilizer for a Pickering emulsion: Influence of the physical properties of particles. *Carbohydrate Polymers*, 239116241. Available from https://doi.org/10.1016/j.carbpol.2020.116241. (April). Elsevier.

Cosmetic Ingredient Review (CIR) (2023). Available from https://www.cir-safety.org/.

Daudt, R. M., Back, P. I., Cardozo, N. S. M., et al. (2015). Pinhão starch and coat extract as new natural cosmetic ingredients: Topical formulation stability and sensory analysis. *Carbohydrate Polymers, 134*, 573–580. Available from https://doi.org/10.1016/j.carbpol.2015.08.038, Elsevier Ltd.

Daudt, R. M., Külkamp-Guerreiro, I. C., Cladera-Olivera, F., et al. (2014). Determination of properties of pinhão starch : Analysis of its applicability as pharmaceutical excipient. *Industrial Crops & Products, 52*, 420–429. Available from https://doi.org/10.1016/j.indcrop.2013.10.052, Elsevier B.V.

European Comission. (2009). European Parliament and of the Council (EC) N° 1223/2009 on cosmetic products.

Fang, K., He, W., Jiang, Y., et al. (2020). Preparation, characterization and physicochemical properties of cassava starch-ferulic acid complexes by mechanical activation. *International Journal of Biological Macromolecules, 160*, 482–488. Available from https://doi.org/10.1016/j.ijbiomac.2020.05.213, Elsevier B.V.

Gladys, D. G., Rose-Monde, M., Micaël, B. E., et al. (2014). Physicochemical characterization of starches from seven improved cassava varieties: Potentiality of industrial utilization. *Journal of Applied Biosciences, 73*, 6002–6011. Available from http://www.m.elewa.org.

Goyal, N., & Jerold, F. (2023). *Biocosmetics: Technological advances and future outlook, . Environmental Science and Pollution Research* (30, pp. 25148–25169). Springer

Science and Business Media Deutschland GmbH. Available from 10.1007/s11356-021-17567-3.

Kanlayavattanakul, M., & Lourith, N. (2015). Biopolysaccharides for skin hydrating cosmetics. *Polysaccharides: Bioactivity and biotechnology* (pp. 1867–1892). Springer International Publishing. Available from http://doi.org/10.1007/978-3-319-16298-0_29.

Larrea-Wachtendorff, D., di Nobile, G., & Ferrari, G. (2020). Effects of processing conditions and glycerol concentration on rheological and texture properties of starch-based hydrogels produced by high pressure processing (HPP. *International Journal of Biological Macromolecules, 159*, 590–597. Available from https://doi.org/10.1016/j.ijbiomac.2020.05.120, Elsevier B.V.

Lehmann, A., Volkert, B., Fischer, S., et al. (2008). Starch based thickening agents for personal care and surfactant systems. *Colloids and Surfaces A: Physicochemical and Engineering Aspects, 331*(1–2), 150–154. Available from https://doi.org/10.1016/j.colsurfa.2008.09.055.

Li, S., Cui, Y., Zhou, Y., et al. (2017). The industrial applications of cassava: Current status, opportunities and prospects. *Journal of the Science of Food and Agriculture, 97*(8), 2282–2290. Available from https://doi.org/10.1002/jsfa.8287, John Wiley and Sons Ltd.

Lochhead, R. Y. (2007). The role of polymers in cosmetics: recent trends. Available from https://pubs.acs.org/sharingguidelines.

Marku, D., Wahlgren, M., Rayner, M., et al. (2012). Characterization of starch Pickering emulsions for potential applications in topical formulations. *International Journal of Pharmaceutics, 428*, 1–7. Available from https://doi.org/10.1016/j.ijpharm.2012.01.031, Elsevier B.V.

Marto, J., Pinto, P., Fitas, M., et al. (2018). Safety assessment of starch-based personal care products: Nanocapsules and pickering emulsions. *Toxicology and Applied Pharmacology, 342*, 14–21. Available from https://doi.org/10.1016/j.taap.2018.01.018, Academic Press Inc.

Matos, M., Timgren, A., Sjöö, M., et al. (2013). Preparation and encapsulation properties of double Pickering emulsions stabilized by quinoa starch granules. *Colloids and Surfaces A: Physicochemical and Engineering Aspects, 423*, 147–153. Available from https://doi.org/10.1016/j.colsurfa.2013.01.060, Elsevier B.V.

Mercado Comum do Sul (MERCOSUL). (1998). Resolution GMC MERCOSUL nº 51/1998.

Mitura, S., Sionkowska, A., & Jaiswal, A. (2020). Biopolymers for hydrogels in cosmetics: Review. *Journal of Materials Science: Materials in Medicine, 31*(6). Available from https://doi.org/10.1007/s10856-020-06390-w.

Nair, B., & Yamarik, A. T. (2002). Cosmetic ingredient review expert panel. Final report on the safety assessment of aluminum starch octenylsuccinate. *International Journal of Toxicology, 21*, 1–7. Available from https://doi.org/10.1080/1091581029009637.

Odeku, O. A. (2013). Potentials of tropical starches as pharmaceutical excipients: A review. *Starch/Staerke, 65*(1–2), 89–106. Available from https://doi.org/10.1002/star.201200076.

Peppas, N. A., Bures, P., Leobandung, W., et al. (2000). Hydrogels in pharmaceutical formulations. *European Journal of Pharmaceutics and Biopharmaceutics, 50*(1), 27–46. Available from https://doi.org/10.1016/S0939-6411(00)00090-4.

Qiu, C., Hu, Y., Jin, Z., et al. (2019). A review of green techniques for the synthesis of size-controlled starch-based nanoparticles and their applications as nanodelivery systems. *Trends in Food Science and Technology*. Available from https://doi.org/10.1016/j.tifs.2019.08.007, Elsevier Ltd.

Rayner, M., Timgren, A., Sjöö, M., et al. (2012). Quinoa starch granules: A candidate for stabilising food-grade Pickering emulsions. *Journal of the Science of Food and Agriculture, 92*(9), 1841–1847. Available from https://doi.org/10.1002/jsfa.5610.

Rincón, A. M., Pérez, R. M. N., de, Reyes, A., et al. (2005). Guapo' (*Myrosma cannifolia*) starch : A natural product with potential use in cosmetic formulations. *International Journal of Cosmetic Science, 27*, 107–114.

Taylor, P., Athawale, V. D., & Rathi, S. C. (2007). 'Graft polymerization : starch as a model substrate graft polymerization. *Journal of Macromolecular Science, Part C Polymer Review*, 37–41. (June 2015).

Tesch, S., Gerhards, C., & Schubert, H. (2002). Stabilization of emulsions by OSA starches. *Journal of Food Engineering, 54*(2), 167–174. Available from https://doi.org/10.1016/S0260-8774(01)00206-0.

Thanyapanich, N., Jimtaisong, A., & Rawdkuen, S. (2021). Functional properties of banana starch (Musa spp.) and its utilization in cosmetics. *Molecules (Basel, Switzerland), 26*(12). Available from https://doi.org/10.3390/molecules26123637, MDPI.

UNION TEPATCOTE. (2009). Regulation no 1223/2009 on cosmetic products. In: Official Journal of the European Union.

U.S. Food & Drug Administration (FDA). (2022). Federal Food, Drug, and Cosmetic Act. Sections 301 - 399d, Short Title and Definitions.

Vieira, G. S., Lavarde, M., Fréville, V., et al. (2020). Combining sensory and texturometer parameters to characterize different type of cosmetic ingredients. *International Journal of Cosmetic Science, 42*(2), 156–166, Blackwell Publishing Ltd.

Vilpoux, O. F., & Siveira Junior, J. F. S. (2023). Global production and use of starch. *Starchy Crops Morphology, Extraction, Properties and Applications* (pp. 43–66). Academic Press.

Wang, X., Huang, L., Zhang, C., et al. (2020). Research advances in chemical modifications of starch for hydrophobicity and its applications: A review. *Carbohydrate*

Polymers, 240(16). Available from https://doi.org/10.1016/j.carbpol.2020.116292.

Wosiacki, G., & Cereda, M. P. (1985). Characterization of pinhão starch. Part I. Extraction and properties of the starch granules. *Starch-Stärke, 37*(7), 224–227.

Yang, J., Gu, Z., Cheng, L., et al. (2021). Preparation and stability mechanisms of double emulsions stabilized by gelatinized native starch. *Carbohydrate Polymers, 262*117926. Available from https://doi.org/10.1016/j.carbpol.2021.117926, Elsevier Ltd.

CHAPTER

12

Spray-dryer and microencapsulation—application of starch as a carrier

Marney Pascoli Cereda[1,2]

[1]Retired from São Paulo State University, Botucatu, São Paulo, Brazil [2]Tecsapp Solutions, Campo Grande, Mato Grosso do Sul, Brazil

12.1 Introduction

Drying is characterized by the removal of humidity to desired values, so that the stability of the product is guaranteed for an adequate period, under preestablished conditions. Drying presents a series of advantages over other forms of conservation, mainly by reducing weight and the possibility of storage at room temperature. The principle of conservation by drying is the reduction of water and water activity (a_W), which hinders or prevents the development of microorganisms and/or reduces the activity of enzymes. The reduction of water activity occurs not only by removing water, but also by increasing the concentration of solutes.

Dehydration to a very low level of humidity also has several drawbacks. Since the material becomes hygroscopic, that is, absorbs moisture, it must be well protected during storage. There are packaging materials suitable for this type of use, but another problem is gas penetration, due to the porous texture of the dehydrated product. The biggest problem occurs with the entry of oxygen, which can cause oxidation reactions even at low water activities. Besides the oxidation of fatty substances, which causes rancidity, color and flavor alterations are common. To avoid or circumvent this problem, it is possible to use nitrogen when closing the final package.

Use of starch as a carrier for spray-dehydrated products is a market that has been largely undervalued by starch companies, especially those of tropical origin, such as cassava. This chapter provides the basics of the spray-drying process to justify the characteristics of starch as a carrier, as well as the main characteristics of starch carriers. The choice of the carrier is a cost/benefit problem, but in general, the best results are obtained with enzymatically produced maltodextrins (for more details on the production of maltodextrins, please refer to Chapter 5).

Spray-drying is widely used in the food, chemical, pharmaceutical, and cosmetic industries to transform liquids into powders. In the food industry, spray-drying prolongs the shelf life and concentrates foods, making storage and transportation easier. A major advantage of spray-drying compared to other drying techniques is that agglomerated powders can be

Starch Industries
DOI: https://doi.org/10.1016/B978-0-323-90842-9.00006-6

produced with excellent functional properties such as flowability, reconstitution behavior, bulk density, and mechanical stability (Eijkelboom et al., 2022). Atomization can also generate microencapsulation. In this case, the carrier agents can protect sensitive food components, such as carotenoids, preserve flavors, and reduce volatility and reactivity (Etzbach et al., 2020).

Regardless of the drying purpose, it is important to choose the carriers with the best performance. Starch can be used in its native form, but derivatives such as maltodextrins, other modified starches, and even cyclodextrins are preferred. Because of their importance, cyclodextrins will be treated in the upcoming Volume 5 of this series, as they can be obtained from cassava bran, which can valorize the residue and reduce its cost.

12.2 Spray-dryer dehydration system

The principle of the spray-drying method is to produce small-sized particles via atomization through a hot gas current and solvent removal (Baltrusch et al., 2022; Nandiyanto & Okuyama, 2011). Water removal by spray-drying is the most widely used practice in the food industry to ensure the microbiological stability of products. It helps to obtain a product with specific functional properties and avoids many of the risks of chemical and/or biological degradation (Bakry et al., 2016). The advantage of a spray-dryer is to allow for the preparation of various particles with a simple process at a high rate of production (Nandiyanto & Okuyama, 2011). According to Domingues et al. (2002), the dehydration by atomization when well conducted, generates a product of greater nutritional value and stability. It is also versatile in its use and can be used for flavoring and coloring. The success of spray-drying can be explained because it is not only efficient, but also economic. Spray-drying is rapid and reproducible, allowing for easy scale-up (Bakry et al., 2016). It has a cheap large-scale production, ease of change in operating conditions, and availability both under laboratory and industrial scales (Baltrusch et al., 2022).

Gava (1978) classifies the types of dehydrators into adiabatic and solid surface heat transfer. Spray-dryer-type dehydrators, or atomizers, are adiabatic, in which heat is conducted by means of hot air. This group also includes cabin dryers, tunnel dryers, fluidized bed dryers, ovens, flash dryers, puff dryers, and foam mat dryers. The author describes atomizers as equipment where drying is done by spraying in a continuous process and where a liquid or paste is transformed into a dry product, characterized by a relatively shorter drying time than in other dryers. The process consists of atomizing the liquid in a chamber in which a flow of hot air circulates. The rapid evaporation of the water allows the temperature of the particles to be kept low, so that the high temperature of the drying air does not affect the product too much. The spray-dryer is used in the food industry to produce powdered milk, instant coffee, and so forth, but also serves the pharmaceutical, detergent, and other industries. Fig. 12.1 presents an overview of the operation scheme of a spray-dryer.

According to Gava (1978) and Masters (1991), dehydration with artificial heat is a function of time, temperature, and ventilation, parameters that should be carefully controlled. For Nandiyanto and Okuyama (2011), optimization can be achieved by managing the geometry of the chamber, the carrier gas flow rate, the pressure, and the temperature process. Gava (1978) indicates that the different phases of atomization interfere in the characteristics of the final powder product. The way of atomization and the properties of the atomized liquid influence the size of the solid particle, its density, appearance, and humidity. The type of contact of the liquid with the hot air and the evaporation influence the density of the powder, appearance, moisture, aroma, and flavor

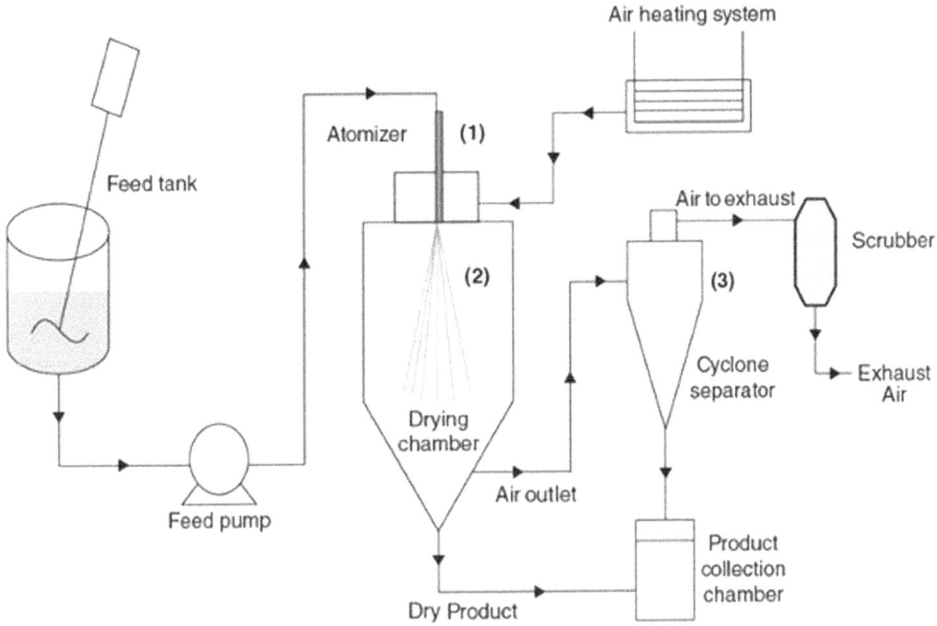

FIGURE 12.1 Schematic representation of the spray-drying mechanism: (1) atomization, (2) droplet-to-particle conversion, (3) particle collection (Santos et al., 2018). *Figure from Santos, D., Maurício, A.C., Sencadas, V., Santos, J.D., Fernandes, M. H. and Gomes, P.S. (2018). Spray drying: An overview.* Biomaterials-physics and chemistry-new edition, *IntechOpen, pp.9–35.*

retention. The technique of separating the powder from the dry air, in turn, influences the particle size of the dehydrated product (Gava, 1978; Nandiyanto & Okuyama, 2011).

Masters (1991) describes the factors involved in the movement of the droplets in the drying chamber and points out that it depends on several factors, the main one being the design of the atomizer. The droplet trajectory follows the design of the atomizer and is affected initially by the swirl around the rotating atomizer, which in turn is caused by the rotation of the atomizer and finally by the drying airflow. This effect is smaller in industrial equipment where the influence of the rotor decreases rapidly with radial distance but is greater in pilot equipment. At the industrial level, the airflow and droplet trajectory are governed mainly by the air disperser. Thus, droplet trajectory depends on the design of the rotary atomizer, feed rate, design, and location of the air disperser. Three types of droplets can be distinguished, those that are ejected from the atomizer, those that are decelerating, and those that are free falling or falling under the influence of the chamber drying airflow.

Among the heat transfer vehicles, air is one of the most widely used, because of its abundance and convenience and because its control of food heating presents no major problems. When air is used, no moisture recovery system is necessary, as is the case with other gases. Air conducts heat to the product, causes water to evaporate, and is also the carrier of the released humid vapor. Five to seven times more air is needed to conduct heat to the product than to transport the removed vapor. The volume of air required to evaporate a given mass of water will depend on the temperature. The air velocity is also dependent on the nature of the product. The most convenient speed is variable depending on the type of dehydrator and can range from 90 to 300 m/min. The evaporation speed of the water in the product is directly proportional to the surface area and porosity of the product but also depends on the air speed.

For Gava (1978) and Nandiyanto and Okuyama (2011), the different steps of drying with spray-dryer are the atomization of the liquid, the contact of the liquid with hot air, droplet-to-particle conversion (solvent evaporation, most of the time water) and separation of the powder product from the drying air.

Atomization is a key parameter in determining particle size. In this step, the initial liquid/slurry is fed into specific atomization equipment for conversion to droplet form. Gava (1978) reminds us that spraying the liquid in the drying chamber can be done by disks or atomizing nozzles. In the first case (centrifugal system), a disk with grooves rotates at high speed and pulverizes the liquid while projecting the droplets radially to the hot airflow that enters through the air disperser located in the upper part of the chamber. The chamber is usually cylindrical at the top and conical at the bottom (Fig. 12.1). Atomization through special nozzles can be done by high-pressure pumps or by pneumatic systems (compressed air). Heating the air can be done by indirect contact, as in pipes with flaps heated by steam, oil, or an electrical system, or by direct burning of gas, oil, or other fuel. The way the hot air encounters the atomized liquid is very important in defining the characteristics of the final product.

Because a single droplet produces one particle, the size of the droplet and the amount of solvent in the droplet influence the final synthesized particle diameter (Nandiyanto & Okuyama, 2011). Thus, for the authors, it is possible to control the outer diameter of the particle, by changing the droplet size and adjusting the droplet concentration. Gava (1978) and Masters (1991) explain that the hot airflow is usually introduced through the air disperser located at the top of the chamber. The sprayed liquid can have the same flow as the hot air, in this case forming a parallel flow where contact occurs between the wetter product and the hotter air. This system is used for products that are more sensitive to heat. In countercurrent flow, the liquid is sprayed in the opposite position to the hot air inlet, so that there is contact of the dryer particle with the hotter air. This type of flow is more efficient in the use of heat and is indicated for thermoresistant products. Some atomizers use a mixed system, combining the two types described above.

When the hot air, at a temperature of 180°C–230°C, meets the humid particle, there is an almost instantaneous dehydration of the droplet and evaporation of the water in the drying chamber, in the form of a mist. The drying air exits through the lower part of the apparatus at a temperature of 60°C–100°C, passing through cyclones to recover the fine particles by gravity and allowing the clean air to exit to the atmosphere through chimneys (Fig. 12.1). The construction of the chamber and the working conditions are adjusted to allow for the evaporation of the water without raising the product temperature too much. The separation of the dry product from the drying air has a great influence on the characteristics of the final powder, due to the mechanical handling used in this separation. Excessive mechanical handling results in finer powder (Masters, 1991).

Fig. 12.2 shows two types of particles that can be formed from the product that is to be dehydrated on contact with hot air, carrier agglomerations with the solid part of the product, and encapsulated particles. The type of sprinkler nozzles is one of the factors responsible for the particle size distribution in the dehydrated product.

According to Gava (1978) and Masters (1991), atomizers use two basic separation systems, the main separation of the powder

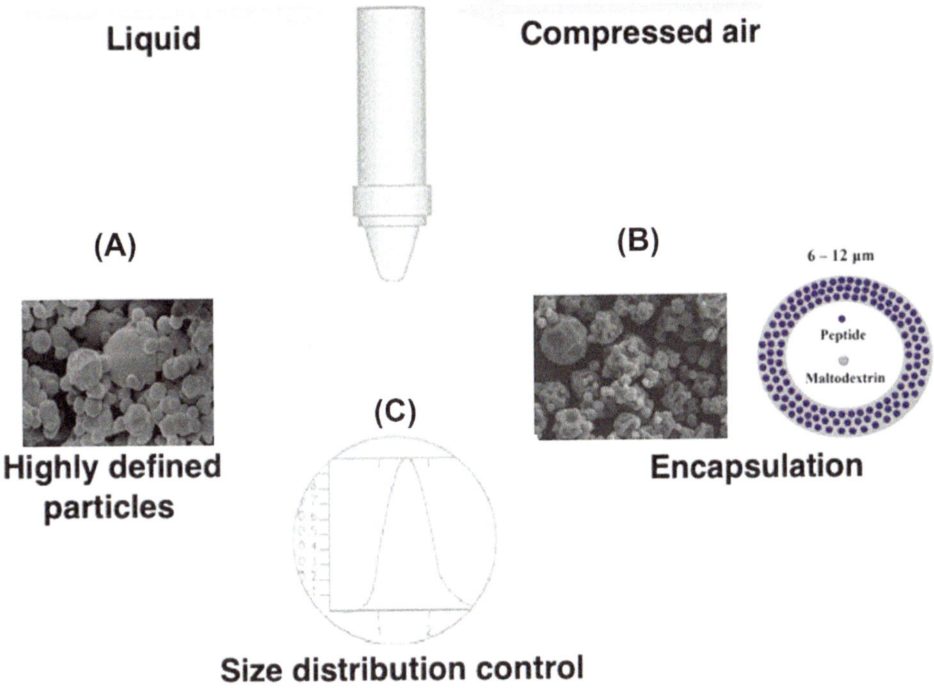

FIGURE 12.2 Conceptual drawing of dry products with maltodextrin carrier particles (A), flaxseed protein hydrolysates microencapsulated with maltodextrin (B) and controlled particle size distribution (C). *Based on Akbarbaglu et al., 2019; Cereda & Daiuto, 2003; Etzbach et al., 2020.*

product that occurs at the bottom or base of the drying chamber and the separation of the finest powder, which occurs only in separation systems, or cyclones (Fig. 12.1). The drying air always carries fine particles that are recovered in separation systems, usually cyclones. In this case, a greater number of cyclones may be necessary, from 3 to 8, depending on the type of atomizer. The finer powder collected in the cyclone may or may not be mixed with the powder obtained at the base of the atomizer.

Because it is a very fast drying, with the use of comparatively low temperatures, atomization is a suitable preservation method for dehydrating heat-sensitive food ingredients (Etzbach et al., 2020). Nevertheless, the incorporation of carriers is an important part of the process.

12.3 Importance of carriers in the drying process

A carrier aims to facilitate the drying process, improve the quality of the final product, and reduce the costs of the process. It is also used in the microencapsulation process, which is discussed in section 12.4.

12.3.1 Use of carriers in fruit juice dehydration with spray-dryer

Atomization dehydration is feasible for most products in suspension or solution, with a minimum solid concentration of 20%, which requires the use of carriers. Its use is a matter of cost and availability. Since atomizers are expensive equipment, in general their use must be continuous to compensate for the costs. This is also an advantage, because the same device can dehydrate, for example, eggs, fruit juice, and milk. For discontinuous use, it is worthwhile to outsource the processing.

According to Brennan et al. (1971), dehydration is one of the most difficult of all available techniques to be applied to fruit juices. Juices are highly sensitive to heat and have high content of hygroscopic substances such as sugars, which can pose problems when it comes to controlling drying time, removing the product from the drying zone, and handling the final product. Etzbach et al. (2020) mentioned that the conversion of sugar-rich juices into powders is often accompanied by problems of stickiness because of the hygroscopicity and thermoplasticity of the products promoted by the high humidity and temperature of the drying air. Stickiness is mainly caused by the low glass transition temperature (T_g) of the sugars present in fruits such as glucose, fructose, and sucrose. Therefore, spray-drying of fruit juices needs the addition of carrier agents to increase the glass transition temperature of the juice.

In juice dehydration, it is likely that only tomato juice can be atomized without the use of additives but using low temperature and special atomizers. For most other juices, spray-dryer dehydration without additives requires the installation of special devices in the drying chamber, such as cooled walls, secondary air coolers, and dispersers, recycling of the dried product, and induction of crystallization in the feeder. To avoid the installation of such devices, it is possible to add carriers in the feed liquid, sometimes in amounts exceeding 75%, calculated on the total solids content (Masters, 1991).

Because of the high proportion of carriers, the characteristics and quality of dehydrated products are strongly influenced by the type of material used. For Etzbach et al. (2020), the most used carrier agents are maltodextrin and gum arabic. Starches as carriers range from the simplest, such as native starches, to the most complex and expensive, such as cyclodextrins. They are used mainly because of their price, performance, and availability

When different sugars such as fructose, glucose, sucrose, lactose, maltose, glucose syrups, and maltodextrins are compared in their performance as carriers for atomization, the best products are those with higher glass transition temperatures (T_g) and melting temperatures (T_m), as already mentioned by Etzbach et al. (2020). These parameters are related to the chain length of the carbohydrates. When used as atomization carriers for orange juice, maltodextrins of low dextrose equivalent (DE) have higher melting temperatures than maltodextrins of higher DE at the same concentrations (Bhandari et al., 1993; Tsourouflis et al., 1976).

The literature mentions the use of high temperatures in the fruit juice drying system, even considering the cooling that occurs with the evaporation process. On the other hand, the drying time is much shorter than in other processes and heating damage can be reduced by controlling the residence time of the dried particles in the hot zone of the dryer. This control is difficult to do because of the extreme hygroscopicity of these products, which, combined with the thermoplastic nature, causes adherence on the dryer walls and difficulties in handling. The two possibilities to reduce these difficulties are the use of additives as a drying carrier and the use of equipment specially designed to facilitate the removal of the product from the drying chamber. Maltodextrins and glucose syrups are the usual carriers for orange, lemon, peach, and apricot juices, as reported by several authors (Bhandari et al., 1993; Gupta, 1978; Lloyd & Nelson, 1984; Masters, 1991).

The presence of acidic substances in the products to be dehydrated is also important. For example, orange juice is more difficult to process than the same materials with lower acidity content. Other components such as soy proteins, pectin, and hemicellulose can act as structural elements in powdered juices, as stated by Mizrahi et al. (1967), Bhandari et al. (1993), and Maltini and Mastrocola (1999). Gum arabic, an encapsulation agent, is useful for preserving flavors in powdered products (Bhandari et al., 1992; Bhandari et al., 1993; Brennan et al., 1971) and can be used in the dehydration of flavored fruit juices.

To dry a certain fruit juice, its initial composition, the way of processing, and the final use of the powdered product should be considered. The choice of technique and additives should be linked to economic considerations. For example, freeze-drying is a very expensive process, although it offers very good results (Maltini & Mastrocola, 1999).

According to Brennan et al. (1971), the selection of the best carrier should take into consideration a series of factors that involve product quality, price, and performance in the process. The authors mention the difficulties to obtain a quality product without the use of a carrier and that they could not dehydrate orange juice with 40% of solids without using a carrier, due to the adherence of the product on the dryer wall. Another problem is that the yield of the process without a carrier was very low, from 1% to 2%, calculated over the total feed solids, even considering the material retained on the walls that was difficult to remove. The product was hygroscopic and difficult to handle.

12.3.2 Starch and derivatives as a carrier in dehydration with a spray-dryer

Among the carriers from starch and derivatives, the following can be mentioned:

- Starches (native or granular): Presence of white starch granules, neutral flavor, with blue coloring with Lugol, pH between 5.5 and 6.0, acidity of 1.0–1.5 mL NaOH.100/g and DE < 2. In this group are available commercial products from corn, cassava, potato, and wheat.
- Dextrins: Dextrins can be separated into two groups, depending on the production process.
- Pyrodextrins or acid dextrins: Cream to yellow powder with a neutral taste. Presence of broken starch granules, with violet to red color reaction with Lugol, pH between 2.0 and 5.0, acidity between 2.0 and 3.0 mL of NaOH.100/g and DE > 2 and < 30. There are dextrins for industrial use (glues and adhesives—see chapter 15) and food use.
- Enzyme dextrins: White powder product, neutral taste, and absence of starch granules, but with violet to red coloration reaction with Lugol. The pH should be between 4.5 and 6.0, and the acidity between 1.0 and 2.0 mL NaOH.100/g and DE > 5 and < 30.
- Maltodextrins (enzymatic): White to slightly creamy powder, neutral to slightly sweet taste, absence of starch granules, but with a violet to yellow coloration with Lugol, pH between 4.5 and 5.5, acidity between 0.5 and 2.0 mL NaOH.100/g and DE > 5 and < 20, with low maltose (< 5%) and glucose (< 3%) contents. Maltodextrins can be considered as pregelatinized starches slightly hydrolyzed by enzymes.
- Cyclodextrins (enzymatic): White to slightly creamy colored powder product, neutral flavor, absence of starch granules, but with reddish coloration or the typical yellow of iodine solution in potassium iodide (Lugol color), with pH between 7.0 and 7.5, acidity between 0.5 and 2.0 mL of NaOH.100/g, and DE > 2 and < 8. Since cyclodextrins have zero DE (no reducing ends), the DE of commercial products is due to the presence of impurities such as dextrins and reducing sugars (glucose and maltose).
- Syrups: Golden-yellow, transparent liquid products (also available in dehydrated form) and yellow color reaction with Lugol. The pH is between 4.5 and 6.0, acidity between 0.5 and 2.0 mL of NaOH.100 g^{-1}, and DE > 20, in the presence of various maltose and glucose molecules and a sweet taste. For production of starch derivatives by hydrolysis (dextrins, maltodextrins, and syrups), see Chapter 5.

Dextrins and native starches are widely used as carriers, but mainly for lower priced dehydrated products where quality is not the most important factor, such as dehydration for animal feed components. Although it is more appropriate to use maltodextrin in products where quality is important, especially to maintain flavors and aromas, it is relatively frequent that in Brazil, many dextrins are sold as if they were maltodextrins. In some cases, they are not products of enzymatic hydrolysis, and it is not even known if they are really food grade. Pyrodextrins have a very acidic pH and are very colorful, interfering with the natural color of dehydrated products. In higher value-added products, mainly for pharmaceutical use and in some fruit juices, cyclodextrins can be used.

Some companies market ready-made mixtures for certain uses, which are called systems. These systems can include mixtures of maltodextrins, glucose syrups, gums, cyclodextrins, among others and are developed to meet specific markets. In the case of Glucidex, the company Roquette Frères, France, reports that it is a non-GMO corn nondemineralized maltodextrin of DE 19 with organic certified, neutral taste, and white power (Glucidex, 2023).

In South America, the availability of tropical fruit juices is large, which requires forms of preservation. Among them, conservation of powdered juices processed by dehydration is the most practical, because it is water removal. The one that guarantees the best quality is freeze-drying, which unfortunately is also the most expensive. Spray-drying is an intermediate solution, and in this case, the use of starch-based carriers is predominant (Azis, 2019).

12.3.2.1 *Use of glucose syrups*

Brennan et al. (1971) reported the evaluation of the use of dextrins, maltodextrins, and low DE glucose syrup, all carriers of the starch family, in fruit juice dehydration. In orange juice, these carriers provided good product flowability and low deposits on the dryer walls, but a panel of tasters found the flavor of the products unacceptable. Liquid glucose syrup with DE 34–43 provided a more acceptable product. The best feed mixture contained 52.50% total solids, being 29.85% from the orange juice and 23.65% from the glucose syrup. The atomizer speed was 31,000 rpm, the airflow rate was 48 m^3/h, and the percentage of insoluble solids in the feed material was 0.07.

The products tested showed good quality and flowability with acceptable aroma and flavor. The highest feed rate was possible at an inlet air temperature of 140°C. At temperature below 130°C, when the feed rate increases much above 15 g/min, deposition occurs on the chamber walls. At inlet air temperature of 150°C, with a feed rate greater than 20 g/min, the deposition on the walls increases greatly.

For a fixed inlet air temperature, there was a tendency for the product moisture to decrease with increasing feed liquid temperature. There appeared to be no correlation between dehydrated product particle size and compacted density, nor between these two properties and inlet or feed air temperature. Color measurements in the reconstituted products revealed reduced yellow and saturation values with increasing inlet air temperature, indicating loss of carotenoids. Further trials using a wider range of inlet air temperatures, from 120°C to 155°C, at a constant feed rate of 15 g/min and feed temperature of 20°C, provided results that corroborated the observations already cited regarding compacted density, particle size, and insoluble solids. There was a tendency for the compacted density to decrease with increasing inlet air temperature. Particle size remained at expected values of 22.5–30.5 μm (Brennan et al., 1971).

12.3.2.2 *Use of maltodextrins*

In Brazil, maltodextrins are widely used for dehydration of products such as fruit juices, chocolate powder, pharmaceutical products, and so forth. However, due to buyers' lack of knowledge, part of the maltodextrins sold in Brazil are in fact dextrins and often pyrodextrins (for more information on the differences between enzymatic dextrins and maltodextrins, see Chapter 5). True enzymatic maltodextrins have almost 10% reducing sugars like glucose and maltose, which is not always found in maltodextrins marketed in Brazil for carrier use.

Zheng, Jin, and Zhang (2007) indicated that carrier hygroscopicity is a factor affecting the quality and durability of a spray-dryer dehydrated product. The authors evaluated several cassava starch derivatives: maltodextrin, cross-linked maltodextrin, octenyl succinic maltodextrin, and cross-linked octenyl succinic maltodextrin, all with DE 10.2. They highlighted the hygroscopicity property that gradually increased as the DE and relative humidity (RH) increased. The hygroscopicity of four types of maltodextrins, at 25°C and RH 85%, were 13.42%, 9.89%, 11.80%, and 8.91%, respectively. It was observed that cross-linking and esterification decreased hygroscopicity by 26.5% and 12.1%, whereas the combination of them reduced it by 33.7%. The surface activities of cassava maltodextrins were estimated from the surface tension measurement. At 5%, the air–water surface tensions of maltodextrin, cross-linked maltodextrin, octenyl succinic maltodextrin, and cross-linked octenyl succinic maltodextrin solutions were 67.24, 64.61, 52.93, and 52.21 mN/m, respectively. The authors concluded that for cassava starch, the influence of esterification on surface activity was more effective than that of cross-linking.

Maltodextrins are especially suitable for dehydration of fruit juices but are more expensive carriers. Bhandari et al. (1993) performed a series of experiments to maximize the amount of fruit juice concentrate relative to the amount of maltodextrin. The results showed that the more fruit juice concentrate is used in the mixture with maltodextrin, the more difficult it is to dry. They also showed that stickiness is closely related to the moisture content of the dried product. In the experiments performed, the authors compared two models of spray-dryers in the preparation of concentrated black currant, apricot, and raspberry juice powders with different maltodextrins as a carrier for dehydration. One of the equipment was from Niro Atomizer, Copenhagen, model Minor Lab, with a centrifugal spreader type atomizer (100 mm diameter) driven by an air turbine at a speed of 35,000 rpm. The other was the Leaflash 100 manufactured by Sante Aoustin, Rouen, France. In this equipment, hot air at a pressure of 1.3 bar was provided by a helical fan and then accelerated in a converging tube before colliding with the feed suspension, where it was atomized and dried. The introduction of cold air at the base of the conical part of the Leaflash type dryer produced rapid cooling of the powder, prior to separation (Bhandari et al., 1993).

Also, according to Bhandari et al. (1993), the control parameters of the experiments were feed rate, initial solids concentration of the feed suspension, atomizer speed, pressure, and temperature of the inlet and outlet airflow of the equipment.

As the use of carrier was considered unavoidable, Bhandari et al. (1993) tried to maximize the proportion of fruit juice to the amount of maltodextrin. The maltodextrin evaluated was from the Roquette company, with DE of 6, 12, and 19 and a moisture content of 4%. The juice concentrate was added under stirring. The solids ranged from 35% to 50% for the Niro equipment and from 30%, 50%, and 60% for Leaflash. The best results were obtained for maltodextrins of DE 6, with juice/maltodextrin ratios of 65/35 for black currant, 60/40 for apricot, and 55/45 for raspberry. The authors considered that the higher the amount of carrier in relation to the fruit juice, the higher the drying yield.

According to the authors, in the case of black currant juice, 65% was the maximum total solids in the feed suspension that could be dried in the Leaflash equipment. For the Niro equipment, this concentration dropped to 50%. The temperature control on the Leaflash wall and the possibility of introducing cold air at the outlet facilitated the recovery of the atomized product.

According to Bhandari et al. (1993), for the various maltodextrins evaluated, the melting temperature or stickiness temperature varied from 100°C to 170°C. The higher this temperature, the more stable the dehydrated product. However, the higher the juice concentration, the lower the melting temperature, which impairs the drying of the product. The maltodextrin of DE 6 was selected because it provides the highest juice concentration, which means a critical temperature of just over 120°C, above which the product can start to become sticky. The lower the DE of the maltodextrin, the higher the melting temperature. The inlet and outlet temperatures in the atomizer are the main variables that control the moisture in the dry product. The nature and concentration of solids in concentrated juices and the type of carrier should also be considered in relation to fruit juices. The ratio of fruit juice to low DE maltodextrins should be increased, as it provides a reduction in the inlet air temperature.

Bhandari et al. (1993) stated that the variation in the ideal proportion of maltodextrin to fruit juice may result from the nature of the sugars present in the juices. In comparative studies with dehydration of aqueous sugar solutions (50% total solids content), results with different behavior were observed for different sugars. The fructose solution required a proportion of at least 50% maltodextrin from DE 12, while the glucose solution required only 35% maltodextrin and sucrose was dehydrated without maltodextrin.

Besides sugars, acids are always present in fruit juices. It has been observed that tartaric, citric, and malic acids, in concentrations above 10% of total solids, cause difficulties in the recovery of the powdered product (Bhandari et al., 1993). When the dehydration of black currant juice was compared to that of apricot juice, with the same total sugars, the fruit juice/maltodextrin ratios were 65/35 and 58.3/41.7, respectively, with the best atomization observed for the apricot juice. This is probably due to the lower fructose and acidity content for this juice, 0.44 versus 0.66 and 0.22 versus 0.32 respectively.

Francisoni et al. (2002) evaluated the effect of Ingredion MOR-REX 1910® maltodextrin concentration and atomization speed on the physical properties of dehydrated passion fruit juice. The spray-dryer operated with an inlet temperature of 190°C and outlet temperature of 90°C, air speed of 25,000 rpm, feed flow rate of 0.18 kg/min, and with varying speeds of the atomizing rotor. The authors concluded that increasing the maltodextrin concentration significantly increased the particle density and the compacted density in the dehydrated juice. At the same time, it reduced the moisture content and the color Lab parameters. The final product became lighter (L%) and the yellow hue lower (b+). On the other hand, the atomization speed did not significantly influence the physical properties, which is why the authors recommended drying at the lowest atomization level and using an intermediate concentration of maltodextrin only to increase the density, because too high a level compromises the color of the product.

Domingues et al. (2002) established the physical properties of pineapple (*Ananas comosus*) juice in spray-dryer dehydration optimized by response surface analysis. Maltodextrin (MOR-REX 1910) was added to the whole pineapple juice at different concentrations. The juice with these treatments was atomized on a semi-industrial scale, operating with the same parameters as for passion fruit juice (Francisoni et al., 2002). Maltodextrin concentration and atomization speed significantly influenced the texture of the juice powder. The authors observed that variation in atomization speed influenced the variation in dehydrated density and that increasing the speed reduced the particle density of the juice. In general, solubility and brightness improved with the reduction in maltodextrin percentage and with the increase in atomization speed. In turn, the reduction in maltodextrin concentration and atomization speed increased the percent moisture and particle density of the pineapple juice.

12.4 Microencapsulation

Microencapsulation is a technique by which solid, liquid, or gaseous active ingredients are packaged within a second material for the purpose of protecting or shielding the active compound from the surrounding environment. Thus, the active compound is designated as the core material, whereas the surrounding material forms the shell. This technique can be employed in a diverse range of fields such as agricultural, chemical, pharmaceutical, cosmetics, printing, and food industry (Raybaudi-Massilia & Mosqueda-Melgar, 2012).

Microencapsulation is a process where an encapsulating agent (the wall) builds a functional barrier around a sensitive compound (the core) to avoid chemical and physical reactions, to maintain the biological, functional, and physicochemical properties of core materials and to provide controlled release dynamics and new properties depending on the carrier material used (Bakry et al., 2016; Baltrusch et al., 2022). For Bakry et al. (2016), the core material is gradually diffused through the capsule walls, thereby offering controlled release properties under desired conditions. Microencapsulation can be defined as a process of building a functional barrier between the core and the wall material to avoid chemical and physical reactions and to maintain the biological, functional, and physicochemical properties of the core materials.

I Ré (1998) identified the major reasons for the use of microencapsulation:

- Protection of the product from the surrounding environment (temperature, moisture, UV radiation, and interaction with other materials).
- Protection of the environment from hazardous or toxic products;
- Decrease of the evaporation or transfer rate of the core material to the outside environment;
- Dry handling—conversion of liquids and sticky solids to free-flowing powders (better mixing, prevent lumping);
- Masking of the undesired properties of the active component, for example, taste or odor masking, or masking the chemical properties such as pH or catalytic activity; and
- Control of the rate of release of the core material under desired conditions, for example, transport of a therapeutic agent selectively to the intended site of action to optimize the biological response, or the release of an active molecule into a selected environment following a predetermined kinetic profile.

In the food sector, the conversion of liquids to dry powders which are easy to handle was one of the first motivations for the use of microcapsules. For food, advantages of microencapsulation are the protection of sensitive food components from other food ingredients during storage and protection against nutritional loss to incorporate unusual or time-release mechanisms into the formulation, to mask or preserve flavors and aromas, and to provide an additional attractiveness for the display and merchandising of food products (greater flexibility and control in developing foods that are more flavorful and nutritious) (I Ré, 1998). According to Potdar et al. (2020), a reduction in flavor is observed during the food processing stage, so it is necessary to replace this loss. Encapsulated flavor helps to compensate by bringing out the flavor in its natural form. Among the many reasons for encapsulating flavor, its functionality and preservation stability are of greatest importance. The encapsulation of flavor compounds attracts the consumer; moreover, these compounds are easily digestible.

Microencapsulation can simplify the food manufacturing process by converting liquids to solid powder, decreasing production costs by allowing for batch processing using low-cost and powder handling equipment. Microcapsules also help fragile and sensitive materials to survive processing and packaging conditions and stabilize the shelf life of the active compound (Raybaudi-Massilia & Mosqueda-Melgar, 2012). Encapsulation can also be used when it is important to add a new ingredient at a certain point in the production process without the need for new feed lines and incorporation steps. For these applications, it is possible to use capsules that will be thermally or shear-activated, taking advantage of or causing changes in process conditions to release the encapsulated material.

Encapsulation is often used to protect an ingredient or product until the time of consumption. Some ingredients are encapsulated to provide functionality during processing or to preserve their functionality during distribution and storage. For Thomas and Atwell (1999), ingredients with leavening agents such as chemical or biological yeast, with almonds or fruits, and finished products such as breakfast cereals, vitamins, flavors, and aromas can be encapsulated.

More recently, microencapsulation became increasingly popular in pharmaceutical, nutraceutical, and functional food industries. It allows to prolong the shelf life of the active compound, to mask the undesirable flavor, color, and taste, and to control the release of bioactive. Functional ingredients, such as omega-3s, probiotics, vitamins, and phytochemicals, are being increasingly incorporated into a wide range of products such as breads, milk, fruit juices, tortillas, chocolate, yoghurt drinks, spreads, peanut butter, eggs, and meat (Timilsena et al., 2020).

For I Ré (1998), following the pharmaceutical, food, and fragrance industries, other industrial activities have expanded the use of microcapsules, such as the parachemical industry for encapsulation of catalysts and release of bactericidal agents and coloring agents and the agrochemical industry for release of pesticides and protection of seeds.

12.4.1 Encapsulation techniques

Besides spray-drying, many other processes exist for encapsulation, such as coating, fluidized bed coating, spray chilling and spray cooling, extrusion coating, liposome entrapment, coacervation, inclusion complexation, and ionotropic gelation (Timilsena et al., 2020). Spray-drying is the most used encapsulation method for food ingredients. It is a well-known process suitable for drying materials due to the very short heat contact time and the high rate of evaporation, resulting in high quality, stable, functional, and low moisture content products (Fernandes et al., 2013). It is a highly automated and cost-effective product, able to process several kinds of materials to obtain product with prespecified properties (Timilsena et al., 2020).

It is a low-cost microencapsulation technology commonly used on an industrial scale, which has the attractive advantages of producing microcapsules in a relatively simple, continuous operation and is inexpensive, compared to other microencapsulation techniques (Bakry et al., 2016).

The mixture to be atomized is prepared by dispersing the material to be encapsulated (core material) in an encapsulating solution or carrier material, in which the material must not be miscible (Beatuz et al., 1985; Constant & Stringheta, 2002). Usually, coarse emulsion is prepared by mixing the core and shell materials in required quantities. It is followed by single-stage or two-stage homogenization that yields a fine emulsion (Timilsena et al., 2020). When the atomized liquid is an emulsion of a core material plus one or more carrier materials, microencapsulation happens (Baltrusch et al., 2022).

The compounds are atomized in a stream of hot air. The atomized particles dry quickly and confine the encapsulated material within the globules formed (Constant & Stringheta, 2002). Atomization of emulsions into a drying chamber at a relatively high temperature leads to very fast water evaporation, allowing for crust formation at fast rate and quasi-instantaneous entrapment of the material to be encapsulated (Bakry et al., 2016). The material that is being coated is termed active material, core material, or also fill, internal phase or payload. The material, which is used to encapsulate the active ingredient, is termed carrier material, wall material, shell, capsule, membrane, matrix, encapsulating agent, or coating material (Potdar et al., 2020; Timilsena et al., 2020).

For Bakry et al. (2016), the microencapsulation by spray-drying involves four stages. It consists of (1) preparation of the dispersion, (2) homogenization of the dispersion, (3) atomization of the emulsion, and (4) dehydration of the atomized particles. Ordinarily, in the first stage, the wall materials are dissolved in distilled water with stirring. The solutions are kept overnight at room or refrigerator temperature to ensure full saturation of the polymer molecules and to avoid any changes due to temperature. The core material is added to the solutions with or without the addition of an emulsifier, depending on the emulsifying properties of the wall materials, and all this before entering the second stage. In the spray-drying process, the initial emulsion droplets are in the range of 0.1–100.0 μm in diameter. The main reason that limits the spray-drying technique in microencapsulation is the limited number of wall materials available and that must have a good solubility in water. Another disadvantage for spray-drying that should be considered is that it produces a fine microcapsule powder, which needs further processing such as agglomeration. In addition, products have a lower oxidative stability due to the high temperatures used during the atomization process.

To obtain high encapsulation efficiency, optimal spray-drying conditions must be used. The feed temperature, air inlet temperature, and air outlet temperature are the main factors that must be optimized. Feed temperature modifies the viscosity and fluidity of the emulsion, and its capacity to be homogeneously sprayed is thus negatively affected. When the feed temperature is increased, viscosity and droplets size should be decreased. However, volatility or deterioration of some heat-sensitive components can occur due to high temperatures (Bakry et al., 2016). The rapid evaporation of water from the encapsulating matrix during drying causes the core to remain at a temperature below 100°C, which allows the process to be used even for thermosensitive substances. However, when the product needs to retain aroma and flavor, there can be losses of components with very low boiling points, due to the complexity of their composition (Constant & Stringheta, 2002).

Microencapsulation efficiency can be increased by increasing the concentration of solids (wall materials) in the solution, which can be related to the formation of a shell around the core material (Bakry et al., 2016).

In general, the microencapsulation process can be classified into macro (> 5 μm) and micro (0.2–5 μm). Emulsion viscosity and particle size distribution have significant effects on microencapsulation by spray-drying (Bakry et al., 2016). The concentration of the main component in the droplet (carrier) influences the prepared particle size. When the main component is added to the initial precursor in small amounts, the concentration ratio of the solvent/main component in the droplet should be high. Because the solvent is evaporated during the spray-drying process, the droplets lose most of their volume, allowing for the production of small particles. On the contrary, with a high concentration of the main component, after the solvent is removed, the generation of large particles is possible (Nandiyanto & Okuyama, 2011).

Capsules are ideally spherical, although their shape is influenced by the original structure of the encapsulated ingredient (Bakan, 1973; King, 1995) and by changing the process conditions (Eijkelboom et al., 2022; Nandiyanto & Okuyama, 2011).

Smooth and spherical particles with limited surface distortions are preferred for a free-flowing powder, which is important for dosing, tableting, blending/mixing, or coating. Powders comprising of particles with irregular shape, in other words, wrinkles and folds, potentially interlock mechanically, leading to inferior flow properties. It is also not uncommon to spray-drying that many small particles (<100 μm) are formed, which negatively influence the flowability (Eijkelboom et al., 2022).

Fig. 12.3 shows the possible forms of microencapsulated dehydrated material particles, which may vary depending on the type of encapsulated material and carrier, but mainly on the models and conditions of the atomizer.

According to Constant and Stringheta (2002), the properties of microcapsules are as follows:

- Structure: The microcapsules formed are ideally spherical in shape but vary predominantly according to the shape of the active material and the encapsulation method employed. Microcapsules can be classified into three basic categories as monocore (also called single-core or reservoir type), polycore (also called multiple-core), and matrix types (Raybaudi-Massilia & Mosqueda-Melgar, 2012). Monocore are microcapsules that have a single hollow chamber within the capsule. In this category, it is possible to find a structure with several different encapsulating walls for the same core (multishell, in Fig. 12.3). Polycore, or multicore, are microcapsules that have different sized chambers within the shell. As for single-core capsules, they can be regular or irregular and can also be grouped in the form of clusters. Matrix type are microparticles that have the active compounds integrated within the matrix of the shell material (Fig. 12.3). It is also possible to find assembly of microcapsules. Although they are formed as monocore capsules, this type of capsule has been classified separately.

  However, the morphology of the internal structure of a microparticle depends mostly on the selected shell materials and the microencapsulation methods that are employed (Raybaudi-Massilia & Mosqueda-Melgar, 2012).
- Release mechanism: The controlled release of the core, carried out in an appropriate manner, is one of the most important properties of microencapsulation. The

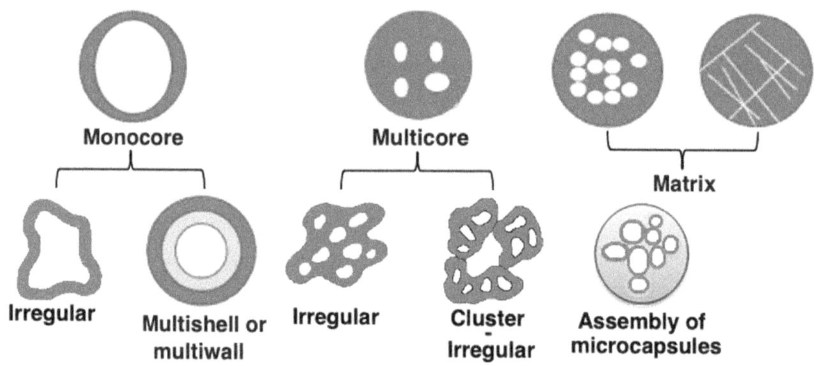

FIGURE 12.3 Different types of microcapsules. *Based on Bakry et al., 2016; Cereda & Daiuto, 2003; Potdar et al., 2020; Raybaudi-Massilia & Mosqueda-Melgar, 2012; Timilsena et al., 2020.*

encapsulating matrix protects the active material from moisture, light, oxygen, and external agents and controls the release of this material.

Santos et al. (2005) used an atomization process to encapsulate paprika oleoresin. They used gum arabic, rice starch porous granules, and gelatin as encapsulating agents. Morphological evaluation by scanning electron microscopy showed that the microcapsules made with gum arabic had a rounded shape, with concavities, without pores or cracks, continuous walls, and the filling distributed in vacuoles on the walls. On the other hand, those produced with porous rice starch granules and gelatin showed a rounded shape, walls formed by the agglomeration of the granules bonded by the gelatin, but with defects in the form of pores or interstices. The particle size distribution was unimodal, and the particles showed an average size of 16 and 20.3 μm for the microcapsules produced with gum Arabic and starch/gelatin, respectively.

12.4.2 Selection of encapsulating agents or wall material

Once identified and defined the release and storage mechanisms, or the process needs, an important step is the selection of the encapsulation material, which must meet all the necessary criteria which are mechanical resistance, compatibility with the product, capsule size, release condition, temperature influence, dissolution form, and so forth. The encapsulation agent must be considered an ingredient since it will be totally incorporated into the final product, food or not. Its nutritional value and its negative points should be carefully evaluated. The ideal encapsulant should have emulsifying properties, be able to form films, have low viscosity and high solid levels, exhibit low hygroscopicity, and be inexpensive. For Timilsena et al. (2020), solubility, interfacial properties, and gelling capacity are other important factors to be considered during the matrix selection process.

Protection of the bioactive ingredients from the environmental stressors (oxygen, radiation, light, moisture, and adverse pH conditions) and release control of actives components from food matrices in the right place at the right time and desired rate are primary objectives of encapsulation. However, recent trends seek to improve the product handling in terms of reduced toxicity, reduced cost, and enhanced bioavailability of a nutrient. The ultimate objective is to extend product shelf life and improve the overall acceptability of the engineered product (Timilsena et al., 2020).

The thickness of the membrane formed over the core can be manipulated to alter the permeability and stability of the microcapsule. The encapsulating agent can represent from 1% to 70% of the weight of the microcapsule; however, in most commercial applications, the weight of this coating varies in a range of 3%–30% (Bakan, 1973; Constant & Stringheta, 2002; Jackson & Lee, 1991). There is a wide variety of encapsulants that can be used in spray-dryer encapsulation (Bakry et al., 2016; Constant & Stringheta, 2002; Thomas & Atwell, 1999; Timilsena et al., 2020), such as:

- Natural polymers:
  - Gums: gum arabic, agar, carrageenan, and sodium alginate.
  - Carbohydrates: starch, maltodextrin, sugar, modified starches, cyclodextrins, and starch syrups.
  - Cellulose: carboxy-methylcellulose, ethylcellulose, methylcellulose, acetylcellulose, and nitrocellulose.
  - Lipids: paraffin wax, tristearin, stearic acid, oils, fats, monoglycerides, diglycerides, and hydrogenated oils.
  - Proteins: gluten, casein, gelatin, albumin, hemoglobin, and peptides.
- Synthetic polymers: nylon, polyethylene, polyester, and others.

Gum arabic, modified starches, and hydrolyzed starches are the most frequently used (Constant & Stringheta, 2002; Reineccius, 1991). Stability of the active ingredient largely depends on the structure and functionality of shell matrix, the processing conditions, and the core–shell interactions (Timilsena et al., 2020).

For food use, the wall material must be soluble in water at an acceptable level, because almost all spray-drying processes in this industry are carried out from an aqueous feed formulation (Bakry et al., 2016). In general, food grade proteins and polysaccharides are used to encapsulate sensitive and bioactive food ingredients including highly unsaturated edible oils (e.g., fish oils), vitamins, enzymes, or various flavors for enhancing their shelf life and/or masking the undesirable flavor or taste. More recently, works focus on enhancing the functionality and health benefits of processed foods and enhancing the efficacy of probiotics and targeted delivery of various enzymes or coenzymes, bioactive peptides, and other therapeutic agents (Timilsena et al., 2020).

Several unstable and sensitive products such as polyunsaturated fats, vitamins, enzymes, and polyphenols can be encapsulated into inert matrices to make them stable and more effective. However, most probiotic strains do not survive well at high temperatures and dehydration during the spray-drying process. The introduction of thermoprotectants such as trehalose and prebiotics can improve the stability of these biomaterials (Timilsena et al., 2020).

In general, an encapsulant should have the following characteristics (Constant & Stringheta, 2002; Shahid & Han, 1993):

- Low viscosity at high concentrations and easy handling during the encapsulation process;
- Ability to form stable emulsions with the core;
- Not react with the encapsulated material;
- Be permeable to the solvent used in the encapsulation process;
- Provide maximum protection to the active material against the external environment;
- To be soluble in food grade solvents;
- To present low hygroscopicity;
- Promote the release of the nucleus at the desired time or in the desired conditions;
- Be economically feasible.

Structural and functional aspects of shell matrices and their compatibility with the core play vital roles in their efficacy and overall acceptability as encapsulants in food industries (Timilsena et al., 2020). It is difficult for an encapsulating agent to present all the cited properties, so in practice, it is common to use mixtures of two or more components. Thus, a blend of 2 or 3 carrier materials can be used to increase encapsulation efficiency as well as cost-effectiveness (Potdar et al., 2020).

The preparation of a microencapsulated product involves several steps. Initially, the need for microencapsulation must be verified. Then the encapsulant must be chosen that allows for the release of the nucleus in the desired form. Finally, the process for preparing the microcapsules must be selected (Constant & Stringheta, 2002).

The encapsulating agent, which is basically a polymeric material capable of forming a film, can be selected from a wide range of natural or synthetic polymers, depending on the active material to be encapsulated, the encapsulation process employed, and the desired characteristics of the final product (Bakan, 1973; Constant & Stringheta, 2002). For example, for flavor retention, Potdar et al. (2020) indicated that gum arabic is having the highest flavor retention ability followed by modified starch having 88% and whey protein isolate 87%.

Generally, water-soluble polymers are used to microencapsulate organic cores, while water-soluble cores are microencapsulated by hydrophobic material (Bakan, 1973; Constant & Stringheta, 2002). A commonly used wall material for liposoluble bioactive compounds is gum arabic, and for hydro soluble compounds, it is maltodextrin. Wall materials such as hydrolyzed starch, octenyl succinic anhydride (OSA), starch and soluble starch are also favorable for spray-drying because of their high solubility in water, low viscosity, bland taste, colorless solutions, and low cost (Ramakrishnan et al., 2018).

12.4.3 Starches and microencapsulation

Starches are preferred for many microencapsulation processes due to their high functionality, versatility, and efficiency in spray-drying microencapsulation. Modified starches offer high oil retention, long shelf life, and high manufacturing efficiency (Thiengnoi et al., 2012).

Initially, the major obstacle in using starch as an encapsulating agent was its susceptibility to browning by Maillard reaction during the encapsulation procedure, resulting in unacceptable taste, color, and aroma (Thomas and Atwell, 1999). Modifications of starches and processes have circumvented this problem. Modified starches offer a better protection than native and waxy starch, as well as a better emulsion stability (Timilsena et al., 2020).

Maltodextrins and dextrins have been used for years as encapsulating agents for flavorings in atomization processes. One of the first successful methods employing drying technology was a process patented by Evans in 1964, which involves the use of dextrins derived from oxidized starches (Thomas & Atwell, 1999).

Akbarbaglu et al. (2019) analyzed the influence of maltodextrin on the retention of antioxidant properties and microstructure of flaxseed protein hydrolysates. They found an increase in production yield because of higher levels in the ratio of maltodextrin. Moreover, moisture content and water activity of powders decreased at higher carrier concentrations, while the solubility of encapsulated powders increased. The authors also encountered a significant effect of maltodextrin on the reduction of hygroscopicity for free hydrolysate, with a reduction from 39.2% to 17.4% in powders produced with a carrier ratio of 3:1. Also, increasing the carrier ratio improved the retention of antioxidant activity in spray-dried hydrolysates.

Etzbach et al. (2020) evaluated the stability of carotenoids and encapsulation efficiency of goldenberry (*Physalis peruviana* L.) powder with different carriers. They encountered a significant reduction in the total carotenoid content of 38.9%–69.7%, depending on the carrier agent used, with the best results being with the use of maltodextrin.

Bakry et al. (2016) described the encapsulation of fish oil with a carbohydrate source (oligosaccharide, dried glucose syrup, and glucose) and protein. They reported that the Maillard reaction increased the encapsulation efficiency and was effective for protecting microencapsulated fish oil and other oils from oxidation. In this case, Maillard reaction is not a problem, as recorded by Thomas and Atwell (1999).

Starch modified with OSA is widely used as a surface-active food additive in microencapsulations of oil-based flavors, nutrients, fragrances, agrichemicals, and pharmaceutical actives, among other biopolymers such as gelatin, protein, and gum arabic (Thomas & Atwell, 1999; Wandrey et al., 2010). Generally, a low-viscosity OSA-treated starch is used to obtain microencapsulates by spray-drying. In drying, this starch achieves better aroma and flavor retention than products encapsulated with gum arabic or conventional dextrins. These encapsulants are easily hydrated, combine with water, and quickly develop the flavoring and aroma agent (Thomas & Atwell, 1999).

There are cases when a more gradual release of flavor is desired, where the encapsulated product must repel water. In this case, a technology patented by Wurzburg, William, and Cole has been used, whereby OSA-treated starch reacts with a polyvalent metal, for example, calcium or magnesium, and results in a more hydrophobic encapsulate. Temperature-sensitive materials, such as vitamins, are encapsulated in OSA-treated starch with a high amylose content, which naturally develops encapsulant characteristics. By using OSA-treated starch, the encapsulates are easy to prepare and do not require high temperature for drying.

Wang et al. (2015) reported that preservation of aromatic additives is often a top concern of food manufacturers because they are usually expensive, volatile, delicate, and hard to handle. Encapsulation provides an effective approach to protect flavors against evaporation and undesirable reactions during storage by coating them with protective carrier materials. Starch has been considered as an excellent wall material for flavor encapsulation over the years. There has been an increasing demand for starches in flavor encapsulation attributed to their significant benefits, as they are fully biodegradable, widely available, inexpensive, and can be easily modified into derivatives with various properties. The authors highlight the used of starch-based ingredients, that is, cyclodextrins, hydrolyzed starches, OSA starches, and porous starches in flavor encapsulation. Cyclodextrins are discussed in Section 12.4.4.

A case of oil encapsulation is reported by Thiengnoi et al. (2012), who evaluated the effects of the type of coating and its concentration on physical properties of agglomerated, spray-dried, and high oil-loaded microcapsules. Spray-drying was used to microencapsulate soybean oil (500 g/kg of dry matter), which served as a model core material in a wall system consisting of soy protein isolate and maltodextrin (1:3 w/w ratio). Spray-drying of the core-in-wall emulsions led to small particles ($\sim 20\ \mu m$) having poor handling and reconstitution properties. Fluidized bed agglomeration of the spray-dried microcapsules using maltodextrin with 10–18 DE value as an aqueous binder solution improved the handling and reconstitution properties of the powders. The system containing maltodextrin with the DE value of 14 presented the highest dispersibility and binder concentrations of 150 g/L.

12.4.4 Microencapsulation with cyclodextrin

Cyclodextrins belong to the group of true encapsulants, as opposed to substances that can act as encapsulants. There are three groups of cyclodextrins, α, β, and γ. In this case, the process takes place at the molecular level and essentially uses β-cyclodextrin as the encapsulating medium (Pagington, 1986; Reineccius, 1991). This molecule, due to its structure, forms a complex with the substances to be encapsulated, so that the material remains in its cavity (more details on the production of cyclodextrins from cassava bran will be presented in the upcoming Volume 5 of this series).

The encapsulating power decreases from γ to α-cyclodextrins, due to the size of their internal cavity. The β-cyclodextrins are intermediates and are FDA approved for food use (for legislation on starch and derivatives, see Chapter 6 in Volume 1 of this series). Microencapsulation in β-cyclodextrin has been recognized as one of the most efficient methods to protect flavors and aromas against oxidation, heat degradation, and evaporation. Inclusion products in β-cyclodextrin molecules often have a longer shelf life than those using common ingredients in spray-drying encapsulation.

The complexes formed by inclusion with β-cyclodextrin have a wide range of application. An increase of stability in oxidative processes of certain oils and fat-soluble vitamins is observed with inclusion. It is also possible to control flavor release, mask odors, and increase the solubility of hydrophobic ingredients (Constant & Stringheta, 2002; Dziezak, 1988). However, the high cost of cyclodextrins restricts their use in foods, making more widespread use unfeasible for many ingredients, even while recognizing the benefits achieved with encapsulation (Constant & Stringheta, 2002; Reineccius, 1991).

The flavor and aroma components are usually made up of volatile substances, which are easily degraded when not protected. The fruit flavor is very sensitive to oxidation by air and can be protected by β-cyclodextrins. Often oils and fats are used to solubilize the flavor components, since they are mostly oil soluble. The complex of cyclodextrins with flavor and aroma compounds remains evenly distributed in the mixtures, making it difficult to separate oily and aqueous phases in the product and eliminating the need for the addition of oils or fats to better disperse the flavors and aromas in the food. Complexations allow the volatility of these compounds to be reduced, increasing storage life, and reducing losses. Many complexes formed with cyclodextrins and terpenoids or phenyl-propane derivatives in dry form remain stable for long periods of time without special care. Thus, flavor and aroma compounds in powder form have been approved and are being produced and consumed in many countries (Alves Prado et al., 2002).

Timilsena et al. (2020) described the encapsulation of ethylene gas in a matrix of α-cyclodextrin using the inclusion complexation method to facilitate its handling and application in fruit ripening.

12.5 Control analyses

Control analyses in spray-drying are very specific and often difficult to evaluate individually. Bakry et al. (2016) described parameters for evaluation of encapsulation efficiency, size particle, stability under different storage conditions, degree of protection provided to the core material, and morphology of microcapsules by scanning microscopy. Some of these methods are described in Chapter 12 of the Volume 1 in this series.

According to the literature, the main evaluation methods are as follows:

- Moisture: determined in an oven with air circulation for 4 hours at 105°C. The results are expressed as percentage on dry basis (Bhandari et al., 1993).
- Compacted density (kg/m^3): A 20 g sample of the powdered product is weighed and placed in a 100 mL graduated beaker. The sample is gently compacted by 10 strokes on a rubber mat from a height of 15 cm (Bhandari et al., 1993). There is equipment to standardize this type of analysis, using the same beating principle, but with greater standardization.
- Particle density: Determined in a pycnometer, using toluene at 20°C (Bhandari et al., 1993).
- Wettability: In this analysis, a 12 g sample of the powdered material is placed on the surface of a container containing water at 20°C and the time in seconds until the sample sinks in water is established (Bhandari et al., 1993; Schubert, 1987).
- Sticking temperature: The stickiness temperature establishes when a dehydrated product that has absorbed water starts to become sticky. It can be measured by visual observation on a polished metal plate heated under a microscope (Kofler) with temperature control, as recommended by Papadakis and Bahu (1992) and Bhandari et al. (1993). It is equivalent to the lowest temperature at which powder adhesion on the surface can be identified. The melting point temperature can be measured visually on the same equipment. It can also be evaluated according to Notter et al. (1959) and Brennan et al. (1971). In this case, a test tube containing a standardized amount of sample powder is immersed in a water bath, in which the temperature gradually increases. The powder is stirred with a rod. The stickiness point corresponds to the temperature at which the rod attached to a reader indicates an increase in the force required for stirring.
- Solubility: The solubility test is very important. Carriers can range from insoluble, such as native starches, to soluble, such as hydrolyzed syrups, to partially soluble products. The literature suggests some

methodologies, but many are too subjective. For fruit juices, water solubility is important, and for pharmaceutical products, it is often important to assess alcohol solubility.

- Image analysis: Samples are observed under an optical microscope at 400X magnification. They can also be observed under an electron microscope to evaluate particle size and morphology. The carriers and insoluble products can be observed in water, but the soluble ones should be observed in glycerol, to maintain the characteristic structure. What is obtained as an image can be seen in Fig. 12.4A, where maltodextrin forms a cluster covered by the dehydrated juice particles. If the product is soluble, by putting water on the slide, the soluble fraction is dissolved. In the case of carriers obtained from starch, Lugol staining can be an important aid, as it indicates the type of carrier used as a function of the reaction color, blue for native starch and violet or red for dextrins (for more information on the use of dye for starch identification, see Chapter 13 in Volume 1 of this series).
- Particle size and distribution: Orr and DallaValle (1959) and Brennan et al. (1971) recommended the use of a microscopic technique where the sample is dispersed in a 20% solution of celoidin in butyl acetate to be examined and photographed. The authors suggest 500 readings for each sample. The average particle size is calculated and expressed in microns (μ), and frequency diagrams are drawn when necessary (Fig. 12.4B). The slide can also be mounted with liquid glycerin. Etzbach et al. (2020) analyzed the particle size distribution by dynamic image analysis using a Camsizer X2 (Retsch Technology, Haan, Germany) operated in XJet mode for air pressure dispersion. The distribution of the particle size should be as narrow as possible to maintain product consistency (Timilsena et al., 2020). For these authors, the size of the microcapsules can be measured using

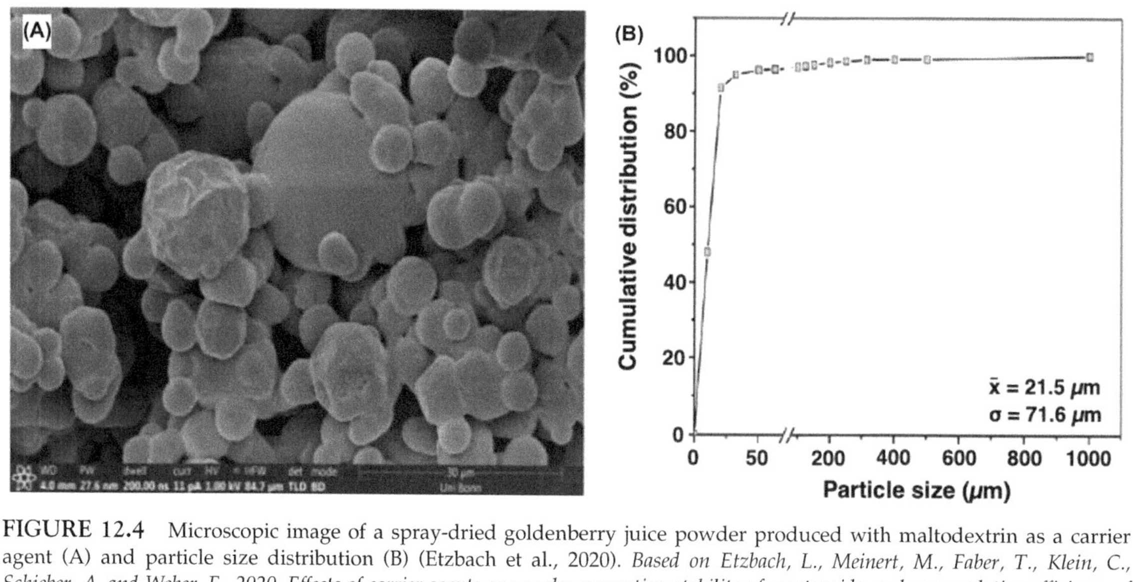

FIGURE 12.4 Microscopic image of a spray-dried goldenberry juice powder produced with maltodextrin as a carrier agent (A) and particle size distribution (B) (Etzbach et al., 2020). *Based on Etzbach, L., Meinert, M., Faber, T., Klein, C., Schieber, A. and Weber, F., 2020. Effects of carrier agents on powder properties, stability of carotenoids, and encapsulation efficiency of goldenberry* (Physalis peruviana *L.) powder produced by co-current spray drying*. Current Research in Food Science, 3,73–81.

techniques such as laser scattering or particle size imaging using microscopy.

- Timilsena et al. (2020) described how to measure the effectiveness of the microencapsulation technique.
- Encapsulation efficiency: The most important parameter to assess the effectiveness of the encapsulation process. Determines the percentage of bioactive substance entrapped inside the inert core during encapsulation process.

$$EE = \frac{Wt - Ws}{Wt} \quad (12.1)$$

where *Wt* is mass of total active ingredient and *Ws* is the mass of surface-active ingredient on the microcapsule.

- Release kinetics: Pattern and kinetics of release of the core material from the capsule. Release can be site-specific, stage-specific, or triggered by changes in pH, temperature, irradiation, or osmotic shock. In the food industry, the most common method is the solvent activated release. In most food applications, the core material is released by simple diffusion or dissolution of the shell matrix due to mechanical force or enzymatic activity.
- Particle size for food applications should be below 100 μm to avoid impacting the mouth feel of the food product.
- Stability: Important to assess the effectiveness of the encapsulation process and quality of resultant product.
- Payload: Percentage of the active ingredient per gram of the powder. Encapsulation process with a higher payload is more economical and the process is considered more effective.

$$PL = \frac{Wt}{Wm} \times 100 \quad (12.2)$$

where *Wt* is the mass of the total active ingredient encapsulated and *Wm* is the total mass of the resultant product.

- Encapsulation yield: Measures the effectiveness of encapsulation

$$EY = \frac{Wt}{Wi} \times 100 \quad (12.3)$$

where *Wt* is the mass of the total active ingredient encapsulated and *Wi* is the total mass of the active ingredient introduced.

12.6 Final considerations

The use of starch and derivatives as carriers in dehydration or microencapsulation by spray-dryer for the food, pharmaceutical, and cosmetics industries, among others, is a reality considering performance, availability, price, and the fact that it is a renewable component. The spray-drying-based microencapsulation method is widely used in various industrial applications. However, more advanced methods, including complex coacervation, are gaining increased attention in recent years and could constitute increasingly important markets for the starch industry (Timilsena et al., 2020).

The main starch used in the spray-dryer market is corn. Few articles are found in the specific literature on the use of starches from underground crops and most of these works focus on cassava, the second most produced starch in the world. Also, little is found on the influence of the type of starch on the characteristics of the products obtained by atomization or encapsulation. This situation is explained by the fact that among the starches marketed in the northern hemisphere, starches from corn, wheat, and potato predominate, with small variations compared to the numerous tropical starches, especially those of underground crops originating from South America.

By tropical starches, we recognize the crops grown between the 20 degree latitude ranges south and north in relation to the equator line. In this range, the starchy crops are of the greatest importance for the diet of the local inhabitants and have therefore been preserved. Regarding starchy raw materials that store starch in underground organs, there are several genera and species, with starch stored in roots, tubers, and rhizomes. In this case, the variability of the characteristics of the starch granules is relevant.

One of the problems encountered in drying as a basis for making tablets or capsules is obtaining a product of adequate bulk density that flows well. If the atomized product is too light, it is difficult to fill the dies of compression machines and pack them into capsules in the desired quantities. However, no references were found in the literature on the influence of starch granule size on spray-dried product quality or encapsulation. The physical characteristics of starch granules, including shape and size, vary according to the botanical source. Corn granules have a polyhedral shape and diameters in the range of 3.0–23.0 μm (Sahai & Jackson, 1996). Cassava starch granules are widely cited in the literature, with a truncated, oval shape and size ranging from 4.0 to 35.0 μm. The granules of taro (*Colocasia esculenta*) are round and oval-shaped, with the diameter ranging from 1.6 to 6.0 μm. Canna (*Canna indica*) starch granules are much larger, 25–45 μm (for more information, see Chapter 10 in Volume 1 of this series).

To cite a specific case about starch as a carrier for atomization, in which the characteristics of tropical starch are considered, Daiuto and Cereda (2006) evaluated two starches extracted from underground crops with potential to be commercial, starches from taro (*Colocasia esculenta*) and canna (*Canna indica*), which present a large difference in granule size. In the evaluation as carriers in processes of drying products by atomization, the bulk density is an important characteristic and should be controlled in the products. The canna and taro starches were selected for representing extremes of granule size and thus allowed to verify the effect of size on the bulk density of dehydrated products by atomization.

For comparison, the authors used the commercial starches of cassava and corn, because in Brazil, they are of common use in atomized products and present intermediate granule sizes. The atomization was done in a Spray Dryer Lab Plant SD 04, operating under 6 $lb.in^2$ pressure, 7.6 mL/min flow, and 1 cm nozzle. The inlet temperature was set at 200°C, but the outlet temperature could not be regulated. The atomized products used boldo extract as a base for the four types of starches. The dried product was evaluated for moisture, water activity, particle size, and bulk density. The results showed that the particle size varied because of the size of the starch granules, which influenced the bulk density of the final product. This was higher (694.27 g/mL) for canna starch and lower (456.13 g/mL) for taro starch. Corn and cassava starches provided very close and intermediate values of 521.51 and 581.48 g/mL, which also represent the pattern of grain size variation (Daiuto & Cereda, 2006).

Besides this article, only one other cites the use of cassava starch, in the area of enzymatic hydrolysates, highlighting the importance of the characteristics of maltodextrins obtained from cassava used in the drying process for the good preservation of the final product. Most publications do not even identify the source of the starch used.

Acknowledgments

The authors would like to thank the Centroflora Anidro do Brasil Group, of Botucatu, SP, Brazil, for drawing their attention to this specific area of application of starches and derivatives and for their support in obtaining information.

The authors would also like to thank the Cargill Foundation, in the person of Cesar Renato Néris, for kindly waiving copyrights.

References

Akbarbaglu, Z., Jafari, S. M., Sarabandi, K., Mohammadi, M., Heshmati, M. K., & Pezeshki, A. (2019). Influence of spray drying encapsulation on the retention of antioxidant properties and microstructure of flaxseed protein hydrolysates. *Colloids and Surfaces B: Biointerfaces*, *178*, 421–429.

Alves Prado, H. F., Gomes, E., & Da Silva, R. (2002). Produção de ciclodextrina glicosiltransferase (CGTase), microbiana e aplicação de ciclodextrinas em alimentos: Revisão. *Boletim da Sociedade Brasileira de Ciências e Tecnologia de Alimentos*, *36*(1), 43–54, 2002.

Azis, A. D. (2019). Effect of vacuum freeze-drying condition and maltodextrin on the physical and sensory characteristics of passion fruit (*Passiflora edulis* sims) extract. In IOP Conference Series: Earth and Environmental Science (Vol. 355, No. 1, p. 012067). IOP Publishing.

Bakry, A. M., Abbas, S., Ali, B., Majeed, H., Abouelwafa, M. Y., Mousa, A., & Liang, L. (2016). Microencapsulation of oils: A comprehensive review of benefits, techniques, and applications. *Comprehensive Reviews in Food Science and Food Safety*, *15*(1), 143–182. Available from https://doi.org/10.1111/1541-4337.12179.

Bakan, J. A. (1973). Microencapsulation of food and related products. *Food Technology,*, *27(11)*, 34–44.

Baltrusch, K. L., Torres, M. D., Domínguez, H., & Florez-Fernandez, N. (2022). Spray-drying microencapsulation of tea extracts using green starch, alginate or carrageenan as carrier materials. *International Journal of Biological Macromolecules*, *203*, 417–429.

Beatuz, Y., Raziel, A., Rosemberg, M., & Kopelman, I. J. (1985). Spray-drying microencapsulation of *Paprika oleoresin*. *Lebensmittel-Wissenschaft & Technologie*, *18*, 28–34.

Bhandari, B. R., Duhoulin, E. D., Richard, H., NoleauI, M. J., & Lebert, A. M. (1992). Flavor encapsulation by spray drying: Application to citral and linaly acetate. *Journal of Food Science*, *57*(11), 217–221.

Bhandari, B. R., Senoussi, A., Dumoulin, E. D., & Lebert, A. (1993). Spray drying of concentrated fruit juices. *Drying Technology*, *11*(5), 1081–1092. Available from https://doi.org/10.1080/07373939308916884.

Brennan, J. G., Herrera, J., & Jowitt, R. (1971). A study of some of the factors affecting the spray drying of concentrated orange juice, on a laboratory scale. *International Journal of Food Science & Technology*, *6*(3), 295–307.

Cereda, M. P., & Daiuto, E. R. (2003). Chap. 16 - Starch as a support in the dehydration through atomization and microencapsulation. In: M. P. Cereda, & O. F. Vilpoux (Eds.), *Tecnologia, usos e potencialidades de tuberosas amiláceas Latino Americanas*, first ed., vol.3. Fundação Cargill, São Paulo.

Constant, P. B. L., & Stringheta, P. C. (2002). Microencapsulação de ingredientes alimentícios. *Boletim da Sociedade Brasileira de Ciência e Tecnologia de Alimentos*, *36*(1), 12–18.

Daiuto, É. R., & Cereda, M. P. (2006). Influência da granulometria de grânulos de amido sobre a densidade aparente de extratos atomizados. *Revista de Ciências Farmacêuticas Básica e Aplicada*, *27*, 51–56.

Domingues, A., Abadio, F.D. B., Oliveira, V.D., Borges, S. V., & Gay, J.. (2002). Characterization of the physical properties of pineapple juice *(Ananás comosus)* powder dried by optimized spray dryer through response surface support analysis. In *Congresso Brasileiro de Ciência e Tecnologia de Alimentos* (Vol. 18, pp. 1717–1720). Porto Alegre: CBCTA (In Portuguese).

Dziezak, J. D. (1988). Microencapsulation and encapsulated ingredients. *Food Technology*, *42*, 136–151.

Eijkelboom, N. M., van Boven, A. P., Siemons, I., Wilms, P. F., Boom, R. M., Kohlus, R., & Schutyser, M. A. (2022). Particle structure development during spray drying from a single droplet to pilot-scale perspective. *Journal of Food Engineering*, *337*111222.

Etzbach, L., Meinert, M., Faber, T., Klein, C., Schieber, A., & Weber, F. (2020). Effects of carrier agents on powder properties, stability of carotenoids, and encapsulation efficiency of goldenberry (*Physalis peruviana* L.) powder produced by co-current spray drying. *Current Research in Food Science*, *3*, 73–81.

Fernandes, R. V. D. B., Borges, S. V., & Botrel, D. A. (2013). Influence of spray drying operating conditions on microencapsulated rosemary essential oil properties. *Food Science and Technology*, *33*, 171–178.

Francisoni, A. D., Campos, F. R., Pereira, D. B., Oliveira, V. D., Borges, S. V., & Gay, J. (2002). Influence of maltodextrin concentration and atomization speed on the physical properties of dehydrated passion fruit juice. *Congresso Brasileiro de Ciência e Tecnologia de Alimentos* (18, pp. 2264–2267). Campinas: CBCTA (In Portuguese).

Gava, A. J. (1978). *Princípios de tecnologia de alimentos*. 1 ed., Nobel, São Paulo.

Glucidex. (2023). Maltodextrins, and dryied glucose syrups, food and dietetic food. *Roquette*, s.d. 28 p. Available from https://www.roquette.com/food-nutrition-maltodextrin-and-dried-glucose-syrup. Consulted on 30/03/2023.

Gupta, A. S. (1978). *Spray drying of orange juice*. US Patent 4.112.130 A.

Ré, M. I. (1998). Microencapsulation by spray drying. *Drying Technology*, *16*(6), 1195–1236.

Jackson, L. S., & Lee, K. (1991). Microencapsulation and the food industry. *Lebensmittel-Wissenschaft & Technologie*, *24* (4), 289–297.

King, A. (1995). Encapsulation of food ingredients. Encapsulation and controlled release of food ingredients. Risch, SJ y Reineccius. In *GA ACS Symposium Series*, Washington DC (pp. p26–39). DOI: https://doi.org/10.1021/bk-1995-0590.ch003

Lloyd, N. E., & Nelson, W. J. (1984). Glucose and fructose-containing sweeteners from starch. In R. L. Whistler, & J. N. Bemiller (Eds.), *Starch: Chemistry and technology* (pp. 611–660). Orlando, FL: Academic Press Inc.

Maltini, E., & Mastrocola, D. (1999). Preparazione di succo integrale crioconcentrato di kiwi fruit. *Industrie delle Bevande*, *28*(159), 6–9.

Masters, K. (1991). Spray drying handbook, 5th edn. Pp. 241–327. London, UK: Longman Scientific and Technical, Microencapsulation: Cutting-edge technique at the service of the food industry. Aditivos e Ingredientes, São Paulo, v.12, jan.-fev., p.46–60, 2001.(In Portuguese) Broadhead, J., Edmond.

Mizrahi, S., Berk, Z., & Cogan, U. (1967). Isolated soybean protein as a banana spray drying aid. *Cereal Science Today*, *12*(8), 322–325.

Nandiyanto, A. B. D., & Okuyama, K. (2011). Progress in developing spray-drying methods for the production of controlled morphology particles: From the nanometer to submicrometer size ranges. *Advanced Powder Technology*, *22*(1), 1–19.

Notter, G. K., Taylor, O. H., & Downes, N. J. (1959). Orange juice powder. Factors affecting storage stability. . *Food Technology*, *13*, 113–118.

Orr, C., Jr., & DallaValle, J. M. (1959). *Particle size determination* (p. 353) Macmillan.

Pagington, J. S. (1986). Beta-cyclodextrin and its uses in the flavour industry. In G. G. Birch, & M. G. Lindley (Eds.), *Developments in food flavours*. Elsevier Applied Science.

Papadakis, S. E., & Bahu, R. E. (1992). The sticky issues of drying. *Drying Technology*, *10*, 817–837.

Potdar, S. B., Landge, V. K., Barkade, S. S., Potoroko, I., & Sonawane, S. H. (2020). Flavor encapsulation and release studies in food. *Encapsulation of active molecules and their delivery system* (pp. 293–321). Elsevier.

Ramakrishnan, Y., Adzahan, N. M., Yusof, Y. A., & Muhammad, K. (2018). Effect of wall materials on the spray drying efficiency, powder properties and stability of bioactive compounds in tamarillo juice microencapsulation. *Powder Technology*, *328*, 406–414.

Raybaudi-Massilia, R. M., & Mosqueda-Melgar, J. (2012). Polysaccharides as carriers and protectors of additives and bioactive compounds in foods. *The Complex Word of Polysaccharides*, *16*, 429–453.

Reineccius, G. (1991). Off-flavors in foods. *Critical Reviews in Food Science & Nutrition*, *29*(6), 381–402.

Sahai, D., & Jackson, D. S. (1996). Structural and chemical properties of native corn starch granules. *Starch/Starke*, *487*(8), 249–255.

Santos, A. B. D., Fávaro-Trindade, C. S., & Grosso, C. R. F. (2005). Preparation and characterization of paprika oleoresin microcapsules obtained by atomization. *Food Science and Technology*, *25*, 322–326. Available from https://doi.org/10.1590/s0101-20612005000200024. (In Portuguese).

Santos, D., Maurício, A. C., Sencadas, V., Santos, J. D., Fernandes, M. H., & Gomes, P. S. (2018). *Spray drying: An overview. Biomaterials-physics and chemistry-new edition* (pp. 9–35). IntechOpen.

Schubert, H. (1987). Food particle technology. Part I: Properties of particles and particulate food systems. *Journal of Food Engineering*, *6*(1), 1–32.

Shahid, F., & Han, X. Q. (1993). Encapsulation of food ingredients. *Critical Review in Food Science and Nutrition*, *33*(6), 501–547.

Thiengnoi, P., Suphantharika, M., & Wongkongkatep, P. (2012). Influence of binder typer and concentration on physical properties of agglomerated, spray-dried, and high oil loaded microcapsules. *Journal of Food, Agriculture & Environment*, *10*(3&4), 141–150.

Thomas, D. J., & Atwell, W. A. (1999). *Starches* (p. 94) Minnesota: Egan Press. (Handbook Series).

Timilsena, Y. P., Haque, M. A., & Adhikari, B. (2020). Encapsulation in the food industry: A brief historical overview to recent developments. *Food and Nutrition Sciences*, *11*(06), 481.

Tsourouflis, S., Flink, J. M., & Karel, M. (1976). Loss of structure in freeze dried carbohydrates solutions: Effect of temperature, moisture content and composition. *Journal of the Science of Food and Agriculture*, *27*(6), 509–519.

Wandrey, C., Bartkowiak, A., & Harding, S. E. (2010). Materials for encapsulation. In N. J. Zuidam, & V. Nedovic (Eds.), *Encapsulation technologies for active food ingredients and food processing* (pp. 31–100). New York: Springer.

Wang, X., Yuan, Y., & Yue, T. (2015). The application of starch-based ingredients in flavor encapsulation. *Starch-Stärke*, *67*(3–4), 225–236. Available from https://doi.org/10.1002/star.201400163.

Zheng, M., Jin, Z., & Zhang, Y. (2007). Effect of cross-linking and esterification on hygroscopicity and surface activity of cassava maltodextrins. *Food Chemistry*, *103*(4), 1375–1379. Available from https://doi.org/10.1016/j.foodchem.2006.10.053.

CHAPTER

13

Starch-based packaging and coating polymers for food

Pricila Veiga-Santos and Lucas Felipe dos Ouros

DPV-Food Technolgy, São Paulo State University - UNESP, Botucatu, Sao Paulo, Brazil

13.1 Introduction

In the food industry, plastic packaging materials are applied in order to facilitate and improve food transportation, protecting against environmental variables (e.g., temperature, humidity, light, oxygen, and radiation), which is a great method to prolong the shelf life. However, these packages are obtained from nonrenewable sources, are single-use products, and are generally harmful to the environment. Searching for more sustainability related to packaging and coating materials, which accounts for 60% of plastic products (Wu et al., 2021), agro-based biomaterials with the ability to form films and coatings have attracted attention for animal-based (ex. chitin, chitosan, and whey protein) and plant-based (pectin, cellulose, and starch) alternatives (Azevedo et al., 2017; Chan-Matú et al., 2021; Mary et al., 2022). In this context, the bio-based packaging materials market value is expected to reach USD 118.85 billion in 2026, with a global production capacity of around 2.34 million metric tons in 2024 (Mordor Intelligence, 2022). Among these materials, starch is one of the most promising substitutes, since it has low cost, high availability, and good properties for film formation (Sánchez Riaño et al., 2022). Starch can be extracted from several raw materials, the most used from cereals (wheat, corn, and rice) and underground reserve organs (potato, cassava, and sweet potato) (for more information about world starch production, see Chapter 3 in Volume 1 of this series). When used as a base for polymers, starch can be applied to pack or coat various products, from fruits to vegetables, to snacks and dry products (Onyeaka et al., 2022).

After extraction, starches can be heated in the presence of water, forming a filmogenic solution, which, depending on the composition, can have a great effect on the formed polymer. Using high amylose content, for example, can be very interesting, since the higher the amylose content, the greater the crystalline domain and, consequently, the mechanical properties of the material (Patkar et al., 2020). However, a negative aspect of the film production process is the heating step, since the heat can break some of the intramolecular bonds and hydrogen bonds with water, decreasing the crystalline regions that would increase the mechanical strength (Corke et al., 2016). More details on the interaction of starch with

DOI: https://doi.org/10.1016/B978-0-323-90842-9.00012-1

water and heat can be found in Volume 1 of this series, with emphasis on Chapters 9 and 10.

13.2 Effects of starch type

There are two types of starch microstructures, linear and branched, which makes starch a heterogeneous material (Onyeaka et al., 2022). The amylose is the linear fraction, composed of unbranched polysaccharide chains, which is amorphous but can be responsible for increasing the crystallization of starch. Amylopectin is the branched structure, with molecular weight greater than amylose and which contribute largely to the recrystallization, or retrogradation, of gelatinized starch during storage (Jiang et al., 2020). More details on the organization of the starch molecule can be found in Volume 1 of this series, with emphasis on Chapters 9 and 10.

Amylose concentration affects polymer formation, where the higher the amylose concentration on the starch-based material, the better the mechanical performance. However, its compact structure can result in poor flexibility and a more difficult gelatinization and film-forming process. These problems could be minimized through higher gelatinization temperatures or by modifying the dispersion solution, for example, ionic liquids, alkali, and inorganic salt solutions (Zou et al., 2021). For more information on amylose content and crystallization of starches, see Chapters 9 and 10 in Volume 1 of this series.

The structural properties of the starch film can influence phase separation, which can happen depending on amylopectin concentration and phosphate groups' presence (Hu et al., 2018).

13.3 Starch polymer formation process

Native starch cannot be used as plastic material, which needs to be plasticized or modified in a deformable thermoplastic material, called thermoplastic starch (Bangar et al., 2021). Among the production methods to develop thermoplastic starch polymers, extrusion is one of the most well-known starch processing procedures, which involves the swelling of starch granules, loss of birefringence, melting, and starch solubilization (Liu et al., 2013; Liu et al., 2014 Onyeaka et al., 2022). Gelatinization of the starch granules occurs only when the water concentration is reduced, thus the shear forces affect the starch and allow for water transfer to the internal molecules (Onyeaka et al., 2022).

Especially in pilot scale or laboratory film and coating formation investigations, the use of casting is a very common procedure, usually with the drying step over air circulation ovens, normally applying temperatures from ambient to around 100°C (Santos et al., 2020), where the lower the temperature, the better we can preserve additives and active or smart compounds. However, casting processes are not the best large-scale method for film production, which can be a hindrance for industrial utilization (González-Seligra et al., 2017).

Extrusion is a well-tested process for starch-based materials production and more viable to industry scale. It consists of applying high pressure, heat, and mechanical shear to starch granules, in the presence of appropriate moisture, triggering chemical and physical reactions. In response, the crystalline structure of starch will be affected, with consequent conversion into thermoplastic starch (González-Seligra et al., 2017).

According to Carvalho et al. (2003), plasticizing agents, such as glycerol, also can reduce amylopectin degradation during extrusion, specially when under decreased screw speed (Davidson et al., 1984; Wen et al., 1990; Willett et al., 1997). However, in the absence of shear stress, Olkku and Rha (1978) found that the factors such as water content and temperature have a higher effect on the extruded starch material. When performed at temperatures below 50°C, van der Waals forces or hydrogen bonds have a big interference on starch stabilization, due to intact crystalline components' existence. When heated above starch gelatinization temperature, greater swelling and dissolution of the crystalline is observed due to the breaking of the bonds between the water molecules and the hydroxyl group of the starch molecules (Lim et al., 2000; Tester & Debon, 2000).

13.4 Coatings and films

There are three ways to use starch in biofilms production as packaging materials. The first is partially replacing plastic polymers. Although this does not make the material biodegradable or sustainable, it increases sustainability and allows for better performance of the starch portion of the material. The second is replacing more than half of the plastic composition; this option follows the improvement of the advantages presented in the first option. The third one is the production of fully biodegradable starch-based films, being in consequence, fully biodegradable and obtained from renewable sources, besides the ecofriendly and already mentioned benefits (nonfossil sources, atoxic nature, easy degradation in nature, abundance of starch sources, and no contribution to CO_2 release). The attributes of the polymer matrix can also be considered (odorless, colorless, permeability to flavoring agents, and to gasses, the last one being an advantage or a disadvantage, depending on the packaged product) (Flieger et al., 2003; Oyom et al., 2022). For more information on biodegradables and biobased plastics, see Chapter 14.

Starch-based materials utilization as packaging and coating for food products is successfully reported, specially considering food and vegetables, allowing for moisture and gas exchange modulation, prolonging senescence, and increasing shelf life as lowering weight, firmness, and chlorophyll loss during storage (Oyom et al., 2022; Thakur et al., 2019). Such barrier selectivity can also be interesting for other products, such as fermented food as cheeses (Veiga-Santos et al., 2022). However, depending on the packed product, gas and humidity permeability can be a problem, which is the present challenge in improving protection performance, such as barrier and mechanical characteristics, making them as competitive as conventional nonbiodegradable materials (e.g., plastics) (Mary et al., 2022).

Each biomaterial will have different functional groups and will require specific strategies to achieve the high performance desired. Starch materials, rich in hydroxyl groups (Mary et al., 2022), present high hydrophilicity, leading to a tendency for poor water solubility, water vapor and gas barrier, and mechanical performance characteristics (Bangar et al., 2021). Such performances have been overcome with additives, process modifications, and blending with other materials.

13.4.1 Additives

Many additives have been successfully added to starch polymers to improve mechanical, optic, gas, and water vapor permeability (Adjouman et al., 2017; Ben et al., 2022; Veiga-Santos et al., 2005). A very simple way to modify starch intermolecular forces and hydrogen bonds in packaging or coating polymers formation is adding additives, such as plasticizers (Bangar et al., 2021). Starch films are very brittle, which is a very limiting factor for their application. The use of plasticizers is a simple and very effective alternative to minimize this problem. Their addition results in decreasing strain stress; increasing hardness and viscosity; increasing elongation; and conferring flexibility and fracture resistance to the polymer chain (Han, 2014). They also can affect other materials' characteristics, such as increasing hydrophilicity, water activity, solubility in water, water absorption kinetic, water vapor permeability, and opacity (Adjouman et al., 2017; Costa et al., 2023; Veiga-Santos et al., 2005).

Although polyols are among the most used additives, Veiga-Santos et al. (2007) successfully substituted them with sugars, such as sucrose and invert sugar, with the advantage of being cheaper, with food grade, easier to access, and more cost-effective.

Another low-cost, highly stable, and efficient alternative is the use of nanofillers, such as montmorillonite nanoclays, which can improve the mechanical and barrier properties of starch materials (Nafchi et al., 2013; Park et al., 2003). Also, an option is the integration of starch with nonbiodegradable materials (Jayakumar et al., 2022; Istiqomah et al., 2022; Kumar et al., 2021) and nanomaterials. Silica nanoparticles, for example, can reduce starch-based polymers' water vapor permeability, helping to extend food shelf life and freshness (Jiménez et al., 2012; Yu et al., 2018). For more information, see Chapter 14.

However, researchers are prioritizing renewable and biodegradable matrixes, aiming for more sustainable blends (Benbettaïeb et al., 2018; da Róz et al., 2016; Bhargava et al., 2020; Istiqomah et al., 2022). In this context, many biobased nanomaterials (e.g., starch, chitosan, gums, and cellulose-based polymers) have been investigated (Al-Tayyar et al., 2020). Additives (Santos et al., 2020; Vaezi et al., 2019) and storage conditions (Onyeaka et al., 2022; Scamparini & Veiga-Santos, 2016), however, can affect starch film and coating optical properties, such as increasing opacity, or make it undergo additive-related chemical reactions, such as crystallization and browning.

13.4.2 Process modifications

Structural modifications had also been successfully applied to starch from different sources for packaging or coating application (Bangar et al., 2021), with previously modified starch, such as dialdehyde starch (Bajer & Burkowska-But, 2022); or using process modifications applied during the polymer formation, such as through reactive extrusion (Da Róz et al., 2016) or ozone gas (La Fuente et al., 2019).

In recent years, studies are pursuing adding functionality to packaging and coatings in addition to conventional protection and satisfying the growing need of consumers to monitor their food during storage. In this sense, active and intelligent (or smart) materials have been developed, where, in brief, active polymers act in interacting within products increasing their shelf life. Smart (or intelligent) polymers interact, indicating shelf life modifications, such as time-temperature indicators, vacuum or gas indicators, humidity, pH, and calorimetric sensors (Firouz et al., 2021).

When polymers affect the atmosphere in contact with the product, they are also considered active. In this context, starch-based coatings can be applied, for example to fruits, by either reducing the rate of gas exchange in the fruit's respiration or by changing the composition of its internal atmosphere (Lima et al., 2022). Table 13.1 presents a few smart starch-based polymers' possibilities.

Besides the great benefits of using renewable, nontoxic, and biobased materials for packages and coatings, as discussed earlier in this work, the sustainable nature of the functional compounds added to active and smart materials must be prioritized (Pandey et al., 2022).

13.5 Active polymers

Many interesting active properties have been applied to biobased polymers, including the development of probiotic starch polymers (Soukoulis et al., 2016) and ethylene absorbers (Jung et al., 2020; Upadhyay et al., 2022).

TABLE 13.1 Smart polymers successfully developed with biobased polymers.

Polymer matrix	"Smart" compound	"Smart" properties	References
Cassava starch	Myoglobin protein	Temperature indicator	Santos et al. (2020)
	Carbohydrates	Temperature indicator	Scamparini and Veiga-Santos (2016)
	Grape and spinach pomace	pH indicator	Veiga-Santos et al. (2011)
	Cassava starch coating	Changed atmosphere	Lima et al. (2022)
Pea starch	Grape skin and metal–organic frameworks crystals	Pork freshness indicator	Lin et al. (2022)

However, this review has focused on antimicrobial and antioxidant materials, which are the most studied.

13.5.1 Antimicrobial active packaging materials

Considering the great susceptibility of food products to cause food-borne pathogen diseases and microbial degradation, polymers that carry antimicrobial components present an excellent extended packed product shelf life opportunity, regarding good spoilage control and safety. Starches can result in successful matrices for antimicrobial polymers or coatings considering their capacity of blending with both water and lipid based natural antimicrobial compounds (Jogee et al., 2021; Oyom et al., 2022; Tadini et al., 2016). However, regarding the great importance of water activity for microbial growth, starch polymers sensitivity for water gain or loss must be considered when planning storage conditions (Adjouman et al., 2017; Costa et al., 2023; Veiga-Santos et al., 2005).

Among the many mechanisms involved in antimicrobial packaging and their mode of action, they can be classified according to their action on microorganisms, by damaging (1) cell membranes or (2) biomolecules inside the cells (e.g., proteins); (3) interfering with metabolic processes; and through (4) oxidative stress (Nizam et al., 2021; Pandey et al., 2022).

According to Zhao et al. (2022), when aiming for sustainability, not only is the polymer matrix material important to consider, but also the search for sustainable and safer antimicrobial components has become a hotspot topic that has attracted attention. However, their effectiveness is highly dependent on the amount of compounds added to the polymer matrix and the amount released to the packed product during shelf life. The antimicrobial release profile is normally controlled by changes in the packaged product, in the polymer or antimicrobial agent itself; or between this agent and the packaged product (Zhong et al., 2020). Kechichian et al., (2010), for example, had followed the efficacy of antimicrobial agents added to cassava starch film polymers, by packing breads with these polymers and controlling their microbial stability during storage.

Nizam et al. (2021) have classified antimicrobial agents based on their chemical composition, between organic (for example, essential oils) and inorganic (for example, metallic particles and nanoparticles). Essential oils have been especially prominent among organic antimicrobial agents, being successfully investigated in starch-based matrices, both in films (Tadini et al., 2016; Alves-Silva et al., 2022) and in coating (Yan et al., 2022) polymers. Sadeghizadeh-Yazdi et al. (2019) noticed also a plasticizing and flavoring effect when adding lemon essential oil into yam and corn starch films.

Although essential oils are considered safe for food application, factors such as their pronounced flavor, their high volatility, and their sensitivity to heat, moisture, and light must be considered (Bai et al., 2022). Besides essential oils, there are many other interesting nontoxic, low-cost, and abundant-in-nature options among organic antimicrobial compounds. A successful example is *Aloe vera*, which when applied to starch polymers have indicated potential to inhibit gram-positive and gram-negative bacteria and molds (Nieto-Suaza et al., 2019; Nizam et al., 2021).

Among inorganic antimicrobial agents applied to starch materials, silver and silver nanoparticles have gained attention due to the facility of being incorporated into starch materials and to the possibility of providing antimicrobial activity and increased mechanical and gas and water vapor permeability resistance. However, it must be considered a possible toxic effect related to prokaryotic cells when these agents are used in high concentrations (Ceballos et al., 2021; Cui et al., 2021; Jogee et al., 2021; Peighambardoust et al., 2019).

There are some aspects that are very important to contribute to the antimicrobial efficacy, such as barrier and mechanical properties, and binding affinity between antimicrobial components and polymeric matrices. Despite the environmental advantages of using biobased and renewable matrices and their good interaction with antimicrobial compounds, when compared to conventional polymers, they normally present poor barrier and mechanical properties (Al-Tayyar et al., 2020; Pandey et al., 2022). In this context, some material modifications (additives, blending, and process modifications), as suggested earlier, can be applied.

Table 13.2 presents some active antimicrobial materials that have been successfully developed for starch-based matrices.

13.5.2 Antioxidant polymers

Traditional antioxidant agents, such as butylhydroxyanisole and butylhydroxytoluene, are usually more commonly applied to packaging polymers due to their already proven effectiveness and their already permitted and regulated use in food. However, the possible toxicity of these synthetic antioxidants is driving a greater demand for natural compounds (Kumar et al., 2021; Veiga-Santos et al., 2022).

Most antioxidant agents have a low molecular weight and show high instability and volatility. This aspect is mitigated by applying these agents in polymers (Peng et al., 2022; Tran et al., 2019), an extra advantage of using starch in the development of antioxidant polymers.

The main form and action of the antioxidant active polymers is to minimize or delay the formation of free radicals in the packaged food, avoiding the chain reaction that harms product quality and decreases its shelf life (Peng et al., 2022; Veiga-Santos et al., 2022). Depending on their mode of action, they are recognized as primary or secondary antioxidants. The primary ones react directly as radical scavengers, generating alkyl and peroxyl radicals, whereas the secondary ones interrupt the generation of new radicals (Mayer et al., 2022).

Many antimicrobial agents also act as antioxidants, as is the case of phenolic compounds and essential oils, and can even be added to polymers seeking both activities (Istiqomah et al., 2022). The oxygen absorbers are another example. When they can minimize oxidation reactions, they also can minimize microorganisms' growth considering their need for an oxygen-rich environment to survive (Promsorn & Harnkarnsujarit, 2022).

Many antioxidant compounds can impart color to the packaging or coating material, either by its own coloration, or by coloration due to some darkening reaction that can occur during storage. Such color addition can be an interesting extra protection (Veiga-Santos et al., 2018), since light is an important catalyst of the oxidation reaction (Fig. 13.1).

Recently, a niche has appeared among the sources for natural antioxidants, which are

TABLE 13.2 Antimicrobial compounds successfully used in biobased polymers.

Polymer matrix	Active compound	References
Cassava starch	Cinnamon and glove	Tadini et al. (2016)
	Peppermint oil and lime oil	Owolabi et al. (2021)
	Clove essential oil	Alves-Silva et al. (2022)
	Thyme essential oil	Sapper et al. (2019)
	Silver nanoparticles and yerba mate extract	Ceballos et al. (2021)
Oxidized esterified tapioca starch	Oregano essential oil filled into mesoporous nanosilica	Yan et al. (2022)
Modified sweet potato starch	Cumin essential oil	Oyom et al. (2022)
Modified moth bean starch	Basil leaves extract	Kumar et al. (2021)
Rice starch	Coconut oil, tea leaf extract	Das et al. (2013)
Green banana starch	Aloe vera and curcumin	Nieto-Suaza et al. (2019)
Starch	Aloe vera	Ortega-Toro et al. (2017)
	Ag, ZnO, and CuO	Peighambardoust et al. (2019)

Adapted from Oyom, W., Zhang, Z., Bi, Y., & Tahergorabi, R. (2022). Application of starch-based coatings incorporated with antimicrobial agents for preservation of fruits and vegetables: A review. Progress in Organic Coatings. *166, 106800. https://doi.org/10.1016/j.porgcoat.2022.106800.*

FIGURE 13.1 Cassava biobased films added with (A) cocoa and coffee compounds and with (B) myoglobin-nitrite after time-temperature exposure.

obtained from solid residues from the food industry. These residues are often rich in antioxidant compounds such as phenolics from grape seeds and peels, red cabbage stirs (Veiga-Santos et al., 2022), and pineapple peels (Kumar et al., 2021).

Table 13.3 presents some antioxidant starch-based polymers with antioxidant compounds that could be applied to starch polymers.

Among the advantages of using antioxidant packaging rather than adding these compounds directly to the food is the possibility of a prolonged release, allowing for a more continuous action during the shelf life of the product. The association of nanomaterials such as nanochitosan with the starch polymer matrix is an effective way to ensure a gradual release of antioxidants into the packaged product (Ma et al., 2022) and can contribute to starch film or coating polymers in many ways.

Nanotechnology can be a great ally to antimicrobial polymers, fighting fungal and bacterial contamination in food, resulting in increased quality, safety, and shelf life (He et al., 2019; Sadeghizadeh-Yazdi et al., 2019). Nanomaterials' application to polymers can increase polymer physicochemical (optical, solubility, etc.) and thermodynamic properties. It is worth mentioning that nanomaterials used in polymers directly in contact with the food must be atoxic and generally recognized as safe, in addition to being stable to adverse conditions used in the food industry, such as pressure, pH, and temperature (Mohamadian et al., 2019; Roselli et al., 2003; Samadi et al., 2018).

The application of nanomaterials can improve the quality of both biodegradable packaging and packed food. Kuswandi and Moradi (2019) have reported that the use of nanomaterials enables a decrease in the polymer gas permeability by up to 90%, which can have a positive impact minimizing degradation mechanisms (such as oxidation, microorganisms proliferation, gas barrier, and respiration of horticultural products), affecting the shelf life of the packed product (Flores et al., 2007; Joye et al., 2016; Khare et al., 2014; Yoksan & Chirachanchai, 2010).

Starch-based polymers' water vapor permeability and moisture absorption can also be decreased with the addition of nanomaterials, such as graphene oxide, which will have a great effect on mechanical strength (Aqlil et al., 2017; Wu et al., 2019).

Promising nanoparticles for biobased packaging and coating are nanoclays, which are

TABLE 13.3 Antioxidant compounds successfully used in starch biobased polymers.

Polymer matrix	Active compound	References
Cassava starch	Coffee and Cocoa	Veiga-Santos et al. (2018)
Cassava starch	Gape and red cabbage pomace	Veiga-Santos et al. (2022)
Cassava starch	Mango and acerola cherry	Souza et al. (2011)
Potato Starch	Sunflower hulls	Menzel et al. (2019)
Potato Starch	Rice straw	Menzel et al. (2020)
Corn starch	α-Tocopherol	Jiménez et al. (2012)
Corn starch	Pineapple peel extract	Kumar et al. (2021)
Potato starch and chitosan nanoparticles	Potato peel	Ma et al. (2022)
Cassava starch	Gallic acid	Promsorn and Harnkarnsujarit (2022)
Dioscorea hispida starch	Lemongrass essential oil	Istiqomah et al. (2022)

proportionally cheap, efficient, and stable. Nanoclays have a thickness of 1 nm, resulting in a largest contact surface of nanometric charges, in turn resulting in a large interface between the biopolymer and its nanofiller, which allows for physical, thermal, and permeability barrier increase (Benbettaïeb et al., 2018). The aforementioned properties are essential for the food packaging materials, as they are necessary for the processing and preservation of the packed foods (Mcglashan & Halley, 2003).

13.6 Smart polymers

While active packaging increases the shelf life of the packed product, smart (or intelligent) packaging reports changes that occur in product during shelf life, or during the processes to be performed at the consumer's home, such as heating or freezing. Intelligent polymers have additives that allow for controlling the internal environment of the package and/or parameters related to the packaged product quality (e.g., temperature, humidity, pH, and microbial contamination). Usually, such materials alert consumers about product quality through easily observed warnings: color changes being among the most studied. Changes in pH can be related to both chemical reactions byproducts and microbial growth metabolites formed during shelf life, resulting in visual changes of the indicators when monitored during storage (Bhargava et al., 2020; Santos et al., 2020).

When reporting pH modifications, chemochromic natural pigments, such as phenolic compounds, are among the most investigated (Lin et al., 2022) due to their ability in changing color depending on the medium pH. Fig. 13.2 demonstrates how cassava starch films added with tannin and anthocyanin can respond to different pH environments.

No acid chemochromic changing responses were noticed for tannin-starch films, as shown in Fig. 13.2(A), indicating that chemochromic color changes are specific for determined pH ranges. The specific condition where a color change will occur will not always indicate a desired or undesired change in pH quality.

In addition to phenolics, other pH indicators were tested on nonstarch biodegradable matrices and show great potential for application in starch-based films and coatings. The use of Alizarin resulted in a change from a light yellow to purple coloration, responding to pH changes between 4 and 10 (Ezati et al., 2020).

Chlorophyll obtained from green tea extract also showed chemochromic potential, resulting in loss of coloration at low pH (pH 3.0) due to degradation of chlorophyll into pheophytin and pheophorbide. In cases of ammonia-like compounds released from the packed product, the green tea chlorogenic acid could oxidize into quinines, with possible color change to brown and yellow (Jamróz et al., 2019).

Smart freshness indicators can also be related to other metabolites, such as carbonic gas, ammonia, hydrogen sulfide, and amines. A carbonic gas indicator polymer was developed using the chemical dyes methyl red, bromothymol blue, and polyethyleneimine as pH dependent indicators, considering that CO_2 level changes may affect pH during storage (Lyu et al., 2019). Other chemical chemochromic dyes have been studied for CO_2 indicator materials such as thymol blue, phenol red, cresol red, or luminescent dyes affected by CO_2 levels (Puligundla et al., 2012).

In many cases, intelligent polymers also act as active polymers (Ezati et al., 2020). Lin et al. (2022) have developed a pork freshness indicator for PVA/Pea starch films, where grape skin anthocyanins act as color pH indicators, expecting to correlate the production of volatile organic amines produced during the meat shelf life with a pH increase. Metal–organic framework crystals were also incorporated into the polymer, expecting that their organic ligands share their positive charge with metal ions present in bacterial cell membranes, acting also as antimicrobial material.

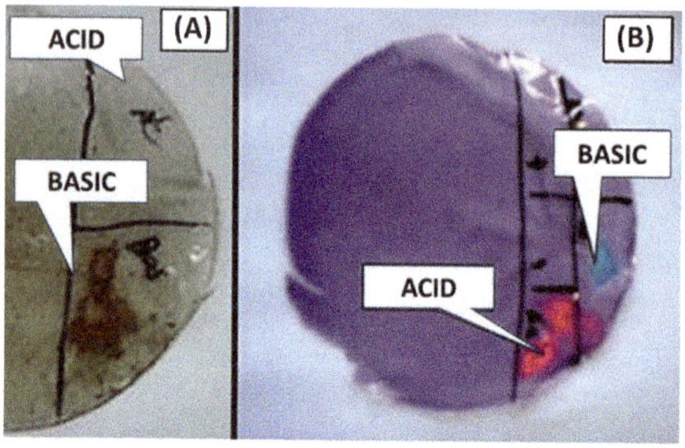

FIGURE 13.2 Cassava starch films rich in (A) tannin and (B) anthocyanin, chemochromic color responses when basic or acid pH solutions were applied on the material surfaces.

Smart polymers can also report changes that the packaged product has undergone during storage or when exposed to a specific situation. An example would be temperature indicators, which can warn of desirable or undesirable exposures that the product has been subjected to. They can be reversible, when the change lasts only when the product is exposed to the temperature, or irreversible, when even after this exposure, it presents a change without returning to its natural state. Irreversible indicators would be more interesting in situations such as to prove a thermal processing performed, like sterilization, or to ensure that a freezing temperature has not been affected during storage or transport (Scamparini & Veiga-Santos, 2016).

13.7 Final considerations

Different starch sources have already been successfully studied for the development of biodegradable coatings and films, and their viability as polymeric matrices for active and intelligent compounds application were demonstrated. Considering the sustainable character, high productivity, and low cost of starch in general, it is undeniable that it ranks among the most promising materials for future generations.

Because starch films and coatings do not have the same mechanical and barrier characteristics as petroleum-based synthetic polymers, aspects like high gas permeability can be pointed out as weaknesses considering protection. However, they can be exploited in a different perspective, being reviewed as strengths for specific applications. For example, a greater gas permeability can allow respiration of fruits. becoming a highly desirable requirement. Further research should therefore not only focus on improving the competitiveness of these materials against the protection requirements of conventional plastic packaging, but also look for applications for which their characteristics have a better protective vocation.

Smart packaging has become a more reliable method for checking the integrity of products than the expiration date. Any variation due to incorrect handling of the product can be recorded on the indicator. Likewise, a product that still fits for consumption does not have to be discarded if it reaches its original expiration date.

Active polymer technology enables a shelf life increase, with the benefit of a gradual release of the active compounds to the product, avoiding their direct addition and therefore allowing for healthier and more green label products.

The combination of smart and active polymers makes it possible to increase the packed product shelf life and communicate its quality to consumers, demonstrating that the company that uses this technology is concerned about not wasting food and food security and intends to increase consumers' convenience, adding value to their product.

A major limiting factor in the large-scale use of these technologies is its cost, for which starch becomes an indispensable ally, contributing both to reduce smart/active packaging costs and making them more sustainable. They also have the advantage that they easily allow for the incorporation of natural smart and active compounds, even when obtained from agroindustrial residues.

References

Adjouman, Y. D., Nindjin, C., Tetchi, F. A., Dalcq, A. C., & Amani, N. G. (2017). Water vapor permeability of edible films based on improved cassava (*Manihot esculenta* Crantz) native starches. *Food Processing and Technology, 8*(3), 665. Available from https://doi.org/10.4172/2157-7110.1000665.

Al-Tayyar, N. A., Youssef, A. M., & Al-Hindi, R. (2020). Antimicrobial food packaging based on sustainable bio-based materials for reducing foodborne Pathogens: A review. *Food Chemistry, 310*. Available from https://doi.org/10.1016/j.foodchem.2019.125915.

Alves-Silva, G. F., Santos, L. G., Martins, V. G., & Cortez-Vega, W. R. (2022). Cassava starch films incorporated with clove essential oil and nanoclay as a strategy to increase the shelf life of strawberries. *International Journal of Food Science & Technology, 57*, 6690–6698. Available from https://doi.org/10.1111/ijfs.16014.

Aqlil, M., Moussemba Nzenguet, A., Essamlali, Y., Snik, A., Larzek, M., & Zahouily, M. (2017). Graphene oxide filled lignin/starch polymer bionanocomposite: Structural, physical, and mechanical studies. *Journal of Agricultural and Food Chemistry, 65*(48), 10571–10581. Available from https://doi.org/10.1021/ACS.JAFC.7B04155/ASSET/IMAGES/MEDIUM/JF-2017-04155R_0009.GIF.

Azevedo, V. M., Borges, S. V., Marconcini, J. M., Yoshida, M. I., Neto, A. R. S., Pereira, T. C., & Pereira, C. F. G. (2017). Effect of replacement of corn starch by whey protein isolate in biodegradable film blends obtained by extrusion. *Carbohydrate Polymers, 157*, 971–980. Available from https://doi.org/10.1016/j.carbpol.2016.10.046.

Bai, M., Zhou, Q., Zhang, J., Li, T., Cheng, J., Liu, Q., Xu, W., & Zhang, Y. (2022). Antioxidant and antibacterial properties of essential oils-loaded β-cyclodextrin-epichlorohydrin oligomer and chitosan composite films. *Colloids and Surfaces B: Biointerfaces, 215*. Available from https://doi.org/10.1016/j.colsurfb.2022.112504.

Bajer, D., & Burkowska-But, A. (2022). Innovative and environmentally safe composites based on starch modified with dialdehyde starch, caffeine, or ascorbic acid for applications in the food packaging industry. *Food Chemistry, 374*. Available from https://doi.org/10.1016/j.foodchem.2021.131639.

Bangar, S. P., Whiteside, W. S., Ashogbon, A. O., & Kumar, M. (2021). Recent advances in thermoplastic starches for food packaging: A review. *Food Packaging and Shelf Life, 30*. Available from https://doi.org/10.1016/j.fpsl.2021.100743.

Ben, Z. Y., Samsudin, H., & Yhaya, M. F. (2022). Glycerol: Its properties, polymer synthesis, and applications in starch based films. *European Polymer Journal, 175*, 111377. Available from https://doi.org/10.1016/j.eurpolymj.2022.111377, ISSN 0014–3057.

Benbettaïeb, N., Debeaufort, F., & Karbowiak, T. (2018). Bioactive edible films for food applications: Mechanisms of antimicrobial and antioxidant activity. *Critical Reviews in Food Science and Nutrition, 59*(21), 3431–3455. Available from https://doi.org/10.1080/10408398.2018.1494132.

Bhargava, N., Sharanagat, V. S., Mor, R. S., & Kumar, K. (2020). Active and intelligent biodegradable packaging films using food and food waste-derived bioactive compounds: A review. *Trends in Food Science & Technology, 105*, 385–401. Available from https://doi.org/10.1016/j.tifs.2020.09.015.

Carvalho, A. J. F., Zambon, M. D., Curvelo, A. A. S., & Gandini, A. (2003). Size exclusion chromatography characterization of thermoplastic starch composites 1. Influence of plasticizer and fibre content. *Polymer Degradation and Stability, 79*(1), 133–138. Available from https://doi.org/10.1016/S0141-3910(02)00265-3.

Ceballos, R. L., Bilderling, C. V., Guz, L., Bernal, C., & Famá, L. (2021). Effect of greenly synthetized silver nanoparticles on the properties of active starch films obtained by extrusion and compression molding. *Carbohydrate Polymers, 261*, 117871. Available from https://doi.org/10.1016/j.carbpol.2021.117871.

Chan-Matú, D. I., Toledo-López, V. M., Vargas, M. L. V., Rincón-Arriaga, S., Rodríguez-Félix, A., & Madera-Santana, T. J. (2021). Preparation and characterization of chitosan-based bioactive films incorporating *Moringa oleifera* leaves extract. *Food Measure, 15*, 4813–4824. Available from https://doi.org/10.1007/s11694-021-01055-w.

Corke, H., Faubion, J., Seetheraman, K., & Wrigley, C. (2016). *Encyclopedia of food grains*. Cambridge, UK: Academic Press.

Costa, B. P., Carpiné, D., Ikeda, M., Pazzini, I. A. E., Alves, F. E., da, S. B., de Melo, A. M., & Ribani, R. H. (2023). Bioactive coatings from non-conventional loquat (*Eriobotrya japonica* Lindl.) seed starch to extend strawberries shelf-life: An antioxidant packaging. *Progress in Organic Coatings, 175*. Available from https://doi.org/10.1016/j.porgcoat.2022.107320.

Cui, C., Ji, N., Wang, Y., Xiong, L., & Sun, Q. (2021). Bioactive and intelligent starch-based films: A review. *Trends in Food Science & Technology, 116*, 854–869. Available from https://doi.org/10.1016/j.tifs.2021.08.024.

Da Róz, A. L., Veiga-Santos, P., Ferreira, A. M., Antunes, T. C. R., Leite, F. L., & De Carvalho, A. J. F. (2016). Water susceptibility and mechanical properties of thermoplastic starch–pectin blends reactively extruded with edible citric acid. *Materials Research, 19*(1). Available from https://doi.org/10.1590/1980-5373-MR-2015-0215.

Das, D. K., Dutta, H., & Mahanta, C. L. (2013). Development of a rice starch-based coating with antioxidant and microbe-barrier properties and study of its effect on tomatoes stored at room temperature. *LWT - Food Science and Technology, 50*(1), 272–278. Available from https://doi.org/10.1016/j.lwt.2012.05.018.

Davidson, V. J., Paton, D., Diosady, L. L., & Larocque, G. (1984). Degradation of wheat starch in a single screw extruder: Characteristics of extruded starch polymers.

Journal of Food Science, *49*(2), 453–458. Available from https://doi.org/10.1111/J.1365-2621.1984.TB12441.X.

Ezati, P., Rhim, J., Moradi, M., Tajik, H., & Molaei, R. (2020). CMC and CNF-based alizarin incorporated reversible pH-responsive color indicator films. *Carbohydrate Polymers*, *246*. Available from https://doi.org/10.1016/j.carbpol.2020.116614.

Firouz, M. S., Mohi-Alden, K., & Omid, M. (2021). A critical review on intelligent and active packaging in the food industry: Research and development. *Food Research International*, *141*, 110113. Available from https://doi.org/10.1016/j.foodres.2021.110113.

Flieger, M., Kantorová, M., Prell, A., Řezanka, T., & Votruba, J. (2003). Biodegradable plastics from renewable sources. *Folia Microbiologica*, *48*(1), 27–44. Available from https://doi.org/10.1007/BF02931273, 2003 48:1.

Flores, S., Famá, L., Rojas, A. M., Goyanes, S., & Gerschenson, L. (2007). Physical properties of tapioca-starch edible films: Influence of filmmaking and potassium sorbate. *Food Research International*, *40*(2), 257–265. Available from https://doi.org/10.1016/J.FOODRES.2006.02.004.

González-Seligra, P., Guz, L., Ochoa-Yepes, O., Goyanes, S., & Famá, L. (2017). Influence of extrusion process conditions on starch film morphology. *LWT*, *84*, 520–528. Available from https://doi.org/10.1016/j.lwt.2017.06.027, ISSN 0023-6438.

Han, J. H. (2014). Edible films and coatings: A review. *Innovations in food packaging*, (pp. 213–255). San Diego, CA, USA: Academic Press. Available from https://www.elsevier.com/books/innovations-in-food-packaging/han/978-0-12-394601-0.

He, X., Deng, H., & Hwang, H. M. (2019). The current application of nanotechnology in food and agriculture. *Journal of Food and Drug Analysis*, *27*(1), 1–21. Available from https://doi.org/10.1016/J.JFDA.2018.12.002.

Hu, H., Zhang, R., Wang, J., Ying, W., Bin., & Zhu, J. (2018). Fully bio-based poly(propylene succinate-co-propylene furandicarboxylate) copolyesters with proper mechanical, degradation and barrier properties for green packaging applications. *European Polymer Journal*, *102*, 101–110. Available from https://doi.org/10.1016/J.EURPOLYMJ.2018.03.009.

Istiqomah, A., Prasetyo, W. E., Firdaus, M., & Kusumaningsih, T. (2022). Valorisation of lemongrass essential oils onto chitosan-starch film for sustainable active packaging: Greatly enhanced antibacterial and antioxidant activity. *International Journal of Biological Macromolecules*, *210*, 669–681. Available from https://doi.org/10.1016/j.ijbiomac.2022.04.223.

Jamróz, E., Kulawik, P., Krzyściak, P., Talaga-Ćwiertnia, K., & Juszczak, L. (2019). Intelligent and active furcellaran-gelatin films containing green or pu-erh tea extracts: Characterization, antioxidant and antimicrobial potential. *International Journal of Biological Macromolecules*, *122*, 745–757. Available from https://doi.org/10.1016/j.ijbiomac.2018.11.008.

Jayakumar, A., Radoor, S., Kim, J. T., Rhim, J. W., Parameswaranpillai, J., Nandi, D., Srisuk, R., & Siengchin, S. (2022). Titanium dioxide nanoparticles and elderberry extract incorporated starch based polyvinyl alcohol films as active and intelligent food packaging wraps. *Food Packaging and Shelf Life*, *34*, 2022. Available from https://doi.org/10.1016/j.fpsl.2022.100967.

Jiang, T., Duan, Q., Zhu, J., Liu, H., & Yu, L. (2020). Starch-based biodegradable materials: Challenges and opportunities. *Advanced Industrial and Engineering Polymer Research*, *3*(1), 8–18. Available from https://doi.org/10.1016/J.AIEPR.2019.11.003.

Jiménez, A., Fabra, M. J., Talens, P., & Chiralt, A. (2012). Effect of re-crystallization on tensile, optical and water vapour barrier properties of corn starch films containing fatty acids. *Food Hydrocolloids*, *26*(1), 302–310. Available from https://doi.org/10.1016/J.FOODHYD.2011.06.009.

Jiménez, A., Fabra, M. J., Talens, P., & Chiralt, A. (2013). Physical properties and antioxidant capacity of starch–sodium caseinate films containing lipids. *Journal of Food Engineering*, *116*(3), 695–702. Available from https://doi.org/10.1016/j.jfoodeng.2013.01.010.11.

Jogee, P. S., Agarkar, G. A., & Rai, M. (2021). Chapter 6 - Starch-based films loaded with nano-antimicrobials for food packaging. In R. Mahendra, & S. Carolina Alves dos (Eds.), *Biopolymer-based nano films* (pp. 99–114). *Elsevier*. Available from https://doi.org/10.1016/B978-0-12-823381-8.00004-1.

Joye, I. J., Davidov-Pardo, G., & Mcclements, D. J. (2016). Nanotechnology in food processing. In. In B. Caballero, P. M. Finglas, & F. Toldrá (Eds.), *Encyclopedia of food and health* (pp. 49–55). Oxford, UK: Academic Press, ISBN 978-0-12-384953-3. Available from https://limo.libis.be/primo-explore/fulldisplay?docid = LIRIAS1639775&context = L&vid = Lirias&search_scope = Lirias&tab = default_tab&fromSitemap = 1.

Jung, J., Deng, Z., & Zhao, Y. (2020). Mechanisms and performance of cellulose nanocrystals Pickering emulsion chitosan coatings for reducing ethylene production and physiological disorders in postharvest 'Bartlett' pears (*Pyrus communis* L.) during cold storage. *Food Chemistry*, *309*, 125693. Available from https://doi.org/10.1016/J.FOODCHEM.2019.125693.

Khare, S., Williams, K., & Gokulan, K. (2014). Nanotechnology. In C. A. Batt, & M. L. Tortorello (Eds.), *Encyclopedia of food microbiology* (2nd ed., pp. 893–900). Oxford, UK: Academic Press, ISBN 978-0-12-384733-1. Available from https://azdoc.tips/documents/encyclopedia-of-food-microbiology-2014-5c15e347d7585.

Kumar, P., Tanwar, R., Gupta, V., Upadhyay, A., Kumar, A., & Gaikwad, K. K. (2021). Pineapple peel extract incorporated poly(vinyl alcohol)-corn starch film for active food packaging: Preparation, characterization and antioxidant activity. *International Journal of Biological Macromolecules*, *187*, 223–231. Available from https://doi.org/10.1016/j.ijbiomac.2021.07.136.

Kuswandi, B., & Moradi, M. (2019). Improvement of food packaging based on functional nanomaterial. In S. Siddiquee, G. J. H. Melvin, & M. M. Rahman (Eds.), *Nanotechnology: Applications in energy, drug and food* (pp. 309–344). Cham, Switzerland: Springer, ISBN 978-3-319-99602-8. Available from https://link.springer.com/book/10.1007/978-3-319-99602-8.

La Fuente, C. I. A., De Souza, A. T., Tadini, C. C., & Augusto, P. E. D. (2019). Ozonation of cassava starch to produce biodegradable films. *International Journal of Biological Macromolecules*, *141*, 713–720. Available from https://doi.org/10.1016/j.ijbiomac.2019.09.028.

Lim, M. H., Wu, H., & Reid, D. S. (2000). The effect of starch gelatinization and solute concentrations on Tg of starch model system. *Journal of Food and Agriculture*, *80*, 1757–1762. Available from https://doi.org/10.1002/1097-0010(20000915)80:12 < 1757::AID-JSFA703 > 3.0.CO;2-Y.

Lima, R. P., De Sousa, A. S. B., Abeli, P., Beaudry, R. M., & Silva, D. M. S. (2022). Setting a safe target internal atmosphere for starch-based coated fruits and vegetables. *Food Control*, *142*, 109270. Available from https://doi.org/10.1016/j.foodcont.2022.109270.

Lin, X., Li, N., Xiao, Q., Guo, Y., Wei, J., Jiao, T., Chen, Q., Chen, Q., & Chen, X. (2022). Polyvinyl alcohol/starch-based film incorporated with grape skin anthocyanins and metal-organic framework crystals for colorimetric monitoring of pork freshness. *Food Chemistry*, *395*, 133613. Available from https://doi.org/10.1016/j.foodchem.2022.133613.

Liu, X., Ma, H., Yu, L., Chen, L., Tong, Z., & Chen, P. (2014). Thermal-oxidative degradation of high-amylose corn starch. *Journal of Thermal Analysis and Calorimetry*, *115*(1), 659–665. Available from https://doi.org/10.1007/S10973-013-3354-2/FIGURES/6.

Liu, X., Wang, Y., Yu, L., Tong, Z., Chen, L., Liu, H., & Li, X. (2013). Thermal degradation and stability of starch under different processing conditions. *Starch - Stärke*, *65* (1–2), 48–60. Available from https://doi.org/10.1002/STAR.201200198.

Lyu, J. S., Choi, I., Hwang, K., Lee, J., Seo, J., Kim, S. Y., & Han, J. (2019). Development of a BTB /TBA + ion-paired dye-based CO2 indicator and its application in a multilayered intelligent packaging system. *Sensors and Actuators B: Chemical*, *282*, 359–365. Available from https://doi.org/10.1016/j.snb.2018.11.073.

Ma, Y., Zhao, H., Ma, Q., Cheng, D., Zhang, Y., Wang, W., Wang, J., & Sun, J. (2022). Development of chitosan/potato peel polyphenols nanoparticles driven extended-release antioxidant films based on potato starch. *Food Packaging and Shelf Life*, *31*, 100793. Available from https://doi.org/10.1016/j.fpsl.2021.100793.

Mary, S. K., Koshy, R. R., Arunima, R., Thomas, S., & Pothen, L. A. (2022). A review of recent advances in starch-based materials: Bionanocomposites, pH sensitive films, aerogels and carbon dots. *Carbohydrate Polymer Technologies and Applications*, *100190*, 2666–8939. Available from https://doi.org/10.1016/j.carpta.2022.100190.

Mayer, J., Metzsch-Zilligen, E., & Pfaendner, R. (2022). Novel multifunctional antioxidants for polymers using eugenol as biogenic building block. *Polymer Degradation and Stability*, *200*, 109954. Available from https://doi.org/10.1016/j.polymdegradstab.2022.109954.

Mcglashan, S. A., & Halley, P. J. (2003). Preparation and characterisation of biodegradable starch-based nanocomposite materials. *Polymer International*, *52*(11), 1767–1773. Available from https://doi.org/10.1002/PI.1287.

Menzel, C., González-Martínez, C., & Chiralt, A. (2019). Antioxidant starch films containing sunflower hull extracts. *Carbohydrate Polymers*, *214*, 142–151. Available from https://doi.org/10.1016/j.carbpol.2019.03.022.

Menzel, C., González-Martínez, C., Vilaplana, F., Diretto, G., & Chiralt, A. (2020). Incorporation of natural antioxidants from rice straw into renewable starch films. *International Journal of Biological Macromolecules*, *146*, 976–986. Available from https://doi.org/10.1016/j.ijbiomac.2019.09.222.

Mohamadian, N., Ghorbani, H., Wood, D. A., & Khoshmardan, M. A. (2019). A hybrid nanocomposite of poly(styrene-methyl methacrylate- acrylic acid) /clay as a novel rheology-improvement additive for drilling fluids. *Journal of Polymer Research*, *26*(2), 1–14. Available from https://doi.org/10.1007/S10965-019-1696-6/TABLES/4.

Mordor Intelligence. (2022). *Biodegradable packaging market - growth, trends, covid-19 impact, and forecasts (2022-2027).* <https://www.mordorintelligence.com/industry-reports/gobal-biodegradable-packaging-solutions-market-industry>.

Nafchi, A. M., Nassiri, R., Sheibani, S., Ariffin, F., & Karim, A. A. (2013). Preparation and characterization of bionanocomposite films filled with nanorod-rich zinc oxide. *Carbohydrate Polymers*, *96*(1), 233–239. Available from https://doi.org/10.1016/J.CARBPOL.2013.03.055.

Nieto-Suaza, L., Acevedo-Guevara, L., Sánchez, L. T., Pinzón, M. I., & Villa, C. C. (2019). Characterization of Aloe vera-banana starch composite films reinforced with curcumin-loaded starch nanoparticles. *Food Structure*, *22*, 100131. Available from https://doi.org/10.1016/j.foostr.2019.100131.

Nizam, N. H. M., Rawi, N. F. M., Ramle, S. F. M., Aziz, A. A., Abdullah, C. K., Rashedi, A., & Kassim, M. H. M. (2021). Physical, thermal, mechanical, antimicrobial and physicochemical properties of starch based film containing aloe vera: A review. *Journal of Materials Research and Technology*, *15*, 1572–1589. Available from https://doi.org/10.1016/j.jmrt.2021.08.138.

Olkku, J., & Rha, C. K. (1978). Gelatinisation of starch and wheat flour starch—A review. *Food Chemistry*, *3*(4), 293–317. Available from https://doi.org/10.1016/0308-8146(78)90037-7.

Onyeaka, H., Obileke, K., Makaka, G., & Nwokolo, N. (2022). Current research and applications of starch based biodegradable films for food packaging. *Polymers*, *14*, 1126. Available from https://doi.org/10.3390/polym14061126.

Ortega-Toro, R., Collazo-Bigliardi, S., Roselló, J., Santamarina, P., & Chiralt, A. (2017). Antifungal starch-based edible films containing Aloe vera. *Food Hydrocolloids*, *72*, 1–10. Available from https://doi.org/10.1016/j.foodhyd.2017.05.023.

Owolabi, I. O., Songsamoe, S., Khunjan, K., & Matan, N. (2021). Effect of tapioca starch coated-rubberwood box incorporated with essential oils on the postharvest ripening and quality control of mangosteen during transportation. *Food Control*, *126*. Available from https://doi.org/10.1016/j.foodcont.2021.108007.

Oyom, W., Xu, L., Liu, Z., Long, H., Li, Y., Zhang, Z., Bi, Y., Tahergorabi, R., & Prusky, D. (2022). Effects of modified sweet potato starch edible coating incorporated with cumin essential oil on storage quality of 'early crisp. *LWT Food Science and Technology*, *153*, 112475.

Oyom, W., Zhang, Z., Bi, Y., & Tahergorabi, R. (2022). Application of starch-based coatings incorporated with antimicrobial agents for preservation of fruits and vegetables: A review. *Progress in Organic Coatings*, *166*, 106800. Available from https://doi.org/10.1016/j.porgcoat.2022.106800.

Pandey, S., Sharma, K., & Gundabala, V. (2022). Antimicrobial bio-inspired active packaging materials for shelf life and safety development: A review. *Food Bioscience*, *48*. Available from https://doi.org/10.1016/j.fbio.2022.101730.

Park, H. M., Lee, W. K., Park, C. Y., Cho, W. J., & Ha, C. S. (2003). Environmentally friendly polymer hybrids Part I Mechanical, thermal, and barrier properties of thermoplastic starch/clay nanocomposites. *Journal of Materials Science*, *38*(5), 909–915. Available from https://doi.org/10.1023/A:1022308705231.

Patkar, M. S., Prasad, R. R., Rahate, K. S., Shetye, P. A., Sinha, R. A., & Chitnis, K. S. (2020). Extraction and characterization of starch from tubers and its application as bioplastic. *World Journal of Environmental Biosciences*, *9*, 1–5. Available from https://doi.org/10.13140/RG.2.2.27746.15040.

Peighambardoust, S. J., Peighambardoust, S. H., Pournasir, N., & Pakdel, P. M. (2019). Properties of active starch-based films incorporating a combination of Ag, ZnO and CuO nanoparticles for potential use in food packaging applications. *Food Packaging and Shelf Life*, *22*, 100420. Available from https://doi.org/10.1016/j.fpsl.2019.100420.

Peng, F., Liu, H., Hu, S., Yue, F., Xiao, D., Guo, L., & Qi, H. (2022). High throughput preparation of antioxidant polysaccharide-based polymers with UV-resistant and antibacterial performance. *Food Hydrocolloids*, *133*, 107936. Available from https://doi.org/10.1016/j.foodhyd.2022.107936.

Promsorn, J., & Harnkarnsujarit, N. (2022). Oxygen absorbing food packaging made by extrusion compounding of thermoplastic cassava starch with gallic acid. *Food Control*, *142*, 109273. Available from https://doi.org/10.1016/j.foodcont.2022.109273.

Puligundla, P., Jung, J., & Ko, S. (2012). Carbon dioxide sensors for intelligent food packaging applications. *Food Control*, *25*(1), 328–333. Available from https://doi.org/10.1016/j.foodcont.2011.10.043.

Roselli, M., Finamore, A., Garaguso, I., Britti, M. S., & Mengheri, E. (2003). Zinc oxide protects cultured enterocytes from the damage induced by *Escherichia coli*. *The Journal of Nutrition*, *133*(12), 4077–4082. Available from https://doi.org/10.1093/JN/133.12.4077.

Sadeghizadeh-Yazdi, J., Habibi, M., Kamali, A. A., & Banaei, M. (2019). Application of edible and biodegradable starch-based films in food packaging: A systematic review and meta-analysis. *Current Research in Nutrition and Food Science*, *7*, 624–637. Available from https://doi.org/10.12944/CRNFSJ.7.3.03.

Samadi, A., Klingberg, H., Jauffred, L., Kjær, A., Bendix, P. M., & Oddershede, L. B. (2018). Platinum nanoparticles: A non-toxic, effective and thermally stable alternative plasmonic material for cancer therapy and bioengineering. *Nanoscale*, *10*(19), 9097–9107. Available from https://doi.org/10.1039/C8NR02275E.

Sánchez Riaño, A. M., Alegría Vivas, E. M., Bermeo Andrade, H. P., Bohórquez Pérez, Y., Villada Castillo, H. S., Daza, L. D., & Valenzuela Real, C. P. (2022). Physicochemical, structural, and thermal characterization of biodegradable film prepared using arracacha thermoplastic starch and polylactic acid. *Journal of Food Measurement and Characterization*, *16*, 3597–3606. Available from https://doi.org/10.1007/s11694-022-01491-2.

Santos, C. T., Veiga-Santos, P., Sestari, P., Sorrini, N. C., & Roça, R. O. (2020). Protein time-temperature sensor for intelligent starch polymers. *Journal of Food Processing and Preservation*, *44*(5), 6. Available from https://doi.org/10.1111/jfpp.14428.

Scamparini, A.R. P., & Veiga-Santos, P. (2016). *Carbohydrate-based irreversible temperature indicator 2004*, Brazil. Patent:

Innovation privilege. registration number: PI0403610, título: "Indicador Irreversível de temperatura À base de carboidratos", Instituição de registro: INPI - Instituto Nacional da Propriedade Industrial. Deposit: 08/23/2004; Grant: 12/06/2016.

Sapper, M., Bonet, M., & Chiralt, A. (2019). Wettability of starch-gellan coatings on fruits, as affected by the incorporation of essential oil and/or surfactants. *LWT, 116*. Available from https://doi.org/10.1016/j.lwt.2019.108574.

Soukoulis, C., Singh, P., Macnaughtan, W., Parmenter, C., & Fisk, I. D. (2016). Compositional and physicochemical factors governing the viability of Lactobacillus rhamnosus GG embedded in starch-protein based edible films. *Food Hydrocolloids, 52*, 876–887. Available from https://doi.org/10.1016/J.FOODHYD.2015.08.025.

Souza, C. O., Silva, L. T., Silva, J. R., López, J. A., Veiga-Santos, P., & Druzian, J. I. (2011). Mango and acerola pulps as antioxidant additives in cassava starch bio-based film. *Journal of Agricultural and Food Chemistry, 59*(6), 2248–2254. Available from https://doi.org/10.1021/jf1040405, 2011.

Tadini, C.C., Keichichian, V., Veiga-Santos, P., & Ditchfield, C. (2016). Biodegradable starch and/or cassava biobased film with natural antimicrobial ingredients and its uses. 2007, Brazil. Patent: Innovation privilege. Registration number: PI07045891, Title: Filme biodegradável à base de amido e/ou fécula contendo ingredientes naturais antimicrobianos e seus usos". Institution of Registration: INPI - National Institute of Industrial Property. Deposit: 04/13/2007; Grant: 03/01/2016.

Tester, R. F., & Debon, S. J. J. (2000). Annealing of starch—a review. *International Journal of Biological Macromolecules, 27*(1), 1–12. Available from https://doi.org/10.1016/S0141-8130(99)00121-X.

Thakur, R., Pristijono, P., Bowyer, M., Singh, S. P., Scarlett, C. J., Stathopoulos, C. E., & Vuong, Q. V. (2019). A starch edible surface coating delays banana fruit ripening. *LWT,, 100*, 341–347. Available from https://doi.org/10.1016/j.lwt.2018.10.055.

Tran, V. V., Moon, J. Y., & Lee, Y. C. (2019). Liposomes for delivery of antioxidants in cosmeceuticals: Challenges and development strategies. *Journal of Controlled Release, 300*, 114–140. Available from https://doi.org/10.1016/j.jconrel.2019.03.003.

Upadhyay, A., Kumar, P., Kardam, S. K., & Gaikwad, K. K. (2022). Ethylene scavenging film based on corn starch-gum acacia impregnated with sepiolite clay and its effect on quality of fresh broccoli florets. *Food Bioscience, 46*, 101556. Available from https://doi.org/10.1016/J.FBIO.2022.101556.

Vaezi, K., Asadpour, G., & Sharifi, H. (2019). Effect of ZnO nanoparticles on the mechanical, barrier and optical properties of thermoplastic cationic starch/montmorillonite biodegradable films. *International Journal of Biological Macromolecules, 124*, 519–529. Available from https://doi.org/10.1016/J.IJBIOMAC.2018.11.142.

Veiga-Santos, P., Antonio, K., De, J., Santos, C. T., Arruda, A. A., Barros, L. B., De., & Gonçalves, L. T. (2022). Pomace-cassava as antioxidant bio-based coating polymers for cheeses. *Polysaccharides, 3*(2), 380–387. Available from https://doi.org/10.3390/polysaccharides3020022.

Veiga-Santos, P., Ditchfield, C., & Tadini, C. C. (2011). Development and evaluation of a novel pH indicator biodegradable film based on cassava starch. *Journal of Applied Polymer Science, 120*, 1069–1079. Available from https://doi.org/10.1002/app.33255.

Veiga-Santos, P., Oliveira, L. M., Cereda, M. P., & Scamparini, A. R. P. (2007). Sucrose and inverted sugar as plasticizer. Effect on cassava starch–gelatin film mechanical properties, hydrophilicity and water activity. *Food Chemistry, 103*(2), 255–262. Available from https://doi.org/10.1016/j.foodchem.2006.07.048.

Veiga-Santos, P., Oliveira, L. M., Cereda, M. P., Alves, A. J., & Scamparini, A. R. P. (2005). Mechanical properties, hydrophilicity and water activity of starch-gum films: Effect of additives and deacetylated xanthan gum. *Food Hydrocolloids, 19*(2), 341–349. Available from https://doi.org/10.1016/j.foodhyd.2004.07.006, ISSN 0268-005X.

Veiga-Santos, P., Silva, L. T., Souza, C. O., Silva, J. R., Albuquerque, E. C. C., & Druzian, J. I. (2018). Coffee-cocoa additives for bio-based antioxidant packaging. *Food Packaging and Shelf Life, 18*, 37–41. Available from https://doi.org/10.1016/j.fpsl.2018.08.005.

Wen, L., Rodis, P., & Wasserman, B. (1990). Starch fragmentation and protein in solubilisation during twin-screw extrusion of corn meal. *Cereal Chemistry, 67*, 268–275.

Willett, J. L., Millard, M. M., & Jasberg, B. K. (1997). Extrusion of waxy maize starch: Melt rheology and molecular weight degradation of amylopectin. *Polymer, 38*(24), 5983–5989. Available from https://doi.org/10.1016/S0032-3861(97)00155-9.

Wu, F., Misra, M., & Mohanty, A. K. (2021). Challenges and new opportunities on barrier performance of biodegradable polymers for sustainable packaging. *Progress in Polymer Science, 117*, 101395. Available from https://doi.org/10.1016/j.progpolymsci.2021.101395.

Wu, Z., Huang, Y., Xiao, L., Lin, D., Yang, Y., Wang, H., Yang, Y., Wu, D., Chen, H., Zhang, Q., Qin, W., & Pu, S. (2019). Physical properties and structural characterization of starch/polyvinyl alcohol/graphene oxide composite films. *International Journal of Biological Macromolecules, 123*, 569–575. Available from https://doi.org/10.1016/J.IJBIOMAC.2018.11.071.

Yan, X., Cheng, M., Zhao, P., Wang, Y., Chen, M., Wang, X., & Wang, J. (2022). Fabrication and characterization of oxidized esterified tapioca starch films encapsulating oregano essential oil with mesoporous nanosilica. *Industrial Crops and Products*, *184*, 115033. Available from https://doi.org/10.1016/j.indcrop.2022.115033.

Yoksan, R., & Chirachanchai, S. (2010). Silver nanoparticle-loaded chitosan–starch based films: Fabrication and evaluation of tensile, barrier and antimicrobial properties. *Materials Science and Engineering: C*, *30*(6), 891–897. Available from https://doi.org/10.1016/J.MSEC.2010.04.004.

Yu, Z., Li, B., Chu, J., & Zhang, P. (2018). Silica in situ enhanced PVA/chitosan biodegradable films for food packages. *Carbohydrate Polymers*, *184*, 214–220. Available from https://doi.org/10.1016/J.CARBPOL.2017.12.043.

Zhao, Y., An, J., Su, H., Li, B., Liang, D., & Huang, C. (2022). Antimicrobial food packaging integrating polysaccharide-based substrates with green antimicrobial agents: A sustainable path. *Food Research International*, *155*, 111096. Available from https://doi.org/10.1016/j.foodres.2022.111096.

Zhong, Y., Godwin, P., Jin, Y., & Xiao, H. (2020). Biodegradable polymers and green-based antimicrobial packaging materials: A mini-review. *Advanced Industrial and Engineering Polymer Research*, *3*(1). Available from https://doi.org/10.1016/j.aiepr.2019.11.002.

Zou, Y., Yuan, C., Cui, B., Liu, P., Wu, Z., & Zhao, H. (2021). Formation of high amylose corn starch/konjac glucomannan composite film with improved mechanical and barrier properties. *Carbohydrate Polymers*, *251*, 117039. Available from https://doi.org/10.1016/j.carbpol.2020.117039.

CHAPTER

14

Biodegradable and biobased plastic materials based on starch

Luc Avérous

BioTeam/ICPEES-ECPM, University of Strasbourg, Strasbourg, France

14.1 Introduction

14.1.1 Generalities

Since the mid-20th century, plastics market has been continuously growing (except in 2020) due to the great versatility, large availability, and unique properties of these exceptional materials. Indeed, they fulfill unique requirements with high performances. Plastics present unbeatable behaviors, using simple and cheap processing techniques with low energy and water consumption, compared to the other materials. The recent society evolution must be largely associated with the large development of fossil-based plastics production and consumption. However, recently, there has been an increased awareness (1) to reduce greenhouse gas emissions and disseminations with long lasting micro/nanoplastics with new challenges to face and (2) to decrease the accumulation of plastic wastes, with different approaches such as the use of alternate resources and the recycling (Edjabou et al., 2015; Eriksen et al., 2019; Garcia & Robertson, 2017). Some waste management options have some disadvantages, which must be mitigated. In this context, the development of biodegradable plastics emerged 25 years ago as a realistic alternative way to deal with plastic wastes and globally the end of life of the plastics (Agarwal, 2020).

However, to reduce the global environmental impact, the management at the end of life is not sufficient, the beginning of the life of these materials must be also considered and improved. In this frame, three main ways or strategies are nowadays developed in parallel with the use of alternative resources such as (1) the use of the products of the recycling steps, (2) the use of CO_2 (trapped directly or not from the atmosphere) to produce polymers, and (3) last but not the least, the development of macromolecules from biomass (i.e., all the materials produced on earth from living organisms, such as plants, animals, fungi, and microorganisms) by direct extraction or by associated fermentation, with or without chemical synthesis. In the latter case, the corresponding polymers are called "biobased polymers."

Finally, some materials present a double positive impact such as the biodegradable and

Starch Industries
DOI: https://doi.org/10.1016/B978-0-323-90842-9.00015-7

biobased polymers, since the beginning and the end of life are considered to reduce the global environmental impact with improved life cycle analysis (LCA). These materials can be associated with a recent active and valuable field called the "circular bioeconomy."

14.1.2 Some definitions

Biodegradable and biobased polymers such as starch can be associated with a large scientific domain at the interface of different scientific fields (polymer science and engineering, chemistry, biology, biotech, physics, etc.). This domain is very particular. Then, to avoid confusions and mistakes, it is essential to provide some very precise definitions, obtained from different sources such as IUPAC recommendations or EU documents (Vert et al., 2012).

- "Plastics": This usual term can be defined as "materials consisting of polymers to which additives or other substances may have been added, and which can function as a main structural component of final products, with the exception of natural polymers that have not been chemically modified" such as or instance native polysaccharides or native proteins.
- "Biopolymer": This misused term must be reserved for substances composed of biomacromolecules, that is, macromolecules formed by living organisms, such as proteins or polysaccharides. Then, biopolymers are no more than a specific category of biobased polymers (Wendels & Avérous, 2021).

Biobased contents can be approached by different ways such as the biogenic carbon atom content, which can be determined by radiocarbon (C^{14}) analysis. Some labels set the minimal biogenic carbon content to 25%. However, different standards must be used to determine the biobased content. The first one in 2004 has been the US norm (ASTM D6866), which gives the standard test methods for determining the biobased content of solid, liquid, and gaseous samples using radiocarbon analysis. This approach is based on C^{14} measurements. The C^{14} content is positive for living organisms. This is due to exchanges between the living organism and its environment. The C^{14} half-life time is 5730 years. Then, after 50,000 years, the C^{14} content cannot be determined. The C^{14} content from fossil resources is then null. In this way, recently (2017) the EU norm EN 16640 (2017) on the determination of the biobased carbon content using the radiocarbon method has been established. However, the biobased (biogenic) carbon content and total biobased content is not equal. For instance, in the case of partially biobased poly(ethylene terephthalate) (PET) (e.g., in "Plant bottle") with a diol obtained from biomass by fermentation, the biobased carbon content is 20%. However, the total biobased content (considering all the atoms) is 31%. Recently, new norms have been established to determine the total biobased content such as the EU norm EN 16785 (2018), which presents two distinct parts and two approaches, with (1) the determination of the biobased content using the radiocarbon analysis and elemental analysis and (2) the determination of the biobased content using the material balance method (From production unit). Biobased polymers represent more than 1% of the production of the plastics in the world with a production higher than 4 million tons in 2021 (Morinval & Averous, 2021).

- "Biodegradable polymers": They are "macromolecules or polymeric substances susceptible to degradation by biological activity by lowering of the molar masses of macromolecules that form the substance" according to IUPAC recommendations (Vert et al., 2012). Biodegradation is a biotic degradation phenomenon mainly caused by the enzymatic process resulting from the action of cells, till the mineralization with the production of a new biomass, water and CO_2 or CH_4, in an aerobic or anaerobic environment, respectively. It can be demonstrated, followed, and measured by standard tests over a specific period of time such as 6 months and must reach after this period a certain level of biodegradation, such as 90%. In connection with the focused applications, different media (compost, soil, fresh water, and marine environment) can be used to analyze the materials' biodegradability. For instance, compostability is associated with the analysis of the biodegradability in compost. Biodegradability and, particularly, compostability of plastics are defined through several standards such as the well-known EN 13432 in Europe, ASTM D6400 (United States), and several international norms as ISO 17088:2012, ISO 14855–2:2018, and ISO 18606. We can notice that the biodegradability of these polymers is an option, which can be or not activated. Then, these polymers can be also recycled as other polymers (materials recycling and chemical recycling) according to their macromolecular architectures. The choice of their ends of life must be based on the results of their respective LCAs, testing the different strategies.

14.1.3 Classification of the biodegradable and biobased polymers

The latest classification of biodegradable and biobased polymers is presented in Fig. 14.1, with different categories. Some acronyms are also given. Initially in 2004, Avérous (2004) established a well-known classification of the biodegradable polymers. In 2021, this classification has been revisited and updated taking into account some recent evolutions on the bioproduction of building blocks (White Biotech) from biomass and then the new potential productions of biobased copolyesters (Morinval & Averous, 2021). These copolyesters were only fossil-based in the past.

Fig. 14.1 presents two main categories with (1) the naturally occurring polymers directly extracted from biomass such as starch. These polymers defined as biopolymers are naturally produced from different resources and (2) the biodegradable and (potentially) biobased polymers from white biotech. This last category is only focused on (co)polyesters architectures. They can be directly obtained by fermentation such as the polyhydroxyalkanoates (PHAs, bacterial polymers) or not. In the latter case, the biobased polymers are synthesized by the chemical process from different building blocks, obtained by fermentation processes (Biotech-Chem approach).

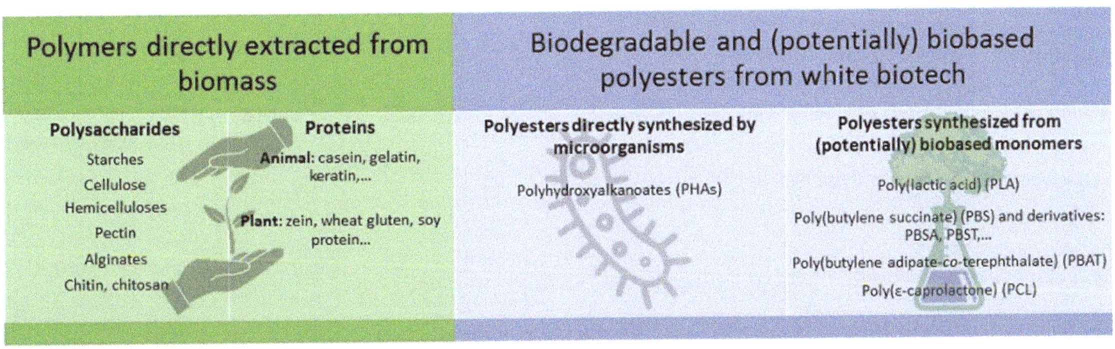

FIGURE 14.1 Classification of biobased and biodegradable polymers into two main categories and three subcategories. Some acronyms are given. *From Morinval, A., & Averous, L. (2021). Systems based on biobased thermoplastics: From bioresources to biodegradable packaging applications.* Polymer Reviews, 62(4), 653–721, *with permissions.*

These macromolecules can be fully biobased (poly(lactic acid), PLA) or partially (poly(butylene adipate-co-terephthalate), PBAT).

After an introduction, this chapter on biodegradable and biobased plastic materials based on starch is structured into different parts. The first part is focused on the presentation of the different main components of the multiphase systems based on starch. The second part is focused on the different morphologies and the corresponding properties of these multiphase materials which can be processed as plastics materials using conventional processing techniques (extrusion, injection, and molding). The last part of this chapter is more focused on some recent applications for these biodegradable and biobased materials focused on starch.

14.2 Starch-based plastics: the different components

Starch-based material has attracted a lot of attention of the academic and industrial communities since the 1980s and has been the topic of several overviews and states of art (Avérous, 2004; Halley & Avérous, 2014; Morinval & Averous, 2021). Starch-based plastics are based on different components. The main component is the starchy material. This component is often associated with cocomponents, which can be (1) other biobased and biodegradable polymers to preserve the final materials biodegradation with a final high biobased content and/or (2) (nano)fillers or reinforcing additives at variable contents to produce biocomposites and nanobiocomposites.

14.2.1 Starch materials (main component)

Starch materials are the main component of the starch-based plastics. From native starches obtained from different resources, a large range of processed materials can be obtained.

14.2.1.1 *Native starches*

Although we can find several nonfood applications (plaster, cement, paper, textile, glue, and plastics), starch stays a major and essential component of the food (more than 60% of the starch consumption in EU). Then, starch is easily available all around the world from different resources (Halley & Avérous, 2014). For more information on starch production and uses in the world, see Chapter 3 in Volume 1 of this series.

Native starches are produced by several plants and act as the energy storage for the plants, during night/day or seasons cycles. They present a granular structure with diameters from 0.5 to hundreds of microns and a specific 3D organization from the hilum with concentric and radial morphologies. It is basically a mix of two polysaccharides based on D-glucose polymers with (1) amylose, linear polymer with a molar mass of 10^5–106 Da, and (2) amylopectin, a highly branched polymers with molar mass of 10^7–109 Da. The latter is generally the main biopolymer (Bowen et al., 2006; Ong et al., 1994; Wu & Sarko, 1978).

Depending on the botanical resource and the local environment, the amylose/amylopectin ratio and residual amounts of lipids and proteins can vary (Ogunsona et al., 2018). These different contents have a major impact on the behavior and the properties of the starchy materials. For instance, the proteins (glutens) can interact during the processing with starch, with the well-known Maillard reaction, which lead to a partial starch degradation with a caramelization of the polysaccharides. It is important to remember that the presence of lipids in starch from roots and tubercles is much lower than in cereal starch, which can interfere with the final properties.

Starch is a semicrystalline material with a complex organization involving different types of crystalline structures, according to the botanical origin and/or the applied process (Lopez-Rubio et al., 2008). The structures of the native starch are largely modified during the processing steps. The structural starch transformations with thermomechanical inputs show different evolutions such as the native granular disruption, the gelatinization, and then the melting, as a function of water content, temperature, shear, and processing time. The relationships between rheological behavior and the corresponding transformation have been largely studied and can be found in the literature with different studies on food or on starch-based plastics (Della Valle et al., 1995, 1996, 1997, 1998; Vergnes & Villemaire, 1987; Xie et al., 2012).

For more information on composition of starch produced from roots and tubers of South American origin, see Chapter 8 of Volume 1 of this series. For information on general starch characteristics and properties, see Chapters 9 and 10 of that same volume.

14.2.1.2 *From native starch to starchy materials*

Native starches present poor properties for material uses. Then, to enhance these properties, native starch must be transformed or modified with different approaches according to targeted final application (Lauer & Smith, 2020).

Native starch can be gelatinized to develop new properties with a disintegration of the granules and the destructuring of its 3D crystalline structure, with chains dispersion under heating input with a high water content (Coffin et al., 1995).

From gelatinized starch, we can obtain foams. A cellular morphology is produced by evaporation of water. These materials can be used to replace PS foams produced by extrusion or by expansion-molding, such as XPS and EPS, respectively, with the production of biodegradable loose fill (continuous extrusion) or expanded trays (heated closed mold as for waffle production) (Kaisangsri et al., 2012).

Native starch melting, like most polysaccharides (as native cellulose for instance), occurs after the degradation temperature. Then native starch cannot be directly processed by plastic processing techniques (Avérous, 2004). Two ways can be developed with (1) the chemical modification based on the hydroxyl groups with the production of, for example, starch ester (starch acetate for instance) with a controlled degree of substitution (DS) till a maximum of 3. The chemical modifications impact the final biodegradability of the corresponding materials. Usually, the final level of biodegradation is decreased after the modification. (2) The formulation with plasticizers to decrease the process temperature below the starch degradation temperature to avoid the product caramelization. Then, plasticizers can be added to the starch to obtain a processable material without degradation. Two types of plasticizers are used and mixed: (1) water, which is a destructuring agent acting also as a volatile plasticizer, and (2) nonvolatile plasticizers as biobased polyols such as glycerol or sorbitol (both can be mixed). After processing, the compound is called thermoplastic starch (TPS) even if the neat thermoplastic character can be discussed. The rheological behavior of these materials is more like a "thermo-mechano-plastic" behavior with a strong dependency of the specific mechanical energy on the viscosity values, at a given temperature. The properties of TPS can be highly variable according to the origins and the content of the plasticizers. A great tunability and modularity is then obtained (Averous, 2000).

TPS is a very versatile material suitable for a variety of fields. However, TPS presents poor mechanical properties; high hydrophily character; and very long postprocessing aging (after the process step, several weeks are needed to obtain a complete stabilization of the properties) compared to conventional thermoplastic materials. To overcome the weaknesses of the material, TPS is often blended or associated with other biobased polymers or/and with (nano)fillers/reinforced to obtain performing multiphase systems with improved properties (Avérous & Fringant, 2001; Kaseem et al., 2012; Koh et al., 2018; Laftah, 2017; Zaaba & Ismail, 2019).

14.2.2 Co-components

Two types of co-components are generally used to produce plastics-based starches: (1) biodegradable and biobased polymers and (2) fillers/reinforcers at different contents. A large variety of morphologies such as foams, blends, multilayers, composites, or nanocomposites can be produced. These different multiphase systems can be mixed or ordered together in supramorphologies to fulfill the adequate requirements for a very large range of environmental applications.

14.2.2.1 Biodegradable and biobased polymers

These polymers are mainly based on different polyesters architectures (Table 14.1). These polyesters are produced (1) from biotech, only as in the case of bacterial polymers, or (2) from the association between biotech and chemistry also called "Biotech-Chem." In the latter case, only the building blocks are produced by biotech. The chains are then synthesized through chemical steps from the building blocks.

14.2.2.1.1 Bacterial polyesters: the case of the polyhydroxyalkanoates

The most basic example of bacterial polymers is the PHAs. It is a large family of polyesters with different architectures. As an intracellular polymer, the microorganisms used PHA as a storage for carbon and energy production. The higher accumulation of PHA occurs in response to stress conditions, till 80% of the dry mass (Ling et al., 2018). Since the end of the 1990s, PHAs can be also bioproduced from transgenic plants such as tobacco or corn plants (Dobrogojski et al., 2018; Tan et al., 2014).

PHAs are the result of different (R)-hydroxy fatty acid-associated monomer units to form a large variety of macromolecular architectures (Table 14.1). Close to 200 different PHAs have been reported (Tan et al., 2014). PHAs are usually identified according to their carbon numbers (C) per repeating unit. From 3 to 5C, we can find the short-chain such as the well-known P(3HB); from 6 to 14C, we have the medium-chain (mcl-PHA); and above 15C, we can find the fewer long-chain length PHA. The chain length impacts their behavior such as the thermal and mechanical properties. Increasing the carbon numbers decreases (1) the crystallinity and the ester density, which results in a decrease of T_m and thermal degradation during the processing steps, and (2) the biodegradability of these polyesters.

P(3HB), also known as PHB, is a good water and gas barrier with high modulus and high crystallinity, for specialty packaging as a monocomponent (El-Hadi et al., 2002; Poirier et al., 1995). However, P(3HB) exhibits several drawbacks such as brittleness and a high melting temperature, close to its degradation temperature. Thermal degradation induces a strong and fast decrease of the chain lengths by the well-known random chain scissions with the formation of crotonic groups. This reaction can be catalyzed with residual amine groups coming for instance from protein byproducts, after the required rupture of the microorganism membrane for the PHAs extraction. Then P(3HB) is usually plasticized with, for example, ester citrate (Citroflex) to decrease the processing temperature (Kurusu et al., 2015). Another strategy

TABLE 14.1 Chemical structures of main biodegradable and biobased polyesters.

Polyhydroxyalkanoates	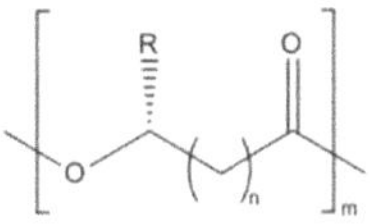
Poly(lactic acid)	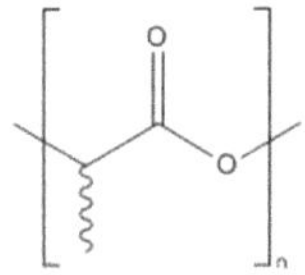
Poly(butylene succinate)	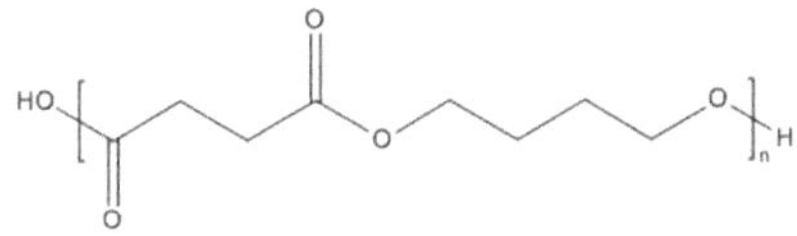
Poly(butylene succinate-co-butylene adipate)	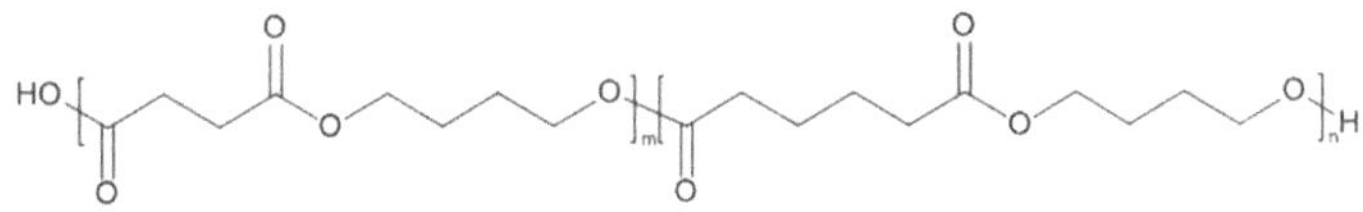
Poly(butylene adipate-co-terephthalate)	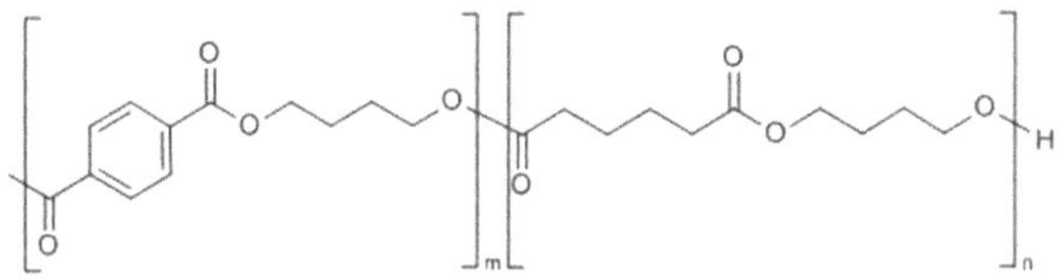
Poly(ε-caprolactone)	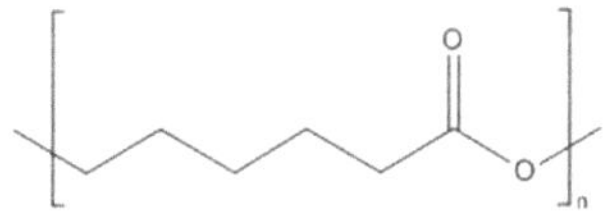

From Morinval, A., Averous, L., 2021. Systems based on biobased thermoplastics: From bioresources to biodegradable packaging applications. Polymer Reviews *62, 653–721, with permissions.*

is based on the development of copolymers such as P(3HB) with (4HB) or (3HV) units as in the case of the Poly(3-hydroxybutyrate-co-3-hydroxyvalerate), also known as PHBV. These copolymers are also synthetized at the industrial level by different companies to reduce their crystallinity, melting temperature, and brittleness (Rivera-Briso & Serrano-Aroca, 2018). Their properties are fine-tuned by varying each unit content (Amass et al., 1998; Shogren, 1997). Such copolymers are more suitable than P(3HB) for usual processing such as extrusion (Malafaya et al., 2007). mcl-PHA can also be used. Some are produced at the industrial level but with an increased cost of the polyester compared to the PHB or the PHBV.

Several companies produce industrially PHAs in the world such as Enmat (China), GreenBio (China), Tianzhu (China), Kaneka (Japan), and Meredian (United States) with a price often higher than 5 Euros/kg in 2020. With all these different small players, considering the rather high price and the PHAs drawbacks as the thermal sensitivity, the PHA market is not till now really mature and stabilized, compared for instance to PLA market.

14.2.2.1.2 Polyesters obtained from Biotech-Chem: the case of the poly(lactic acid)s

The chemical structure of PLA is presented Table 14.1. It is one of the nicest examples of association between biotech and chemistry (Arrieta et al., 2014; Nova-Institut, IKT Stuttgart, OWS, HYDRA Marine Sciences, 2021; Quecholac-Piña et al., 2020; Saadi et al., 2012). Two enantiomers are used to synthetize PLA: L- and D-lactic acid. Lactic acid can be chemically produced from fossil resources or biobased from the lacto-fermentation of sugars, starchy materials, or lignocellulosic biomass (Castillo Martinez et al., 2013; Jem & Tan, 2020). The bioproduction has several advantages compared to synthetic production of lactic acid with gains in terms of environmental impact (John et al., 2007).

Finally, PLA can be produced using different pathways to obtain high molar masses such as (1) the direct condensation polymerization from lactic acid, followed by a chemical coupling of the short chains, (2) the azeotropic dehydration condensation, and (3) the polymerization through lactide formation. From the lactides, high molar masses are obtained by ring opening polymerization (ROP). The latter pathway is the most used at the industrial level since it presents several advantages such as the control of the final L/D ratio. PLAs have strong variations of the properties such as the degree of crystallinity, T_g and T_m values according to the L/D ratio, and the copolymer structure (random, alternate, and block) (Bigg, 2005). For instance, stereoregular P(D)LA and P(L)LA are semicrystalline, while atactic P(D, L) LA is often completely amorphous when D content is lower than 96 wt.%. We have also the case of the stereocomplex based on the mix 50/50 of P(D)LA and P(L)LA with higher T_g and T_m compared to the corresponding neat P (D)LA or P(L)LA. Mechanical, thermal, and barrier properties vary depending on the PLA microstructure and the formulation. In this way, PLA can be reinforced or plasticized. Martin and Avérous (2001) showed that the most efficient plasticizers for PLA were oligomeric lactic acid (OLA) and low molar mass polyethylene glycol. In this frame, OLA is now currently commercialized by Condensia (Spain) for the industrial application of plasticized and softer PLA. PLA plasticization favors chain mobility, easing crystallization.

During the last two decades, several rigid packaging applications such as cups to drink, bowls, pots for yogurt, and food trays have been commercialized worldwide, particularly in Europe (Rai et al., 2021). One of the first examples was the production of yogurt pots for the German market by Danone (France), at the end of the nineties. PLA textiles have been also largely developed more particularly in Asia and United States. PLA foams have also been developed by Toyota (Japan) for inner applications of cars (Ema et al., 2006; Lee et al., 2008).

PLA is the biobased polyester with the highest worldwide production, with several hundred KTons per year, with some big players such as NatureWorks (Cargill-PTT, Thailand/United States), Teijin (Japan), Futerro (Belgium), Total Corbion (Belgium/Netherland/Thailand), Evonik (Germany), or Zhejiang Hisun (China) (European Bioplastics, 2020).

In all the cases, the PLA demand is increasing, as it is the cheapest biobased and biodegradable polyester. The price was below 2$/kg in 2020 (Oever et al., 2017). Moreover, LCAs tend to show that PLA have environmental benefits compared to fossil-based and nonbiodegradable equivalents like PS or PET (Madival et al., 2009; Vink et al., 2003). Besides, PLA could become cheaper than fossil-based PS and PET in the near future, which encourages enthusiastic market forecasts. However, common P(D,L)LA can present strong drawbacks such as rather low barrier properties, its intermediate T_g (around 60°C), and a strong decrease of the modulus after the T_g.

This behavior can be an issue for the transport and the storage (Logistic network) during summer periods. PLA packaging must be then transported in climatized containers. It is also a drawback for aseptic hot filling or treatment in the context of food packaging applications. Another drawback is that PLA is biodegradable only in industrial composts and anaerobic digesters. For instance, PLA is not really biodegradable in the case of home compost. It is clearly a strong limitation.

As a transition with the next part, we can notice more and more the production of several PLA-based renewable and sustainable blends as Ecovio (BASF, Germany), which is a blend between PBAT and PLA.

14.2.2.1.3 (Co)Polyesters obtained by Biotech-Chem associations: the case of the polybutylene derivates

This large category of polyesters and copolyesters is based on several associated building blocks, where we find a common diol: the 1,4-butanediol (BDO). Biobased BDO can be synthetized from the chemical succinic acid platform (Litsanov et al., 2014) by metal-catalyzed hydrogenation (Delhomme et al., 2009; Le & Nishimura, 2019). However, several companies such as Genomatica (United States) and Novamont (Italy) have developed direct biobased BDO production processes, which allows for direct conversion of carbohydrates into BDO by liquid fermentation (Burgard et al., 2016; Dani et al., 2017; Yim et al., 2011).

The main polymers based on polybutylene derivates are the poly(butylene succinate) (PBS), poly(butylene succinate-*co*-butylene adipate) (PBSA), and the PBAT. Nowadays, the trend is to develop these biodegradable polymers with the highest biobased content with the increasing production of biobased building blocks from Biotech. Then the polymers are synthesized from diol and di-carboxyl acids, using conventional polymer chemistry by, for example, polycondensation.

PBAT is an aromatic copolyester (Table 14.1), synthetized by polycondensation between three building blocks: BDO and two carboxylic acids with the adipic acid and the terephthalic acid (TA) (Zhao et al., 2010). PBAT synthesis can be catalyzed by different organometallic catalysts under high vacuum and high temperature (Jian et al., 2020). For fully biobased PET, the production of biobased TA is currently under strong investigations at academic and industrial levels. TA could be, for instance, obtained by oxidation of biobased *p*-xylene obtained from different biomass (Pang et al., 2016). Indeed, several pathways for the synthesis of TA from biomass are based on terpenes. They are currently studied for industrial scaling up (Akanuma et al., 2014; Collias et al., 2014; Smith, 2015; Volanti et al., 2019). Currently, commercial fossil-based adipic acid is mainly obtained by oxidation of cyclohexanol and cyclohexanone with nitric acid. To replace this tough process, researchers have developed many pathways for the synthesis of biobased adipic acid (Vyver & Román-Leshkov, 2013) such as (1) a chemical path based on 5-hydroxymethylfurfural, a chemical platform obtained from different mono- and polysaccharides, (2) the hydrodeoxygenation of glucaric acid, obtained by oxidation of D-glucose, and (3) a chemical pathway based on γ-valerolactone synthetized by catalytic hydrogenation of levulinic acid, which is obtained from cellulose and hemicellulose. Then, in this large context, at the industrial level and till now, PBAT can be only partially biobased.

The largest PBAT producers are (1) Novamont (Italy) with the biobased Origo-Bi and (2) BASF (Germany) with the Ecoflex with capacities of production of about 100 and 74 KTons/year, respectively. We can also find some new smaller players in China.

PBAT's properties strongly depend on the ratio between the different monomers.

PBAT presents high flexibility linked to a relatively low crystallinity, low ester density, and some π stacking between the aromatic rings (Jiang & Li, 2001; Ren et al., 2009; Signori et al., 2009; Son et al., 2018). PBAT is only biodegradable in industrial and home composts and in the soil. PBAT cannot be considered as biodegradable in fresh or marine waters.

Several applications based on neat PBAT include, for instance, bags for shopping or organic wastes and mulch films in agriculture (Bordes et al., 2009; Yamamoto et al., 2005). The applications can be extended with the development of multiphase systems. Except Ecovio (BASF, Germany), another good example of PBAT-commercial blend is the Mater-Bi (e.g., PBAT/TPS blends from Novamont S.p.A., Italy).

PBS is a biodegradable and semicrystalline nonaromatic polyester (Table 14.1) obtained by polycondensation of BDO and succinic acid. Succinic acid was usually obtained by hydrogenation of maleic anhydride from fossil resources. Succinic acid can be produced from biomass by bioproduction (Song & Lee, 2006). Then, succinic acid is largely industrially produced using white biotechnology processes, by fermentation of, for example, carbohydrates by bacteria or fungi (Bechthold et al., 2008; Zeikus et al., 1999). There are several commercial biobased succinic acid in the world, such as the Biosuccinium (Roquette Frères S.A., France). Originally PBS is brittle, but some diisocyanate chain extenders are commonly used in order to improve by coupling the molar masses, with a strong increase of the elongation at break values (Gigli et al., 2016). PBS is biodegradable in only industrial compost and not in home compost. It is a strong limitation for the uses. However, PBS grades are approved by the Food and Drug Administration for food contact. Then, it is used in many packaging applications, such as shopping bags, or in food packaging, coffee capsules, tea bags, and even for agriculture in mulch films.

PBSA is a popular linear aliphatic copolyester from PBS (Table 14.1). PBSA is synthetized by melt polycondensation or transesterification of succinic acid, adipic acid, and BDO (Nikolic & Djonlagic, 2001). All these building blocks can be biobased. The main properties of PBSA are like those of PBS, but its lower crystallinity and ester density bring more flexibility. This crystallinity reduction increases the biodegradability of PBSA compared to PBS, which can be managed in home and industrial composts and in soil (as PBAT). It can be easily processed and used in various packaging applications (Gigli et al., 2016).

The largest PBS and PBSA producer is MCPP (Mitsubishi, Japan) with the BioPBS with a production capacity of 20KTons/year. The price of the grades is currently relatively high at around 4€/kg in 2020. However, the expected production increases should decrease this cost (Oever et al., 2017).

14.2.2.1.4 From fossil to biobased transition: the specific case of the poly(ε-caprolactone)

Poly(ε-caprolactone) (PCL) (Table 14.1) is well known by the students and the researchers on polymer science since it is one of the first reported syntheses of biodegradable polymers by Natta, Hill, and Carothers in 1930 (Natta et al., 1934). As for PLA, different pathways for PCL synthesis can be used with (1) the polycondensation of hydroxycarboxylic acids (6-hydroxycaproic acid) or (2) the well-known ROP of ε-caprolactone. The latter is preferred because it leads to high molar mass and lower dispersity (Đ) (Labet & Thielemans, 2009). ROP of ε-caprolactone can be catalyzed by various metallic compounds (Labet & Thielemans, 2009). ε-caprolactone is usually industrially obtained from a petrochemical process from benzene. However, researchers have shown the possibility to produce fully biobased ε-caprolactone, with the conversion of 5-hydroxymethylfurfural into 1,6-hexanediol (as for the biobased adipic acid production).

In this chemical process, the oxidation of 1,6-hexanediol to ε-caprolactone stays a critical step. Recently, the enzymatic oxidation of 1,4-hexanediol to 6-hydroxyhexanoic acid has been reported and then could be used to produce ε-caprolactone (Pyo et al., 2020). To date, no industrial production of biobased PCL has been reported. However, several companies work on this topic (Thaore et al., 2018).

PCL is a soft semicrystalline aliphatic polyester with a very low T_m (around 60°C) and a low T_g (around −60°C). Such a low T_m can be a barrier to the use of PCL for applications requiring heat resistance. Nevertheless, PCL is stable until 190°C. This low T_m can be also an advantage because this polyester can be shaped at low temperature in hot water. PCL is biodegradable in industrial and home composts.

Different suppliers act on this market such as Perstorp (Sweden) with Capa grades. PCL is of major interest for the development of different biodegradable multiphase systems (Averous, 2000; Liu et al., 2004; López et al., 2019; Mi et al., 2017; Mohamed & Yusoh, 2015; Siparsky et al., 1997; Woodruff & Hutmacher, 2010), mainly for the manufacturing of bags or food packaging. PCL is also currently used in different other applications, for instance, as a plasticizer (PVC) or a polyol for the synthesis of different macromolecular architectures.

14.2.2.2 *Fillers and reinforcers*

The main difference between fillers and reinforcements is linked to the main action of the corresponding added particles. In the case of reinforcer, the main advantage and effect are linked to the evolution of the mechanical properties with, for instance, a strong increase of the Young's modulus. However, the difference between these two terms is often very thin.

The description of the different fillers, nanofillers, and reinforcements to develop biocomposites (with micro-/macrofillers/reinforcers) and nanobiocomposites (with nanofillers) by dispersion of macro/micro/nano rigid objects into a biobased and biodegradable polymer matrix has been the main topic of books and overviews (Avérous & Pollet, 2012; Avérous, 2004; Halley & Avérous, 2014; Morinval & Averous, 2021). This part briefly presents some examples linked to this field.

14.2.2.2.1 Main examples of biobased fillers

The main examples of biobased fillers are based on lignocellulose structures obtained from soft or hard woods, or from annual plants such as wheat or rice straws. Lignocellulosic fibers present a very specific 3D structure, as a super and complex composite combining different biopolymers such as cellulose, hemicellulose, and lignins, with some lower molecules as the aromatic tannins, for instance. Lignocellulosic fibers are among the most common biobased reinforcements for the elaboration of biocomposites (Biagiotti et al., 2004; Dixit et al., 2017; Faruk et al., 2012; Mohammed et al., 2015; Puglia et al., 2005; Singh et al., 2018; Yang et al., 2019). Except the case of wood fibers, the most produced technical lignocellulosic fibers around the world come from bagasse, bamboo, jute, and kenaf. They bring high modulus and mechanical resistances.

Except mechanical properties, the lignocellulosic fibers exhibit different attractive properties, such a low density, easy recyclability, and low mold shrinkage at lower cost compared to common synthetic fibers as glass, Kevlar, or carbon fibers (Satyanarayana et al., 2009).

14.2.2.2.2 Examples of usual nanofillers

Nanofillers are usually classified according to their shapes (spherical, acicular, layered, or elongated) and their origin, with inorganic vs organic resources (e.g., the nanocellulose). Fig. 14.2 presents the main nanofillers, which have been extensively used into biodegradable and biobased polymer matrix (Bari et al., 2016; Chivrac et al., 2010; Olivato et al., 2015, 2017; Shamsuri & Jamil, 2020). Among them, nanoclays have been largely tested and used and more particularly

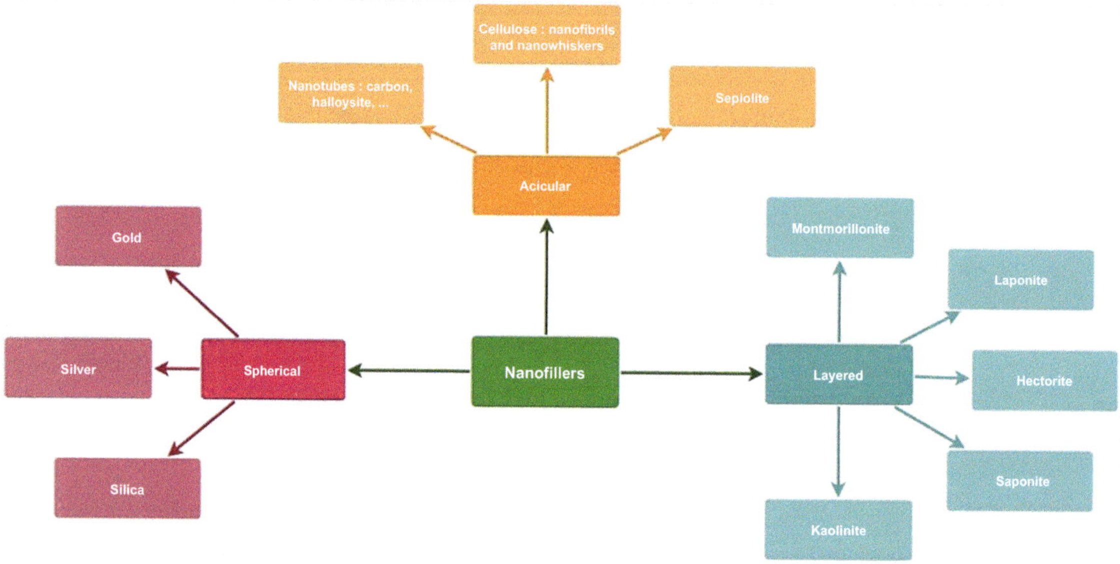

FIGURE 14.2 Presentation of the main nanofillers according to their shapes to produce nanobiocomposites. *From Morinval, A., & Averous, L. (2021). Systems based on biobased thermoplastics: From bioresources to biodegradable packaging applications.* Polymer Reviews, *62(4), 653–721, with permissions.*

the layered nanofillers from nanoclays. The corresponding nanobiocomposites exhibit numerous improved properties compared to the neat matrix, such as higher stiffness and barrier properties, when they are correctly distributed into the matrix. Due to their geometry and high potential interface area with the matrix, competitive performances can be reached even at very low content (lower than 6 wt.%) and compared to conventional composites, without drastically affecting the global biodegradation level or compostability of the corresponding multiphase systems (Giannelis, 1996).

Layered silicates have been largely described and used. The incorporation of these layered nanofillers into a polymer matrix can lead to distinct morphologies according to the level of exfoliation of the nanoclays, from microcomposites (without exfoliation) to fully exfoliated nanocomposites (Avérous & Pollet, 2012). Intermediate morphology can be obtained with the intercalated nanocomposites. One of the most used layered nanoclays is the montmorillonite (MMT) with different important mining deposits in the world. This common nanoclay has been incorporated into various biobased or not matrixes (Avérous & Pollet, 2012; Biswas & Ray, 2001; Bruzaud & Bourmaud, 2007; Chieng et al., 2010; Chiu, 2016; Chivrac et al., 2006; Habibi et al., 2019; Lal et al., 2017; Lee et al., 2002; Madhumitha et al., 2018; Mangiacapra et al., 2006; Paul et al., 2005; Tan et al., 2019; Vandewijngaarden et al., 2016; Wang et al., 2005). To achieve a better affinity with the polymer matrix, by reducing the interfacial tension between the filler and the matrix, organomodified MMT (OMMT) is prepared and even commercialized. For instance, OMMT with a cationic starch has been successfully incorporated into a TPS matrix, resulting in an efficient exfoliated morphology, improving strongly the mechanical properties (Chivrac et al., 2008).

The most used commercial OMMTs are sold under the trademark Cloisite (Southern Clay Products, United States). They are obtained by organomodification of MMT with different ammonium-based cations, which (1) decrease MMT's polarity at the surface, (2) improve the spacing between the layer improving the exfoliation phenomenon, and (3) improve the chemical affinity and then dispersion in the matrix (Adrar et al., 2017; Bordes et al., 2008).

14.3 Starch-based plastics: elaboration and properties of different multiphase systems

Despite the behaviors exhibited by the different biobased and biodegradable polymers, it remains difficult to use them alone to compete to more conventional plastics and to fulfill industrial requirements. An effective strategy is to design starch-based materials combining the different components previously presented. In this part, several morphologies such as blends, (nano)composites, and multilayers are reported with their corresponding challenges. This part briefly presents different examples of these different multiphase systems with their advantages and issues.

14.3.1 Biodegradable and renewable blends based on starch

The blends and the blending are one of the basic and easiest ways to obtain multiphase materials combining the properties of the different polymers. Then, the lack of miscibility and compatibility between the components is a major and common issue in blends, resulting in low material properties. TPS has been associated with all the different biobased and biodegradable polymers previously described with a range of final properties, largely reported in several reviews (Avérous, 2004; Morinval & Averous, 2021).

For partially or fully immiscible blends, the compatibilization is a crucial step to obtain a correct mix with adequate properties. The reduction of the interfacial tension at the interface between the blend components, resulting in a more continuous morphology, with good interfacial adhesion, and thus better materials properties.

The blends compatibilization can be achieved by ex situ or in situ compatibilization (Muthuraj et al., 2018). For instance, PBAT grafted with maleic anhydride improved the tensile strength and the elongation at break of TPS/PBAT blends. In addition to mechanical properties, barrier properties can also be improved by compatibilization. Indeed, a higher dispersion of polymers increases the "tortuosity factor" of the different associated phases, causing a decrease of gas diffusion through the blend membrane. For instance, oxygen permeability of TPS/PLA blends has been largely reduced, thanks to grafted PCL-based compatibilizers (Collazo-Bigliardi et al., 2019).

The use of biodegradable compatibilized blends is already widely implemented and several commercial starch plastics exist, such as the family of starch-based blends Mater-Bi, mainly based on renewable polyesters as cocomponents with a large range of applications, for instance, in packaging or agriculture (Imre & Pukánszky, 2013).

14.3.2 Renewable biocomposites and nanobiocomposites based on starch

According to the size of the fillers or the reinforcements, two main heterogeneous systems can be obtained with (1) the biocomposites based on micro-/macrofillers and (2) the nanobiocomposites with nanofillers. In the latter case, at least one dimension such as the length or the diameter of the fillers is below the micron.

14.3.2.1 Biocomposites with lignocellulosic fibers

Not only to increase mechanical properties but also to maintain or to increase the biobased content, lignocellulosic fibers have been largely incorporated into biodegradable polymers since several decades (Gallos et al., 2017; Le Digabel et al., 2004). In this way, TPS can be reinforced, for instance, with coffee and rice husk fibers (Collazo-Bigliardi et al., 2018), wood fibers (Cheng, 2019), or pineapple fibers (Smitthipong et al., 2015).

Compared to the neat matrix, these cellulose reinforcements can bring different global advantages such as (1) higher mechanical properties linked to strong matrix/filler interactions between the different polysaccharides (cellulose/amylose and amylopectin), (2) higher thermal resistance, due to the transition shift of T_g and an increase in the rubber plateau, (3) a reduced water sensitivity due to fiber–matrix interactions and to the higher hydrophobic character of the cellulose linked to its high crystallinity, and (4) a reduced post-processing aging, due to the formation of a 3D network between the different matrix–filler carbohydrates based on hydrogen bonds. However, lignocellulosic fibers can bring some shortcomings such as some microbial and fungal attack and degradation (Güven et al., 2016).

14.3.2.2 Nanobiocomposites

These multiphasic materials are also known as "bionanocomposites" in some publications. Beyond the mechanical, thermal, and barrier properties, nanofillers such as nanoclay (e.g., Cloisite) can bring antimicrobial activity (Rhim et al., 2006). This effect is not really linked to the filler but to the associated organomodifier. This behavior is linked to the well-known disruption of the bacterial membranes by the quaternary ammonium groups (Avérous & Pollet, 2012; Rhim et al., 2009). However, some nanofillers are very well known for such an activity, as the silver-based nanofillers (nanosilver).

The effect of nanoclays on the biodegradability of polymers is largely discussed. Some studies concluded that nanoclays could enhance the biodegradability kinetic of certain polymers. However, an inhibition can be caused by OMMT due to their specific antimicrobial behavior depending on the associated organomodifier (Rhim et al., 2013). The effect of nanoclays on the biodegradability is highly dependent on the testing conditions.

Thanks to the enhancement of mechanical, thermal, and barrier properties, nanobiocomposites could find their place in the packaging market (Wyser et al., 2016), but raise a significant concern due to potential nanofiller effect with potential negative impact on human healthiness and the environment (Bandyopadhyay & Ray, 2019). Then the use of nanobiocomposites based on starch is quasi-inexistent on the market. This is due to the regulation linked to the used of nanofillers and the negative perception by the public of such systems.

14.3.3 Biodegradable and renewable multilayers based on starch

To replace blends with additional properties, multilayer films are widely used in packaging. They allow to associate the behaviors, such as the barrier properties, of several polymers while keeping the possibility to choose the polymer in contact, for example, food contact, with the packed article(s), controlling then the container–content interactions. Skin layers can be also chosen according to printability and/or esthetic criteria. Most multilayer films are made up of 2–5 layers, but this number can be higher, till 7 layers, for some specific packaging applications with some compatibility or recycled layers and interlayers (Morris, 2016). The main processes for multilayers manufacturing are coextrusion, lamination, or coating (Anukiruthika et al., 2020). Coextrusion is a family of techniques, of which most widespread ones are flat, tube dies and blown systems. For packaging application, the most used is a flat die associated with several extruders and a repartition box to obtain the initial multilayer structure and a rolls-based system at the exit to produce flat multilayer films with controlled thicknesses of the whole films and of each layer.

Several issues are encountered when elaborating biodegradable and biobased multilayers based on TPS. However, these problems are more or less similar to the ones encountered with conventional polymers, such as encapsulation and interfacial instabilities, for instance (Martin & Avérous, 2002). Several factors, including extrusion rate, die geometry, and the gaps between the viscosities of the different components in contact, impact the occurrence of such instabilities. Besides, all the polymers are not extruded at the same temperature. For instance, we can notice that TPS usually presents a high viscosity in the molten phase and then it is difficult to obtain the viscosity ratio close to one at the given temperature and shearing with the associated cocompound (such as a biodegradable and biobased polyester). Besides, TPS presents a high thermal sensitivity and can degrade at temperature higher than 150°C. The corresponding process windows is narrow to avoid the polysaccharide degradation and to obtain a perfect superposition of layers with well controlled and stable thicknesses.

Among the biodegradable and biobased multilayer film morphology, three-layer films with TPS as a core layer and polyester in the outer layers have been largely studied for decades and largely patented and used as packaging materials (Martin et al., 2001). In such systems, TPS brings an oxygen barrier behavior and the biodegradable polyesters in the outer layers bring water barrier properties, preserving the core layers from water uptake and bringing some mechanical properties. Usually, the polyester layers are very thin since the cost of the polyesters are largely higher than the TPS one. In some case, compatibility layers can be added between the TPS and the polyester layers. This layer can be based (1) on compatibilized blends and/or (2) on copolymers systems from starch and the polyesters. In the latter case, the hydroxyl groups from the polysaccharides are used to develop a grafting chemistry.

From such multilayer structures, films can be obtained. However, at higher thickness, plates can be also produced to fabricate trays, for instance, with an additional thermoforming step. In this case, the core layer can be also based on starch-based foam to bring low density and thermal insulation, for instance.

14.4 Starch-based plastics: some applications

Starch-based plastics based on biodegradable and biobased polymers are principally used in short-term application, which means mainly packaging and agriculture. Mainly common starches are used from large productions such as starches from corn, potatoes, wheat, or cassava. In some rare and specific cases, amylomais with high amylose content can be used. According to the targeted applications, cheapest nonfood starch grades can be also used, to reduce the materials cost.

14.4.1 Agriculture items from starch-based plastics

Starch-based plastic can be found in different biodegradable and biobased items for agriculture for short- or medium-term uses such as biodegradable mulch films, planting pots, planting stakes, or clips (e.g., for tomatoes cultivation) (Cameron & Styles, 2009). Mater-Bi grades based on starch with biodegradable polyesters are largely used for these different applications. When reinforcement and high modulus are needed, lignocellulosic fibers are added such as in the case of some planting pots, stakes, or clips.

14.4.2 Packaging from starch-based plastics

Packaging is the first plastic market, but it is also the first and the most common application for biodegradable polymers (Din et al., 2020). This domain packaging is vast and presents two subdomains: food and nonfood packaging, which need relevant and specific properties. Food packaging is associated with different food contact constraints with, for example, limited migration according to the national or international regulations in the frame of container food interactions and/or barrier properties (oxygen and water vapor).

14.4.2.1 Food packaging

Different types of food packaging have been elaborated from starch with the adequate requirements (Halley & Avérous, 2014). The main types of biodegradable food packaging include films wrapped with vegetables or fruits, for instance. Sealed films can be also used in the cases of food trays or pouches. One of the most basic biodegradable films is based on three components with starch, flour, and water (Aung, 1993). Such starchy films are then edible (for more information on starchy films see Volume 13 of this series). Such edible packaging are of growing interest in the context of the global source reduction of waste and single uses plastics in the frame of the recent evolutions and regulations in Europe (Umaraw & Verma, 2017). They often combine different biodegradable polymers (Gilleland et al., 2003; Han, 2008). Several biodegradable and biobased polymers can be used to manufacture bags for the packaging of food items such as fruit and vegetable in the supermarket to replace HDPE bags. Some of these bags are produced from Mater-Bi blends based on starch. Food packaging films based on multiphase systems can combine the complementary barrier properties of several biodegradable polymers, as the starch-cellulose laminates (Zagrodzki, 2018), or/and the blends from TPS, PLA, and PCL (Castillo et al., 2013). Until now, most of biodegradable and biobased multiphase films are used for short-life foods such as fresh fruits or vegetable.

Thermoformed or injected food containers and trays are used in commercial applications, including the packaging of sensitive foods such as biscuits, pastries, meats, or fishes where controlled barriers and food contacts are key points. Biodegradable food containers are often used for the packaging of takeaway, home-delivered, and fast foods. Such packages are already widely available and can be based on starch and PBS (Ozasa et al., 2006) or starch with cellulose pulp (Bowden & Johnston, 2011). For instance, cutlery items and cups for fast food or picnic can be fabricated by injection molding techniques combining starch, PLA and PBAT (Scheer, 2007), or starch and wood fibers (Bowden & Johnston, 2011). Several biodegradable and biobased polymers can be used to manufacture bags for the packaging of food items such as fruit and vegetable in the supermarket to replace PE bags. Some of these bags are produced from Mater-Bi blends based on starch.

14.4.2.2 Nonfood packaging

Several biodegradable and biobased polymers can be used to manufacture bags for the packaging of nonfood articles, such as shopping, trash, and compost bags, or to pack various other nonfood products such as, for instance, fertilizers for agriculture. These different bags can be produced from Mater-Bi blends by Novamont (Italy) or from Sphere (France) production. They are based on starch-based blends, and mainly produced in Europe (Morinval & Averous, 2021).

Packaging for cosmetics and parapharmaceutical or pharmaceutical uses can be based on biodegradable and biobased materials. It is the case, for instance, of tubes for product like toothpaste, creams, or shampoo, in this context, Essel Propack (India)-patented starch-based blends for extrusion and injection molding processes (Abhyankar & Banerjee, 2018).

Biodegradable and biobased polymers and starch-based plastics are used to pack goods such as detergents (Dituro et al., 2013) or to fabricate cushioning systems (Hornstein & Landrum, 1998; Kannankeril et al., 2014). Recently, Nova-Institute (Germany) has published the "BioSinn" report (Bauchmüller et al., 2021), which shows several examples as (1) dishwater tabs films, which could be replaced by multilayer films based on starch or (2) seed coatings based on starch.

References

Abhyankar, C.R., & Banerjee, M.K. (2018). *Polymer composition for manufacturing biodegradable articles and process thereof*. EP2718076B1.

Adrar, S., Habi, A., Ajji, A., & Grohens, Y. (2017). Combined effect of epoxy functionalized graphene and organomontmorillonites on the morphology, rheological and thermal properties of poly (butylenes adipate-co-terephtalate) with or without a compatibilizer. *Applied Clay Science,, 146*, 306–315. Available from https://doi.org/10.1016/j.clay.2017.06.009.

Agarwal, S. (2020). Biodegradable polymers: Present opportunities and challenges in providing a microplastic-free environment. *Macromolecular Chemistry and Physics,, 221*, 2000017. Available from https://doi.org/10.1002/macp.202000017.

Akanuma, Y., Selke, S. E. M., & Auras, R. (2014). A preliminary LCA case study: Comparison of different pathways to produce purified terephthalic acid suitable for synthesis of 100% bio-based PET. *The International Journal of Life Cycle Assessment, 19*, 1238–1246. Available from https://doi.org/10.1007/s11367-014-0725-2.

Amass, W., Amass, A., & Tighe, B. (1998). A review of biodegradable polymers: Uses, current developments in the synthesis and characterization of biodegradable polyesters, blends of biodegradable polymers and recent advances in biodegradation studies. *Polymer International, 47*, 89–144. Available from https://doi.org/10.1002/(SICI)1097-0126(199810)47:2 < 89::AID-PI86 > 3.0.CO;2-F.

Anukiruthika, T., Sethupathy, P., Wilson, A., Kashampur, K., Moses, J. A., & Anandharamakrishnan, C. (2020). Multilayer packaging: Advances in preparation techniques and emerging food applications. *Comprehensive Reviews in Food Science and Food Safety,, 19*, 1156–1186. Available from https://doi.org/10.1111/1541-4337.12556.

Arrieta, M. P., López, J., Rayón, E., & Jiménez, A. (2014). Disintegrability under composting conditions of plasticized PLA–PHB blends. *Polymer Degradation and Stability,, 108*, 307–318. Available from https://doi.org/10.1016/j.polymdegradstab.2014.01.034.

Aung, D. (1993). *Biodegradable packaging film*. WO1993006013A2.

Avérous, L. (2004). Biodegradable multiphase systems based on plasticized starch: A review. *Journal of Macromolecular Science Part C- Polymer Reviews,, 44*, 231–274. Available from https://doi.org/10.1081/MC-200029326.

Averous, L. (2000). Properties of thermoplastic blends: Starch–polycaprolactone. *Polymer, 41*, 4157–4167. Available from https://doi.org/10.1016/S0032-3861(99)00636-9.

Avérous, L., & Fringant, C. (2001). Association between plasticized starch and polyesters: Processing and performances of injected biodegradable systems. *Polymer Engineering and Science, 41*, 727–734. Available from https://doi.org/10.1002/pen.10768.

Bandyopadhyay, J., & Ray, S. S. (2019). Are nanoclay-containing polymer composites safe for food packaging applications?—An overview. *Journal of Applied Polymer Science, 136*, 47214. Available from https://doi.org/10.1002/app.47214.

Bari, S. S., Chatterjee, A., & Mishra, S. (2016). Biodegradable polymer nanocomposites: An overview. *Polymer Reviews,, 56*, 287–328. Available from https://doi.org/10.1080/15583724.2015.1118123.

Bauchmüller, V., Carus, M., Chinthapalli, R., Dammer, L., Hark, N., Partanen, A., Ruiz, P., & Lajewski, S. (2021). *BioSinn - Products for which biodegradation makes sense*. Nova-Institut für politische und ökologische Innovation GmbH, Hürth.

Bechthold, I., Bretz, K., Kabasci, S., Kopitzky, R., & Springer, A. (2008). Succinic acid: A new platform chemical for biobased polymers from renewable resources. *Chemical Engineering & Technology,, 31*, 647–654. Available from https://doi.org/10.1002/ceat.200800063.

Biagiotti, J., Puglia, D., & Kenny, J. M. (2004). A review on natural fibre-based composites-Part I: Structure, processing and properties of vegetable fibres. *Journal of Natural Fibers,, 1*, 37–68. Available from https://doi.org/10.1300/J395v01n02_04.

Bigg, D. M. (2005). Polylactide copolymers: Effect of copolymer ratio and end capping on their properties. *Advances in Polymer Technology,, 24*, 69–82. Available from https://doi.org/10.1002/adv.20032.

Biswas, M., & Ray, S. S. (2001). Recent progress in synthesis and evaluation of polymer-montmorillonite nanocomposites. *New polymerization techniques and synthetic methodologies* (pp. 167–221). Berlin, Heidelberg: Springer Berlin Heidelberg.

Bordes, P., Pollet, E., & Averous, L. (2009). Nano-biocomposites: Biodegradable polyester/nanoclay systems. *Progress in Polymer Science,, 34*, 125–155. Available from https://doi.org/10.1016/j.progpolymsci.2008.10.002.

Bordes, P., Pollet, E., Bourbigot, S., & Avérous, L. (2008). Structure and properties of PHA/clay nano-biocomposites prepared by melt intercalation. *Macromolecular Chemistry and Physics,, 209*, 1473–1484. Available from https://doi.org/10.1002/macp.200800022.

Bowden, J.A., & Johnston, C.C. (2011). *Biodegradable or compostable containers*. US7967904B2.

Bowen, S. E., Gray, D. A., Giraud, C., Majzoobi, M., Testa, C. E. M., Pérez, L. A. B., & Hill, S. E. (2006). Lipid oxidation and amylopectin molecular weight changes occurring during storage of extruded starch samples. *Journal of Cereal Science,, 43*, 275–283. Available from https://doi.org/10.1016/j.jcs.2005.08.010.

Bruzaud, S., & Bourmaud, A. (2007). Thermal degradation and (nano)mechanical behavior of layered silicate reinforced poly(3-hydroxybutyrate-co-3-hydroxyvalerate) nanocomposites. *Polymer Testing,, 26*, 652–659. Available from https://doi.org/10.1016/j.polymertesting.2007.04.001.

Burgard, A., Burk, M. J., Osterhout, R., Van Dien, S., & Yim, H. (2016). Development of a commercial scale process for production of 1,4-butanediol from sugar. *Current Opinion in Biotechnology,, 42*, 118–125. Available from https://doi.org/10.1016/j.copbio.2016.04.016.

Cameron, G., & Styles, P. (2009). *Biodegradable plant pots*. US20090249688A1.

Castillo, H.S. V., Porras, D.P. N., & Niño, J.P. C. (2013). *Biodegradable films obtained from cassava starch and their manufacture process*. WO2013042083A1.

Castillo Martinez, F. A., Balciunas, E. M., Salgado, J. M., Domínguez González, J. M., Converti, A., Oliveira, R. P., & de, S. (2013). Lactic acid properties, applications and production: A review. *Trends in Food Science & Technology, 30*, 70–83. Available from https://doi.org/10.1016/j.tifs.2012.11.007.

Cheng, W. (2019). Preparation and properties of lignocellulosic fiber/CaCO3/thermoplastic starch composites. *Carbohydrate Polymers,, 211*, 204–208. Available from https://doi.org/10.1016/j.carbpol.2019.01.062.

Chieng, B. W., Ibrahim, N. A., & Wan Yunus, W. M. Z. (2010). Effect of organo-modified montmorillonite on poly(butylene succinate)/poly(butylene adipate-co-terephthalate) nanocomposites. *Express Polymer Letters,, 4*, 404–414. Available from https://doi.org/10.3144/expresspolymlett.2010.51.

Chiu, F.-C. (2016). Fabrication and characterization of biodegradable poly(butylene succinate-co-adipate) nanocomposites with halloysite nanotube and organo-montmorillonite as nanofillers. *Polymer Testing, 54*, 1–11. Available from https://doi.org/10.1016/j.polymertesting.2016.06.018.

Chivrac, F., Kadlecová, Z., Pollet, E., & Avérous, L. (2006). Aromatic copolyester-based nano-biocomposites: Elaboration, structural characterization and properties. *Journal of Polymers and the Environment, 14*, 393–401. Available from https://doi.org/10.1007/s10924-006-0033-4.

Chivrac, F., Pollet, E., Schmutz, M., & Avérous, L. (2010). Starch nano-biocomposites based on needle-like sepiolite clays. *Carbohydrate Polymers, 80*, 145–153. Available from https://doi.org/10.1016/j.carbpol.2009.11.004.

Chivrac, F., Pollet, E., Schmutz, M., & Avérous, L. (2008). New approach to elaborate exfoliated starch-based nanobiocomposites. *Biomacromolecules, 9*, 896–900. Available from https://doi.org/10.1021/bm7012668.

Coffin, D. R., Fishman, M. L., & Cooke, P. H. (1995). Mechanical and microstructural properties of pectin/starch films. *Journal of Applied Polymer Science, 57*, 663–670. Available from https://doi.org/10.1002/app.1995.070570602.

Collazo-Bigliardi, S., Ortega-Toro, R., & Chiralt, A. (2019). Using grafted poly(ε-caprolactone) for the compatibilization of thermoplastic starch-polylactic acid blends. *Reactive & Functional Polymers, 142*, 25–35. Available from https://doi.org/10.1016/j.reactfunctpolym.2019.05.013.

Collazo-Bigliardi, S., Ortega-Toro, R., & Chiralt Boix, A. (2018). Reinforcement of thermoplastic starch films with cellulose fibres obtained from rice and coffee husks. *Journal of Renewable Materials, 6*, 599–610. Available from https://doi.org/10.32604/JRM.2018.00127.

Collias, D. I., Harris, A. M., Nagpal, V., Cottrell, I. W., & Schultheis, M. W. (2014). Biobased terephthalic acid technologies: A literature review. *Industrial Biotechnology, 10*, 91–105. Available from https://doi.org/10.1089/ind.2014.0002.

Dani, M., Ruggiero, G., Perini, D., & Bianchi, A. (2017). *Process for the production of 1,4-butanediol*. US20170029852A1.

Delhomme, C., Weuster-Botz, D., & Kühn, F. E. (2009). Succinic acid from renewable resources as a C4 building-block chemical—a review of the catalytic possibilities in aqueous media. *Green Chemistry: An International Journal and Green Chemistry Resource: GC, 11*, 13–26. Available from https://doi.org/10.1039/B810684C.

Della Valle, G., Boché, Y., Colonna, P., & Vergnes, B. (1995). The extrusion behaviour of potato starch. *Carbohydrate Polymers, 28*, 255–264. Available from https://doi.org/10.1016/0144-8617(95)00111-5.

Della Valle, G., Buleon, A., Carreau, P. J., Lavoie, P.-A., & Vergnes, B. (1998). Relationship between structure and viscoelastic behavior of plasticized starch. *Journal of*

Rheology, *42*, 507–525. Available from https://doi.org/10.1122/1.550900.

Della Valle, G., Colonna, P., Patria, A., & Vergnes, B. (1996). Influence of amylose content on the viscous behavior of low hydrated molten starches. *Journal of Rheology*, *40*, 347–362. Available from https://doi.org/10.1122/1.550747.

Della Valle, G., Vergnes, B., Colonna, P., & Patria, A. (1997). Relations between rheological properties of molten starches and their expansion behaviour in extrusion. *Journal of Food Engineering*, *31*, 277–295. Available from https://doi.org/10.1016/S0260-8774(96)00080-5.

Din, M. I., Ghaffar, T., Najeeb, J., Hussain, Z., Khalid, R., & Zahid, H. (2020). Potential perspectives of biodegradable plastics for food packaging application-review of properties and recent developments. *Food Additives & Contaminants: Part A*, *37*, 665–680. Available from https://doi.org/10.1080/19440049.2020.1718219.

Dituro, J., Dauksevicz, E., Furze, B., & Juckett, M. (2013). Biodegradable package for detergent. US20130053293A1.

Dixit, S., Goel, R., Dubey, A., Shivhare, P. R., & Bhalavi, T. (2017). Natural fibre reinforced polymer composite materials - A review. *Polymers from Renewable Resources*, *8*, 71–78. Available from https://doi.org/10.1177/204124791700800203.

Dobrogojski, J., Spychalski, M., Luciński, R., & Borek, S. (2018). Transgenic plants as a source of polyhydroxyalkanoates. *Acta Physiologiae Plantarum/Polish Academy of Sciences, Committee of Plant Physiology Genetics and Breeding*, *40*, 162. Available from https://doi.org/10.1007/s11738-018-2742-4.

Edjabou, M. E., Jensen, M. B., Götze, R., Pivnenko, K., Petersen, C., Scheutz, C., & Astrup, T. F. (2015). Municipal solid waste composition: Sampling methodology, statistical analyses, and case study evaluation. *Waste Management (New York, N.Y.)*, *36*, 12–23. Available from https://doi.org/10.1016/j.wasman.2014.11.009.

El-Hadi, A., Schnabel, R., Straube, E., Müller, G., & Henning, S. (2002). Correlation between degree of crystallinity, morphology, glass temperature, mechanical properties and biodegradation of poly (3-hydroxyalkanoate) PHAs and their blends. *Polymer Testing*, *21*, 665–674. Available from https://doi.org/10.1016/S0142-9418(01)00142-8.

Ema, Y., Ikeya, M., & Okamoto, M. (2006). Foam processing and cellular structure of polylactide-based nanocomposites. *Polymer*, *47*, 5350–5359. Available from https://doi.org/10.1016/j.polymer.2006.05.050.

Environmental silicate nano-biocomposites. In L. Avérous, & E. Pollet (Eds.), *Green energy and technology*. London: Springer. Available from https://doi.org/10.1007/978-1-4471-4108-2.

Eriksen, M. K., Damgaard, A., Boldrin, A., & Astrup, T. F. (2019). Quality assessment and circularity potential of recovery systems for household plastic waste. *Journal of Industrial Ecology*, *23*, 156–168. Available from https://doi.org/10.1111/jiec.12822.

European Bioplastics. (2020). *Bioplastics market development update 2020* [WWW Document]. https://docs.european-bioplastics.org/conference/Report_Bioplastics_Market_Data_2020_short_version.pdf (accessed 6.24.21).

Faruk, O., Bledzki, A. K., Fink, H.-P., & Sain, M. (2012). Biocomposites reinforced with natural fibers: 2000–2010. *Progress in Polymer Science*, *37*, 1552–1596. Available from https://doi.org/10.1016/j.progpolymsci.2012.04.003.

Gallos, A., Paës, G., Allais, F., & Beaugrand, J. (2017). Lignocellulosic fibers: A critical review of the extrusion process for enhancement of the properties of natural fiber composites. *RSC Advances*, *7*, 34638–34654. Available from https://doi.org/10.1039/C7RA05240E.

Garcia, J. M., & Robertson, M. L. (2017). The future of plastics recycling. *Science (New York, N.Y.)*, *358*, 870–872. Available from https://doi.org/10.1126/science.aaq0324.

Giannelis, E. P. (1996). Polymer layered silicate nanocomposites. *Advanced Materials*, *8*, 29–35.

Gigli, M., Fabbri, M., Lotti, N., Gamberini, R., Rimini, B., & Munari, A. (2016). Poly(butylene succinate)-based polyesters for biomedical applications: A review. *European Polymer Journal*, *75*, 431–460. Available from https://doi.org/10.1016/j.eurpolymj.2016.01.016.

Gilleland, G.M., Turner, J.L., Patton, P.A., & Harrison, M.D. (2003). *Highly flexible starch-based films*. US6649188B2.

Güven, O., Monteiro, S. N., Moura, E. A. B., & Drelich, J. W. (2016). Re-emerging field of lignocellulosic fiber – Polymer composites and ionizing radiation technology in their formulation. *Polymer Reviews*, *56*, 702–736. Available from https://doi.org/10.1080/15583724.2016.1176037.

Habibi, S., Saket, M., Nazockdast, H., & Hajinasrollah, K. (2019). Fabrication and characterization of exfoliated chitosan–gelatin–montmorillonite nanocomposite nanofibers. *The Journal of The Textile Institute*, *110*, 1672–1677. Available from https://doi.org/10.1080/00405000.2019.1613029.

Halley, P. J., & Avérous, L. (Eds.), (2014). *Starch polymers: From genetic engineering to green applications*. Elsevier.

Han, J.H. (2008). *Physical and mechanical properties of pea starch edible films containing beeswax emulsions*. CA2552179A1.

Hornstein, R.R., & Landrum, G.F. (1998). *Molded biodegradable packaging. US5826725A*.

Imre, B., & Pukánszky, B. (2013). Compatibilization in biobased and biodegradable polymer blends. *European Polymer Journal*, *49*, 1215–1233. Available from https://doi.org/10.1016/j.eurpolymj.2013.01.019.

Jem, K. J., & Tan, B. (2020). The development and challenges of poly (lactic acid) and poly (glycolic acid). *Advanced Industrial and Engineering Polymer Research*, *3*, 60–70. Available from https://doi.org/10.1016/j.aiepr.2020.01.002.

Jian, J., Xiangbin, Z., & Xianbo, H. (2020). An overview on synthesis, properties and applications of poly(butylene-adipate-co-terephthalate)−PBAT. *Advanced Industrial and Engineering Polymer Research*, *3*, 19−26. Available from https://doi.org/10.1016/j.aiepr.2020.01.001.

Jiang, Y., & Li, Y. (2001). Effects of chitosan coating on post-harvest life and quality of longan fruit. *Food Chemistry*, *73*, 139−143. Available from https://doi.org/10.1016/S0308-8146(00)00246-6.

John, R. P., Nampoothiri, K. M., & Pandey, A. (2007). Fermentative production of lactic acid from biomass: An overview on process developments and future perspectives. *Applied Microbiology and Biotechnology*, *74*, 524−534. Available from https://doi.org/10.1007/s00253-006-0779-6.

Kaisangsri, N., Kerdchoechuen, O., & Laohakunjit, N. (2012). Biodegradable foam tray from cassava starch blended with natural fiber and chitosan. *Industrial Crops and Products*, *37*, 542−546. Available from https://doi.org/10.1016/j.indcrop.2011.07.034.

Kannankeril, C., Conroy, P., & Shankaranarayan, N. (2014). *Cushioning assembly*. US20140117071A1.

Kaseem, M., Hamad, K., & Deri, F. (2012). Thermoplastic starch blends: A review of recent works. *Polymer Science, Series A*, *54*, 165−176. Available from https://doi.org/10.1134/S0965545X1202006X.

Koh, J. J., Zhang, X., & He, C. (2018). Fully biodegradable poly(lactic acid)/starch blends: A review of toughening strategies. *International Journal of Biological Macromolecules*, *109*, 99−113. Available from https://doi.org/10.1016/j.ijbiomac.2017.12.048.

Kurusu, R. S., Siliki, C. A., David, É., Demarquette, N. R., Gauthier, C., & Chenal, J.-M. (2015). Incorporation of plasticizers in sugarcane-based poly(3-hydroxybutyrate) (PHB): Changes in microstructure and properties through ageing and annealing. *Industrial Crops and Products*, *72*, 166−174. Available from https://doi.org/10.1016/j.indcrop.2014.12.040, Special issue derived from International Conference on Bio-based Materials and Composites.

Labet, M., & Thielemans, W. (2009). Synthesis of polycaprolactone: A review. *Chemical Society Reviews*, *38*, 3484. Available from https://doi.org/10.1039/b820162p.

Laftah, W. A. (2017). Starch based biodegradable blends: A review. *International Journal of Engineering Research*, *6*, 19.

Lal, S., Perwez, A., Rizvi, M. A., & Datta, M. (2017). Design and development of a biocompatible montmorillonite PLGA nanocomposites to evaluate in vitro oral delivery of insulin. *Applied Clay Science*, *147*, 69−79. Available from https://doi.org/10.1016/j.clay.2017.06.031.

Lauer, M. K., & Smith, R. C. (2020). Recent advances in starch-based films toward food packaging applications: Physicochemical, mechanical, and functional properties. *Comprehensive Reviews in Food Science and Food Safety*, *19*, 3031−3083. Available from https://doi.org/10.1111/1541-4337.12627.

Le Digabel, F., Boquillon, N., Dole, P., Monties, B., & Averous, L. (2004). Properties of thermoplastic composites based on wheat-straw lignocellulosic fillers. *Journal of Applied Polymer Science*, *93*, 428−436. Available from https://doi.org/10.1002/app.20426.

Le, S. D., & Nishimura, S. (2019). Highly selective synthesis of 1,4-butanediol via hydrogenation of succinic acid with supported Cu−Pd alloy nanoparticles. *ACS Sustainable Chemistry & Engineering*, *7*, 18483−18492. Available from https://doi.org/10.1021/acssuschemeng.9b04447.

Lee, S.-R., Park, H.-M., Lim, H., Kang, T., Li, X., Cho, W.-J., & Ha, C.-S. (2002). Microstructure, tensile properties, and biodegradability of aliphatic polyester/clay nanocomposites. *Polymer*, *43*, 2495−2500. Available from https://doi.org/10.1016/S0032-3861(02)00012-5.

Lee, S. T., Kareko, L., & Jun, J. (2008). Study of thermoplastic PLA foam extrusion. *Journal of Cellular Plastics*, *44*, 293−305. Available from https://doi.org/10.1177/0021955X08088859.

Ling, C., Qiao, G.-Q., Shuai, B.-W., Olavarria, K., Yin, J., Xiang, R.-J., Song, K.-N., Shen, Y.-H., Guo, Y., & Chen, G.-Q. (2018). Engineering NADH/NAD+ ratio in *Halomonas bluephagenesis* for enhanced production of polyhydroxyalkanoates (PHA). *Metabolic Engineering*, *49*, 275−286. Available from https://doi.org/10.1016/j.ymben.2018.09.007.

Litsanov, B., Brocker, M., Oldiges, M., & Bott, M. (2014). *Succinic acid. Bioprocessing of renewable resources to commodity bioproducts* (pp. 435−472). John Wiley & Sons.

Liu, L., Li, Y., Liu, H., & Fang, Y. (2004). Synthesis and characterization of chitosan-graft-polycaprolactone copolymers. *European Polymer Journal*, *40*, 2739−2744. Available from https://doi.org/10.1016/j.eurpolymj.2004.07.016.

López, O. V., Ninago, M. D., Lencina, M. M. S., Ciolino, A. E., Villar, M. A., & Andreucetti, N. A. (2019). Starch/poly(ε-caprolactone) graft copolymers synthetized by γ-radiation and their application as compatibilizer in polymer blends. *Journal of Polymers and the Environment*, *27*, 2906−2914. Available from https://doi.org/10.1007/s10924-019-01568-y.

Lopez-Rubio, A., Flanagan, B. M., Gilbert, E. P., & Gidley, M. J. (2008). A novel approach for calculating starch crystallinity and its correlation with double helix content: A combined XRD and NMR study. *Biopolymers*, *89*, 761−768. Available from https://doi.org/10.1002/bip.21005.

Madhumitha, G., Fowsiya, J., Roopan, S. M., & Thakur, V. K. (2018). Recent advances in starch−clay nanocomposites. *International Journal of Polymer Analysis and Characterization*, *23*, 331−345. Available from https://doi.org/10.1080/1023666X.2018.1447260.

Madival, S., Auras, R., Singh, S. P., & Narayan, R. (2009). Assessment of the environmental profile of PLA, PET and PS clamshell containers using LCA methodology. *Journal of Cleaner Production, 17*, 1183–1194. Available from https://doi.org/10.1016/j.jclepro.2009.03.015.

Malafaya, P. B., Silva, G. A., & Reis, R. L. (2007). Natural–origin polymers as carriers and scaffolds for biomolecules and cell delivery in tissue engineering applications. *Advanced Drug Delivery Reviews, 59*, 207–233. Available from https://doi.org/10.1016/j.addr.2007.03.012.

Mangiacapra, P., Gorrasi, G., Sorrentino, A., & Vittoria, V. (2006). Biodegradable nanocomposites obtained by ball milling of pectin and montmorillonites. *Carbohydrate Polymers, 64*, 516–523. Available from https://doi.org/10.1016/j.carbpol.2005.11.003.

Martin, O., & Avérous, L. (2002). Comprehensive experimental study of a starch/polyesteramide coextrusion: Starch/polyesteramide coextrusion. *Journal of Applied Polymer Science, 86*, 2586–2600. Available from https://doi.org/10.1002/app.11191.

Martin, O., & Avérous, L. (2001). Poly(lactic acid): Plasticization and properties of biodegradable multiphase systems. *Polymer, 42*, 6209–6219. Available from https://doi.org/10.1016/S0032-3861(01)00086-6.

Martin, O., Schwach, E., Avérous, L., & Couturier, Y. (2001). Properties of biodegradable multilayer films based on plasticized wheat starch. , *Starch - Starke, 53*, 372–380. Available from https://doi.org/10.1002/1521-379X(200108)53:8 < 372::AID-STAR372 > 3.0.CO;2-F.

Mi, H.-Y., Jing, X., Napiwocki, B. N., Hagerty, B. S., Chen, G., & Turng, L.-S. (2017). Biocompatible, degradable thermoplastic polyurethane based on polycaprolactone-block-polytetrahydrofuran-block-polycaprolactone copolymers for soft tissue engineering. *Journal of Materials Chemistry B, 5*, 4137–4151. Available from https://doi.org/10.1039/C7TB00419B.

Mohamed, R. M., & Yusoh, K. (2015). A review on the recent research of polycaprolactone (PCL). *Advanced Materials Research, 1134*, 249–255. Available from https://doi.org/10.4028/http://www.scientific.net/AMR.1134.249.

Mohammed, L., Ansari, M. N. M., Pua, G., Jawaid, M., & Islam, M. S. (2015). A review on natural fiber reinforced polymer composite and its applications. *International Journal of Polymer Science, 2015*, 1–15. Available from https://doi.org/10.1155/2015/243947.

Morinval, A., & Averous, L. (2021). Systems based on biobased thermoplastics: From bioresources to biodegradable packaging applications. *Polymer Reviews, 62*, 653–721. Available from https://doi.org/10.1080/15583724.2021.2012802.

Morris, B. A. (2016). *The science and technology of flexible packaging: Multilayer films from resin and process to end use.* William Andrew.

Muthuraj, R., Misra, M., & Mohanty, A. K. (2018). Biodegradable compatibilized polymer blends for packaging applications: A literature review. *Journal of Applied Polymer Science, 135*, 45726. Available from https://doi.org/10.1002/app.45726.

Natta, F. J., van, Hill, J. W., & Carothers, W. H. (1934). Studies of polymerization and ring formation. XXIII.1 ε-caprolactone and its polymers. *Journal of the American Chemical Society, 56*, 455–457. Available from https://doi.org/10.1021/ja01317a053.

Nikolic, M. S., & Djonlagic, J. (2001). Synthesis and characterization of biodegradable poly(butylene succinate-co-butylene adipate)s. *Polymer Degradation and Stability, 74*, 263–270. Available from https://doi.org/10.1016/S0141-3910(01)00156-2.

Nova-Institut, IKT Stuttgart, OWS, HYDRA Marine Sciences (2021). *Biodegradable Polymers in Various Environments According to Established Standards and Certification Schemes* [WWW Document]. https://renewable-carbon.eu/publications/product/biodegradable-polymers-in-various-environments-according-to-established-standards-and-certification-schemes-graphic-png/ (accessed 5.21.21).

Oever, M.V., Molenveld, K., Zee, M., & van der, Bos, H. (2017). *Bio-based and biodegradable plastics : Facts and figures : Focus on food packaging in the Netherlands.* Wageningen Food & Biobased Research, Wageningen.

Ogunsona, E., Ojogbo, E., & Mekonnen, T. (2018). Advanced material applications of starch and its derivatives. *European Polymer Journal, 108*, 570–581. Available from https://doi.org/10.1016/j.eurpolymj.2018.09.039.

Olivato, J. B., Marini, J., Pollet, E., Yamashita, F., Grossmann, M. V. E., & Avérous, L. (2015). Elaboration, morphology and properties of starch/polyester nano-biocomposites based on sepiolite clay. *Carbohydrate Polymers, 118*, 250–256. Available from https://doi.org/10.1016/j.carbpol.2014.11.014.

Olivato, J. B., Marini, J., Yamashita, F., Pollet, E., Grossmann, M. V. E., & Avérous, L. (2017). Sepiolite as a promising nanoclay for nano-biocomposites based on starch and biodegradable polyester. *Materials Science and Engineering: C, 70*, 296–302. Available from https://doi.org/10.1016/j.msec.2016.08.077.

Ong, M. H., Jumel, K., Tokarczuk, P. F., Blanshard, J. M. V., & Harding, S. E. (1994). Simultaneous determinations of the molecular weight distributions of amyloses and the fine structures of amylopectins of native starches. *Carbohydrate Research, 260*, 99–117. Available from https://doi.org/10.1016/0008-6215(94)80025-1.

Ozasa, A., Hashimoto, A., Shinohara, R., & Tanaka, S. (2006). *Biodegradable molded articles, process for producing the same and compositions for foam molding.* EP1321289B1.

Pang, J., Zheng, M., Sun, R., Wang, A., Wang, X., & Zhang, T. (2016). Synthesis of ethylene glycol and terephthalic acid from biomass for producing PET. *Green Chemistry: An International Journal and Green Chemistry Resource: GC, 18*, 342–359. Available from https://doi.org/10.1039/C5GC01771H.

Paul, M.-A., Delcourt, C., Alexandre, M., Degée, P., Monteverde, F., & Dubois, P. (2005). Polylactide/montmorillonite nanocomposites: Study of the hydrolytic degradation. *Polymer Degradation and Stability, 87*, 535–542. Available from https://doi.org/10.1016/j.polymdegradstab.2004.10.011.

Poirier, Y., Nawrath, C., & Somerville, C. (1995). Production of polyhydroxyalkanoates, a family of biodegradable plastics and elastomers, in bacteria and plants. *Bio/Technology, 13*, 142–150. Available from https://doi.org/10.1038/nbt0295-142.

Puglia, D., Biagiotti, J., & Kenny, J. M. (2005). A review on natural fibre-based composites-Part II: Application of natural reinforcements in composite materials for automotive industry. *Journal of Natural Fibers, 1*, 23–65. Available from https://doi.org/10.1300/J395v01n03_03.

Pyo, S.-H., Park, J. H., Srebny, V., & Hatti-Kaul, R. (2020). A sustainable synthetic route for biobased 6-hydroxyhexanoic acid, adipic acid and ε-caprolactone by integrating bio- and chemical catalysis. *Green Chemistry: An International Journal and Green Chemistry Resource: GC, 22*, 4450–4455. Available from https://doi.org/10.1039/D0GC01454K.

Quecholac-Piña, X., Hernández-Berriel, M., del, C., Mañón-Salas, M. D. C., Espinosa-Valdemar, R. M., & Vázquez-Morillas, A. (2020). Degradation of plastics under anaerobic conditions: A short review. *Polymers, 12*, 109. Available from https://doi.org/10.3390/polym12010109.

Rai, P., Mehrotra, S., Priya, S., Gnansounou, E., & Sharma, S. K. (2021). Recent advances in the sustainable design and applications of biodegradable polymers. *Bioresource Technology, 325*, 124739. Available from https://doi.org/10.1016/j.biortech.2021.124739.

Ren, J., Fu, H., Ren, T., & Yuan, W. (2009). Preparation, characterization and properties of binary and ternary blends with thermoplastic starch, poly(lactic acid) and poly(butylene adipate-co-terephthalate). *Carbohydrate Polymers, 77*, 576–582. Available from https://doi.org/10.1016/j.carbpol.2009.01.024.

Rhim, J.-W., Hong, S.-I., & Ha, C.-S. (2009). Tensile, water vapor barrier and antimicrobial properties of PLA/nanoclay composite films. *LWT - Food Science and Technology, 42*, 612–617. Available from https://doi.org/10.1016/j.lwt.2008.02.015.

Rhim, J.-W., Hong, S.-I., Park, H.-M., & Ng, P. K. W. (2006). Preparation and characterization of chitosan-based nanocomposite films with antimicrobial activity. *Journal of Agricultural and Food Chemistry, 54*, 5814–5822. Available from https://doi.org/10.1021/jf060658h.

Rhim, J.-W., Park, H.-M., & Ha, C.-S. (2013). Bio-nanocomposites for food packaging applications. *Progress in Polymer Science, 38*, 1629–1652. Available from https://doi.org/10.1016/j.progpolymsci.2013.05.008, Progress in Bionanocomposites: from green plastics to biomedical applications.

Rivera-Briso, A., & Serrano-Aroca, Á. (2018). Poly(3-hydroxybutyrate-co-3-hydroxyvalerate): Enhancement strategies for advanced applications. *Polymers, 10*, 732. Available from https://doi.org/10.3390/polym10070732.

Saadi, Z., Rasmont, A., Cesar, G., Bewa, H., & Benguigui, L. (2012). Fungal degradation of poly(l-lactide) in soil and in compost. *Journal of Polymers and the Environment, 20*, 273–282. Available from https://doi.org/10.1007/s10924-011-0399-9.

Satyanarayana, K. G., Arizaga, G. G. C., & Wypych, F. (2009). Biodegradable composites based on lignocellulosic fibers—An overview. *Progress in Polymer Science, 34*, 982–1021. Available from https://doi.org/10.1016/j.progpolymsci.2008.12.002.

Scheer, F. (2007). *Bio based biodegradable polymer compositions and use of same*. WO2007063361A1.

Shamsuri, A. A. M., & Jamil, S. N. A. (2020). A short review on the effect of surfactants on the mechanico-thermal properties of polymer nanocomposites. *Applied Science, 10*, 4867. Available from https://doi.org/10.3390/app10144867.

Shogren, R. (1997). Water vapor permeability of biodegradable polymers. *Journal of Polymers and the Environment, 5*, 91–95. Available from https://doi.org/10.1007/BF02763592.

Signori, F., Coltelli, M.-B., & Bronco, S. (2009). Thermal degradation of poly(lactic acid) (PLA) and poly(butylene adipate-co-terephthalate) (PBAT) and their blends upon melt processing. *Polymer Degradation and Stability, 94*, 74–82. Available from https://doi.org/10.1016/j.polymdegradstab.2008.10.004.

Singh, H., Inder Preet Singh, J., Singh, S., Dhawan, V., & Kumar Tiwari, S. (2018). A brief review of jute fibre and its composites. *Materials Today: Proceedings, 5*, 28427–28437. Available from https://doi.org/10.1016/j.matpr.2018.10.129.

Siparsky, G. L., Voorhees, K. J., Dorgan, J. R., & Schilling, K. (1997). Water transport in polylactic acid (PLA), PLA/polycaprolactone copolymers, and PLA/polyethylene glycol blends. *Journal of Environmental Polymer Degradation, 5*, 125–136. Available from https://doi.org/10.1007/BF02763656.

Smith, P. B. (2015). Bio-based sources for terephthalic acid. *Green polymer chemistry: Biobased materials and biocatalysis, ACS symposium series* (pp. 453–469). American Chemical Society.

Smitthipong, W., Tantatherdtam, R., & Chollakup, R. (2015). Effect of pineapple leaf fiber-reinforced thermoplastic starch/poly(lactic acid) green composite: Mechanical, viscosity, and water resistance properties. *Journal of Thermoplastic Composite Materials*, *28*, 717–729. Available from https://doi.org/10.1177/0892705713489701.

Son, S. Y., Kim, J.-H., Song, E., Choi, K., Lee, J., Cho, K., Kim, T.-S., & Park, T. (2018). Exploiting π–π stacking for stretchable semiconducting polymers. *Macromolecules*, *51*, 2572–2579. Available from https://doi.org/10.1021/acs.macromol.8b00093.

Song, H., & Lee, S. Y. (2006). Production of succinic acid by bacterial fermentation. *Enzyme and Microbial Technology*, *39*, 352–361, The Asia-Pacific Biochemical Engineering Conference (APBioChEC 2005). Available from https://doi.org/10.1016/j.enzmictec.2005.11.043.

Tan, G.-Y., Chen, C.-L., Li, L., Ge, L., Wang, L., Razaad, I., Li, Y., Zhao, L., Mo, Y., & Wang, J.-Y. (2014). Start a research on biopolymer polyhydroxyalkanoate (PHA): A review. *Polymers*, *6*, 706–754. Available from https://doi.org/10.3390/polym6030706.

Tan, L., He, Y., & Qu, J. (2019). Structure and properties of polylactide/poly(butylene succinate)/ORGANICALLY MODIFIED MONTMORILLonite nanocomposites with high-efficiency intercalation and exfoliation effect manufactured via volume pulsating elongation flow. *Polymer*, *180*, 121656. Available from https://doi.org/10.1016/j.polymer.2019.121656.

Thaore, V., Chadwick, D., & Shah, N. (2018). Sustainable production of chemical intermediates for nylon manufacture: A techno-economic analysis for renewable production of caprolactone. *Chemical Engineering Research and Design*, *135*, 140–152. Available from https://doi.org/10.1016/j.cherd.2018.05.026.

Umaraw, P., & Verma, A. K. (2017). Comprehensive review on application of edible film on meat and meat products: An eco-friendly approach. *Critical Reviews in Food Science and Nutrition*, *57*, 1270–1279. Available from https://doi.org/10.1080/10408398.2014.986563.

Vandewijngaarden, J., Wauters, R., Murariu, M., Dubois, P., Carleer, R., Yperman, J., D'Haen, J., Ruttens, B., Schreurs, S., Lepot, N., Peeters, R., & Buntinx, M. (2016). Poly(3-hydroxybutyrate-co-3-hydroxyhexanoate)/organomodified montmorillonite nanocomposites for potential food packaging applications. *Journal of Polymers and the Environment*, *24*, 104–118. Available from https://doi.org/10.1007/s10924-016-0751-1.

Vergnes, B., & Villemaire, J. P. (1987). Rheological behaviour of low moisture molten maize starch. *Rheologica Acta*, *26*, 570–576. Available from https://doi.org/10.1007/BF01333742.

Vert, M., Doi, Y., Hellwich, K.-H., Hess, M., Hodge, P., Kubisa, P., Rinaudo, M., & Schué, F. (2012). Terminology for biorelated polymers and applications (IUPAC Recommendations 2012). *Pure and Applied Chemistry. Chimie Pure et Appliquee*, *84*, 377–410. Available from https://doi.org/10.1351/PAC-REC-10-12-04.

Vink, E. T. H., Rábago, K. R., Glassner, D. A., & Gruber, P. R. (2003). Applications of life cycle assessment to NatureWorks™ polylactide (PLA) production. *Polymer Degradation and Stability*, *80*, 403–419. Available from https://doi.org/10.1016/S0141-3910(02)00372-5.

Volanti, M., Cespi, D., Passarini, F., Neri, E., Cavani, F., Mizsey, P., & Fozer, D. (2019). Terephthalic acid from renewable sources: Early-stage sustainability analysis of a bio-PET precursor. *Green Chemistry: An International Journal and Green Chemistry Resource: GC*, *21*, 885–896. Available from https://doi.org/10.1039/C8GC03666G.

Vyver, S. V. D., & Román-Leshkov, Y. (2013). Emerging catalytic processes for the production of adipic acid. *Catalysis Science & Technology*, *3*, 1465–1479. Available from https://doi.org/10.1039/C3CY20728E.

Wang, S. F., Shen, L., Tong, Y. J., Chen, L., Phang, I. Y., Lim, P. Q., & Liu, T. X. (2005). Biopolymer chitosan/montmorillonite nanocomposites: Preparation and characterization. *Polymer Degradation and Stability*, *90*, 123–131. Available from https://doi.org/10.1016/j.polymdegradstab.2005.03.001.

Wendels, S., & Avérous, L. (2021). Biobased polyurethanes for biomedical applications. *Bioactive Materials*, *6*, 1083–1106. Available from https://doi.org/10.1016/j.bioactmat.2020.10.002.

Woodruff, M. A., & Hutmacher, D. W. (2010). The return of a forgotten polymer-Polycaprolactone in the 21st century. *Progress in Polymer Science*, *35*, 1217–1256. Available from https://doi.org/10.1016/j.progpolymsci.2010.04.002.

Wu, H.-C. H., & Sarko, A. (1978). The double-helical molecular structure of crystalline a-amylose. *Carbohydrate Research*, *61*, 27–40. Available from https://doi.org/10.1016/S0008-6215(00)84464-X.

Wyser, Y., Adams, M., Avella, M., Carlander, D., Garcia, L., Pieper, G., Rennen, M., Schuermans, J., & Weiss, J. (2016). Outlook and challenges of nanotechnologies for food packaging: Outlook and challenges of nanotechnologies for food packaging. *Packaging Technology and Science*, *29*, 615–648. Available from https://doi.org/10.1002/pts.2221.

Xie, F., Halley, P. J., & Avérous, L. (2012). Rheology to understand and optimize processibility, structures and properties of starch polymeric materials. *Progress in Polymer Science, Topical Issue on Polymer Physics*, *37*, 595–623. Available from https://doi.org/10.1016/j.progpolymsci.2011.07.002.

Yamamoto, M., Witt, U., Skupin, G., Beimborn, D., & Müller, R.-J. (2005). Biodegradable aliphatic-aromatic polyesters: "Ecoflex®,". In A. Steinbüchel (Ed.), *Biopolymers Online*. New York, NY: John Wiley & Sons.

Yang, J., Ching, Y., & Chuah, C. (2019). Applications of lignocellulosic fibers and lignin in bioplastics: A review. *Polymers, 11*, 751. Available from https://doi.org/10.3390/polym11050751.

Yim, H., Haselbeck, R., Niu, W., Pujol-Baxley, C., Burgard, A., Boldt, J., Khandurina, J., Trawick, J. D., Osterhout, R. E., Stephen, R., Estadilla, J., Teisan, S., Schreyer, H. B., Andrae, S., Yang, T. H., Lee, S. Y., Burk, M. J., & Van Dien, S. (2011). Metabolic engineering of *Escherichia coli* for direct production of 1,4-butanediol. *Nature Chemical Biology, 7*, 445–452. Available from https://doi.org/10.1038/nchembio.580.

Zaaba, N. F., & Ismail, H. (2019). A review on tensile and morphological properties of poly (lactic acid) (PLA)/thermoplastic starch (TPS) blends. *Polymer-Plastics Technology and Materials, 58*, 1945–1964. Available from https://doi.org/10.1080/25740881.2019.1599941.

Zagrodzki, A. (2018). *Multi-layer biodegradable laminated structures intended for the production especially of food packaging and food packaging obtained therefrom*. EP3266608A1.

Zeikus, J. G., Jain, M. K., & Elankovan, P. (1999). Biotechnology of succinic acid production and markets for derived industrial products. *Applied Microbiology and Biotechnology, 51*, 545–552. Available from https://doi.org/10.1007/s002530051431.

Zhao, P., Liu, W., Wu, Q., & Ren, J. (2010). Preparation, mechanical, and thermal properties of biodegradable polyesters/poly(lactic acid) blends. *Journal of Nanomaterials, 2010*, 1–8. Available from https://doi.org/10.1155/2010/287082.

CHAPTER

15

Starch glues and adhesives

Marney Pascoli Cereda[1,2]

[1]Retired from São Paulo State University, Botucatu, São Paulo, Brazil [2]Tecsapp Solutions, Campo Grande, Mato Grosso do Sul, Brazil

15.1 Introduction

The most different types of adhesives are used in everyday life, at home, in commerce, and in industry. The great number of adhesives, with different adhesive capacities, becomes possible due to the variety of production processes and materials used in their manufacture. The level of effort required to keep the parts (substrates) glued allows for defining the adhesive to be used, as well as the number of layers to be used.

The simplest adhesives are mixtures of water with starch, which may include flours. After being boiled in water, they are used to bond wood, fabric, or paper. Among the starch-based adhesives, the most widely used are dextrin derivatives. In the competition for light adhesives, dextrin derivatives can compete with gums extracted from vegetables, such as arabic, xanthan, and guar gums. From these simpler adhesives, more elaborate products have been developed, with a mixture of natural or synthetic polymers.

All these adhesives should be evaluated for how their adhesion properties change over time and with temperature. The most widely adopted tests for adhesives are those described by the American Society for Testing and Materials (ASTM International, 2021), which also describes manufacturing processes.

In this chapter, the word substrate has been adopted to characterize the surface that receives a body or other surface, adhered by the use of an adhesive, or glue. The word film was also adopted to designate the presence of adhesive on the surface of the substrate.

15.2 Glues and adhesives—concepts

The American Society for Testing and Materials (ASTM International, 2021) defines glue as a product capable of forming a film between surfaces in order to hold materials together by surface interaction, resisting separation. Also, according to ASTM, the two most important phenomena involved in gluing are adhesion and cohesion. Cohesion is the property of molecules or atoms of the same material to stick together due to mutual attraction. Adhesion is the property of different molecules and atoms to stick together (Fig. 15.1).

The mechanical properties of adhesives are mainly evaluated in peel tests, which evaluate

Starch Industries
DOI: https://doi.org/10.1016/B978-0-323-90842-9.00009-1

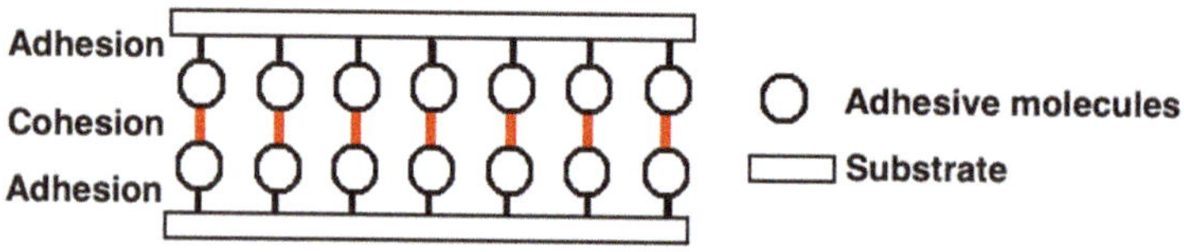

FIGURE 15.1 Representation of adhesion and cohesion in a gluing process.

the adhesive failures. A cohesive failure occurs when the adhesive film breaks and leaves adhesive residue on both substrates. In an extreme case, the adhesive completely detaches from the substrate during peel-off and the entire film remains on one of the substrates, indicating an adhesive failure (Kim et al., 2005).

15.2.1 Different types of adhesion

Adhesion can be classified into specific and mechanical. Specific adhesion, also called chemical adhesion, consists of the bond strength between the adhesive and the substrate, which is greater when the chemical affinity between the two is greater. Mechanical adhesion, on the other hand, occurs when the adhesive penetrates the pores of the substrate, which function as a kind of mechanical fixing hooks (Voiutskii, 1963).

For Silva (2008), the main mechanisms that explain the adhesion phenomenon are chemical and electrical bonding, adsorption and diffusion, and mechanical anchoring. The chemical bonding mechanism suggests that covalent chemical bonds can be formed along the adhesive–substrate interface. This bonding occurs when reactive sites present on the substrate are attacked by the adhesive. In the case of wood bonding with reactive polyurethanes, the breakdown energy associated with this type of force is approximately 600 kJ/mol (Silva, 2008).

The electrical mechanism is a type of chemical bonding that occurs mainly in metal bonding. It is based on the electronegativity difference between adhesive and substrate, and the bonding force is attributed to electron transfer along the interface creating positive and negative charges in an electrical double layer. This type of charge-bonding force is the second most intense intermolecular interaction, with a breakdown energy on the order of 200 kJ/mol (Silva, 2008).

Adsorption is attributed to chemical adhesion by van der Waals interactions between the surface molecules of the adhesive and those of the substrate. For De Bruyne (1957), the forces acting at the threshold between the solid and liquid phases are sufficient to ensure good adhesion. Silva (2008) added that adhesion involving polymers can be divided into two stages. The first stage comprises the migration of the macromolecules of the adhesive from the liquid phase to the material surface that will receive the object to be bonded. This movement at the molecular level occurs by macro-Brownian motion and allows the polar groups of both materials to come closer together. The second stage of adhesion occurs with the establishment of the adsorption equilibrium. The attractive van der Waals forces begin to act when the distance between the adhesive and substrate molecules becomes less than 5 Å, while adhesion requires more interaction between the adhesive and the object to be bonded.

Diffusion requires Brownian motion to occur, which in turn is based on the chain structure and flexibility of the molecules, which are essential characteristics of the polymers that constitute it. According to Silva (2008), the bonding force between the adhesive and the substrate is caused by the diffusion of polymer chains or their individual segments over undulations and spaces formed by irregularities on the substrate surface. The author also points out that on substrates where the roughness can be observed without the need for a microscope, the adhesion mechanism is usually called mechanical anchoring. On smooth substrates, this phenomenon also occurs, but on a scale that is only visible under a microscope.

Adhesion depends on several factors, such as the formation of the adhesive joint resulting from the pressure applied at the time of bonding, the temperature, the roughness of the substrate, the chemical structure of the adhesive and the substrate, and the type of attractive intermolecular forces (Satas, 1989).

The adhesion of a polymer to any substrate increases with contact time. Initially, it occurs rapidly and then more and more slowly, tending to a limit value at infinity. The adsorption and electrical bonding theories alone are not sufficient to provide a satisfactory interpretation of this behavior. By the diffusion theory, over time, the polymer continues to diffuse, more slowly through the substrate, managing in the end to occupy the smaller scale surface imperfections of the substrate. For Silva (2008), equilibrium is reached more quickly in crystalline adhesives with high melting temperature. Adhesives made with elastomeric materials, such as pressure-sensitive adhesives (PSA), take time to reach equilibrium, which causes a continuous increase in adhesion. Therefore, the tests to be performed with elastomeric adhesives should consider the time after application (Silva, 2008).

When it comes to favoring starch-based adhesives, viscosity is mandatorily included as a parameter to be considered.

15.2.2 Viscoelasticity

A viscoelastic material exhibits the characteristics of both a liquid and a solid. For Silva (2008), viscoelastic materials are those that, when subjected to a stress, present at the same time responses characterized as elastic and viscous. This behavior is characteristic of most polymeric materials, such as starch pastes, and is a consequence of the high molar mass of the polymer chains and the fact that they are partially or totally in random intertwined ball-shaped conformation.

Adhesives are polymeric materials, with a viscoelastic behavior. When a stress is applied to a polymer, a part of the energy is dissipated by the movements of chains (viscous flow). These viscous losses are time- (or speed or frequency) and temperature-dependent (Guillemenet et al., 2002). For example, in the case of two irregular and rough surfaces to be bonded, the use of a hard adhesive without viscoelastic properties will not allow for the good penetration of the adhesive material between the irregularities of the material to be bonded. In this case, the bonding will not be efficient. In the case of an adhesive with good viscoelastic properties, it will be able to penetrate more easily between the irregularities of the material to be bonded, improving the bonding efficiency (Budnick Converting, 2023).

However, as already mentioned, the gluing efficiency depends on time and temperature. A very general rule is that an adhesive will reach 50% total adhesion in 20 minutes, 80% within 24 hours, and 100% in 72 hours (Budnick Converting, 2023).

Many substances have been used as adhesive, as long as they have plasticity and adherence characteristics. Among them are starch-based pastes and their derivatives and gums and mucilage.

15.3 Starch and derivatives used to make adhesives

Fabrício and Mählmann (2014) justified the use of starch as a raw material for making adhesives as being due to the growing importance of using biodegradable polymers, the growing need for sustainable development, the desire to reduce dependence on nonrenewable resources, and changing policies and attitudes in waste management. Most biodegradable polymers on the market are based on renewable raw materials of agricultural or forestry origin (Murphy & Bartle, 2004).

Several raw materials can be used in the production of starch-based adhesives, but the most common are based on commercial starches: potato, cassava, corn, and wheat (See Chapter 3 of Volume 1 in this series). The most used for adhesive production is corn starch.

The use of starch meets the proposals for cultivated and biodegradable renewable raw materials through the use of agricultural resources (Lörcks, 1998) and the need to reduce dependence on petroleum (Vroman & Tighzert, 2009).

Adhesives based on starch and one of its derivatives, dextrin, are of great importance in industrial production in the packaging industry. Starch and dextrin are used mainly for gluing paper products. Most corrugated boxes are bonded with starch-based adhesives, but other porous substrates can easily be bonded with these versatile adhesives (ASI-Adhesives & Sealants Industry, 2005). They are also the primary adhesives used in paper bonding.

Adhesives obtained from starch provide good adhesion to paper and wood due to their cellulosic origin. Starch adhesives have excellent heat resistance, low cost, stable quality, and good adhesion to cellulose and other porous substrates, besides being biodegradable and nontoxic. These adhesives are widely used in the production of cardboard boxes, paper bags, hanging files, wallpaper, cigarette gluing, bottle labeling, envelopes, bookbinding, and other paper packages that require some type of gluing.

Developing starch-based adhesives enables the use of a raw material from renewable sources that combines biodegradability with low cost and high availability. Starch-based adhesives have the characteristic of being easily removable with water and at the same time being chemically stable. They are flexible when applied in thin layers and also quite strong, but have the disadvantage of being slow-drying. Pre-cooked starch glues, which are easy to prepare, can be found on the market (Rodrigues et al., 2014).

15.3.1 Glue from native starches

The basic raw material for adhesive production in powder or syrup form is native starches. It turns out that these starches have high viscosity, which requires using low solids content so that the paste (gel) can be transported. In addition, the paste based on native starch presents viscosity variation with time (ASI-Adhesives & Sealants Industry, 2005).

To produce gum, which is already a type of adhesive, it is necessary to break up the starch granules. Heating in water is the simplest method to break these granules, which first swell and then collapse, releasing a thick suspension. The temperature at which this thickening of the suspension occurs is called the gelatinization temperature. For most starches in pure water, gelatinization occurs between 57°C and 72°C (for more information on starch properties, see Chapter 10 of Volume 1 in this series). It is important to remember that the gelatinization of a starch does not occur at a fixed temperature, but over a range of temperatures. Table 15.1 indicates the onset temperatures identified in the scientific literature for the main commercial starches used in the manufacture of adhesives.

TABLE 15.1 Comparison of the most used commercial starches for adhesive production, in characteristics that can affect their processing and performance.

Sources	**Cereals**	**Underground organs**	
Raw materials	**Corn**	**Cassava**	**Potato**
Granule diameter (μm)	2–30	4–35	5–100
Onset temperature (°C)	62–77	56–66	58–63
Amylose (%)	19–32	16–29	18–30

From Chapter 10 in Volume 1 of this series.

Because it gives rise to a colloidal suspension, starch cannot be considered a true solution (ASI-Adhesives & Sealants Industry, 2005). This colloidal suspension is already a type of adhesive obtained from native starch.

To produce glue, the heating process can be done with the addition of salts that facilitate swelling. Salts such as calcium, magnesium, or zinc chloride are added to a suspension of starch in water, which is heated by a controlled system and stirred for the time necessary to obtain the characteristics required for the desired adhesive. The most common production system uses caustic soda (NaOH), which is added to the starch suspension. Once the reaction is complete, the product is neutralized with acids (ASI-Adhesives & Sealants Industry, 2005).

At the end of the process, the starch paste has a high viscosity, which distinguishes it as an adhesive. If the starch concentration is above 7%, the cooked paste is very viscous and difficult to pump. Above 15% starch solids, the cooked paste forms a rubbery looking mass on cooling. Starch suspensions with a high amylose/amylopectin ratio will also have higher viscosity (ASI-Adhesives & Sealants Industry, 2005).

Pregelatinized starch exhibits the same behavior, but it is produced by physical modification, which allows for the formation of a paste that is easily dispersed in cold water. The process consists of heating the starch paste to a temperature above the gelatinization temperature, and then drying it quickly before retrogradation (for more on the retrogradation property of starch pastes, see Chapter 10 of Volume 1 in this series).

Various equipment can be used to generate the starch paste made from pregelatinized starch, such as spray drier and drum dryer, but the main process equipment used in the industries to produce this type of adhesive is the heated-drum drier or hot roll (for details on obtaining modified starch, see Chapter 6).

Natural or native starches subjected to heating followed by drying are marketed in powder form (pre-gel) or as high viscosity, low solid content pastes that do not have a stable viscosity over time. Therefore, several treatments have been developed to produce liquid adhesives that do not retrograde and have a more stable viscosity and rheology suitable for many applications. These treatment methods include alkaline treatment, acid treatment, and oxidation (ASI-Adhesives & Sealants Industry, 2005).

15.3.2 Adhesives from modified starches

Modified starches may present several advantages and are therefore widely used as thickening, stabilizing, or emulsifying agents. However, compared to native starches, they show worse water retention characteristics and shorter shelf life, which reduces the technical advantages, when compared to acacia gum and its possible substitutes (World Bank, 2007).

While native starch contains only hydroxyl groups and has limited scope, chemically modified starch shows superior water resistance properties for adhesive applications. Esterified starches, starches with ester substituents, can be utilized to prepare biobased adhesives with improved water resistance (Watcharakitti et al., 2022). Starches modified by oxidation or acidification are more suitable for making adhesives than native starches. Both are modified starches that are widely marketed and simple to produce. However, the production of these modified starches without adequate control may lead to a nonstandard adhesive (to learn more about the processes for obtaining these starches, see Chapter 6).

Acid-modified starches are obtained by heating the suspension in water at 49°C–54°C, with a low amount of aqueous acid solution, followed by neutralization with a base. This treatment tends to give a thinner suspension with the same solid concentration, when compared to native starch. Starches prepared in this way are often referred to as thin-boiling or high-fluidity starches. Acid modification is often used when the application requires higher solids content and a low, controlled viscosity (ASI-Adhesives & Sealants Industry, 2005).

Oxidation of starch is usually achieved by treatment with aqueous alkaline hypochlorite solution. A starch suspension at pH 8–10 is treated with hypochlorite, with the chlorine concentration set at 5%–10% on the weight of the starch, for long enough to produce the desired viscosity. The acid released during the reaction must be neutralized to maintain the pH at the previously established value. Dry oxidized starch is usually whiter than native starch, which is why adhesives made with these starches are often used for printing papers. Oxidized or chlorinated starches behave like the acid-modified ones; however, oxidized starch has better adhesion when used in the preparation of adhesives (ASI-Adhesives & Sealants Industry, 2005).

Fabrício and Mählmann (2014) described the preparation of adhesive using combinations of corn and cassava starches previously oxidized with sodium hypochlorite (NaOCl). Starches from the same botanical sources subjected to treatment with phosphoric acid (H3PO4) were also evaluated.

15.3.3 Adhesives based on dextrins and the process to obtain them

Dextrins are starch derivatives, consisting of oligosaccharides of 6–8 glucose units (Souza & Magalhães, 2010) (for more on enzymatic hydrolysis of starch, read Chapter 5). Regardless of the botanical origin of the starch, the production of dextrin for glue is essentially done by breaking down the starch molecules by heat (ASI-Adhesives & Sealants Industry, 2005). In this case, dextrins can also be called pyrodextrins.

For starch to be suitable for production of adhesives, it requires alkaline, acid, and oxidation treatments. To accelerate and improve the quality of the pyrodextrins, it is possible, in addition to heat, to add mineral or organic acids, or even enzymes. As an adhesive, dextrins stand out for their high tensile strength (ASI-Adhesives & Sealants Industry, 2005).

15.3.3.1 Classic processes or use of dry heat

The physicochemical process to obtain dextrins involves the use of heat, 110°C–120°C, on starch moistened with a dilute acid solution (1000 parts starch, 250 parts water, 2 parts HCl). This process produces low yields and an inferior quality starch (ASI-Adhesives & Sealants Industry, 2005).

The gelatinization temperature can be reduced by adding an alkali to the starch suspension, such as sodium hydroxide. If sufficient alkali is added, the starch can gelatinize at room temperature (ASI-Adhesives & Sealants Industry, 2005) (to learn more about cold gelatinization of starch with NaOH, it is suggested to look at Chapter 10 in Volume 1 of this series).

The dextrin obtained by heat treatment, used as a base for adhesives, is the pyrodextrin. In production by roasting, a direct-fired or steam pan is used (more suitable as it does not create overheating breakdowns). The roasting point is generally established by color (ASI-Adhesives & Sealants Industry, 2005). There are three types of pyrodextrin marketed as adhesives: white dextrins, "Canarian" dextrins, and "British gums." The difference is determined by the roasting time, temperature, and amount of catalyst used.

Pyrodextrins are also obtained by adding acid as a catalyst during starch roasting. With heat and acid treatment, the starch molecules are hydrolyzed, or broken into small fragments, and then polymerized again into highly branched, moderately sized polymer molecules. Although other starches can also be used, corn starch is most often used because of its low cost and ready availability (ASI-Adhesives & Sealants Industry, 2005).

Unlike starch, dextrins are water-soluble. The degree of water solubility is regulated by the severity of the heat and acid treatment. The degree of solubility is one of the determinants of dextrin classification. At the end of the process, dextrins are in the form of a very fine powder, ranging in color from pure white to brown (ASI-Adhesives & Sealants Industry, 2005).

Among the pyrodextrins, British gums are the dextrins with the highest average molecular weight. Although they are the strongest adhesives among the dextrins, their maximum solid content used is approximately 25%. Yellow dextrins are the ones with the lowest average molecular weight, available with a viscosity of only 1700 cps at 60% concentration. White dextrins have the lightest coloration of the three, and their average molecular weight is intermediate between the other two (ASI-Adhesives & Sealants Industry, 2005).

The solubility range of dextrin-based adhesives is important and allows this starch derivative to be used to formulate a variety of suspensions of very different viscosities, creating a wide range of gel strength. The same additives and processes used to make native starch-based adhesives are generally used to formulate dextrin adhesives. In general, dextrins are preferable raw materials to native starch because they require less or no cooking and much less water to form the paste. This means that during the drying time of the adhesive the likelihood of mold growth is reduced when use occurs in warmer climates.

As adhesives, dextrins are used in many of the same applications as native starch adhesives. They can also be used as a core-binding material in the manufacture of foundry cores, with the main function of providing good mold strength and preventing core deformation before it is burned in the furnace (ASI-Adhesives & Sealants Industry, 2005).

15.3.3.2 *Production of dextrins by the enzymatic process*

Dextrin and its derivatives are modifications of starch, which can add value to starch, adjusting it to its applications. For more details on physical, chemical, and enzymatic modifications, please refer to Chapter 5.

However, today's consumer does not trust chemically modified starches and prefers products from physical and enzymatic processes. Although food consumption is not the focus of this chapter, the use of enzymes to produce dextrins for adhesive purposes has several advantages. We highlight the lower risks of accidents during processing on the factory floor, the lower temperatures, and especially the advantage of obtaining more uniform products, since enzymes present specificities of action.

Souza and Magalhães (2010) described α-amylases as one of the main enzymes used in industry. Such enzymes hydrolyze the starch molecules into polymers composed of glucose units. Amylases have potential application in a wide number of industrial processes such as food, fermentation, and pharmaceutical industries. The α-amylases can be obtained from plants, animals, and microorganisms. However, enzymes from fungal and bacterial sources have dominated applications in industrial sectors. The production of α-amylase is essential for conversion of starches into oligosaccharides. Starch is an important constituent of the human diet and is a major storage product of many economically important crops such as wheat, rice, maize, tapioca, and potato. Starch conversion by enzymes is used in the production of maltodextrin, modified starches, or glucose and fructose syrups. A large number of microbial α-amylases have applications in different industrial sectors such as food, textile, paper, and detergent industries. The production of α-amylases has generally been carried out using submerged fermentation, but solid-state fermentation systems appear as a promising technology. The properties of each α-amylases such as thermostability, pH profile, pH stability, and calcium independency are important in the development of the fermentation process.

When the process is enzymatic, the use of enzymes allows to bypass the high viscosity, especially when using starches extracted from raw materials that store starch in underground organs (for more details on the properties of these starches, we suggest reading Volume 1 in this series). Concentration allows the syrup to be adjusted so that it can be stored at room temperature.

15.4 Applications of starch-based adhesives (ASI-Adhesives & Sealants Industry, 2005)

The starch and dextrin-based adhesives represent a significant part of the industrial use, of which the paper, cardboard, and weaving sectors stand out, but with use also in cosmetics, pharmaceuticals, and food, depending only on meeting the specific standards of each one.

Depending on the concentration, native starch pastes with high viscosity and low solids content show viscosity variation with storage time, with a risk of retrogradation. Several treatments have been developed to obtain liquid adhesives that do not retrograde or at least retrograde little, so that their viscosity and rheology are more suitable for various applications. These methods include alkaline treatment, acid treatment, and oxidation, as seen above.

It should be remembered that in both modified starches, oxidized and acid-modified, partial hydrolysis occurs, which is evidenced by the determination of reducing ends (DE), carboxyl index, and other characteristics (for more details on modified starches, please read Chapter 12 of Volume 1 in this series, which deals with specific analyses for modified starches).

Because they are whiter than their native starches, oxidized starches are often used for sizing and coating printing papers. These oxidized or chlorinated starches behave similar to acid-modified starches. However, the oxidized starch has higher permanent adhesion (TACK) and adhesive character and is therefore often used in adhesive applications (ASI-Adhesives & Sealants Industry, 2005).

The paper and cardboard sector, especially in the packaging box industries, requires specific properties from adhesives. Rodrigues et al. (2014) reported, in a technical article, the experience with adjustments in the formulation of starch-based adhesives for the manufacture of corrugated cardboard, one of the main applications of dextrin-based adhesives. According to the authors, the concentration of the adhesive and its application point, as well as the gluing process, influence the quality of the cardboard.

Corrugated board is formed by the association of core and covers, which can form several layers, and which are bonded together by adhesives that are applied on top of the waves of paper core. For a good gluing process to occur that produces a good quality corrugated board, it is necessary to control three elements: paper, glue, and heat. To achieve a good bond, the paper quality and some other factors must be constantly monitored during the process. For papers with resin applications, it is recommended to work with a solid content above 28%, because this condition avoids the increase of humidity, prevents loss of speed in the forming machine (corrugator), and provides better adhesion. With this, it is possible to reduce glue application and increase machine speed. At the same time, it is possible to count on new equipment developed to control and make the temperature and pressure more precise (Rodrigues et al., 2014).

However, Rodrigues et al. (2014) explained that this same equipment works with higher pressure on the drying table, with values around 14 kgf/cm^2. Thus, lower solid contents are needed in the glue formulation, requiring rheological control additives, called gluing aids. These auxiliaries contribute to the reduction of solids in the adhesive by increasing the anchoring effect of the adhesive to the paper fibers, allowing for higher operation speed, and ensuring the quality of the corrugated paper sheet. The authors also prove the improving effect of the use of rheological modifiers as additives to the starch.

15.4.1 Adhesive manufacturing processes

The main processes for making adhesives from starch are discussed below.

15.4.1.1 *The Stein-Hall process*

The first large-scale commercial process for producing starch adhesives was the Stein-Hall process, introduced in the 1930s to produce an adhesive for corrugated boxes. The Stein-Hall system combines the properties of an 8%–9% cooked starch paste (Upper Tank System) and a 25% starch paste (Lower Tank) in a two-component, two-process container. Sodium hydroxide is used so that gelatinization occurs at room temperature, and borax is used to increase the quality of permanent adhesion adhesives (TACK). Cross-linking starches can be used to confer water resistance (ASI-Adhesives & Sealants Industry, 2005).

The two phases are divided into two tanks (ASI-Adhesives & Sealants Industry, 2005):

- Upper tank (carrier starch)
  - In 182 kg of water, add by stirring 45 kg of starch, followed by 8 kg of caustic soda in 4 liters of water.
  - Heat with hot steam to 71°C, stirring for 15 minutes, and then dilute in 227 kg of cold water.
  - Add a preserving agent or other additive if necessary.
- Lower tank (suspended starch)
  - In 690 kg of water, add 8 kg of Borax and let it dissolve for 2–3 minutes and add 227–270 kg of starch. Stir to homogenize.
  - Gradually empty the upper tank completely into the lower tank, mix efficiently, and stir until the viscosity is consistent.
  - Add a preserving agent or other additive if necessary.

Units have been adjusted for the metric system. The quoted temperatures, times, and concentrations are for corn starch, and in the case of starches extracted from underground organs, an adjustment must be made; otherwise the texture, viscosity, and adhesion may be greatly reduced by heat.

The cooked, high-viscosity starch paste serves as a support to prevent the starch granules that remain ungelatinized in the paste from settling to the bottom of the tank. When the adhesive is applied and heated above its gelatinization temperature, the granules finish their gelatinization process and a very viscous, sticky paste is produced. In corrugated boxboard manufacturing, the starch in suspension must gelatinize immediately when it comes in contact with the heated-drum drier and develop permanent adhesion instantly upon reaching 60°C 70°C. Starch solids can range from 20% to 30% (ASI-Adhesives & Sealants Industry, 2005).

15.4.1.2 *Single container or "carrierless" process*

All the operations described for the two-tank system can also be performed in a one-tank system, without having to transport the prepared adhesive to another tank. In this process, a caustic soda solution is added to a starch paste to swell the starch granules uniformly and controlled until they reach the required adhesive viscosity. Acid is then added to stop this swelling process before gelatinization is complete. Borax is added at the end of the process (ASI-Adhesives & Sealants Industry, 2005).

According to ASI-Adhesives & Sealants Industry (2005), the preparation of adhesives in a single tank presents a series of advantages such as lower heat requirement to achieve the bonds between the granules, since they are already partially swollen. It also requires less free water, providing higher viscosity, which allows to increase the machine speed and reduce the occurrence of warped paper, besides improving the storage conditions, with better shear stability. However, it also requires certain precautions, such as the need to carefully control the concentrations and temperature of the starch paste and caustic soda solution, to stop the swelling of the granules at the correct stage, and to ensure that the required viscosity is achieved and maintained.

15.5 Adhesive performance analysis methods

The analysis methods described are intended for technical evaluation of adhesives, although they can be used in research and laboratory. Of the adhesive properties, adhesion and cohesion were selected. They have already been highlighted in the introduction of this chapter as being the most important for this type of evaluation.

15.5.1 Characterization of the rheological behavior

According to Silva (2008), the characterization of the rheological behavior can be obtained with viscosity measurements in Brookfield viscometer and by linear viscoelastic parameters in rotational rheometers.

The same author reminds that the analysis of rheological behavior in a rotational rheometer is done in oscillatory mode, with a frequency sweep between 0.1 and 500 Hz, with a strain determined at temperatures of 40°C and 150°C. The choice of strain should be made after strain sweeps at frequencies of 0.1 and 500 Hz and selection of a strain that matches in the linear region for both frequencies. Evaluations at 40°C are done using the parallel plate geometry and at 150°C using Couette. In the case of using a Brookfield viscometer, the analysis is done at a temperature ranging between 130°C and 160°C, with a 5°C interval between them, using shear sensor 29 and rotation according to the experiment's viscosity, to obtain a torque within the equipment's precision range.

15.5.2 Adhesion

Adhesion is evaluated by the peel test at varying angles from 90 to 180 degrees, according to ASTM D3330M. The peel test consists of measuring the force required to peel a 25 mm wide flexible substrate from a rigid substrate bonded with the adhesive under evaluation. Other variations of the peel method refer to the peel angle, as illustrated in Fig. 15.2.

Silva (2008) evaluated adhesion using the 180 degree peel test according to ASTM D3330M. A polyester film and a stainless-steel plate type L304 were used as flexible and rigid substrates, respectively.

15.5.3 Cohesion

Silva (2008) stated that the main cohesion test for PSA is the shear adhesion test, also called holding power. In this test, a tape with the adhesive applied is glued to a standard surface and subjected to a suspended weight. The result is given by the time elapsed between

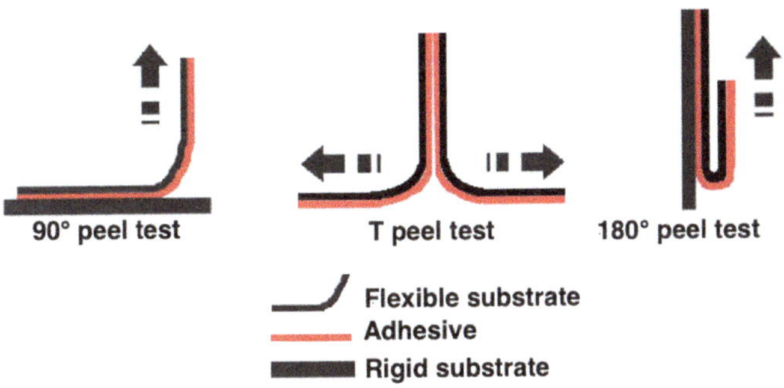

FIGURE 15.2 Schematic representation of the peel test and its various assessment angles (Heinzmann et al., 2016). *Adapted from Heinzmann, C., Weder, C. & de Espinosa, L. M. (2016). Supramolecular polymer adhesives: Advanced materials inspired by nature.* Chemical Society Reviews, *45(2), 342–358.*

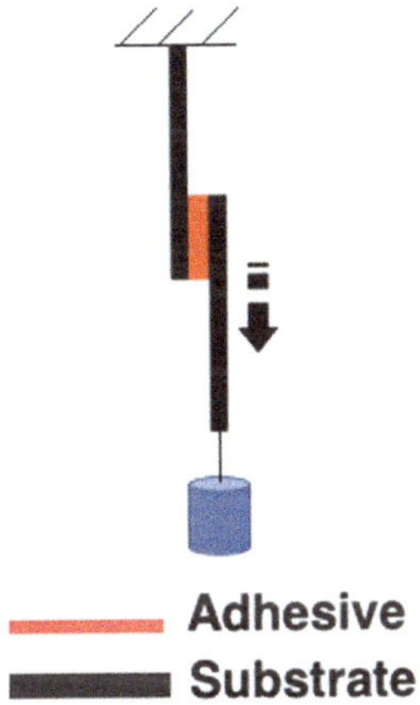

FIGURE 15.3 Shear adhesion, or holding power, test. *Adapted from Heinzmann, C., Weder, C. & de Espinosa, L. M. (2016). Supramolecular polymer adhesives: Advanced materials inspired by nature.* Chemical Society Reviews, *45(2), 342–358.*

when the weight is suspended and when it falls.

According to ASTM D3654, Method A, the cohesion by the "shear" or "holding power" test uses a weight of 2 kgf. In the "shear" test, a vertically arranged 1 inch (2.54 cm) area of bonding is subjected to an applied shear stress with the suspension of a weight at its end (Fig. 15.3). The time measured between the suspension of the weight and its fall through peeling from the substrate represents the cohesion of the adhesive. Variations in this test refer to the use of different weights and the temperature of the test. One particular case is called shear adhesive failure temperature, and it consists of subjecting the device to a temperature gradient and measuring the temperature at which adhesion breakdown occurs (Silva, 2008).

Other tests can be used such as the permanent adhesion test (TACK), which can be measured by two tests, the "rolling ball," which follows ASTM D3121, and the "Loop TACK," which follows ASTM D6195, method A, both considered by Silva (2008) as the main evaluations for adhesives.

The "rolling ball" test consists of measuring the distance traveled by a ball of known weight, over the adhesive-coated substrate, after it has traveled down an inclined surface of defined geometry (Fig. 15.4).

The Loop TACK test (Fig. 15.5) consists of measuring the force required to peel off a 1-square inch surface (6.4516 cm^2) immediately after it has been bonded, without pressure being applied for bonding (Silva, 2008).

Although relatively simple, the apparatus required for these tests are more commonly used in industry, while more sophisticated equipment is used in research, such as rheometers that can measure the viscosity of an adhesive very accurately, but do not provide the information that companies need to improve equipment performance, productivity, or product quality.

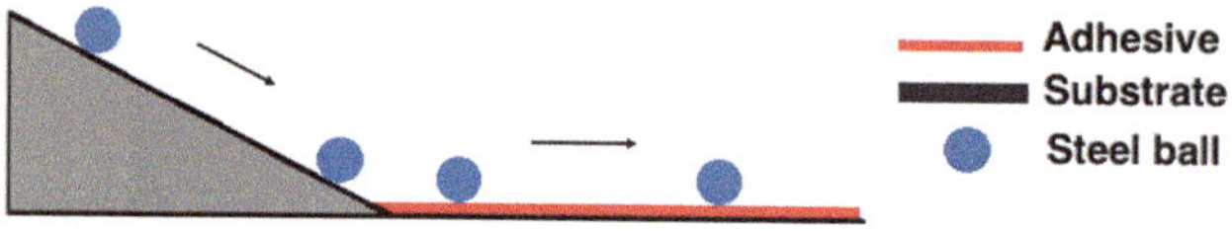

FIGURE 15.4 Description of the method used for the Rolling Ball Adherence Test.

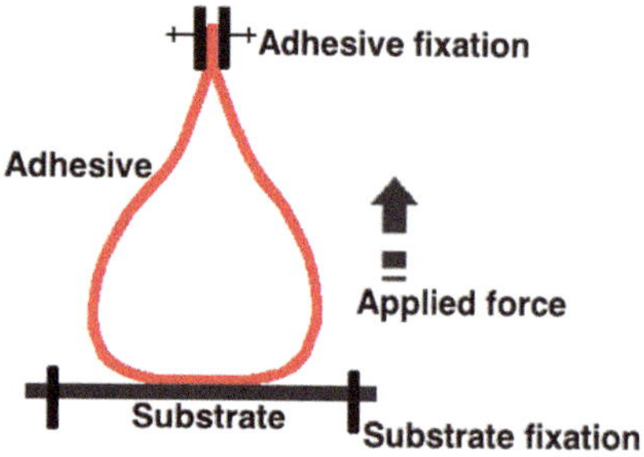

FIGURE 15.5 Loop TACK test.

15.6 Final considerations

As reported in the introduction, the technical area of production of starch-based adhesives or its derivatives is little disclosed, even in specialized books. From the publications about the industry, the most important are the technical standards and results published by suppliers. Important information remains in the producing companies, to prevent it from reaching the competitors.

The research has presented laboratory results that often do not meet the needs of the industries. Many formulations and practical guidelines for the preparation of adhesives are available, all based on cereals, such as corn and wheat, which is understandable because they are the main starches produced in more industrialized countries. In Europe, there are also references to the use of wheat flour-based adhesives (D'Amico et al., 2012).

A trend toward developing or adjusting formulations compatible with preserving or improving environmental conditions has also been observed. An example is the elimination of formaldehyde-emission in urea-oxidized adhesive formulations (Zhao et al., 2018). Also, in the starch-based wood adhesive sector, improvements have been found by increasing or replacing formulation components, as cited by Li et al. (2015).

Of the few papers that mention the use of tropical starches, only cassava starch is highlighted. However, a cereal starch-based formulation can be adapted for use with starches extracted from underground organs, because there is no requirement for specific properties for use as adhesives, as these properties are destroyed by the action of acids or enzymes.

It should also not be forgotten that although starch adhesives are available at a low price, they face competition from other types of materials, such as gums (gum arabic), which have the advantage of gelatinizing at room temperature, but also in hot water (Bobbio & Bobbio, 1989) and do not manifest the phenomenon of retrogradation. With the gums, it is possible to adjust the consistency of the adhesives online with the usage process.

The use of mixtures of starch with gums is also possible. Rodrigues et al. (2014) reported that when the concentration of corn dextrin-based adhesive was increased to 28% under industrial conditions, they obtained better results with the addition of xanthan gum-based additive to optimize molding machine speed. However, as for starch, xanthan gum is susceptible to bacterial attack, which requires the use of a biocide in the adhesive formulation, leading to increased costs and extra regulatory procedures (Rheological Modifiers).

Gums are natural polymers, important not only because they are readily available and renewable, but also because of their gel-forming properties (Paula et al., 2002; Souza, 2014).

Gums have high molecular weight and colloidal properties capable of forming highly viscous suspensions, with thickening, gelling, emulsifying, stabilizing, and binding functions (Towle & Whistler, 1973). They are considered soluble cellulose derivatives that fall under the concept of mucilaginous polysaccharides (Azeez, 2005). Souza (2014) stated that gums, as high molecular weight polymers, can be extracted from algae, seeds, animals, or from microorganisms via fermentation. Among the different gums and mucilage, it is possible to mention alginates, agar, carrageenan, arabic, guar, xanthan, gellan, yam extract, konjac, and pectins.

For food uses, these gums and mucilage have been more widely available on the market and at more convenient prices and could be used together with starch in the production of adhesives, allowing for better adjustment to factory production conditions.

The specific case of gum arabic is interesting because it was one of the most traditional glues, used on stamps and valued for being water-soluble and not damaging the stamps for use in philately. Its extraction in Sub-Saharan Africa (Sahel zone) is generally done by extractivism as an exudate of the stems and branches of trees of the Acacia species, the most commercially exploited species (Al-Assaf et al., 2007).

For Souza (2014), Brazil has a dependent relationship with imported gum arabic, which is the oldest and best known of all tree gum exudates. In the case of gums extracted from vegetables, the exudates are generally collected as air-dried droplets (Fitwi, 2000). The gums thus purified are in the form of a white powder that is completely odorless and soluble in water. Unlike several other plant-based gums, gum arabic up to 50% concentration dissolves easily in water. It has a low viscosity and forms a colorless, tasteless solution that is largely nonreactive when mixed with other chemical compounds.

According to Souza (2014), gum arabic is imported to Brazil at a high price, but it is still frequently used in the glue industries. Besides gum arabic, other gums can be cited, such as guar and xanthan.

Finally, Rodrigues et al. (2014) called attention to the need for adjustments in adhesive formulations, as well as in quality testing, due to changes that have occurred in traditional markets. Among these changes, the authors cited the increased variation in the standard of pulp used, due to the increasing use of recycled paper and cardboard. This reuse of paper and packaging comes from measures taken to avoid waste or environmental burden.

References

Al-Assaf, S., Phillips, G. O., Aoki, H., & Sasaki, Y. (2007). Characterization and properties of *Acacia senegal* (L.) Willd. var. senegal with enhanced properties (Acacia (sen) SUPER GUM™): Part 1—Controlled maturation of *Acacia senegal* var. senegal to increase viscoelasticity, produce a hydrogel form and convert a poor into a good emulsifier. *Food Hydrocolloids*, *21*(3), 319–328.

ASI-Adhesives & Sealants Industry. (2005). Packaging end user: Starch- and dextrin-based adhesives. Available from <https://www.adhesivesmag.com/articles/84472-packaging-enduser-starch-and-dextrin-based-adhesives>. (Consulted in 03/15/2023).

ASTM International. (2021). Available from <https://webstore.ansi.org/sdo/astm?gclid = Cj0KCQjwgtWDBhDZARIsADEKwgP8gr36DF-o0Z_Q9tagibxJeubAqDIb3GPnSgOHX7T2sy5GkcaPX4-XUaAijDEALw_wcB>. (Consulted in 04/13/2021).

Azeez, O. S. (2005). Production of gum from cashew tree latex. *Leonardo Electronic Journal of Practices and Technologies*, 17–22.

Bobbio, F. O., & Bobbio, P. A. (1989). *Introdução à química de alimentos* (p. 231) *Introdução à química de alimentos*, -231.

Budnick Converting. (2023). What is viscoelasticity, and why should I care? Available from <https://www.budnick.com/learning-center/blog/what-is-viscoelasticity-and-why-should-i-care-1>.

D'Amico, S., Hrabalova, M., Müller, U., & Berghofer, E. (2012). Influence of ageing on mechanical properties of wood to wood bonding with wheat flour glue. *European Journal of Wood and Wood Products*, *70*(5), 679–688. Available from https://doi.org/10.1007/s00107-012-0595-x.

De Bruyne, N. A. (1957). How glue sticks. *Nature*, *180*, 262–266.

Fabrício, R. F., & Mählmann, C. M. (2014). Desenvolvimento de adesivo à base de amido de milho e de mandioca.

Tecno-Lógica, *18*(1), 1–12. Available from https://doi.org/10.17058/tecnolog.v18i1.3836.

Fitwi, G. (2000). *The status of gum arabic and resins in ethiopia, report of the meeting of the network for natural gums and resins in Africa (NAGRA)*, Kenya College of Communications Technology, Nairobi, Kenya, p. 14–22.

Guillemenet, J., Bistac, S., & Schultz, J. (2002). Relationship between polymer viscoelastic properties and adhesive behaviour. *International Journal of Adhesion and Adhesives*, *22*(1), 1–5.

Heinzmann, C., Weder, C., & de Espinosa, L. M. (2016). Supramolecular polymer adhesives: Advanced materials inspired by nature. *Chemical Society Reviews*, *45*(2), 342–358.

Kim, D. J., Kim, H. J., & Yoon, G. H. (2005). Effect of substrate and tackifier on peel strength of SIS (styrene-isoprene-styrene)-based HMPSAs. *International Journal of Adhesion and Adhesives*, *25*(4), 288–295.

Li, Z., Wang, J., Li, C., Gu, Z., Cheng, L., & Hong, Y. (2015). Effects of montmorillonite addition on the performance of starch-based wood adhesive. *Carbohydrate Polymers*, *115*, 394–400. Available from https://doi.org/10.1016/j.carbpol.2014.08.106.

Lörcks, J. (1998). Properties and applications of compostable starch-based plastic material. *Polymer Degradation and Stability*, *59*(1–3), 245–249.

Murphy, R., & Bartle, I. (2004, May). Biodegradable polymers and sustainability: Insights from life cycle assessment. In *Summary Report presented at the National Non-Food Crops Centre seminar, London* (Vol. 25).

Paula, H. C., Gomes, F. J., & de Paula, R. C. (2002). Swelling studies of chitosan/cashew nut gum physical gels. *Carbohydrate Polymers*, *48*(3), 313–318.

Rodrigues, D., Chagas, R., & Aguiar, F. E. C. (2014). Importância da qualidade do adesivo na produção de papelão corrugado. *O Papel*, *4*, 59–61.

Satas, D. (1989). ed. *Handbook of pressure sensitive adhesive technology* (1, p. 940). New York: Van Nostrand Reinhold.

Silva, S.A. D. (2008). *Estudo do comportamento reológico dos adesivos hot melt PSA e sua relação com a composição e as propriedades adesivas*.

Souza, A.C. R.D. (2014). *Contribuição para uso da goma de cajueiro como um insumo industrial*.

Souza, P. M. D., & Magalhães, P. D. O. (2010). Application of microbial α-amylase in industry-A review. *Brazilian Journal of Microbiology*, *41*, 850–861.

Towle, G. A., & Whistler, R. L. (1973). Hemicelluloses and gums. In L. P. Miller (Ed.), *Phytochemistry: The process and products of photosynthesis* (1, pp. 198–248). New York: Van Nostrand Reinhold.

Voiutskii, S. S. (1963). *Autohesion and adhesion of high polymers*. New York: Interscience Publishers.

Vroman, I., & Tighzert, L. (2009). Biodegradable polymers. *Materials*, *2*(2), 307–344.

Watcharakitti, J., Win, E. E., Nimnuan, J., & Smith, S. M. (2022). Modified starch-based adhesives: A review. *Polymers*, *14*(10), 2023.

World Bank. (2007). Policy note: Export marketing of gum arabic from Sudan.

Zhao, X. F., Peng, L. Q., Wang, H. L., Wang, Y. B., & Zhang, H. (2018). Environment-friendly urea-oxidized starch adhesive with zero formaldehyde-emission. *Carbohydrate Polymers*, *181*, 1112–1118. Available from http://doi.org/10.1016/j.carbpol.2019.115668.

CHAPTER

16

Starch applied for pharmaceutical use

Maria Aparecida Vieira Teixeira Garcia[1], *Aina Liz Alves Cesar*[2], *Cleverson Fernando Garcia*[3] *and André Augusto Gomes Faraco*[4]

[1]Food Department, School of Pharmacy, Federal University of Minas Gerais, Belo Horizonte, Minas Gerais, Brazil [2]Una, Bom Despacho, Minas Gerais, Brazil [3]Department of Chemistry, Federal Center of Technological Education of Minas Gerais, Belo Horizonte, Minas Gerais, Brazil [4]Department of Pharmaceutical Sciences, Faculty of Pharmacy, Federal University of Minas Gerais, Belo Horizonte, Minas Gerais, Brazil

16.1 Introduction

Starch is one of the most important natural polymeric carbohydrates used as excipients in the pharmaceutical industry, as it is cheap, nontoxic, biodegradable, renewable, and biocompatible with many other materials for industrial applications (Builders & Arhewoh, 2016; Guo et al., 2021; Lemos et al., 2021).

It can represent about 30%–90% of the dry weight of the material *in natura* depending on the plant origin: cereal grains (40%–90% of dry weight), vegetables (30%–50% of dry weight), tubers (65%–85% of dry weight), and immature or unripe fruits (40%–70% of dry weight dry) (Storpirtis, 2006).

Although starches can be found in abundance in a wide variety of plants, only some starches are approved by official pharmaceutical regulatory agencies and pharmaceutical compendiums (e.g., British Pharmacopoeia/BP; United States Pharmacopeia and the National Formulary/USP-NF; Japanese Pharmacopoeia/JP, European Pharmacopoeia/PhEur, Brazilian Pharmacopoeia/BP 6) to be used as adjuvants in pharmaceutical formulations as long as they have the appropriate degree of purity: corn/maize starches (BP; JP; PhEur; USP-NF; BP 6); rice starches (BP; JP; PhEur; BF 6); wheat starches (BP; JP; PhEur; USP-NF; BP 6); pea starches (PhEur); and, for the starchy roots and tubers crops of South American origin, potato starches (BP; JP; PhEur; USP-NF; BP 6) and cassava starches (BP; USP-NF; BP 6) (ANVISA, 2019; Builders & Arhewoh, 2016; Garcia et al., 2020; Kunle & Kunle, 2019; Rowe et al., 2009). For more information on standards for the use of starches, see the Chapter 6 of Volume 1 in this series.

It should be noted that maize (corn) starch is the most used when compared to other starches, probably because of the low cost due to its large world production and the greater availability of pharmaceutical grade raw materials in the market. The second most popular type of starch

Starch Industries
DOI: https://doi.org/10.1016/B978-0-323-90842-9.00004-2

used in the pharmaceutical industry is cassava starch, which is the second most produced and the most exported (for more information, see Chapter 3 of Volume 1 in this series). The physicochemical properties of these starches are well known and meet the requirements for use as excipients (Swarbrick, 2013; Charoenthai, Sangangam, Puttipipatkhachorn, (2018); Charoenthai, Wickramanayaka, Sungthongjeen, et al., 2018; Kunle & Kunle, 2019).

Cassava (*Manihot esculenta Crantz*), also known as yucca (Central America), mandioca or manioca (Brazil), tapioca (Asia), manioc (France), and cassada or cassava (Africa and Southeast Asia), is an important carbohydrate in tropical countries worldwide. It could be a suitable alternative in tablet formulations once it has been shown to deform mainly by plastic flow during compression in a similar manner to corn starch (Odeku, 2013). These starches give a stringy cohesive paste when gelatinized. The packing and cohesive properties of cassava starch could promote closer packing of particles and could be useful in the production of capsules.

Other starchy crops have been characterized and evaluated as excipients such as sweet potato (*Ipomoea batatas*), arrowroot (*West Indian arrowroot: Maranta arundinacea*), Indian arrowroot (*Curcuma angustifolia*), yams (Dioscorea species including *D. alata*, *D. dumetorum*, *D. oppositifolia*, *D. rotundata*, *D. esculenta*, and *D. cayenensis*), Tacca (*Tacca leontopetaloides*), Queensland arrowroot, or canna (*Canna indica*), yam bean (*Pachyrhizus erosus L. Urban*), and cocoyam, which are the edible aroid root crops belonging to the family Araceae, which include *Colocasia esculenta*, *Xanthosoma sagittifolium*, and finally, sorghum (*Sorghum bicolor* L. Moench). However, the use of these starches has not been approved for insertion in Pharmacopoeias yet. These botanical sources of starch are available in South America and in many tropical countries and have the potential to starch extraction. However, it is also important to consider some requirements such as agricultural productivity and dry matter content. To better know these cultures, it is recommended to read the Volume 2 in this series.

Some studies point out the potential use of sweet potato roots as an alternative binding agent to produce tablets with higher bond strength than corn starch, indicating that it could be an alternative as an excipient (Nwachukwu & Ubieko, 2020; Odeku et al., 2005).

Starch is mainly used as excipients in solid systems, for example, in powders evolved into granules and their complementary forms, such as capsules, tablets, coated tablets, and dragees, which correspond to approximately 80% of the production of the pharmaceutical industry (Builders & Arhewoh, 2016; Garcia et al., 2020; Lemos et al., 2021). In semisolid dosage products (ointments, creams, gels, and pastes) and in products with liquid dosage forms (solutions, drops, syrups, suspensions, and elixirs), starches are rarely used due to their insolubility in water. Reported examples of the use of starch in these forms are rice and wheat starches that can be applied in topical preparations and starch pastes used in ointments, usually in the presence of large amounts of glycerin (Rowe et al., 2009).

Excipients are components of the formulation that do not have pharmacological activity. In most formulations, excipients are at a higher concentration than the active ingredient. In this way, they significantly contribute to the functionality and processability of pharmaceutical preparations: they allow to solubilize, suspend, thicken, preserve, emulsify, modify the dissolution, favor compressibility, and correct organoleptic characteristics of the drug, allowing for acquisition of different preparations or pharmaceutical forms. In addition, they can be used as drug release systems, contributing to the pharmaceutical, pharmacokinetic, and pharmacodynamic phases of a drug, such as its stability, transport, biological availability of the drug, and its action. They also help in the acceptability of the drug by the patient, facilitating the administration and preservation of the active ingredient (Allen et al., 2013; van der Merwe et al., 2020).

In addition to the possibility of using pure starches, the pharmaceutical excipient industries also use starch derivatives called coprocessed excipients: a type of excipient that mixes one or more substances with the starch granule to obtain better flow, agglutination, disintegration, and compressibility properties in the final product. Furthermore, in order to improve the process of fabrication, some properties from starch can be used in problematic formulations that have some bad features like short disintegration time, difficulty of compression, and poor agglutination. Additionally other properties from starch can be performed to facilitate its utilization such as various physical, chemical, enzymatic, and/or genetic modifications. Depending on the modification, it may be possible to improve flow and compressibility properties of the starches (Lemos et al., 2021).

Some of the products from these modifications have been approved by the FDA. Two of them, sodium starch glycolate and pregelatinized starch (chemically and physically modified starch, respectively), can be used either as an isolated excipient or as a matrix for drug delivery systems (DDS) (FDA, 2022).

Among the modified starches, pregelatinized starch is the most used since it is the easiest to produce and the cheapest. In addition, its properties for obtaining tablets are superior when compared to conventional starch, without modification. To read more about starch modifications, it is suggested to read Chapter 6.

With the increasing number of drug moieties and DDS, new excipients with different functionalities have been studied. Several works present new starches, chemical reagents, or starch derivatives to be used in the development of drugs that have not been approved for commercial use yet. These restrictions happen to protect the consumer, guaranteeing its safety and the environment and controlling production costs.

Next, we review different types of *in natura* and modified starches applied for pharmaceutical use.

16.2 Starch

Starch is found in plants in the form of partially crystalline granules. It is the main energy reserve of plants, obtained via a photosynthetic route by a varied class of vegetables, such as cereals, seeds of vegetables, and tuberous sources. It is a polysaccharide made up of anhydroglucose units linked together primarily through α-D-glycosidic bonds. Its two main polymer chains are amylose and amylopectin (Fig. 16.1) linked together primarily through α-D-glucosidic bonds (Sivamaruthi et al., 2022; Waterschoot et al., 2014). Amylose is essentially a linear polymer in which the anhydroglucose units are predominantly linked through $\alpha(1\rightarrow4)$ glucosidic bonds, featuring a reducing end. The amylopectin molecule is branched and formed by anhydroglucose units joined at α(1he and α(1nd glycosidic bonds (polymers of α[1gl] linked glucopyranose with α[1li] branches). Amylopectin contains approximately 1 million glucose units, making it one of the largest biomolecules. The branches, in turn, are found every 24–30 units, but, due to the structural profile, only one end has a reducing character. To read more about starch structure, it is suggested to read Chapter 9 in Volume 1 of this series.

Unmodified (*in natura*) starches contain more amylopectin than amylose. In general, the percentage composition by mass of the starch granule from vegetable sources can vary around 80% ± 5% amylopectin and 20% ± 5% amylose. Other components that are part of the starch are lipid (1%), protein (0.4%) and traces of phosphate. The amount of these components

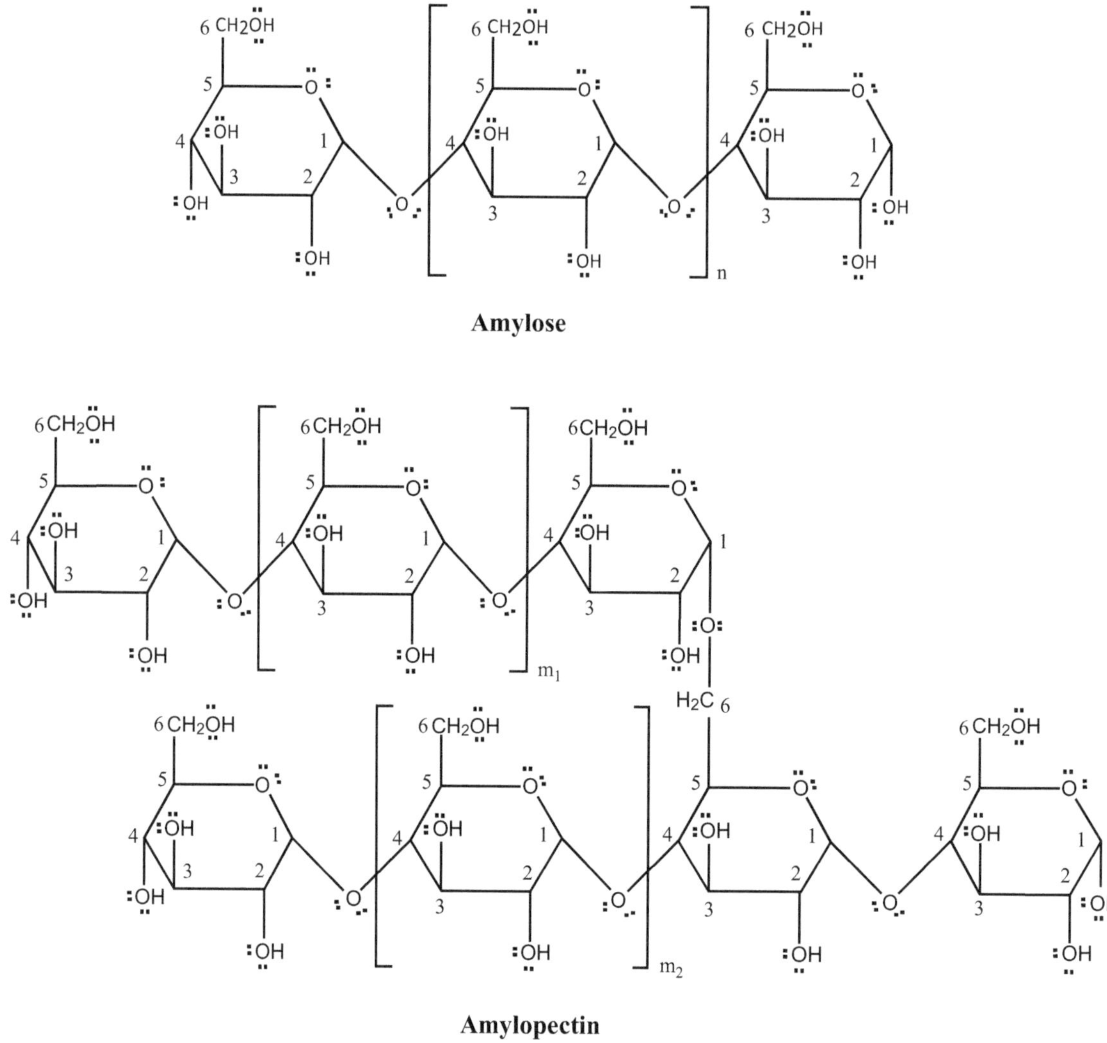

FIGURE 16.1 Representation of the structural formulas of amylose and amylopectin.

may vary depending on the plant source. Note that starches, such as corn and potato starches, can be genetically modified so that they have a higher concentration of amylose (>30%) or amylopectin (99%) (Garcia et al., 2020; Sivamaruthi et al., 2022; Waterschoot et al., 2014; see also Chapter 10 in Volume 1 of this series).

The amylose content can be distinct within the same botanical variety because of differences in geographic origin and culture conditions (Gao et al., 2014). It can be present in the form of free amylose or forming complexes with a large number of linear compounds, such as long chain fatty acids, fatty alcohols, and monoglycerides (amylose-lipids), which despite being detected in native starch, possibly are formed to a greater extent during hydrothermal treatment or gelatinization. Also, amylose can form inclusion complexes with other polar and nonpolar compounds, forming molecular inclusion complexes by wrapping around the ligand and forming a left-handed helix (Fredriksson et al., 1998; Tetlow & Bertoft, 2020). To read more about starch modifications, it is suggested to read Chapter 6.

Larger amylose molecules are concentrated in the center of the granule and probably participate in double helices with amylopectin, while smaller molecules present at the periphery can be leached out of the granule. The structure of starch granules consists of concentric shells alternating amorphous and crystalline rings. The crystalline region is made up of the double helices of the parallel amylopectin chains and is more compact, while the amorphous region, less ordered, contains the branching points of the amylopectin side chains and possibly some amylose (Cozzolino et al., 2014; Osborne, 1998). Despite its limited role in crystal formation, amylose can influence the organization of double helices, interfering with the packing density of amylopectin chains, and may participate in the formation of crystalline lamellae. So, while the amylopectin represents predominantly the crystalline domain of native starch which is characterized by an ordered, tightly packed parallel glucan chain, the amylose represents the amorphous region (Builders & Arhewoh, 2016; Kuakpetoon & Ya-Jane, 2007; Tetlow & Bertoft, 2020).

Starch is insoluble in water (Xu et al., 2005). When in the presence of cold water, the granules swell slightly (10%–20%) due to the diffusion and absorption of water in the amorphous regions. Water absorption and solubility are related to the ratio between crystalline and amorphous components of its granules (Wani et al., 2016).

Once starches are heated in excess water, at a particular temperature, the granular structure breaks down followed by granular swelling, in a phenomenon called gelatinization (Bogracheva et al., 2002). The crystalline structure is disrupted due to the relaxation of hydrogen bonds, and the water molecules interact with the hydroxyl groups of amylose and amylopectin, causing an increase in granule size and leaching of amylose from the intergranular phase to the aqueous phase occurs (Garcia et al., 1997; Han et al., 2019; Hoover, 2001).

The set of changes involving the breakdown of granular structure, swelling, hydration, and solubilization of starch molecules is defined as the end of gelatinization (Brouillet-Fourmann et al., 2003; Tharanathan, 2002). When cooled, the gel can undergo a phenomenon called retrogradation: the chains begin to reassociate with the formation of crystalline areas. It is observed that the gel becomes more opaque as the retrogradation takes place. The strong interaction of the chains with each other promotes the exit of water from the system, and this expulsion is called syneresis.

The structural characteristics of starch, such as its amylose content and distribution of amylopectin chain lengths, may be closely related to its crystallinity and its physicochemical properties—as, for example, with the events associated with gelatinization and retrogradation, such as swelling of the granule, leaching of amylose and/or amylopectin, and loss of radial structure (birefringence)—and its technological application. Starch sources with varying contents of amylose and amylopectin levels and other noncarbohydrate constituents result in different functional properties (Table 16.1). This difference can affect functional properties such as viscosity, shear resistance, gelling power, solubility, stability, swelling, and retrogradation. Thus, in certain applications, starch from one source cannot be automatically interchangeable, unless there is a previous study indicating that this act will not cause damage or significant differences in the pharmaceutical form in question (Builders & Arhewoh, 2016; Rowe et al., 2009). Starch pastes from tubers and roots, for example, have weaker and less consistent gels, but are highly viscous,

TABLE 16.1 Properties of native starches from different plant matrices used in the pharmaceutical industry.

Properties	Maize	Potato	Wheat	Cassava	Rice	Sweet potato
Amylose content	24%–28%	20%–23%	24%–28%	17%–20%	18%	9.7%–15.1%
Gelatinization temperature (°C)	64.4–80.4	59.6–75.7	57.5–73.9	61.9–81.4	71.1–87.6	64.6–84.6
Mean particle diameter (μm)	13 (2–32)	46 (10–100)	23 (1–45)	13 (4–35)	6 (3–10)	13.1 (0.85–44.69)

translucent, long, cohesive, fibrous, and with a low tendency to retrograde. Cassava starch exhibits lower gelatinization temperature and higher water-binding capacity, viscosity, and shear resistance, making it a suitable option in food, feed, chemicals, and pharmaceutical products (Wang et al., 2022). For more information about starch properties, it is suggested to read Chapter 10 in Volume 1 of this series.

The botanical origin of starch can interfere in properties such as shape, size, and structure. The size of the granules can vary between four categories: large (>25 mm), medium (10–25 mm), small (5–10 mm), and very small (5 mm). Also, starch granules can be round, lenticular, or polygonal shape (Builders & Arhewoh, 2016; Daiuto & Cereda, 2006; Sivamaruthi et al., 2022.). To know more about characterization of underground starchy crops, it is suggested to read Chapter 8 in Volume 1 of this series.

From a pharmaceutical point of view, the size of the granules is a relevant parameter, as it can interfere in the production process and, consequently, in the final quality of the drug, which can affect tabletability, compressibility, dissolution rate, and particle flowability. One can mention, for example, the fluidity, which is better in larger particles. This parameter must be adjusted properly, as it interferes with the flow of the powder mixture and the filling of the matrices at the time of tablet production. This, in turn, interferes with the variation of the mass of the tablet during its production and consequently in its dose (Builders & Arhewoh, 2016; Palanisamy et al., 2020; Shi et al., 2019; Allen et al., 2013).

The rate and extent of digestion and absorption of starch in the human body can be influenced by several factors, including variation in the amylose—amylopectin ratio, physicochemical properties such as gelatinization and retrogradation, and the way in which it is processed. Starches can be classified as follows: (1) rapidly digestible starch; (2) slowly digestible starch, and (3) resistant starch. Some studies point to the use of resistant starch as materials with potential biomedical applications as DDS. Starch could form the V-type inclusion complex with polyphenols which was similar to the inclusion complex formed by starch and lipids. These interactions would lead to a reduction in starch. To know more about food uses and nutritional applications of resistant starches, it is suggested to read Chapter 15 in Volume 1 of this series.

Starches are stable if protected from moisture and are considered chemically and microbiologically inert under normal storage conditions. These facts, and the absence of toxicity, facilitate their use in the pharmaceutical industry as they can be handled as required. In this way, starches are used as excipients to produce solid pharmaceutical forms. On the other hand, their dispersions are susceptible to being metabolized by microorganisms, which is one of the factors for its low use in pharmaceutical liquid preparations, requiring the use of preservatives. From a regulatory point of view, starch is a substance registered in the American and Brazilian Pharmacopeias as an inactive pharmaceutical excipient for use in solid forms, mainly being able to be used in suspensions when it requires adsorption of an active substance in inert particles.

16.3 Modified starch

Starches may suffer physical, chemical, enzymatic, and genetic modifications to present novel characteristics that will improve functional properties, such as improved flow, diluting, binding and disintegrating properties, solubility, formation of stable gels, improved compressibility, and dissolution rate (Tester et al., 2004).

The physical changes correspond to pregelatinization processes, heat treatment, or mechanical processes, in the dry or wet state, including fractionation. Thus, pharmaceutical grades of fully pregelatinized starch use no additives and are prepared by spreading an aqueous suspension of ungelatinized starch on hot drums where gelatinization and subsequent drying takes place. Partially pregelatinized starch is produced by subjecting moistened starch to mechanical pressure. The resultant material is ground, and the moisture content is adjusted to specifications. Gelatinized starches are among the derivatives that have acquired greater importance as excipients over the last few years.

Chemical transformations are responsible for the structural modification of part of the α-D-glucopyranose units of the starch, conferring properties such as higher viscosity. Examples of chemical transformations are etherification, esterification, oxidation of hydroxyl carbonyl groups to carbonyl or carboxylic groups, cross-linked reactions, and hydrolysis (Kaur et al., 2007).

In enzymatic modifications, enzymes are used to catalyze hydrolytic processes responsible for the formation of oligosaccharides and maltodextrins, reducing viscosity and ensuring greater gel stability, heat resistance, and water solubility, among other characteristics.

Modified starches generally have markedly altered physicochemical properties, compared to their native starches, depending on the degree of substitution and the type of functional groups introduced. In drugs, modified starches can be used as excipients to improve drug stability or as DDS, for example, in oral sustained-release dosage forms or as subcutaneous implants. Functionalities such as the ability to release drugs in biological fluids can be optimized through modifications, in view of their effects on the fluidity and compressibility of starch (Lawal, 2018; Charoenthai, Sanga-ngam, Puttipipatkhachorn, (2018); Charoenthai, Wickramanayaka, Sungthongjeen, et al., 2018; Garcia et al., 2020). For more information about starch modification, see Chapters 5 and 6.

16.4 Use of native and modified starches in the pharmaceutical industry

In the pharmaceutical industry, the main raw materials used are the active drug substances and excipients, the choice of which affects the type of formulation (solid, liquid, or semisolid) and, consequently, the pharmaceutical form (Allen et al., 2013).

As stated earlier, starch is present in several tropical crops such as roots, tubers, grains, cereals, and fruits. Among these, we can highlight corn starch as the most worldwide exploited by the pharmaceutical industry as an excipient in the preparation of medicines. The second most popular type is cassava/tapioca starch. Some studies suggest that they may be interchangeable as they have a similar plastic deformation behavior during compression for tablet manufacture (Allen et al., 2013; Odeku, 2013).

The use of starches in pharmaceutical formulations is cost-effective: in addition to being cheap, obtained from renewable and available sources, they are innocuous, which is an advantage when the formulations are administered orally. Also, they do not have the need for intense purification, as with other natural polymers, such as cellulose and gums

(Demirgöz et al., 2000; Rodrigues & Emeje, 2012).

In this way, starches in their natural or modified form are used in the production of drugs. They are mostly used as binders, filler diluents, and disintegrants in tablet formulation. They can also be used as a sustained-release agent in a matrix system or as a subcutaneous implant to maximize the bioavailability of drugs and reduce their side effects (Garcia et al., 2020).

Many attempts have been made to modify cassava starch by physical and chemical processes for use as pharmaceutical excipients: as a carrier for solid dispersion (for more on this topic, see Chapter 12), a suspending agent, a direct compression filler, a matrix forming agent for controlled release tablet, a film coating agent, and a carrier for mucoadhesive microsphere are reported. The natural and cross-linked cassava starches produced tablets with very low crushing strength, while the acid-modified starches produced tablets with higher crushing strength. Acid-modified and acid-modified cross-linked cassava starches were shown to be useful as fillers in direct compression tablet preparation (Charoenthai, Sanga-ngam, Puttipipatkhachorn, (2018); Charoenthai, Wickramanayaka, Sungthongjeen, et al., 2018).

Some modified cassava starches (carboxymethyl, acid-modified, cross-linked, grafted, enzyme-catalyzed, pregelatinized, and hydroxypropyl) have a potential to be used as pharmaceutical excipients: as a carrier for solid dispersion, a suspending agent, a direct compression filler, a matrix forming agent for controlled release tablet, a film coating agent, and a carrier for mucoadhesive microsphere (Charoenthai, Sanga-ngam, Puttipipatkhachorn, (2018); Charoenthai, Wickramanayaka, Sungthongjeen, et al., 2018). Additionally, carboxymethyl cassava starch has been reported to be used in medicines and cosmetics and in the food industry. This modified form is advantageous, as it has a lower tendency to retrogradation, lowered gelatinization temperature, increased solubility in cold water with clear gel, and higher storage stability. Lin et al. (2012) demonstrated the potential of nanosized flake carboxymethyl cassava to improve the solubility of poorly water-soluble drugs.

Pregelatinized starch has physical properties that are important for tablet production, such as better compressibility and flowability. In the first case, due to the greater number of free hydroxyl groups, there is a favoring of interactions by hydrogen bonding between these molecules and other excipients, resulting in better consolidation of the tablet. Thus, there is a plastic deformation of the particles maintaining the physical structure in a compact form. Depending on the process chosen to make this type of starch, particles with a spherical shape can be obtained, which would allow for better flow. This format contributes to the constant powders of a formulation being carried into the matrices during the tablet formulation process. This good flow results in a better quality of the tablet obtained regarding its homogenization in relation to its weight.

In addition, pregelatinized starch can give rise to other products called coprocessed excipients, in which there is a small amount of it in these preparations. The presence in these cases of pregelatinized starch is less than the main ingredient, such as sucrose, (Di-Pac and Nu-Tab); dextrose (EMdex and Maltrim); and lactose monohydrate (StarLac).

16.4.1 Excipients: definition and characteristics

In a pharmaceutical formulation, three important aspects must be considered: the active substance, the pharmaceutical form, and the excipients used. Excipients in a formulation are expected to be physically and chemically stable. Most formulations contain excipients at a higher concentration than the active drug. Consequently, the excipients contribute significantly to the functionality and processability of the formulation.

In this way, pharmaceutical products are composed of therapeutically active substances and substances without pharmacological activity called excipients/vehicle/adjuvant. The meaning of these words can be correlated with the function of these substances in pharmaceutical products: excipient/ excipere in Latin, which means "to receive" the active principle; vehicle, which refers to the function of transporting the drug to the body's absorption site; and adjuvant/adjuvare in Latin, which means to assist the active principle. In this chapter, all these substances with no pharmacological activity will be called excipients.

Excipients in a pharmaceutical form are substances that have no pharmacological activity, which should not interfere negatively in the dosage, stability, and bioavailability of the drug. However, depending on the absorption of water by the excipients, there may be changes in molecular mobility within the system, which could accelerate chemical reactions and contribute to degradation reactions or drug interactions (Crowley & Martini, 2001).

Despite not having pharmacological activity, it is important to maintain control over the quality of the excipients to ensure the efficacy and safety of the final product. Due to the increase in regulatory pressure related to purity, safety, and standardization of the excipients, an international body, the International Pharmaceutical Excipients Council (IPEC), with representation from the United States, Europe, and Japan was formed. The IPEC has made efforts to harmonize requirements for purity and functionality testing.

Excipients can be divided according to their performance/function in the pharmaceutical product as follows:

1. Used to perform important and specific technological functions such as solubilizing, suspending, thickening, preserving, emulsifying, modifying the dissolution, favoring compressibility, enabling the acquisition of various preparations or pharmaceutical forms in the correct dosage, and guaranteeing the dosage, stability, and bioavailability of the product active principle. Its application is practically mandatory in almost all formulas.
2. Used to make drug administration easier and improve patient compliance: absorbents, wetting agents, flavorings, sweeteners, and dyes. They are optionally added.
3. Drug release systems/drug vehicles.

16.4.2 Use of starch as an excipient

Nowadays, there are several starch-based excipients approved by The United States Food and Drug Administration (FDA) to produce pharmaceutical forms: native starch, sodium starch glycolate (chemically modified starch), and pregelatinized starch (physically modified starch). The physicochemical characteristics and functional properties of these polymers are responsible for making them attractive to be used as pharmaceutical excipients (Builders & Arhewoh, 2016; FDA, 2022). Thus, in addition to the properties that will be discussed below, starch as an excipient has advantages in terms of cost, safety, and product quality. The choice of the most suitable type of excipient is related to the way in which the drug is used, the application site, and the drug delivery system (Builders & Arhewoh, 2016; Garcia et al., 2020).

The choice of starch (*in natura*, coprocessed, or modified) influences its physicochemical characteristics such as whiteness, morphology, particle size, shape, surface area, porosity, density, soft, smooth dryness as well as gelling, viscosity, flow, and disintegration properties, among others. In this way, depending on the type of starch-based excipient chosen, important factors in the production of medicines such as fluidity, compactability, dilution and disintegration potential, and formation of stable gels in hot and cold water can be changed, defining the method of production of the drug, such as direct or indirect compression into tablets (Builders & Arhewoh, 2016).

Starch is basically used in solid forms (capsules, tablets, coated tablets, and dragees), which represent about 80% of the total dosage forms used in therapeutic administration (Jivraj et al., 2000). In these cases, the excipient fills in the volume and mass necessary for the product to have an adequate size for handling, in addition to being able to act as a disintegrant, diluent, and binder and to prevent the powder from adhering to punches and dies and to facilitate the flow of the product (Prista et al., 1996). The amount of starch used will influence its role as an excipient, for example, concentrations of 3%–25% w/w may be assigned to its disintegrating action (typically 15% w/w) or 3%–10% w/w as a nonstick and tablet lubricant (Prista et al., 1996).

Considering the solid forms, starch is more commonly used in tablets. Tablets can be obtained by means of three methods: dry granulation, wet granulation, and direct compression. The composition of the mixture and the intrinsic characteristics of the raw material, such as the compressibility and particle size (of the granules or powders) and the stability of the drugs during the manufacturing process, will define the most suitable method to produce the drug (Allen et al., 2013). A large variety of starch types and its granulometric parameters, obtained from the plants already mentioned, enable the use of the three types of tableting methods (Garcia et al., 2020; Prista et al., 1996).

Direct compression has advantages compared to dry and wet granulation in terms of cost and production time, as the process requires only two steps: mixing and compression of the components, consisting of the simplest form of tablet manufacturing. In the direct compression process, starch and its modified derivatives can act as diluents, binders, and disintegrators (e.g., Sta-Rx 1500, Cellutab, and Primojel), aiming at improving the characteristics of compressibility, disaggregation capacity, and flow (Prista et al., 1996; Sivamaruthi et al., 2022; Swarbrick, 2013).

However, the production of satisfactory dosage forms from certain active ingredients has limitations. The size and shape of granules are an important parameter on direct compression and, as previously described in this section, it can vary according to the origin of the starch. In this method, the starch grain must have homogeneous size distribution to ensure good flowability and avoid segregation during drug production (Allen et al., 2013; Prista et al., 1996; Sivamaruthi et al., 2022). The adjustment of these parameters (fluidity and particle size) will guarantee the reproducibility of the quality between the batches produced (Jivraj et al., 2000).

One way to adjust the starch granule size is through spray dryer atomization, as described by Daiuto and Cereda (2006). Through this technique, it is also possible to reduce the water activity, which is of interest in the industry, since with the reduction of the amount of free water, there is an increase in the microbiological stability of the product (Daiuto & Cereda, 2006). For more information, see Chapter 12.

Wet and dry granulation processes do not require differentiated particle shape, unlike the direct compression. These processes can use poorly shaped particles, but before being compressed, they must go through a previous granulation, whose objective is to standardize the size of the particles. This has an impact on the longer production process with more steps and increased financial cost for tablet production (Garcia et al., 2020).

Depending on the design of the pharmaceutical form (concentration) and processes used, starch acts as an excipient in the drug, performing the following functions.

16.4.2.1 *Binders, diluents, and disintegrants in tablet formulation*

Starches can be used, with a wide variety of active substances and with other excipients, as diluent, disintegrant, and binder (Rowe et al., 2009). Table 16.2 presents the classification of starch, its properties, and application in the pharmaceutical industry as excipients. Among the starches used in the development of pharmaceutical formulas, we can highlight the use of native and modified cassava starch as diluents, binders, and disintegrants in tablets and capsules.

16.4.2.1.1 Disintegrant

Disintegrants are used to promote the breakdown of the solid oral pharmaceutical form, with the aim of facilitating its dispersion and dissolution, to increase the absorption of intestinal actives.

Due to their compression characteristics, native starches present difficulty in tablet compaction because of elastic deformation, causing defects in tablets such as chipping and capping. Thus, among native starches, we can mention potato starch, which produces hard tablets with rapid disintegration, but which is not as recommended as modified starch (Klein, 2009). The use of starch as a disintegrant generally requires a prior granulation step to avoid problems with insufficient flow and segregation. Separation of the powders from the mixture occurs because of the fine particles of the starch that leads to a stratification of the powders, and, consequently, lack of uniformity in the produced tablet. In the granulation process, the particles (fine starch particles) are agglutinated with the other raw materials in the formulation. For this purpose, starch is used in the concentration between 5% to 25% (w/w), with greater application around 15% (Pagtalunan et al., 2022; Rowe et al., 2009).

Starch derivatives can also be used as disintegrants, more specifically as superdisintegrants. For example, sodium starch glycolate acts through swelling, allowing for faster disintegration of the pharmaceutical form, thus increasing its dissolution rate. In general, superdisintegrants are used in the concentration range of 1%–10% w/w. The increased dissolution capacity is a consequence of the chemical modification of the starch structure and binding of hydrophilic groups, which increases the swelling capacity of the molecule (Lawal, 2018). This type of excipient is very common in orodispersible tablets, which require rapid dissolution in the oral cavity for the absorption of the active substance to occur through the sublingual route (Allen et al., 2013; Pagtalunan et al., 2022).

Adjei et al. (2017) studied the disintegrant potential of cassava native starches. The tablets produced were tested for disintegration, and when compared to tablets containing corn starch, they showed faster disintegration times. This disintegration time is inversely proportional to the concentration of cassava starch. In addition, dissolution tests for immediate release tablets were also performed (≥70% release in 45 min), demonstrating profiles similar to those containing corn starch. This is yet another study that proves that corn and cassava starches can be interchangeable.

16.4.2.1.2 Binders

Binders are substances used to promote adhesion of powder particles to granulates intended for compression.

Starches are used in the form of freshly prepared pastes, as a binder for wet granulation, at a concentration of 5%–10% w/w, depending on the type of starch. To determine the required amount, parameters such as friability and hardness of the tablet, disaggregation rate, and dissolution rate of the drug should be considered; therefore, formulation optimization studies are necessary (Rowe et al., 2009).

When used as a binding agent in a paracetamol tablet formulation, cassava starch exhibited stronger binding properties than gelatin.

TABLE 16.2 Types of starches, their properties, and application in the pharmaceutical industry as binders, filler diluents, disintegrants, and glidants.

Starch type	Modification	Properties	Applications
Starch *in natura*	*Not applicable*	• Granules with limited solubility • Insoluble in cold water • Highly flexible polymer	• Diluents • Disintegrants • Binders
Physical modification			
Pregelatinized	Chemically and/or mechanically processed to rupture all or part of the starch granules	• Excellent wettability • Dispersible in cold water • High viscosity, and moisture sorption and swelling	• Tablet and capsule diluent • Binder and disintegrant direct compression tablets • Filler and disintegrant hard gelatin capsules • Flow aid in powder blend • Hydrophilic matrices: extended release
Chemically modified starch			
Sodium starch glycolate 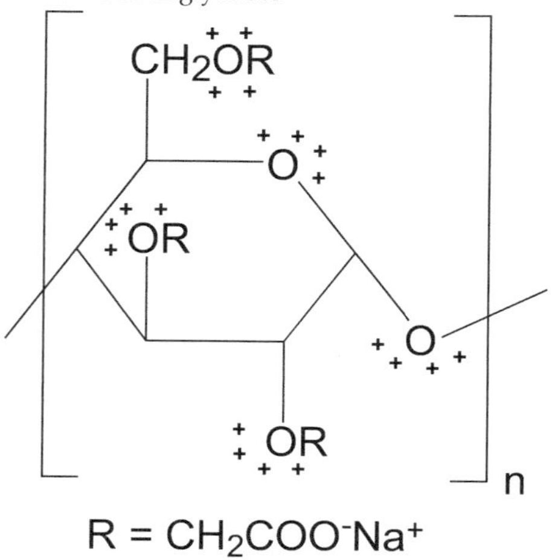 	Carboxymethylation in two stages 1. Formation of alkoxide groups in basic medium 2. Formation of CMS from the reaction between alkoxides and monochloroacetic acid or its sodium salt	• Stabilized starch in aqueous media • Reduced the tendency to retrogradation • Increased swelling capacity • Good flow properties	Superdisintegrant

Acetylated starch 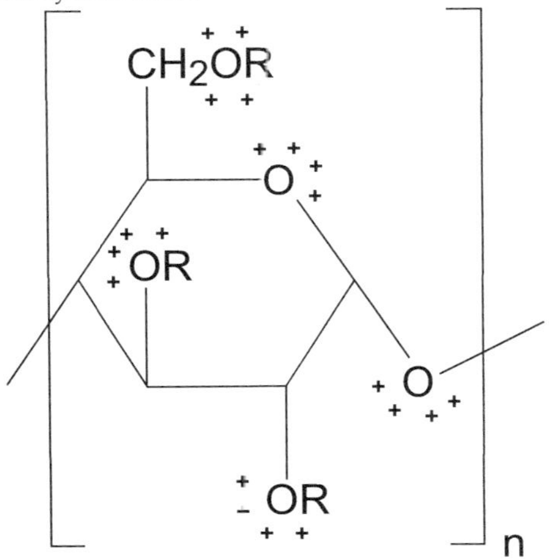 $R = COCH_3$	Esterification of native starch with acetic anhydride	• Increased granular size, solubility, swelling power, and water • Absorption capacity • Readily deformed plastically • Resistant to retrogradation due to reassociation	• Diluent, tablet matrix • Binder when a tablet with low crushing strength and fast disintegration is desired • Good flow and compressional properties in the tablet formulations

(Continued)

TABLE 16.2 (Continued)

Starch type	Modification	Properties	Applications
Cross-linked phosphorylated starch 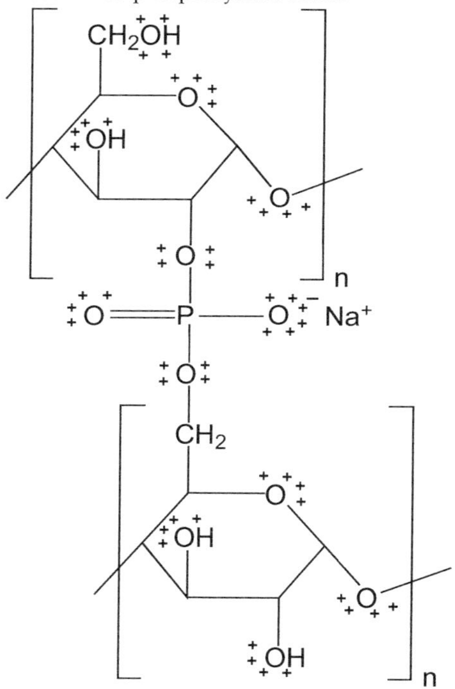	Starch network structures are frequently reinforced by applying crosslinking	• Control granular swelling • Tolerant to high temperature, shear, and acidic conditions • More opaque gels • Less water retention capacity during aging • Structural integrity to stabilize and reinforce structures, to improve resistance to high temperature, low pH, and high shear	• Disintegrant agent • Matrices in controlled release formulations

Converted starches 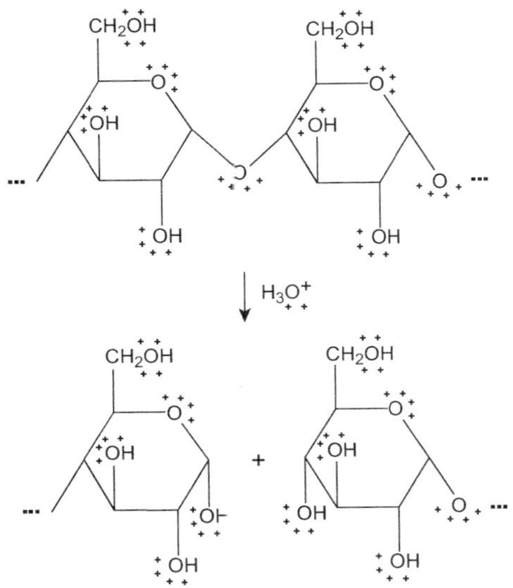	Acid hydrolysis (generally hydrochloric acid is used)	• Increases water solubility • Controls gel strength • Reduced viscosity and swelling capacity • Increased solubility • Greater gel strength	Filler binder: optimal flow and compressibility properties in formulations
Coprocessed starches	Mixes of one or more substances with the starch	• Better synergism in functionality of excipients • Better flow, agglutination, disintegration, and compressibility properties in the final product • Enhance the functionality of modified starches, with incorporation of physically and chemically modified starches or the use of a combination of native and modified starch	Capsules and directly compressed tablets: Better flow compressibility and disintegration potential in direct compression formulation

Adapted from Garcia, M. A. V. T., Garcia, C. F., & Faraco, A. A. G. (2020). Pharmaceutical and biomedical applications of native and modified starch: A review. Starch - Stärke, *72(7-8), 1900270. https://doi.org/10.1002/star.201900270.*

Also, the binding capacity of cassava starch in paracetamol tablets was much greater (in terms of friability, disintegration time, and hardness) than that of industrial maize starch (Chalapathi et al., 2010). Cassava starches can be used as binders in uncoated ibuprofen tablets under the following operating conditions: punch setting of 24 and binder concentration of 2% w/w. Under such conditions, less friable tablets were produced, and reduced amounts of materials were used (Chitedze et al., 2012).

Pregelatinized starch is the most common derivative of starch used in the pharmaceutical industry, spreading an aqueous suspension of ungelatinized starch on hot drums where gelatinization and subsequent drying takes place. Partially pregelatinized starch is produced by subjecting moistened starch to mechanical pressure. The resultant material is ground, and the moisture content is adjusted to specifications (Lawal et al., 2015; Mounsey & O'Riordan, 2008). Due to the changes undergone during the gelatinization process, its flow and compression characteristics are improved. In this way, pregelatinized starch can be used as a binder for tablets in compression processes via dry granulation or direct compression. In these specific cases, it can act as a self-lubricant. It is noteworthy, however, that this does not dispense with the use of another lubricant in the formulation (e.g., magnesium stearate 0.25% w/w or stearic acid) (Lawal, 2018; Rowe et al., 2009). At this point, it is worth mentioning that tablets formed with pregelatinized starches have lower tensile strength values and greater hardness than native starches.

It should be noted that, except for the case described, the use of starch as a lubricant is unusual and not the best choice for this function. When used as a lubricant, its concentration must be obtained through tests, since it can vary according to the formulation. The lack or excess of starch can impair the flow of powders through the feeder of a compressor, for example, negatively interfering in the compaction capacity of the powder mixture and in the filling of the matrices (Builders & Arhewoh, 2016; Rashid et al., 2011; Sohail et al., 2021).

The use of native and modified cassava starch as binders occurs in the form of starch paste in the wet granulation process, generally resulting in tablets with greater tensile strength, lower friability, longer disintegration time, and lower drug dissolution rate (Charoenthai, Sanga-ngam, Puttipipatkhachorn, 2018; Charoethai, Wickra-manayaka, Sungthongjeen, et al., 2018). As an example, we can highlight that the binding capacity of cassava starch in paracetamol tablets was much greater (in terms of friability, disintegration time, and hardness) than that of industrial maize starch (Chalapathi et al., 2010).

16.4.2.1.3 Diluents

Diluents are the excipients that provide mass to solid formulations that contain a small dose of the active ingredient (mainly tablets), making the manufacturing process more reliable and reproducible. As a diluting agent, starch generally is used in concentrations greater than 30% w/w (Allen et al., 2013; Rowe et al., 2009).

In addition to the possibility of using pure starch in the dosage forms, pharmaceutical excipient industries also use starch derivatives called coprocessed excipients. These excipients were developed to encompass several parameters related to fluidity, compressibility, and disintegration, among other functions. They provide a synergy of functionality improvements and mask the undesirable properties of individual excipients (Rowe et al., 2009; Shi et al., 2019).

These excipients are produced with a view to use direct compression as a tablet production method and to increase the tablet production speed, since wet or dry compression takes longer. For this reason, coprocessed excipients seem to be more attractive to the pharmaceutical industries (Bhatia et al., 2022). In addition, when used, they can accelerate the disintegration process of the pharmaceutical form. They also allow for the production of tablets and capsules with low weight and good content uniformity (Rowe et al., 2009).

Examples of these substances that are nowadays registered with the FDA as coprocessed starches include corn starch 10% w/w, alpha-lactose monohydrate 70% w/w, and microcrystalline cellulose 20% w/w; corn starch 20% w/w and mannitol 80% w/w; and corn starch 15% w/w and alpha-lactose monohydrate 85% w/w (Garcia et al., 2020). These coprocessed starches are mostly prepared using the spray dryer method, since this method allows obtaining particles with uniform granulometry (Lawal, 2018)

16.4.2.2 *Drug delivery systems*

The use of starch as an excipient has evolved significantly in recent years, including its application in novel DDS. Unlike traditional drugs, which have immediate drug release, DDS allow the active substances to have a delayed or prolonged release. DDS have a modified active release profile, being able to maintain constant and/or prolonged concentrations of therapeutic agents, and consequently, increase the bioavailability of the active substances. The goal of these systems is to load homogeneously and optimize the effect of the drug (Villanova et al., 2010).

Sustained-release systems are characterized by the slow releasing of the drug on a predetermined rate to maintain a constant plasma concentration, leading to optimization of treatment through increased bioavailability. They also allow for a reduction in the frequency and number of administered doses, which leads to a decrease in toxicity and adverse reactions and consequently increase the patient compliance when compared to conventional release pharmaceutical forms. In addition, controlled release systems can protect the active substances from degradation reactions and chemical interactions among the product ingredients (Allen et al., 2013; Aulton, 2016; Lemos et al., 2021).

Delayed release systems can protect the drug from the gastric fluid, so that the release of the active principle occurs only in the small intestine or colon, where it will be better absorbed. This type of formulation prevents the inactivation of the drug by enzymes or acidic pH and reduces stomach discomfort. In addition, DDS allow for the release of the active directly at the site of action, when dealing with intestinal diseases (Allen et al., 2013; Aulton, 2016).

Gangopadhyay et al. (2022) studied a modified retrograded acetylated corn starch system for targeted delivery of budesonide into the colon for the treatment of ileocolonic irritable bowel disease. The retrogradation is a structural modification that makes the starch resistant to the degradation it could undergo in the upper gastrointestinal tract, making this system more effective in drug delivery compared to the conventional formulation (Gangopadhyay et al., 2022). Recife et al. (2017) evaluated the high amylose starch as an excipient for controlled release for diclofenac sodium tablets similar to the previous study. It was demonstrated that in enteric medium, the release of the drug was greater, demonstrating, again, to be resistant to the upper gastrointestinal tract.

Another study was carried out using octenyl succinate derivatives of cassava starch, synthesized by microwave irradiation in the solid state using a single-mode reactor. The modified starch was used to formulate tablets by wet granulation. In vitro studies demonstrated that the degree of starch substitution was directly related to the formulation's ability to sustain release over a longer period. The release kinetics data indicate that the incorporated drug appears to be released through the coupled mechanism of diffusion and erosion, following the model described by Higuchi model and Korsemeyer Peppas (Athira & Jyothi, 2015).

There are several other forms of prolonged release systems such as micro- and nanostructured drugs, matrix systems (tablets or capsules), reservoir systems (osmotic pump), transdermal drugs, and gastroresistant systems (as described above) (Allen et al., 2013; Aulton, 2016).

Matrix systems are used as solid systems for the controlled release of drugs, due to the possibility of incorporating high drug doses and capable of sustaining the therapeutic action over time and/or releasing the drug at the specific target. They are advantageous because they are versatile and effective, have a low cost, and use conventional equipment and techniques for their production. Matrix systems can be obtained through coatings, matrices, chemical reactions, or physical interaction. In these systems, the release of the molecularly dispersed drug is controlled by the diffusion through the matrix in hydrophobic matrices and by dissolution in hydrophilic matrices (Allen et al., 2013; Lopes et al., 2005; Pezzini et al., 2007). Examples of these extended release systems are controlled release tablets (Lemieux et al., 2009), films (Kaewkroek et al., 2022), micro- or nanoparticles (Kim et al., 2015; Tuovinen et al., 2004), and subcutaneous implant (Esfahani et al., 2022).

Pregelatinized starches, in addition to being used as a diluent in capsules and as disintegrants and binders to produce tablets, promoting a uniform filling of dies to ensure correct dosage with excellent disintegration/dissolution properties, have also shown promising results as hydrophilic matrices for extended drug release. Cassava starch has been pregelatinized and used as a polymer in combination with sodium alginate for the formulation of floating gastroretentive metronidazole microspheres. Buoyancy and rate of drug release appeared to be modulated by the concentration of cassava starch in the polymer blend. The results showed that pregelatinized cassava could be useful in the formulation of floating gastroretentive metronidazole microspheres (Odeku et al., 2017).

In addition to the systems previously cited as excipients in controlled released drugs, we can also highlight hydroxypropylated starches (which improve the shelf-life, freeze-thaw stability, cold-storage stability, clarity, and texture of starch paste and lower the pasting temperature); oxidized starches (which retard retrogradation, increase hydrophilicity, increase viscosity, and reduce gelatinization temperature); excipient for some DDS; and sustained and controlled drug release (van der Merwe et al., 2020).

Recently, new release systems using starch have been developed including polymeric starch films, nanoparticles, hydrogels, molecular encapsulation, and scaffolds. To be safe and effective, it must be ensured that these systems do not release a drug accidentally, are easy to produce, and, if necessary, have the possibility of sterilization (Garcia et al., 2020; Santander-Ortega et al., 2010).

Next, we present some examples of DDS using starch.

16.4.2.2.1 Polymeric starch films

Starches can form biofilms by techniques such as solution casting, internal mixing, extrusion, injection molding, and compression molding. These films can maintain the therapeutic drug concentration with minimum side effects. Film-coated pellets are an excellent and potential oral colon targeting drug delivery system. Succinate starches, for example, which have low gelatinization temperature and tendency to swell in cold water, have been used as matrix material for drug delivery and encapsulation material because of their good film properties and ability to form gels.

Chen et al. (2007) developed a prototype of a gastroresistant tablet using resistant maize starch as coating films, aiming at the delayed release of the active ingredient in the colon. Resistant starch was prepared by cross-linked and pregelatinized maize starch. It was demonstrated through in vitro drug release studies that the pharmaceutical form remained practically intact, starting to release the active substances after 8 hours of incubation, that is, the tablet passed through the upper gastrointestinal tract without releasing a significant amount of the active, and then, the release was held over a period of 36 hours.

Wannaphatchaiyong et al. (2019) studied tapioca starch films for buccal delivery of lidocaine and lidocaine hydrochloride, and the drug release profiles indicated 8 hours drug release. Another raw material for buccal films is phthalated cassava starch, which can be used in the formulation of buccal mucoadhesive films with a controlled drug release profile (Surini et al., 2018). Some other examples are gastroretentive expandable films to treatment of gastric diseases (Kaewkroek et al., 2022), starch-based biofilms for the treatment of ocular diseases that can maintain the therapeutic drug concentration for prolonged periods (Tuovinen et al., 2004).

Cassava starch–konjac glucomannan blend films were used as matrices for the sustained-release drug. The study showed that the blend film stored at a relative humidity in the range of 11%–53% was appropriate for the sustained release of the less water-soluble drug theophylline (Nair & Jyothi, 2013).

16.4.2.2.2 Nano- and microparticles

Nanoparticles are excellent carriers, as they have prolonged release characteristics and can ensure greater intracellular absorption due to their small size (preparation, characterization, and utilization of starch nanoparticles). Nanoscale starch particles could be prepared from starch granules, which have unique physical properties. These new particles have great potential for applicability in cosmetics, medicines, as well as various composites. Starch has been used as a delivery carrier for tumor-targeted drugs and transdermal nanoparticles (Rodrigues & Emeje, 2012; Santander-Ortega et al., 2010).

Particularly, cassava starch acetate–polyethylene glycol–gelatin nanocomposites were successfully used for a controlled release of an anticancer drug (Raj & Prabha, 2015). Amphoteric cassava starch nanoparticles (CA-CANPs) are prepared by a water/oil microemulsion crosslinking method using (3-chloro-2-hydroxypropyl) trimethyl ammonium chloride as a cationic reagent and $POCl_3$ as an anionic reagent. The CA-CANPs have the possibility of being a targeted controlled release carrier with pH responsiveness for antitumor drugs (Xie et al., 2022).

Shahrodin et al. (2015) called cassava starch nanocrystals (CSNs) the residues obtained from the treatment of native cassava starch (NCS) by sulfuric acid solution, with a variation of 27–333 g H2SO4/L. Hydrolysis was carried out by stirring for 5 days with continuous stirring speed of 300–400 rpm, with constant temperature of 37°C. The generated residue was washed with distilled water until neutral pH to make sure that no acid residues remain in the CSN solution. The precipitate was dried in an oven over night at 60°C. The highest yield was 1.1%. The morphology test analysis by transmission electron microscopy indicated that the samples have been destroyed and degraded to be nanoparticles with a size range of 5–20 nm. X-ray diffraction and 13°C nuclear magnetic resonance were used to indicate the type of crystallinity for both NCS and CSN. These modified starches can be used to fabricate a robust pulsatile delivery system with rapid drug release after lag time (Charoenthai, Sanga-ngam, Puttipipatkhachorn, (2018); Charoenthai, Wickramanayaka, Sungthongjeen, et al., 2018).

Nanosized flake carboxymethyl cassava starch (CMCS) was first synthesized under ultrasonic irradiation. Then acetylsalicylic acid was selected as a water insoluble drug model to prepare solid dispersions using three different kinds of excipients. The results indicated that the nanoscale CMCS was a good carrier for solid dispersion to improve the solubility of poorly water-soluble drugs (Lin et al., 2012).

Also, it is worth mentioning that systems that use starch as a drug carrier can protect bioactive compounds from thermal or oxidative degradation, as well as adverse environmental conditions often encountered during the drug manufacturing process (Allen et al., 2013; Pei et al., 2017; Rodrigues & Emeje, 2012).

16.4.2.2.3 Hydrogels

Another example of extended release DDS is polysaccharide hydrogels, which are three-dimensional polymer networks. They have a porous structure, which allows for loading of drugs into the gel network and enables precise control of drug delivery. Like other modified-release systems, they are capable of increased bioavailability, making them suitable for delivery and controlled release in the small intestine, and apparently, they are more efficient to protect sensitive substances from adverse processing conditions than the existing conventional encapsulation techniques (Marinopoulou et al., 2019). Examples are hybrid gels based on a high amylose starch (Xu et al., 2019) and hydrogels through a complex formed by native maize starch and myristic acid (Konstantakos et al., 2019).

Pregelatinized maize starch hydrogels were used in DDS. In this case, the drug release mechanism is controlled by diffusion through polymer chains, and it is determined by the structural organization, diffusion capability, and gel strength by the processes of polymer swelling and gel layer erosion (Michailova et al., 2001).

Hydrogels can also be developed to respond to external stimuli such as pH, temperature, light, and magnetic field. They can be used as DDS due to magnetic response sensitiveness for targeted drug delivery or as theragnostic agents, which consist of devices capable of simultaneously diagnosing and treating (Fang et al., 2022; Fragal et al., 2022). Magnetic hydrogels based on potato starch beads were evaluated for the treatment of colon cancer. The authors demonstrated that in addition to having magnetic properties, the release of actives occurred at different pH levels, and this was directly related to the proportion of starch. The release was lower in an acid medium due to the lower swelling capacity of the system; the opposite occurred in a basic medium, that is, the presence of starch was relevant to prevent the release of the actives in the stomach environment (Mohammadi et al., 2021). Another example is magnetic iron oxide nanoparticles coated with cross-linked thiolated potato starch to carry an antituberculosis drug. These magnetic nanoparticles showed characteristics of sustained drug release, good biocompatibility, and mucoadhesive properties. Saikia et al. (2014) also observed that there was an improvement in cell viability, proving that this system has great potential for targeted treatment of lung diseases.

16.4.2.2.4 Molecular encapsulation

Cyclodextrin (CD) is a degradation product of the starch produced by CD glycosyltransferase enzyme (CGTase), an alfa-amylase produced by many microorganisms. The CDs are cyclic oligosaccharides in the shape of a truncated cone that have a hydrophilic exterior and a hydrophobic interior, capable of binding molecules forming the so-called inclusion complexes. This "molecular encapsulation" is widely used in the medicine, cosmetics, and food industry, as it allows for solubilization of hydrophobic substances to protect and deliver the drug to the targeted site in a controlled delivery way (Szejtli, 1998). CDs are applicable in the formulation of starch delivery systems such as: PVA-Starch films incorporated with β-CD inclusion complex (Chen et al., 2020) and cationic starch and PVA membrane as a vehicle for β-CD/curcumin inclusion complex (Gularte et al., 2020). For more information on CDs, see Chapter 12 and the upcoming Volume 5 in this series.

Amylose presents a unique behavior among natural polymers, because, when under suitable conditions, it can interact spontaneously with several molecules to form helical inclusion complexes. These systems act as potential nanocarriers by confining the guest molecules inside the helical cavity of the amylose. The complexed guest molecules present greater stability in addition to remaining protected from the gastric environment, being released in the small intestine, thus increasing their bioavailability (Lalush et al., 2005; Yang et al., 2009). These DDS have already been developed and studied for actives such as ibuprofen (Yang et al., 2013), nimesulide, and tretinoin (Marinopoulou et al., 2020) and a spray-dried system powder, which contain many active substances (Marinopoulou et al., 2021).

16.4.2.2.5 Scaffolds

A good alternative to solve the hydrophilic nature of starches is the grafting of hydrophobic side chains onto the hydrophilic starch backbone (scaffolds). The large amount of hydroxyl groups in the polysaccharide backbone allows for incorporation of different specific ligands to obtain polyfunctional colloidal systems. Recently, several studies (Salgado et al., 2004; Santos et al., 2009) demonstrate the applicability of starch-based scaffolds in the healing of wounds and burns. This is possible because starch can accelerate the healing process (Chella Perumal et al., 2022) and can also be used in tissue engineering (Prasopdee et al., 2022). In addition, this system can be used to transport and deliver drugs, such as antiinflammatory drugs (Malafaya et al., 2001) and colonic DDS (Pu et al., 2011).

16.4.2.2.6 Other systems

Other types of systems where starch is used are 3D printed drug release systems. They can be used in personalized therapies following a quick and easy methodology that allowed for preparation of tablets with different geometries loaded with a nonsoluble drug (González et al., 2022).

A natural polymer-based sponge was prepared by Long et al. (2022) using a combination of oxidized starch, microfibrillated cellulose, and carboxymethylchitosan. This system was proposed to aid in the treatment of chronic wounds. In addition, it was modified with silver and collagen nanoparticles that increased its antimicrobial effect in vitro. In vivo experiments demonstrated that the system could promote wound repair by inhibiting inflammation and promoting angiogenesis and cell proliferation (Long et al., 2022).

16.4.3 Other industrial applications of starch

Other uses for starch that can be highlighted are the use of rice starch in the treatment of diarrhea in children and starch with a high amylose content (resistant starch), which can be used as insoluble fibers in clinical nutrition and in colonic segmentation applications (Rowe et al., 2009).

Starch can also be used in the field of cosmetology, as seen in Chapter 11. An example is the use of rice starch as a powder, which, when applied to the skin, produces a soft, mattifying effect, which also helps minimize the appearance of fine lines, wrinkles, and blemishes. The same type of starch is part of the composition of dry shampoo formulations, being responsible for absorbing the oiliness of the hair fiber while providing a desirable volume. Finally, we can mention the use of corn starch as a substitute for talc, mainly in topical formulations for children. Such a demand arose after recent alerts from the FDA indicating the contamination of talc with asbestos, which is supposedly linked to the occurrence of cancer (Builders & Arhewoh, 2016; Garrido-Miranda et al., 2018; COSMETICS, 2020).

The demand for sustainable cosmetics has been growing over the years, and starch together with other natural polymers [e.g., poly(hydroxyalkanoate), poly(butylene succinate-co-adipate), or poly(butylene adipate-co-terephthalate)] has shown to be promising in, for example, formulations of ecofriendly beauty masks, which are generally produced from polymeric fibers derived from petroleum. Studies evaluate the use of these polymeric blends containing starch and point out several advantages such as the fact that the mask can be sold dry, which avoids the use of preservatives, reinforcing the environmental appeal. In addition, they are biodegradable and, after being moistened, they present bioadhesiveness, become flexible (adjusting easily to the curvature of the face), and allow for a quick release of the loaded active to confer specific functions to the mask (Coltelli et al., 2020; Garrido-Miranda et al., 2018).

16.5 Final considerations

This chapter aimed to disclose the different types of starch and their potential applications in the pharmaceutical production sector. Cassava starch is the main starchy crop of South American origin commercially used in the drug manufacture as pharmaceutical excipients. It is the second most used type of starch, the other being potato starch. The most used is corn starch, which originates from central America. It is noteworthy that both have a similar plastic deformation behavior, that is, during the production of tablets, these two types of starch can be interchanged without prejudice to the quality characteristics of the final product.

The physical and chemical properties of starches (*in natura*, modified, or coprocessed) and their availability and low immunogenicity are the properties that allow for their use as excipients in the pharmaceutical application. Starches can be used as binders, filler diluents, and disintegrants in tablet formulations. Also, they can be used in the formulation of DDS, as they have high drug loading efficiency and controlled drug delivery. It should be emphasized, however, that the choice of starch as an excipient depends not only on its physicochemical characteristics, but also on its interaction with other excipients and drugs. This choice will only be made during the development of the formulation and in the stability study of the drug.

Industries are interested in identifying new species that produce native starches, with special characteristics to meet the special needs of drug formulators and the demands of new formulations. These can replace chemically modified starches and open new markets for their use in pharmaceutical formulations. There is a great potential for the development of tropical starches as pharmaceutical excipients. It is possible to observe the versatility of these materials. Some of these crops grow in abundance and contain starch as one of their major components (40%–80% w/w db). However, it is true that new studies must subsidize their uses and prove their efficacy and safety of use in formulations before being insert in Pharmacopoeias. Therefore, it becomes necessary that these starches are fully characterized before pharmaceutical grade excipients are produced and used for commercial purposes.

16.6 Pharmaceutical glossary

- Active pharmaceutical ingredient: Also called drug, or simply active principle, is the pharmacologically active component intended for use in medicine.
- Adjuvants: An ingredient in a medicine that increases or modifies the activity of the

other ingredients. Examples are binders, disintegrants, and diluting.

- Bioavailability: Indicates the speed and extent of absorption of an active principle in a dosage form, based on its concentration/time curve in the systemic circulation or its excretion in the urine.
- Delayed release: Drug release occurs at some point after a certain period has elapsed since administration. Gastroresistant preparations are considered delayed release forms, as they are designed to resist gastric fluid and release the active principle into the intestinal fluid.
- Extended release: Allows for a reduction in the frequency of dose when compared to the medicine presented in the form of immediate release. It is obtained through a special formulation design and/or manufacturing method.
- Immediate release: Type of release in which the asset is released immediately after administration.
- Pharmaceutical form: Final state of presentation of the drug after one or more pharmaceutical operations performed with or without the addition of excipients. The characteristics of each pharmaceutical form are appropriate for a given route of administration.
- Pharmacodynamics: It is the study of a drug's molecular, biochemical, and physiologic effects or actions. It comes from the Greek words "pharmakon," meaning "drug," and "dynamikos," meaning "power."
- Pharmacokinetics: It is the study of how the body interacts with administered substances for the entire duration of exposure (medications for the sake of this article). This is closely related to but distinctly different from pharmacodynamics, which examines the drug's effect on the body more closely. The four main parameters generally examined by this field include absorption, distribution, metabolism, and excretion.
- Route of administration: Site of the body through which the drug is administered.

References

Adjei, F., Osei, Y., Kuntworbe, N., & Ofori-Kwakye, K. (2017). Evaluation of the disintegrant properties of native starches of five new cassava varieties in paracetamol tablet formulations. *Journal of pharmaceutics, 2017*. Available from https://doi.org/10.1155/2017/2326912.

Allen Jr., L. V., Popovich, N. G., & Ansel, H. C. (2013). *Formas Farmacêuticas e Sistemas de Liberação de Fármacos* (9. ed ed.). Artmed.

ANVISA. (2019). Farmacopeia Brasileira. In A. N. D. V. Sanitária (Ed.), (6ª ed., Vol. Volume II). Brasília.

Athira, G., & Jyothi, A. (2015). Octenyl succinate cassava starch as an excipient for controlled release of theophylline: Microwave-assisted synthesis, characterization and in-vitro drug release studies. *International Journal of Pharmaceutical Sciences and Research, 6*, 200–211. Available from https://doi.org/10.13040/IJPSR.0975-8232.6(1)0.200-11.

Aulton, M. M. E. (2016). *Delineamento de formas farmacêuticas* (4ª ed.). Guanabara Kooganp.

Bhatia, V., Dhingra, A. K., Chopra, B., & Guarve, K. (2022). Co-processed excipients: Recent advances and future perspective. *Journal of Drug Delivery Science and Technology, 71*(5–6)103316. Available from https://doi.org/10.1016/j.jddst.2022.103316.

Bogracheva, T. Y., Wang, Y., Wang, T., & Hedley, C. (2002). Structural studies of starches with different water contents. *Biopolymers, 64*(5). Available from https://doi.org/10.1002/bip.10190.

Brouillet-Fourmann, S., Carrot, C., & Mignard, N. (2003). Gelatinization and gelation of corn starch followed by dynamic mechanical spectroscopy analysis | Request PDF. *Rheologica Acta, 42*(1), 110–117. Available from https://doi.org/10.1007/s00397-002-0261-z.

Builders, P. F., & Arhewoh, M. I. (2016). Pharmaceutical applications of native starch in conventional drug delivery. *Starch - Stärke, 68*, 864–873. Available from https://doi.org/10.1002/star.201500337.

Chalapathi, V., Yuvaraj, T. V., & Jaganathan, A. (2010). Formulation of paracetamol tablets using a novel binder isolated from *Manihot esculenta*. L and its evaluation. *International Journal of ChemTech Research*, 2(1), 406–411.

Charoenthai, N., Sanga-ngam, T., & Puttipipatkhachorn, S. (2018). Use of modified tapioca starches as pharmaceutical excipients. *Pharmaceutical Sciences Asia, 45*(4), 195–204. Available from https://doi.org/10.29090/psa.2018.04.018.0048.

Charoenthai, N., Wickramanayaka, A., Sungthongjeen, S., & Puttipipatkhachorn, S. (2018). Use of cassava starch nanocrystals to make a robust rupturable pulsatile release pellet. *Journal of Drug Delivery Science and Technology*, *47*(47), 283–290. Available from https://doi.org/10.1016/j.jddst.2018.07.026.

Chella Perumal, P., Cui, B., Zhang, H., Gunasekaran, V., Adeniran, L., Jayaraman, S., ... Long, Q. (2022). A critical review on starch-based electrospun nanofibrous scaffolds for wound healing application. *International Journal of Biological Macromolecules*, *222*. Available from https://doi.org/10.1016/j.ijbiomac.2022.09.274.

Chen, L., Li, X., Pang, Y., Li, L., Zhang, X., & Long, Y. (2007). Resistant starch as a carrier for oral colon-targeting drug matrix system. *Journal of Materials Science. Materials in Medicine*, *18*(11). Available from https://doi.org/10.1007/s10856-007-3009-6.

Chen, Z., Zong, L., Chen, C., & Xie, J. (2020). Development and characterization of PVA-Starch active films incorporated with β-cyclodextrin inclusion complex embedding lemongrass (Cymbopogon citratus) oil. *Food Packaging and Shelf Life*, *26*(100565). Available from https://doi.org/10.1016/j.fpsl.2020.100565.

Chitedze, J., Monjerezi, M., Saka, J. D., & Steenekamp, J. (2012). Binding effect of Cassava starches on the compression and mechanical properties of Ibuprofen tablets. *Journal of Applied Pharmaceutical Science*, *2*(4–10). Available from https://doi.org/10.7324/JAPS.02.4.004.

Coltelli, M., Panariello, L., Morganti, P., Danti, S., Baroni, A., Lazzeri, A., ... Donnarumma, G. (2020). Skin-compatible biobased beauty masks prepared by extrusion. *Journal of functional biomaterials*, *11*(2). Available from https://doi.org/10.3390/jfb11020023.

COSMETICS. (2020). Starch: the alternative to talc and microplastics for sensory enhancementAvailable from https://www.cosmeticsdesign-europe.com/Headlines/Promotional-Features/Starch-the-alternative-to-talc-and-microplastics-for-sensory-enhancement. (2020) Accessed 19.10.22.

Cozzolino, D., Degner, S., & Eglinton, J. (2014). A review on the role of vibrational spectroscopy as an analytical method to measure starch biochemical and biophysical properties in cereals and starchy foods. *Foods (Basel, Switzerland)*, *3*(4). Available from https://doi.org/10.3390/foods3040605.

Crowley, P., & Martini, L. G. (2001). Drug-excipient interactions. *Pharmaceutical Technology*, *13*, 26–34.

Daiuto, É. R., & Cereda, M. P. (2006). Influência da granulometria de grânulos de amido sobre a densidade aparente de extratos atomizados [Research Article]. *Revista de Ciências Farmacêuticas Básica e Aplicada*, *27*(1), 51–56, Retrieved 2020-03-12, disponível em. Available from http://hdl.handle.net/11449/69329.

Demirgöz, D., Elvira, C., Mano, J. F., Cunha, A. M., Piskin, E., & Reis, R. L. (2000). Chemical modification of starch based biodegradable polymeric blends: Effects on water uptake, degradation behaviour and mechanical properties. *Polymer Degradation and Stability*, *70*(2), 161–170. Available from https://doi.org/10.1016/S0141-3910(00)00102-6.

Esfahani, G., Häusler, O., & Mäder, K. (2022). Controlled release starch-lipid implant for the therapy of severe malaria. *International Journal of Pharmaceutics*, *622*. Available from https://doi.org/10.1016/j.ijpharm.2022.121879.

Fang, K., Zhang, Y., Yin, J., Yang, T., Li, K., & He, W. (2022). Hydrogel beads based on carboxymethyl cassava starch/alginate enriched with MgFe2O4 nanoparticles for controlling drug release. *International Journal of Biological Macromolecules*, *220*, 573–588. Available from https://doi.org/10.1016/j.ijbiomac.2022.08.081.

FDA. (2022). *Inactive ingredients in approved drug products search*. Food and Drug Administration Retrieved 20 september from. Available from https://www.fda.gov/drugs/drug-approvals-and-databases/inactive-ingredients-approved-drug-products-search-frequently-asked-questions.

Fragal, E., Fragal, V., Silva, E., Paulino, A., da Silva Filho, E., Mauricio, M., ... Muniz, E. (2022). Magnetic-responsive polysaccharide hydrogels as smart biomaterials: Synthesis, properties, and biomedical applications. *Carbohydrate polymers*, *292*. Available from https://doi.org/10.1016/j.carbpol.2022.119665.

Fredriksson, H., Silverio, J., Andersson, R., Eliasson, A.-C., & Åman, P. (1998). The influence of amylose and amylopectin characteristics on gelatinization and retrogradation properties of different starches. *Carbohydrate PolymersVolume*, *35*(3–4), 119–134. Available from https://doi.org/10.1016/S0144-8617(97)00247-6.

Gangopadhyay, A., Bose, A., Rout, S. S., & Mohapatra, R. (2022). Application of dual modified corn starch as a polymer for the colon targeted direct compressible budesonide tablet. *Journal of Drug Delivery Science and Technology*, *74*(103556). Available from https://doi.org/10.1016/j.jddst.2022.103556.

Gao, H., Cai, J., Han, W., Huai, H., Chen, Y., & Wei, C. (2014). Comparison of starches isolated from three different Trapa species. *Food Hydrocolloids*, *37*, 174–181. Available from https://doi.org/10.1016/j.foodhyd.2013.11.001.

Garcia, M. A. V. T., Garcia, C. F., & Faraco, A. A. G. (2020). Pharmaceutical and biomedical applications of native and modified starch: A review. *Starch - Stärke*, *72*(7-8), 1900270. Available from https://doi.org/10.1002/star.201900270.

Garcia, V., Colonna, P., Bouchet, B., & Gallant, D. J. (1997). Structural changes of cassava starch granules after heating at intermediate water contents. *Starch - Stärke*, *49*(5),

171–179. Available from https://doi.org/10.1002/star.19970490502.

Garrido-Miranda, K., Rivas, B., Pérez, M., Sanfuentes, E., & Peña Farfal, C. (2018). Antioxidant and antifungal effects of eugenol incorporated in bionanocomposites of poly-(3-hydroxybutyrate)-thermoplastic starch. *Lebensmittel-Wissenschaft und-Technologie, 98*(9). Available from https://doi.org/10.1016/j.lwt.2018.08.046.

González, K., Larraza, I., Berra, G., Eceiza, A., & Gabilondo, N. (2022). 3D printing of customized all-starch tablets with combined release kinetics. *International Journal of Pharmaceutics, 622*. Available from https://doi.org/10.1016/j.ijpharm.2022.121872.

Gularte, M., Quadrado, R., Pedra, N., Soares, M., Bona, N., Spanevello, R., & Fajardo, A. (2020). Preparation, characterization and antitumor activity of a cationic starch-derivative membrane embedded with a β-cyclodextrin/curcumin inclusion complex. *International Journal of Biological Macromolecules, 148*. Available from https://doi.org/10.1016/j.ijbiomac.2020.01.104.

Guo, Y., Qiao, D., Zhao, S., Zhang, B., & Xie, F. (2021). Starch-based materials encapsulating food ingredients: Recent advances in fabrication methods and applications. *Carbohydrate polymers, 270*. Available from https://doi.org/10.1016/j.carbpol.2021.118358, Retrieved 10/15/2021, from.

Han, H., Hou, J., Yang, N., Zhang, Y., Chen, H., Zhang, Z., ... Guo, S. (2019). Insight on the changes of cassava and potato starch granules during gelatinization. *International Journal of Biological Macromolecules, 126*. Available from https://doi.org/10.1016/j.ijbiomac.2018.12.201.

Hoover, R. (2001). Composition, molecular structure, and physicochemical properties of tuber and root starches: A review. *Carbohydrate Polymers, 45*, 253–267. Available from https://doi.org/10.1016/S0144-8617(00)00260-5.

Jivraj, I., Martini, L., & Thomson, C. (2000). An overview of the different excipients useful for the direct compression of tablets. *Pharmaceutical Science & Technology Today, 3*(2). Available from https://doi.org/10.1016/s1461-5347(99)00237-0.

Kaewkroek, K., Petchsomrit, A., Wira, S. A., & Wiwattanapatapee, R. (2022). Development of starch/chitosan expandable films as a gastroretentive carrier for ginger extract-loaded solid dispersion. *Saudi Pharmaceutical Journal: The Official Publication of the Saudi Pharmaceutical Society, 30*(2). Available from https://doi.org/10.1016/j.jsps.2021.12.017.

Kaur, L., Singh, J., & Liu, Q. (2007). Starch – A potential biomaterial for biomedical applications. *Nanomaterials and Nanosystems for Biomedical Applications*. Springer. Available from https://doi.org/10.1007/978-1-4020-6289-6_5.

Kim, H.-Y., Park, S. S., & Lim, S.-T. (2015). Preparation, characterization and utilization of starch nanoparticles. *Colloids and Surfaces B: Biointerfaces, 126*, 607–620.

Klein, S. (2009). Polysaccharides in oral drug delivery—recent applications and future perspectives. *ACS Symposium Series, 1017*, 13–30. Available from https://doi.org/10.1021/bk-2009-1017.ch001.

Konstantakos, S., Marinopoulou, A., Papaemmanouil, S., Emmanouilidou, M., Karamalaki, M., Kolothas, E., ... Karageorgiou, V. (2019). Preparation of model starch complex hydrogels. *Food Hydrocolloids, 96*, 365–372. Available from https://doi.org/10.1016/j.foodhyd.2019.05.046.

Kuakpetoon, D., & Ya-Jane, W. (2007). Internal structure and physicochemical properties of corn starches as revealed by chemical surface gelatinization. *Carbohydrate Research, 342*, 2253–2263. Available from https://doi.org/10.1016/j.carres.2007.06.010.

Kunle, O. O., & Kunle, O. O. (2019). Starch Source and Its Impact on Pharmaceutical Applications. https://doi.org/10.5772/intechopen.89811.

Lalush, I., Bar, H., Zakaria, I., Eichler, S., & Shimoni, E. (2005). Utilization of amylose-lipid complexes as molecular nanocapsules for conjugated linoleic acid. *Biomacromolecules, 6*(1). Available from https://doi.org/10.1021/bm049644f, Retrieved 2005 Jan-Feb, from.

Lawal, M. V., Odeniyi, M. A., & Itiola, O. A. (2015). Material and rheological properties of native, acetylated, and pregelatinized forms of corn, cassava, and sweet potato starches. *Starch - Stärke, 67*(11–12), 964–975. Available from https://doi.org/10.1002/star.201500044.

Lawal, M. (2018). Modified starches as direct compression excipients – Effect of physical and chemical modifications on tablet properties: A review. *Starch - Stärke, 71* (1800040). Available from https://doi.org/10.1002/star.201800040.

Lemieux, M., Gosselin, P., & Mateescu, M. (2009). Carboxymethyl high amylose starch as excipient for controlled drug release: Mechanistic study and the influence of degree of substitution. *International Journal of Pharmaceutics, 382*(1–2). Available from https://doi.org/10.1016/j.ijpharm.2009.08.030.

Lemos, P. V. F., Marcelino, H. R., Cardoso, L. G., Souza, C. O., & Druzian, J. I. (2021). Starch chemical modifications applied to drug delivery systems: From fundamentals to FDA-approved raw materials. *International Journal of Biological Macromolecules, 184*. Available from https://doi.org/10.1016/j.ijbiomac.2021.06.077.

Lin, X., Gao, W., Li, C., Chen, J., Yang, C., & Wu, H. (2012). Nano-sized flake carboxymethyl cassava starch as excipient for solid dispersions. *International Journal of Pharmaceutics, 423*(2), 435–439. Available from https://doi.org/10.1016/j.ijpharm.2011.12.014.

Long, L., Hu, C., Liu, W., Wu, C., Lu, L., Yang, L., & Wang, Y.-b. (2022). Microfibrillated cellulose-enhanced carboxymethyl chitosan/oxidized starch sponge for chronic diabetic wound repair. *Materials Science and Engineering: C, 135*(112669). Available from https://doi.org/10.1016/j.msec.2022.112669.

Lopes, C., Lobo, J., & Costa, P. (2005). Formas farmacêuticas de liberação modificada: Polímeros hidrifílicos. *Revista Brasileira de Ciências Farmacêuticas, 41*, 143–154. Available from https://doi.org/10.1590/S1516-93322005000200003.

Malafaya, P., Elvira, C., Gallardo, A., San, R. J., & Reis, R. (2001). Porous starch-based drug delivery systems processed by a microwave route. *Journal of Biomaterials Science. Polymer Edition, 12*(11). Available from https://doi.org/10.1163/156856201753395761.

Marinopoulou, A., Christofilos, D., Arvanitidis, J., & Raphaelides, S. (2019). Study of molecular inclusion complex formation of amylose with indomethacin. *Starch - Stärke, 71*1800295. Available from https://doi.org/10.1002/star.201800295.

Marinopoulou, A., Christofilos, D., Arvanitidis, J., & Raphaelides, S. (2020). Interaction of tretinoin and nimesulide with amylose matrices. *Starch - Stärke, 73*(2000054). Available from https://doi.org/10.1002/star.202000054.

Marinopoulou, A., Zoumaki, M., Goulas, A., Dimitris, P., Raphaelides, S., Antonios, A., Tzikas, A., & Karageorgiou, V. (2021). Functional characteristics and physical properties of spray dried starch inclusion complexes with drugs. *Starch - Starke, 74*. Available from https://doi.org/10.1002/star.202100176.

Michailova, V., Titeva, S., Kotsilkova, R., Krusteva, E., & Minkov, E. (2001). Influence of hydrogel structure on the processes of water penetration and drug release from mixed hydroxypropylmethyl cellulose/thermally pregelatinized waxy maize starch hydrophilic matrices. *International Journal of Pharmaceutics, 222*(1). Available from https://doi.org/10.1016/s0378-5173(01)00706-2.

Mohammadi, R., Saboury, A., Javanbakht, S., Foroutan, R., & Shaabani, A. (2021). Carboxymethylcellulose/polyacrylic acid/starch-modified Fe3O4 interpenetrating magnetic nanocomposite hydrogel beads as pH-sensitive carrier for oral anticancer drug delivery system. *European Polymer Journal, 153*110500. Available from https://doi.org/10.1016/j.eurpolymj.2021.110500.

Mounsey, J. S., & O'Riordan, E. D. (2008). Influence of pregelatinised maize starch on the rheology, microstructure and processing of imitation cheese. *Journal of Food Engineering, 84*(1), 57–64. Available from https://doi.org/10.1016/j.jfoodeng.2007.04.017.

Nair, S., & Jyothi, A. (2013). Cassava starch–konjac glucomannan biodegradable blend films: In vitro study as a matrix for controlled drug delivery. *Starch – Starke, 65*. Available from https://doi.org/10.1002/star.201200070.

Nwachukwu, N., & Ubieko, E. (2020). Disintegrant properties of native starches obtained from cassava, sweet potato and corn in ibuprofen tablet formulations. *Journal of Drug Delivery and Therapeutics, 10*, 264–273. Available from https://doi.org/10.22270/jddt.v10i5.4324.

Odeku, O. A. (2013). Potentials of tropical starches as pharmaceutical excipients: A review. *Starch – Starke, 65*. Available from https://doi.org/10.1002/star.201200076.

Odeku, O. A., Aderogba, A. A., Ajala, T. O., Akin-Ajani, O. D., & Okunlola, A. (2017). Formulation of floating metronidazole microspheres using cassava starch (*Manihot esculenta*) as polymer. *Journal of Pharmaceutical Investigation, 47*(5), 445–451. Available from https://doi.org/10.1007/s40005-017-0319-7, Retrieved 2017-03-25, from.

Odeku, O., Awe, O. O., Popoola, B., Odeniyi, M., & Itiola, O. A. (2005). Compression and mechanical properties of tablet formulations: Containing corn, sweet potato, and cocoyam starches as binders. *Pharmaceutical Technology, 29*, 82–90.

Osborne, B. G. (1998). NIR measurements of the development of crystallinity in stored bread crumb. *Analusis, 26* (4), 55–57. Available from https://doi.org/10.1051/analusis:199826040055.

Pagtalunan, E., Ramos, J., & Guzman, G. (2022). Application of hydroxypropylated crosslinked starch from the grains of *Oryza sativa* L. as potential disintegrant in compressed solid dosage form. *Journal of Pharmacy & Pharmacognosy Research; a Journal of Science and its Applications, 10*, 701–716. Available from https://doi.org/10.56499/jppres22.1377_10.4.701.

Palanisamy, C. P., Cui, B., Zhang, H., Jayaraman, S., & Kodiveri Muthukaliannan, G. (2020). A comprehensive review on corn starch-based nanomaterials: properties, simulations, and applications. *Polymers, 12*(9). Available from https://doi.org/10.3390/polym12092161.

Pei, X., Zhai, K., Liang, X., Deng, Y., Tan, Y., Wang, P., & Xu, K. (2017). Interfacial activity of starch-based nanoparticles at the oil-water interface. *Langmuir: The ACS Journal of Surfaces and Colloids, 33*(15). Available from https://doi.org/10.1021/acs.langmuir.7b00035.

Pezzini, B. R., Silva, M. A. S., & Ferraz, H. G. (2007). Formas farmacêuticas sólidas orais de liberação prolongada: Sistemas monolíticos e multiparticulados. *Revista Brasileira de Ciências Farmacêuticas, 4*(43), 491–502. Available from https://doi.org/10.1590/S1516-93322007000400002.

Prasopdee, T., Sinthuvanich, C., Chollakup, R., Uttayarat, P., & Smitthipong, W. (2022). The albumin/starch scaffold and its biocompatibility with living cells. *Materials Today Communications, 27*(2021)102164. Available from https://doi.org/10.1016/j.mtcomm.2021.102164.

Prista, N. N., Alves, A. C., & Morgado, R. M. R. (1996). *Técnica farmacêutica e farmácia galênica* (3 ed ed., Vol. 1). Fundação Calouste Gulbenkian.

Pu, H., Chen, L., Li, X., Xie, F., Yu, L., & Li, L. (2011). An oral colon-targeting controlled release system based on resistant starch acetate: Synthesization, characterization, and preparation of film-coating pellets. *Journal of Agricultural and Food Chemistry*, *59*(10). Available from https://doi.org/10.1021/jf2005468.

Raj, V., & Prabha, G. (2015). Synthesis, characterization and in vitro drug release of cisplatin loaded Cassava starch acetate–PEG/gelatin nanocomposites. *Journal of the Association of Arab Universities for Basic and Applied Sciences*, *21*, 10–16. Available from https://doi.org/10.1016/j.jaubas.2015.08.001.

Rashid, I., Al-Remawi, M., Leharne, S. A., Chowdhry, B. Z., & Badwan, A. (2011). A novel multifunctional pharmaceutical excipient: Modification of the permeability of starch by processing with magnesium silicate. *International Journal of Pharmaceutics*, *411*(Issues 1–2), 18–26. Available from https://doi.org/10.1016/j.ijpharm.2011.03.011.

Recife, A., Meneguin, A., Cury, B., & Evangelista, R. (2017). Evaluation of retrograded starch as excipient for controlled release matrix tablets. *Journal of Drug Delivery Science and Technology*, *40*, 83–94. Available from https://doi.org/10.1016/j.jddst.2017.06.003.

Rodrigues, A., & Emeje, M. (2012). Recent applications of starch derivatives in nanodrug delivery. *Carbohydrate Polymers*, *87*(2), 987–994. Available from https://doi.org/10.1016/j.carbpol.2011.09.044.

Rowe, R. C., Sheskey, P. J., & Quinn, M. E. (2009). *Handbook of Pharmaceutical Excipients*. Pharmaceutical Press.

Saikia, C., Hussain, A., Ramteke, A., Sharma, H., & Maji, T. (2014). Crosslinked thiolated starch coated Fe3O4 magnetic nanoparticles: Effect of montmorillonite and crosslinking density on drug delivery properties. *Starch - Starke*, *66*(760–771). Available from https://doi.org/10.1002/star.201300277.

Salgado, A. J., Coutinho, O. P., & Reis, R. L. (2004). Novel starch-based scaffolds for bone tissue engineering: Cytotoxicity, cell culture, and protein expression. *Tissue Engineering*, *10*(3–4). Available from https://doi.org/10.1089/107632704323061825.

Santander-Ortega, M., Stauner, T., Loretz, B., Ortega-Vinuesa, J., Bastos-González, D., Wenz, G., ... Lehr, C. (2010). Nanoparticles made from novel starch derivatives for transdermal drug delivery. *Journal of Controlled Release: Official Journal of the Controlled Release Society*, *141*(1). Available from https://doi.org/10.1016/j.jconrel.2009.08.012.

Santos, M., Pashkuleva, I., Alves, C., Gomes, M., Fuchs, S., Unger, R., ... Kirkpatrick, C. (2009). Surface-modified 3D starch-based scaffold for improved endothelialization for bone tissue engineering. *Journal of Materials Chemistry*, *19*(24), 4091–4101. Available from https://doi.org/10.1039/B819089E.

Shi, L., Hopfer, H., Ziegler, G. R., & Kong, L. (2019). Starch-menthol inclusion complex: Structure and release kinetics. *Food Hydrocolloids*, *97*, 105183. Available from https://doi.org/10.1016/j.foodhyd.2019.105183.

Sivamaruthi, B. S., Nallasamy, P. k, Suganthy, N., Kesika, P., & Chaiyasut, C. (2022). Pharmaceutical and biomedical applications of starch-based drug delivery system: A review. *Journal of Drug Delivery Science and Technology*, *77*, 103890. Available from https://doi.org/10.1016/j.jddst.2022.103890.

Sohail, A. M., Zafar, S., Yousef, B., Alyassin, Y., Ali, R., AlAsiri, A., ... Pitt, K. (2021). A review of emerging technologies enabling improved solid oral dosage form manufacturing and processing. *Advanced Drug Delivery Reviews*, *178*. Available from https://doi.org/10.1016/j.addr.2021.113840.

Storpirtis, S. (2006). Carbohidratos en alimentos regionales iberoamericanos. *Revista Brasileira de Ciências Farmacêuticas*, *42*, 615–616. Available from https://doi.org/10.1590/S1516-93322006000400018.

Surini, S., Gotalia, F., & Putri, K. S. S. (2018). Formulation of mucoadhesive buccal films using pregelatinized cassava starch phthalate as a film-forming polymer. *International Journal of Applied Pharmaceutics*, *10*(1), 225–229. Available from https://doi.org/10.22159/ijap.2018.v10s1.50, Retrieved 2018-12-20, from.

Swarbrick, J. (2013). *Encyclopedia of pharmaceutical technology* (6). CRC press.

Szejtli, J. (1998). Introduction and general overview of cyclodextrin chemistry. *Chemical Reviews*, *98*(5). Available from https://doi.org/10.1021/cr970022c.

Shahrodin, N. S., Rahmat, A. R., & Arsad, A. (2015). Synthesis and characterization of cassava starch nanocrystals by hydrolysis method. *Advanced Materials Research*, *1113*, 446–452. Available from https://doi.org/10.4028/http://www.scientific.net/amr.1113.446.

Tester, R., Karkalas, J., & Qi, X. (2004). Starch—composition, fine structure and architecture. *Journal of Cereal Science*, 151–165. Available from https://doi.org/10.1016/j.jcs.2003.12.001.

Tetlow, I. J., & Bertoft, E. (2020). A review of starch biosynthesis in relation to the building block-backbone model. *International Journal of Molecular Sciences*, *21*(19), 7011.

Tharanathan, R. (2002). Food-derived carbohydrates—structural complexity and functional diversity. *Critical Reviews in Biotechnology*, *22*(1). Available from https://doi.org/10.1080/07388550290789469.

Tuovinen, L., Ruhanen, E., Kinnarinen, T., Rönkkö, S., Pelkonen, J., Urtti, A., ... Jarvinen, K. (2004). Starch acetate microparticles for drug delivery into retinal pigment epithelium-in vitro study. *Journal of Controlled Release: Official Journal of the Controlled Release Society*,

98(3). Available from https://doi.org/10.1016/j.jconrel.2004.05.016.

van der Merwe, J., Steenekamp, J., Steyn, D., & Hamman, J. (2020). The role of functional excipients in solid oral dosage forms to overcome poor drug dissolution and bioavailability. *Pharmaceutics*, *12*(5). Available from https://doi.org/10.3390/pharmaceutics12050393.

Villanova, J. C. O., Oréfice, R. L., & Cunha, A. S. (2010). Aplicações farmacêuticas de polímeros. *Polímeros*, *20*(1), 51–64. Available from https://doi.org/10.1590/S0104-14282010005000009.

Wang, Z., Mhaske, P., Farahnaky, A., Kasapis, S., & Majzoobi, M. (2022). Cassava starch: Chemical modification and its impact on functional properties and digestibility, a review. *Food Hydrocolloids*, *129*107542. Available from https://doi.org/10.1016/j.foodhyd.2022.107542.

Wani, I., Sogi, D., Hamdani, A., Gani, A., Bhat, N., & Shah, A. (2016). Isolation, composition, and physicochemical properties of starch from legumes: A review. *Starch - Starke*, *68*. Available from https://doi.org/10.1002/star.201600007.

Wannaphatchaiyong, S., Heng, P., Suksaeree, J., Boonme, P., & Pichayakorn, W. (2019). Lidocaine loaded gelatin/gelatinized tapioca starch films for buccal delivery and the irritancy evaluation using chick chorioallantoic membrane. *Saudi Pharmaceutical Journal: The Official Publication of the Saudi Pharmaceutical Society*, *27*(8). Available from https://doi.org/10.1016/j.jsps.2019.09.005.

Waterschoot, J., Gomand, S., Fierens, E., & Delcour, J. (2014). Production, structure, physicochemical and functional properties of maize, cassava, wheat, potato and rice starches. *Starch - Starke*, *67*. Available from https://doi.org/10.1002/star.201300238.

Xie, X., Zhang, Y., Zhu, Y., & Lan, Y. (2022). Preparation and drug-loading properties of amphoteric cassava starch nanoparticles. *Nanomaterials (Basel, Switzerland)*, *12*(4). Available from https://doi.org/10.3390/nano12040598.

Xu, J., Tan, X., Chen, L., Li, X., & Xie, F. (2019). Starch/microcrystalline cellulose hybrid gels as gastric-floating drug delivery systems. *Carbohydrate Polymers*, *215*, 151–159. Available from https://doi.org/10.1016/j.carbpol.2019.03.078.

Xu, Y. X., Dzenis, Y., & Hanna, M. A. (2005). Water solubility, thermal characteristics and biodegradability of extruded starch acetate foams. *Industrial Crops and Products*, *21*, 361–368. Available from https://doi.org/10.1016/j.indcrop.2004.05.004.

Yang, L., Zhang, B., Yi, J., Liang, J., Liu, Y., & Zhang, L. M. (2013). Preparation, characterization, and properties of amylose-ibuprofen inclusion complexes. *Starch - Starke*, *65*. Available from https://doi.org/10.1002/star.201200161.

Yang, Y., Gu, Z., & Zhang, G. (2009). Delivery of bioactive conjugated linoleic acid with self-assembled amylose-CLA complex. *Journal of Agricultural and Food Chemistry*, *57*(15). Available from https://doi.org/10.1021/jf9016306.

Index

Note: Page numbers followed by "*f*" and "*t*" refer to figures and tables, respectively.

B

D

N

T